Introduction to the Electron Theory of Metals

The electron theory of metals describes how electrons are responsible for the bonding of metals and subsequent physical, chemical and transport properties. This textbook gives a complete account of electron theory in both periodic and non-periodic metallic systems.

The author presents an accessible approach to the theory of electrons, comparing it with experimental results as much as possible. The book starts with the basics of one-electron band theory and progresses to cover up-to-date topics such as high-T_c superconductors and quasicrystals. The relationship between theory and potential applications is also emphasized. The material presented assumes some knowledge of elementary quantum mechanics as well as the principles of classical mechanics and electromagnetism.

This textbook will be of interest to advanced undergraduates and graduate students in physics, chemistry, materials science and electrical engineering. The book contains numerous exercises and an extensive list of references and numerical data.

UICHIRO MIZUTANI was born in Japan on March 25, 1942. During his early career as a post-doctoral fellow at Carnegie–Mellon University from the late 1960s to 1975, he studied the electronic structure of the Hume-Rothery alloy phases. He received a doctorate of Engineering in this field from Nagoya University in 1971. Together with Professor Thaddeus B. Massalski, he wrote a seminal review article on the electron theory of the Hume-Rothery alloys (*Progress in Materials Science*, 1978). From the late 1970s to the 1980s he worked on the electronic structure and transport properties of amorphous alloys. His review article on the electronic structure of amorphous alloys (*Progress in Materials Science*, 1983) provided the first comprehensive understanding of electron transport in such systems. His research field has gradually broadened since then to cover electronic structure and transport properties of quasicrystals and high-T_c superconductors. It involves both basic and practical application-oriented science like the development of superconducting permanent magnets and thermoelectric materials.

He became a professor of Nagoya University in 1989 and was visiting professor at the University of Paris in 1997 and 1999. He received the Japan Society of Powder and Powder Metallurgy award for distinguished achievement in research in 1995, the best year's paper award from the Japan Institute of Metals in 1997 and the award of merit for Science and Technology of High-T_c Superconductivity in 1999 from the Society of Non-Traditional Technology, Japan.

INTRODUCTION TO THE ELECTRON THEORY OF METALS

UICHIRO MIZUTANI

Department of Crystalline Materials Science, Nagoya University

CAMBRIDGE UNIVERSITY PRESS

PUBLISHED BY THE PRESS SYNDICATE OF THE UNIVERSITY OF CAMBRIDGE
The Pitt Building, Trumpington Street, Cambridge, United Kingdom

CAMBRIDGE UNIVERSITY PRESS
The Edinburgh Building, Cambridge CB2 2RU, UK
40 West 20th Street, New York, NY 10011–4211, USA
10 Stamford Road, Oakleigh, VIC 3166, Australia
Ruiz de Alarcón 13, 28014 Madrid, Spain
Dock House, The Waterfront, Cape Town 8001, South Africa

http://www.cambridge.org

Japanese edition © Uchida Rokakuho 1995 (Vol. 1, pp. 1–260); 1996 (Vol. 2, pp. 261–520)
English edition © Cambridge University Press 2001

This book is in copyright. Subject to statutory exception
and to the provisions of relevant collective licensing agreements,
no reproduction of any part may take place without
the written permission of Cambridge University Press.

First published 2001

Printed in the United Kingdom at the University Press, Cambridge

Typeface Monotype Times NR 11/14 pt   *System* QuarkXPress™   [SE]

A catalogue record for this book is available from the British Library

ISBN 0 521 58334 9 hardback
ISBN 0 521 58709 3 paperback

Contents

Preface	*page* xi
1 Introduction	**1**
1.1 What is the electron theory of metals?	1
1.2 Historical survey of the electron theory of metals	3
1.3 Outline of this book	8
2 Bonding styles and the free-electron model	**10**
2.1 Prologue	10
2.2 Concept of an energy band	10
2.3 Bonding styles	13
2.4 Motion of an electron in free space	16
2.5 Free electron under the periodic boundary condition	18
2.6 Free electron in a box	20
2.7 Construction of the Fermi sphere	21
Exercises	28
3 Electrons in a metal at finite temperatures	**29**
3.1 Prologue	29
3.2 Fermi–Dirac distribution function (I)	29
3.3 Fermi–Dirac distribution function (II)	34
3.4 Electronic specific heat	37
3.5 Low-temperature specific heat measurement	40
3.6 Pauli paramagnetism	44
3.7 Thermionic emission	50
Exercise	53
4 Periodic lattice, and lattice vibrations in crystals	**54**
4.1 Prologue	54
4.2 Periodic structure and reciprocal lattice vectors	54
4.3 Periodic lattice in real space and in reciprocal space	57
4.4 Lattice vibrations in one-dimensional monatomic lattice	64

4.5	Lattice vibrations in a crystal	66
4.6	Lattice waves and phonons	69
4.7	Bose–Einstein distribution function	69
4.8	Lattice specific heat	72
4.9	Acoustic phonons and optical phonons	77
4.10	Lattice vibration spectrum and Debye temperature	80
4.11	Conduction electrons, set of lattice planes and phonons	81
	Exercises	83

5 Conduction electrons in a periodic potential — 86

5.1	Prologue	86
5.2	Cosine-type periodic potential	86
5.3	Bloch theorem	88
5.4	Kronig–Penney model	93
5.5	Nearly-free-electron model	97
5.6	Energy gap and diffraction phenomena	103
5.7	Brillouin zone of one- and two-dimensional periodic lattices	105
5.8	Brillouin zone of bcc and fcc lattices	106
5.9	Brillouin zone of hcp lattice	113
5.10	Fermi surface–Brillouin zone interaction	116
5.11	Extended, reduced and periodic zone schemes	121
	Exercises	125

6 Electronic structure of representative elements — 126

6.1	Prologue	126
6.2	Elements in the periodic table	126
6.3	Alkali metals	126
6.4	Noble metals	130
6.5	Divalent metals	132
6.6	Trivalent metals	135
6.7	Tetravalent metals and graphite	137
6.8	Pentavalent semimetals	141
6.9	Semiconducting elements without and with dopants	143

7 Experimental techniques and principles of electronic structure-related phenomena — 148

7.1	Prologue	148
7.2	de Haas–van Alphen effect	148
7.3	Positron annihilation	155
7.4	Compton scattering effect	160
7.5	Photoemission spectroscopy	162
7.6	Inverse photoemission spectroscopy	169
7.7	Angular-resolved photoemission spectroscopy (ARPES)	172

7.8	Soft x-ray spectroscopy	176
7.9	Electron-energy-loss spectroscopy (EELS)	181
7.10	Optical reflection and absorption spectra	184
	Exercises	188

8 Electronic structure calculations — 190

8.1	Prologue	190
8.2	One-electron approximation	190
8.3	Local density functional method	195
8.4	Band theories in a perfect crystal	199
8.5	Tight-binding method	200
8.6	Orthogonalized plane wave method	203
8.7	Pseudopotential method	204
8.8	Augmented plane wave method	207
8.9	Korringa–Kohn–Rostoker method	211
8.10	LMTO	215
	Exercises	223

9 Electronic structure of alloys — 224

9.1	Prologue	224
9.2	Impurity effect in a metal	224
9.3	Electron scattering by impurity atoms and the Linde law	226
9.4	Phase diagram in Au–Cu alloy system and the Nordheim law	228
9.5	Hume-Rothery rule	232
9.6	Electronic structure in Hume-Rothery alloys	235
9.7	Stability of Hume-Rothery alloys	240
9.8	Band theories for binary alloys	245

10 Electron transport properties in periodic systems (I) — 249

10.1	Prologue	249
10.2	The Drude theory for electrical conductivity	249
10.3	Motion of electrons in a crystal: (I) – wave packet of electrons	254
10.4	Motion of electrons in a crystal: (II)	257
10.5	Electrons and holes	261
10.6	Boltzmann transport equation	264
10.7	Electrical conductivity formula	267
10.8	Impurity scattering and phonon scattering	270
10.9	Band structure effect on the electron transport equation	271
10.10	Ziman theory for the electrical resistivity	275
10.11	Electrical resistivity due to electron–phonon interaction	280
10.12	Bloch–Grüneisen law	284
	Exercises	291

11 Electron transport properties in periodic systems (II) — 293

11.1 Prologue — 293
11.2 Thermal conductivity — 293
11.3 Electronic thermal conductivity — 296
11.4 Wiedemann–Franz law and Lorenz number — 299
11.5 Thermoelectric power — 302
11.6 Phonon drag effect — 307
11.7 Thermoelectric power in metals and semiconductors — 309
11.8 Hall effect and magnetoresistance — 312
11.9 Interaction of electromagnetic wave with metals (I) — 317
11.10 Interaction of electromagnetic wave with metals (II) — 321
11.11 Reflectance measurement — 324
11.12 Reflectance spectrum and optical conductivity — 325
11.13 Kubo formula — 328
Exercises — 333

12 Superconductivity — 334

12.1 Prologue — 334
12.2 Meissner effect — 335
12.3 London theory — 338
12.4 Thermodynamics of a superconductor — 341
12.5 Ordering of the momentum — 343
12.6 Ginzburg–Landau theory — 344
12.7 Specific heat in the superconducting state — 346
12.8 Energy gap in the superconducting state — 347
12.9 Isotope effect — 347
12.10 Mechanism of superconductivity–Fröhlich theory — 349
12.11 Formation of the Cooper pair — 351
12.12 The superconducting ground state and excited states in the BCS theory — 353
12.13 Secret of zero resistance — 358
12.14 Magnetic flux quantization in a superconducting cylinder — 359
12.15 Type-I and type-II superconductors — 360
12.16 Ideal type-II superconductors — 362
12.17 Critical current density in type-II superconductors — 364
12.18 Josephson effect — 368
12.19 Superconducting quantum interference device (SQUID) magnetometer — 373
12.20 High-T_c superconductors — 376
Exercises — 382

13 Magnetism, electronic structure and electron transport properties in magnetic metals — 383

- 13.1 Prologue — 383
- 13.2 Classification of crystalline metals in terms of magnetism — 383
- 13.3 Orbital and spin angular momenta of a free atom and of atoms in a solid — 386
- 13.4 Localized electron model and spin wave theory — 390
- 13.5 Itinerant electron model — 395
- 13.6 Electron transport in ferromagnetic metals — 400
- 13.7 Electronic structure of magnetically dilute alloys — 403
- 13.8 Scattering of electrons in a magnetically dilute alloy – "partial wave method" — 405
- 13.9 Scattering of electrons by magnetic impurities — 410
- 13.10 s–d interaction and Kondo effect — 414
- 13.11 RKKY interaction and spin-glass — 418
- 13.12 Magnetoresistance in ferromagnetic metals — 420
- 13.13 Hall effect in magnetic metals — 428
- Exercises — 431

14 Electronic structure of strongly correlated electron systems — 432

- 14.1 Prologue — 432
- 14.2 Fermi liquid theory and quasiparticle — 433
- 14.3 Electronic states of hydrogen molecule and the Heitler–London approximation — 434
- 14.4 Failure of the one-electron approximation in a strongly correlated electron system — 438
- 14.5 Hubbard model and electronic structure of a strongly correlated electron system — 441
- 14.6 Electronic structure of 3d-transition metal oxides — 444
- 14.7 High-T_c cuprate superconductors — 447
- Exercise — 450

15 Electronic structure and electron transport properties of liquid metals, amorphous metals and quasicrystals — 451

- 15.1 Prologue — 451
- 15.2 Atomic structure of liquid and amorphous metals — 452
- 15.3 Preparation of amorphous alloys — 462
- 15.4 Thermal properties of amorphous alloys — 464
- 15.5 Classification of amorphous alloys — 466
- 15.6 Electronic structure of amorphous alloys — 467
- 15.7 Electron transport properties of liquid and amorphous metals — 472

15.8	Electron transport theories in a disordered system	474
15.8.1	Ziman theory for simple liquid metals in group (V)	475
15.8.2	Baym–Meisel–Cote theory for amorphous alloys in group (V)	479
15.8.3	Mott s–d scattering model	482
15.8.4	Anderson localization theory	483
15.8.5	Variable-range hopping model	486
15.9	Electron conduction mechanism in amorphous alloys	488
15.10	Structure and preparation method of quasicrystals	494
15.11	Quasicrystals and approximants	495
15.12	Electronic structure of quasicrystals	500
15.13	Electron transport properties in quasicrystals and approximants	502
15.14	Electron conduction mechanism in the pseudogap systems	507
15.14.1	Mott conductivity formula for the pseudogap system	507
15.14.2	Family of quasicrystals and their approximants	509
15.14.3	Family of amorphous alloys in group (IV)	510
15.14.4	Family of "unusual" pseudogap systems	512
	Exercises	515
	Appendix 1 Values of selected physical constants	516
	Principal symbols (by chapter)	517
	Hints and answers	539
	References	569
	Materials index	577
	Subject index	579

Preface

This book is an English translation of my book on the electron theory of metals first published in two parts in 1995 and 1996 by Uchida Rokakuho, Japan, the content of which is based on the lectures given for advanced undergraduate and graduate students in the Department of Applied Physics and in the Department of Crystalline Materials Science, Nagoya University, over the last two decades. Some deletions and additions have been made. In particular, the chapter concerning electron transport properties is divided into two in the present book: chapters 10 and 11. The book covers the fundamentals of the electron theory of metals and also the greater part of current research interest in this field. The first six chapters are aimed at the level for advanced undergraduate students, for whom courses in classical mechanics, electrodynamics and an introductory course in quantum mechanics are called for as prerequisites in physics. It is thought to be valuable for students to make early contact with original research papers and a number of these are listed in the *References* section at the end of the book. Suitable review articles and more advanced textbooks are also included. Exercises, and hints and answers are provided so as to deepen the understanding of the content in the book.

It is intended that this book should assist students to further their training while stimulating their research interests. It is essentially meant to be an introductory textbook but it takes the subject up to matters of current research interest. I consider it to be very important for students to catch up with the most recent research developments as soon as possible. It is hoped that this book will be found helpful to graduate students and to specialists in other branches of physics and materials science. It is also designed in such a way that the reader can find interest in learning some more practical applications which possibly result from the physical concepts treated in this book.

I am pleased to acknowledge the valuable discussions that I have had with many colleagues throughout the world, which include Professors T. B.

Massalski, K. Ogawa, M. Itoh, T. Fukunaga, H. Sato, T. Matsuda and H. Ikuta, also Drs E. Belin-Ferré, J. M. Dubois and T. Takeuchi. I would like to thank them all for their interest and helpfulness. With regard to the actual production of this book, the situation is more straightforward. In this regard, I would especially like to thank Professor M. Itoh, Shimane University and Professor K. Ogawa, Yokohama City University, for allowing me to include some of their own thoughts in my textbook. I am also grateful to Dr Brian Watts of Cambridge University Press for his advice on form and substance, and assistance with the English of the book at the final stage of its preparation.

Uichiro Mizutani
Nagoya

Chapter One

Introduction

1.1 What is the electron theory of metals?

Each element exists as either a solid, or a liquid, or a gas at ambient temperature and pressure. Alloys or compounds can be formed by assembling a mixture of different elements on a common lattice. Typically this is done by melting followed by solidification. Any material is, therefore, composed of a combination of the elements listed in the periodic table, Table 1.1. Among them, we are most interested in solids, which are often divided into metals, semiconductors and insulators. Roughly speaking, a metal represents a material which can conduct electricity well, whereas an insulator is a material which cannot convey a measurable electric current. At this stage, a semiconductor may be simply classified as a material possessing an intermediate character in electrical conduction. Most elements in the periodic table exist as metals and exhibit electrical and magnetic properties unique to each of them. Moreover, we are well aware that the properties of alloys differ from those of their constituent elemental metals. Similarly, semiconductors and insulators consisting of a combination of several elements can also be formed. Therefore, we may say that unique functional materials may well be synthesized in metals, semiconductors and insulators if different elements are ingeniously combined.

A molar quantity of a solid contains as many as 10^{23} atoms. A solid is formed as a result of bonding among such a huge number of atoms. The entities responsible for the bonding are the electrons. The physical and chemical properties of a given solid are decided by how the constituent atoms are bonded through the interaction of their electrons among themselves and with the potentials of the ions. This interaction yields the electronic band structure characteristic of each solid: a semiconductor or an insulator is described by a filled band separated from other bands by an energy gap, and a metal by

Table 1.1. Periodic table of the elements

Legend: atomic number, Symbol, atomic weight, outer electron configurations in the ground state

1	2	3	4	5	6	7	8	9	10	11	12	13	14	15	16	17	18
1 **H** 1.008 $1s$																	2 **He** 4.003 $1s^2$
3 **Li** 6.941 $2s$	4 **Be** 9.012 $2s^2$											5 **B** 10.81 $2s^2 2p$	6 **C** 12.01 $2s^2 2p^2$	7 **N** 14.01 $2s^2 2p^3$	8 **O** 16.00 $2s^2 2p^4$	9 **F** 19.00 $2s^2 2p^5$	10 **Ne** 20.18 $2s^2 2p^6$
11 **Na** 22.99 $3s$	12 **Mg** 24.31 $3s^2$											13 **Al** 26.98 $3s^2 3p$	14 **Si** 28.09 $3s^2 3p^2$	15 **P** 30.97 $3s^2 3p^3$	16 **S** 32.07 $3s^2 3p^4$	17 **Cl** 35.45 $3s^2 3p^5$	18 **Ar** 39.95 $3s^2 3p^6$
19 **K** 39.10 $4s$	20 **Ca** 40.08 $4s^2$	21 **Sc** 44.96 $4s^2 3d$	22 **Ti** 47.88 $4s^2 3d^2$	23 **V** 50.94 $4s^2 3d^3$	24 **Cr** 52.00 $4s 3d^5$	25 **Mn** 54.94 $4s^2 3d^5$	26 **Fe** 55.85 $4s^2 3d^6$	27 **Co** 58.93 $4s^2 3d^7$	28 **Ni** 58.69 $4s^2 3d^8$	29 **Cu** 63.55 $4s 3d^{10}$	30 **Zn** 65.39 $4s^2 3d^{10}$	31 **Ga** 69.72 $4s^2 4p$	32 **Ge** 72.59 $4s^2 4p^2$	33 **As** 74.92 $4s^2 4p^3$	34 **Se** 78.96 $4s^2 4p^4$	35 **Br** 79.90 $4s^2 4p^5$	36 **Kr** 83.80 $4s^2 4p^6$
37 **Rb** 85.47 $5s$	38 **Sr** 87.62 $5s^2$	39 **Y** 88.91 $5s^2 4d$	40 **Zr** 91.22 $5s^2 4d^2$	41 **Nb** 92.91 $5s 4d^4$	42 **Mo** 95.94 $5s 4d^5$	43 **Tc** — $5s 4d^6$	44 **Ru** 101.1 $5s 4d^7$	45 **Rh** 102.9 $5s 4d^8$	46 **Pd** 106.4 $4d^{10}$	47 **Ag** 107.9 $5s 4d^{10}$	48 **Cd** 112.4 $5s^2 4d^{10}$	49 **In** 114.8 $5s^2 5p$	50 **Sn** 118.7 $5s^2 5p^2$	51 **Sb** 121.8 $5s^2 5p^3$	52 **Te** 127.6 $5s^2 5p^4$	53 **I** 126.9 $5s^2 5p^5$	54 **Xe** 131.3 $5s^2 5p^6$
55 **Cs** 132.9 $6s$	56 **Ba** 137.3 $6s^2$	Lanthanide	72 **Hf** 178.5 $6s^2 5d^2 4f^{14}$	73 **Ta** 180.9 $6s^2 5d^3$	74 **W** 183.9 $6s^2 5d^4$	75 **Re** 186.2 $6s^2 5d^5$	76 **Os** 190.2 $6s^2 5d^6$	77 **Ir** 192.2 $5d^9$	78 **Pt** 195.1 $6s 5d^9$	79 **Au** 197.0 $6s 5d^{10}$	80 **Hg** 200.6 $6s^2 5d^{10}$	81 **Tl** 204.4 $6s^2 6p$	82 **Pb** 207.2 $6s^2 6p^2$	83 **Bi** 209.0 $6s^2 6p^3$	84 **Po** — $6s^2 6p^4$	85 **At** — $6s^2 6p^5$	86 **Rn** — $6s^2 6p^6$
87 **Fr** — $7s$	88 **Ra** 226.0 $7s^2$	Actinide															

Lanthanide:

57 **La** 138.9 $6s^2 5d$	58 **Ce** 140.1 $6s^2 4f^2$	59 **Pr** 140.9 $6s^2 4f^3$	60 **Nd** 144.2 $6s^2 4f^4$	61 **Pm** — $6s^2 4f^5$	62 **Sm** 150.4 $6s^2 4f^6$	63 **Eu** 152.0 $6s^2 4f^7$	64 **Gd** 157.3 $6s^2 5d 4f^7$	65 **Tb** 158.9 $6s^2 5d 4f^8$	66 **Dy** 162.5 $6s^2 4f^{10}$	67 **Ho** 164.9 $6s^2 4f^{11}$	68 **Er** 167.3 $6s^2 4f^{12}$	69 **Tm** 168.9 $6s^2 4f^{13}$	70 **Yb** 173.0 $6s^2 4f^{14}$	71 **Lu** 175.0 $6s^2 5d 4f^{14}$

Actinide:

89 **Ac** 227.0 $7s^2 6d$	90 **Th** 232.0 $7s^2 6d^2$	91 **Pa** 231.0 $7s^2 6d 5f^2$	92 **U** 238.0 $7s^2 6d 5f^3$	93 **Np** 237.0 $7s^2 5f^5$	94 **Pu** — $7s^2 5f^6$	95 **Am** — $7s^2 5f^7$	96 **Cm** — $7s^2 6d 5f^7$	97 **Bk** —	98 **Cf** —	99 **Es** —	100 **Fm** —	101 **Md** —	102 **No** —	103 **Lr** —

overlapping continuous bands. The resulting electronic structure affects significantly the observed electron transport phenomena. The electron theory of metals in the present book covers properties of electrons responsible for the bonding of solids and electron transport properties manifested in the presence of external fields or a temperature gradient.

Studies of the electron theory of metals are also important from the point of view of application-oriented research and play a vital role in the development of new functional materials. Recent progress in semiconducting devices like the IC (Integrated Circuit) or LSI (Large Scale Integrated circuit), as well as developments in magnetic and superconducting materials, certainly owe much to the successful application of the electron theory of metals. As another unique example, we may refer to amorphous metals and semiconductors, which are known as non-periodic solids having no long-range order in their atomic arrangement. Amorphous Si is now widely used as a solar-operated battery for small calculators.

It may be worthwhile mentioning what prior fundamental knowledge is required to read this book. The reader is assumed to have taken an elementary course of quantum mechanics. We use in this text terminologies such as the wave function, the uncertainty principle, the Pauli exclusion principle, the perturbation theory etc., without explanation. In addition, the reader is expected to have learned the elementary principles of classical mechanics and electromagnetic dynamics.

The units employed in the present book are mostly those of the SI system, but CGS units are often conventionally used, particularly in tables and figures. Practical units are also employed. For example, the resistivity is expressed in units of Ω-cm which is a combination of CGS and SI units. Important units-dependent equations are shown in both SI and CGS units.

1.2 Historical survey of the electron theory of metals

In this section, the reader is expected to grasp only the main historical landmarks of the subject without going into details. The electron theory of metals has developed along with the development of quantum mechanics. In 1901, Planck [1][†] introduced the concept of discrete energy quanta, of magnitude $h\nu$, in the theory of a "black-body" radiation, to eliminate deficiencies of the classical Rayleigh and Wien approaches. Here h is called the Planck constant and ν is the frequency of the electromagnetic radiation expressed as the ratio of the speed of light c over its wavelength λ. In 1905, Einstein [2] explained the

---

[†] Numbers in square brackets are references (see end of book, p. 569).

photoelectric effect (generation of current by irradiation) by making assumptions similar to those of Planck. He assumed the incident light to be made up of energy portions (or "photons" as named later) having discrete energies in multiples of $h\nu$ but that it still behaves like waves with the corresponding frequency. The assumption about a relationship between wave-like and particle-like behavior of light had not been easily accepted at that time.

In 1913, Bohr [3] proposed the electron shell model for the hydrogen atom. He assumed that an electron situated in the field of a positive nucleus was restricted to only certain allowed orbits and that it could "fall" from one orbit to another thereby emitting a quantity of radiation with an energy equal to the difference between the energies of the two orbits. In 1914, Franck and Hertz [4] found that electrons in mercury vapor accelerated by an electric field would cause emission of monochromatic radiation with the wavelength 253.6 nm only when their energy exceeds 4.9 eV. This was taken as a demonstration for the correctness of Bohr's postulate.[1]

There is, however, a difficulty in the semiclassical theory of an atom proposed by Bohr. According to the classical theory, an electron revolving round a nucleus would lose its energy by emitting radiation and eventually spiral into the nucleus. An enormous amount of effort was expended to resolve this paradox in the period of time between 1913 and 1926, when the quantum mechanical theory became ultimately established. In 1923, Compton [5] discovered that x-rays scattered from a light material such as graphite contained a wavelength component longer than that of the incident beam. A shift of wavelength can be precisely explained by considering the conservation of energy and momentum between the x-ray photons and the freely moving electrons in the solid. This clearly demonstrated that electromagnetic radiation treated as particles can impart momenta to particles of matter and it created a need for constructing a theory compatible with the dual nature of radiation having both wave and particle properties.

In 1925, Pauli [6] postulated a simple sorting-out principle by thoroughly studying a vast amount of spectroscopic data including those associated with the Zeeman effect described below. Pauli found the reason for Bohr's assignment of electrons to the various shells around the nuclei for different elements in the periodic table. Pauli's conclusion, which is now known as the "exclusion principle", states that not more than two electrons in a system (such as an atom) should exist in the same quantum state. This became an important basis

---

[1] Radiation with $\lambda = 253.6$ nm is emitted upon the transition from the 6s6p $^3P_1$ excited state to the 6s^2 $^1S_0$ ground state in mercury. According to Bohr's postulate, some excited atoms would fall into the ground state thereby emitting radiation with the wavelength $\lambda = 253.6$ nm. Insertion of $\lambda = 253.6$ nm into $\Delta E = hc/\lambda$ exactly yields the excitation energy of 4.9 eV.

1.2 Historical survey of the electron theory of metals

in the construction of quantum mechanics. Another important idea was set forth by de Broglie [7] in 1924. He suggested that particles of matter such as electrons, might also possess wave-like characteristics, so that they would also exhibit a dual nature. The de Broglie relationship is expressed as $\lambda = h/p = h/mv$, where p is the momentum of the particle and λ is the wavelength. A wavelength is best associated with a wave-like behavior and a momentum is best associated with a particle-like behavior. According to this hypothesis, electrons should exhibit a wave-like nature. Indeed, Davisson and Germer [8] discovered in 1927 that accelerated electrons are diffracted by a Ni crystal in a similar manner to x-rays. The formulation of quantum mechanics was completed in 1925 by Heisenberg [9]. Our familiar Schrödinger equation was established in 1926 [10].

The beginning of the electron theory of metals can be dated back to the works of Zeeman [11] and J. J. Thomson [12] in 1897. Zeeman studied the possible effect of a magnetic field on radiation emitted from a flame of sodium placed between the poles of an electromagnet. He discovered that spectral lines became split into separate components under a strong field. He supposed that light is emitted as a result of an electric charge, really an electron, vibrating in a simple harmonic motion within an atom and could determine from this model the ratio of the charge e to the mass m of a charged particle.

At nearly the same time, J. J. Thomson demonstrated that "cathode rays" in a discharge tube can be treated as particles with a negative charge, and he could independently determine the ratio $(-e)/m$. Soon, the actual charge $(-e)$ was separately determined and, as a result, the electron mass calculated from the ratio $(-e)/m$ turned out to be extremely small compared with that of an atom. In this way, it had been established by 1900 that the negatively charged particles of electricity, which are now known as electrons, are the constituent parts of all atoms and are responsible for the emission of electromagnetic radiation when atoms become excited and their electrons change orbital positions.

The classical theory of metallic conductivity was presented by Drude [13] in 1900 and was elaborated in more detail by Lorentz [14] originally in 1905. Drude applied the kinetic theory of gases to the freely moving electrons in a metal by assuming that there exist charged carriers moving about between the ions with a given velocity and that they collide with one other in the same manner as do molecules in a gas. He obtained the electrical conductivity expression $\sigma = ne^2\tau/m$, which is still used as a standard formula. Here, n is the number of electrons per unit volume and τ is called the relaxation time which roughly corresponds to the mean time interval between successive collisions of the electron with ions. He also calculated the thermal conductivity in the same manner and successfully provided the theoretical basis for the Wiedemann–Frantz law already established in 1853. It states that the ratio of the electrical and thermal conductivities of any metal is a universal constant at a given temperature.

Lorentz later reinvestigated the Drude theory in a more rigorous manner by applying Maxwell–Boltzmann statistics to describe the velocities of the electrons. However, a serious difficulty was encountered in the theory. If the Boltzmann equipartition law $\frac{1}{2}mv^2 = \frac{3}{2}k_B T$ is applied to the electron gas, one immediately finds the velocity of the electron to change as $\sqrt{T}$. According to the Drude model, the mean free path is obviously temperature independent, since it is calculated from the scattering cross-section of rigid ions. This results in a resistivity proportional to $\sqrt{T}$, provided that the number of electrons per unit volume n is temperature independent.[2] However, people at that time had been well aware that the resistivity of typical metals increases linearly with increasing temperature well above room temperature. In order to be consistent with the equipartition law, one had to assume n to change as $1/\sqrt{T}$ in metals. This was not physically accepted.

The application of the equipartition law to the electron system was apparently the source of the problem. Indeed, the true mean free path of electrons is found to be as long as 20 nm for pure Cu even at room temperature (see Section 10.2).[3] Another serious difficulty had been realized in the application of the Boltzmann equipartition law to the calculation of the specific heat of free electrons, which resulted in a value of $\frac{3}{2}R$. The well-known Dulong–Petit law holds well even for metals in which free electrons are definitely present. This means that the additional specific heat of $\frac{3}{2}R$ is somehow missing experimentally. We had to wait for the establishment of quantum mechanics to resolve the failure of the Boltzmann equipartition law when applied to the electron gas.

Quantum mechanics imposes specific restrictions on the behavior of electron particles. The Heisenberg uncertainty principle [15] does not permit an exact knowledge of both the position and the momentum of a particle and, as a result, particles obeying the quantum mechanics must be indistinguishable. In 1926, Fermi [16] and Dirac [17] independently derived a new form of statistical mechanics based on the Pauli exclusion principle. In 1927, Pauli [18] applied the newly derived Fermi–Dirac statistics to the calculation of the paramagnetism of a free-electron gas.

In 1928, Sommerfeld [19] applied the quantum mechanical treatment to the electron gas in a metal. He retained the concept of a free electron gas originally introduced by Drude and Lorentz, but applied to it the quantum mechanics

---

[2] The resistivity ρ is given by $\rho = mv/n(-e)^2 \Lambda$, where m is the mass of electron, v is its velocity, n is the number of electrons per unit volume, Λ is the mean free path for the electron and $(-e)$ is the electronic charge (see Section 10.2).

[3] By applying quantum statistics to the electron gas, we will find (in Section 10.2) the true electron velocity responsible for electron conduction in typical metals to be of the order of 10^6 m/s and temperature independent. Instead, the mean free path is shown to be temperature dependent.

coupled with the Fermi–Dirac statistics. The specific heat, the thermionic emission, the electrical and thermal conductivities, the magnetoresistance and the Hall effect were calculated quite satisfactorily by replacing the ionic potentials with a constant averaged potential equal to zero. The Sommerfeld free-electron model could successfully remove the difficulty associated with the electronic specific heat derived from the equipartition law.

The Sommerfeld model was, however, unable to answer why the mean free path of electrons reaches 20 nm in a good conducting metal like silver at room temperature. Indeed, electrons in a metal are moving in the presence of strong Coulomb potentials due to ions. Therefore the success based on the concept of free-electron behavior was received at that time with a great deal of surprise. The ionic potential is periodically arranged in a crystal. In 1928, Bloch [20] showed that the wave function of a conduction electron in the periodic potential can be described in the form of a plane wave modulated by a periodic function with the period of the lattice, no matter how strong the ionic potential. The wave function is called the Bloch wave. The Bloch theorem provided the basis for the electrical resistivity; the entity that is responsible for the scattering of electrons is not the strong ionic potential itself but the deviation from its periodicity. Based on the Bloch theorem, Wilson [21] in 1931 was able to describe a band theory, which embraces metals, semiconductors and insulators. The main frame of the electron theory of metals had been matured by about the middle of the 1930s. We can see it by reading the well-known textbooks by Mott and Jones [22] and Wilson [23] published in 1936.

Before ending this section, the most notable achievements since the 1940s in the field of the electron theory of metals may be briefly mentioned. Bardeen and Brattain invented the point-contact transistor in 1948–49 [24]. For this achievement, the Nobel prize was awarded to Bardeen, Brattain and Shockley in 1956. Superconductivity is a phenomenon in which the electrical resistivity suddenly drops to zero at its transition temperature T_c. The theory of superconductivity was established in 1957 by Bardeen, Cooper and Schrieffer [25]. The so called BCS theory has been recognized as one of the greatest accomplishments in the electron theory of metals since the advent of the Sommerfeld free-electron theory. Naturally, the higher the superconducting transition temperature, the more likely are possible applications. A maximum superconducting transition temperature had been thought to be no greater than 30–40 K within the framework of the BCS theory. However, a new material, which undergoes the superconducting transition above 30 K, was discovered in 1986 [26] and has received intense attention from both fundamental and practical points of view. This was not an ordinary metallic alloy but a cuprate oxide with a complex crystal structure. More new superconductors in this family have been discovered successively and the superconducting transition temperature T_c has increased to be above 90 K in 1987, above 110 K in 1988 and almost 140 K in 1996. The electronic properties manifested by these superconducting oxides have become one of the most exciting and challenging topics in the field of the electron theory of metals.

Originally, the electron theory of metals was constructed for crystals where the existence of a periodic potential was presupposed. Subsequently, an electron theory treatment of a disordered system, where the periodicity of the ionic potentials is heavily distorted, was also recognized to be significantly important. Liquid metals are typical of such disordered systems. More recently, amorphous metals and semiconductors have received considerable attention not only from the viewpoint of fundamental physics but also from many possible practical applications. In addition to these disordered materials, a non-periodic yet highly ordered material known as a quasicrystal was discovered by Shechtman *et al.* in 1984 [27]. The icosahedral quasicrystal is now known to possess two-, three- and five-fold rotational symmetry which is incompatible with the translational symmetry characteristic of an ordinary crystal. The electron theory should be extended to these non-periodic materials and be cast into a more universal theory.

1.3 Outline of this book

Chapters 2 and 3 are devoted to the description of the Sommerfeld free-electron theory. The free-electron model and the concept of the Fermi surface are discussed in Chapter 2. The Fermi–Dirac distribution function is introduced in Chapter 3 and is applied to calculate the electronic specific heat and the thermionic emission. Pauli paramagnetism is also discussed as another example of the application of the Fermi–Dirac distribution function.

Before discussing the motion of electrons in a periodic lattice, we have to study how the periodic lattice can be described in both real and reciprocal space. Fundamental properties associated with both the periodic lattice and lattice vibrations in both real and reciprocal space are dealt with in Chapter 4. In Chapter 5, the Bloch theorem is introduced and then the energy spectrum of conduction electrons in a periodic lattice potential is given in the nearly-free-electron approximation. The mechanism for the formation of an energy gap and its relation to Bragg scattering are described. The concept of the Brillouin zone and its construction are then shown. The Fermi surface and its interaction with the Brillouin zone are considered and the definitions of a metal, a semiconductor and an insulator are given.

In Chapter 6, the Fermi surfaces and the Brillouin zones in elemental metals and semimetals in the periodic table are presented. The reader will discover how the Fermi surface–Brillouin zone interaction in an individual metal results in its own unique electronic band structure. In Chapter 7, the experimental techniques and the principles involved in determining the Fermi surface of metals are introduced. The behavior of conduction electrons in a magnetic field is also treated in this chapter. In Chapter 8, electronic band structure calculation techniques are introduced. The electron theory in alloys is treated in Chapter 9.

Transport phenomena of electrons in crystalline metals are discussed in both Chapters 10 and 11. The derivation of the Boltzmann transport equation and its application to the electrical conductivity are discussed in Chapter 10. In Chapter 11, other transport properties including thermal conductivity, thermoelectric power, Hall coefficient and optical properties are discussed within the framework of the Boltzmann transport equation. At the end of Chapter 11, the basic concept of the Kubo formula is introduced. Superconducting phenomena are presented in Chapter 12, including the introduction of basic theories such as the London theory and BCS theory. The superconducting properties of high-T_c-superconducting materials are also briefly discussed. In Chapter 13, we focus on the electronic structure and electron transport phenomena in magnetic metals and alloys. For example, the resistivity minimum phenomenon known as the Kondo effect, which is observed when a very small amount of magnetic impurities is dissolved in a non-magnetic metal, is described.

The chapters up to 13 are based on the one-electron approximation. But its failure has been recognized to be crucial in the high-T_c-superconducting cuprate oxides and related materials. The materials in this family have been referred to as strongly correlated electron systems. The electronic structure and electron transport properties of a strongly correlated electron system have been studied extensively in the last decade. Its brief outline is, therefore, introduced in Chapter 14. Finally, the electron theory of non-periodic systems, including liquid metals, amorphous metals and quasicrystals is discussed in Chapter 15.

Exercises are provided at the end of most chapters. The reader is asked to solve them since this will certainly assist in the understanding of the chapter content and ideas. Hints and answers are given at the end of the book. References pertinent to each chapter are listed at the end of the book. Several modern textbooks on solid state physics that include the electron theory of metals are also listed [28–32].

Chapter Two
Bonding styles and the free-electron model

2.1 Prologue

The electron theory of metals pursues the development of ideas that lead to an understanding of various properties manifested by different kinds of materials on the basis of the electronic bondings among constituent atoms. Here the concept of the energy band plays a key role and is introduced in Section 2.2. Condensed matter is often classified in terms of bonding mechanisms; metallic bonding, covalent bonding, ionic bonding and van der Waals bonding. After their brief introduction in Section 2.3, we focus on metallic bonding and discuss the Sommerfeld free-electron model in Sections 2.4–2.6. The construction of the Fermi sphere is discussed in Section 2.7.

2.2 Concept of an energy band

Let us first briefly consider the electron configurations in a free atom. The central-field approximation is useful to describe the motion of each electron in a many-electron atom, since the repulsive interaction between the electrons can be included on an average as a part of the central field. Because of the spherical symmetry of the field, the motion of each electron can be conveniently described in polar coordinates r, θ and ϕ centered at the nucleus. All three variables r, θ and ϕ are needed to describe electron motion in three-dimensional space. In quantum mechanics, the three degrees of freedom lead to three different quantum numbers, by which the stationary state or the quantum state of an electron is specified; the principal quantum number n, which takes a positive integer, the azimuthal or orbital angular momentum quantum number ℓ, which takes integral values from zero to $n-1$, and the magnetic quantum number m, which can vary in integral steps from $-\ell$ to ℓ, including zero. Furthermore, the spin quantum number s, which takes either $\frac{1}{2}$ or $-\frac{1}{2}$, is needed

2.2 Concept of an energy band

to describe the spin motion of each electron. The letters s, p, d, f, ..., are often used to signify the states with $\ell = 0, 1, 2, 3, \ldots$, each preceded by the principal quantum number n.

Because of the Pauli exclusion principle, no two electrons are assigned to the same quantum state. For the lowest energy state of the atom, the electrons must be assigned to states of the lowest energy possible. The first two electrons are accommodated in the quantum states $n = 1$, $\ell = 0$, $m = 0$ and $s = \pm\frac{1}{2}$, which is denoted as $(1s)^2$. Here, the superscript denotes the number of electrons in the 1s state. The third and fourth electrons have to occupy the next lowest energy level with the quantum state $n = 2$, $\ell = 0$, $m = 0$ and $s = \pm\frac{1}{2}$ or $(2s)^2$. The next six electrons, from the fifth up to the tenth electron, are accommodated in the quantum states $n = 2$, $\ell = 1$, $m = \pm 1$ and 0 with $s = \pm\frac{1}{2}$ or $(2p)^6$. The next higher energy level corresponds to the quantum state $n = 3$, $\ell = 0$, $m = 0$ and $s = \pm\frac{1}{2}$ or $(3s)^2$. We can continue this process up to the last electron, the number of which is equal to the atomic number of a given atom. The electron configurations for all elements in the periodic table can be constructed in this manner and are listed in Table 1.1.

An isolated Na atom is positioned in the periodic table with atomic number 11. Since it possesses a total of 11 electrons, its electron configuration (its ground state) can be expressed as $(1s)^2(2s)^2(2p)^6(3s)^1$ with four different orbital energy levels 1s, 2s, 2p and 3s. Now we consider a system consisting of a molar quantity of 10^{23} identical Na atoms separated from each other by a distance far larger than the scale of each atom. All energy levels including those of the outermost 3s electrons must be degenerate, i.e., identical in all 10^{23} atoms, as long as the neighboring wave functions do not overlap with each other.

What happens when the interatomic distance is uniformly reduced to an atomic distance of a few-tenths nm? Figure 2.1 illustrates the probability density of the 1s, 2s, 2p and 3s electrons of two free Na atoms separated by 0.37 nm corresponding to the nearest neighbor distance in sodium metal. It is clear that the 3s wave functions overlap substantially so that some of the 3s electrons belong to both atoms, but the 1s, 2s and 2p wave functions remain still isolated from each other. This means that the degenerate 3s energy levels begin to be "lifted" (i.e., begin possessing slightly different energies), but other levels are still degenerate, when the interatomic distance is reduced to the order of the lattice constant of sodium metal.

As is shown schematically in Fig. 2.2, the energy levels for the 10^{23} 3s electrons are split into quasi-continuously spaced energies when the interatomic distance is reduced to a few-tenths nm. The quasi-continuously spaced energy levels thus formed are called an energy band. Since each level accommodates two electrons with up and down spins, the 3s band must be half-filled by 3s

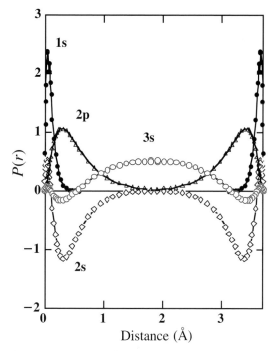

Figure 2.1. 1s, 2s, 2p and 3s wave functions for a free Na atom. Identical wave functions are shown in duplicate both at the origin and 3.7 Å (or 0.37 nm) corresponding to the interatomic distance in Na metal. $P(r)$ represents r times the radial wave function $R(r)$. $P(r)=rR(r)$ is used as a measure of the probability density, since the probability of finding electrons in the spherical shell between r and $r+dr$ is defined as $4\pi r^2|R(r)|^2 dr$. The wave functions are reproduced from D. R. Hartree and W. Hartree, *Proc. Roy. Soc.* (London) **193** (1947) 299.

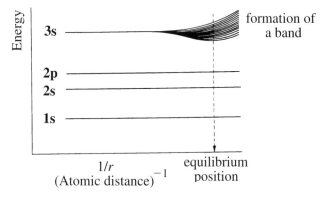

Figure 2.2. Schematic illustration for the formation of an energy band. The energy levels for a huge number of Na free atoms are degenerate when their interatomic distances are very large. The outermost 3s electrons form an energy band when the interatomic distance becomes comparable to the lattice constant of sodium metal.

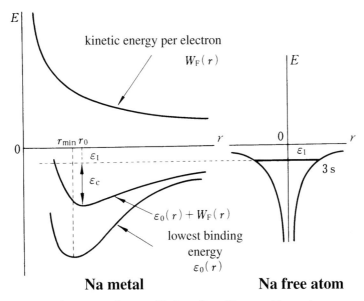

Figure 2.3. Cohesive energy in metallic bonding. Na metal is used as an example. The curve $\varepsilon_0(r)$ represents the lowest energy of electrons with the wave vector $k=0$ (see the lowest curve for the 3s electrons in Fig. 2.2), while the curve W_F represents an average kinetic energy per electron. ε_I represents the ionization energy needed to remove the outermost 3s electron in a free Na atom to infinity and ε_0 is the cohesive energy. The position of the minimum in the cohesive energy gives an equilibrium interatomic distance r_0.

electrons. The 3p level is unoccupied in the ground state of a free Na atom. But the 3p states in sodium metal also form a similar band and mix with the 3s band without a gap between them. As can be understood from the argument above, the energy distribution of the outermost electrons (the valence electrons) spreads into a quasi-continuous band when a solid is formed. This is referred to as the electronic band structure or valence band structure of a solid.

2.3 Bonding styles

We discussed in the preceding section how a piece of sodium metal is formed when a large number of Na atoms are brought together. Now we look into more details of the 3s-band structure shown in Fig. 2.2. The lowest energy level ε_0 obtained after lifting the 10^{23}-fold degeneracy is shown in Fig. 2.3 as a function of interatomic distance r [1,2]. It is seen that the energy ε_0 takes its minimum at $r=r_{min}$. Because of the Pauli exclusion principle, only two electrons with up and down spins among the 10^{23} 3s electrons can occupy this lowest energy level and the next 3s electron must go to the next higher level. As mentioned in the preceding section, one-half of the 3s-band is filled with electrons. This implies that a large amount of kinetic energy is furnished to the electrons. As will be seen in Section 2.7, an average kinetic energy W_F per electron is given by equation (2.24). It increases with decreasing interatomic distance, as shown in Fig. 2.3 (see Exercise 2.2). A distance r_0, at which the value of $\varepsilon_0 + W_F$ takes its minimum, corresponds to the equilibrium interatomic distance observed in sodium metal.

The reason why sodium metal can exist as a solid at ambient temperature arises from the fact that the value of $\varepsilon_0 + W_F$ is lower than the ionization energy ε_I of a free Na atom. The quantity $\varepsilon_c = |\varepsilon_0 + W_F| - |\varepsilon_I|$ is called the cohesive energy and takes its minimum at $r = r_0$. In other words, the 3s electrons can lower their total energy when they form an energy band and gain cohesive energy by overlapping their wave functions. As a result, each 3s electron no longer belongs to any particular atom but moves about almost freely in the system. The freely moving electrons in a band are called valence electrons or simply free electrons. They are responsible for the electron conduction in a metal. In this sense, these electrons are also called conduction electrons.

The remaining ten electrons associated with Na atoms are composed of two 1s electrons, two 2s electrons and six 2p electrons. They are still bound to the nucleus of each given Na atom and maintain their own degenerate energy levels in a free atom. All these bound electrons are called core electrons. The sum of the charges due to the nucleus and the core electrons results in a net charge equal to $+e$ centered at the nucleus. This assembly constitutes a positive ion. Hence, sodium metal is viewed as a solid containing 3s valence electrons moving freely in the potential due to the periodic array of positive Na^+ ions. The net charge of all valence electrons is just equal and opposite in sign to that of the positive ions to maintain charge neutrality. As emphasized above, such a uniform distribution of the valence electrons in the presence of positive ionic potential fields lowers the total energy and thus gains a finite cohesive energy to stabilize a solid. The formation of a solid in this style is called metallic bonding.

Apart from metallic bonding, there are three other bonding styles: ionic bonding, covalent bonding and van der Waals bonding. Typical examples of ionic bonding are the crystals NaCl and KCl. They are made up of positive and negative ions, which are alternately arranged at the lattice points of two interpenetrating simple cubic lattices. The electron configurations for both K^+ and Cl^- ions in a KCl crystal are equally given as $(1s)^2(2s)^2(2p)^6(3s)^2(3p)^6$. Figure 2.4 shows the overlap of the 3p wave functions associated with K^+ and Cl^- free ions separated by a distance equal to 0.315 nm. It can be seen that the overlap

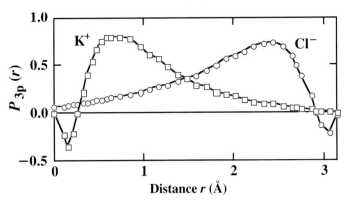

Figure 2.4. 3p wave functions for K^+ and Cl^- ions, both being separated by 3.15 Å (or 0.315 nm) corresponding to the interatomic distance in a KCl crystal. $P(r) = rR(r)$, where $R(r)$ is the radial wave function (see caption to Fig. 2.1). The overlap of wave functions is small at the midpoint. The bonding is due mainly to the electrostatic interaction of oppositely charged ions.

of 3p electron wave functions is less significant relative to that in metallic bonding. The cohesive energy in ionic bonding is gained mainly by the electrostatic interaction arising from the Coulomb force exerted by oppositely charged ions.

Representative elements characteristic of covalent bonding are C, Si and Ge. Figure 2.5 shows the 3s and 3p wave functions of two free Si atoms separated by the nearest neighbor distance of 0.235 nm in solid Si. The overlap of wave functions is substantial and is apparently similar to that of the outermost electron wave functions in metallic bonding. A clear difference from the metallic bonding style cannot be realized, as far as Fig. 2.5 is concerned. The most salient feature of covalent bonding is found in the directional bonding characteristic between the neighboring atoms, illustrated schematically in Fig. 2.6.

Inert gases like He, Ne and Ar are electrically neutral and extremely stable as gases at ambient temperatures and pressures. Inert gases, except for He, solidify at low temperatures. For example, the melting points for Ne and Ar are 24.56 and 83.81 K, respectively. Helium does not solidify even at absolute zero under normal pressures because of a large zero-point motion. More than 25 atmospheric pressures are needed for its solidification at about 2 K. The van der Waals force, which is much weaker than the Coulomb force, is responsible for the bonding of these gases. Indeed, the cohesive energy in solid Ne and Ar is 0.5 and 1.85 kcal/mol, respectively, which is very small relative to that in other bonding styles, for example 26 kcal/mol for Na metal, 98.9 kcal/mol for Fe, 107 kcal/mol for Si and 178 kcal/mol for NaCl [3].

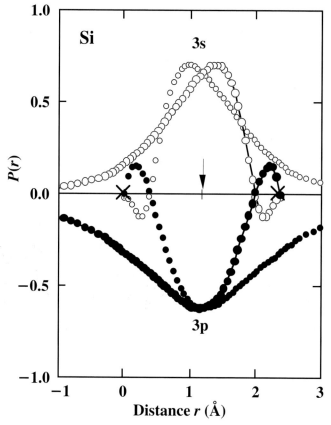

Figure 2.5. 3s and 3p wave functions of a free Si atom. Two identical wave functions are shown: one at the origin and the other at 2.35 Å (or 0.235 nm) corresponding to the interatomic distance in solid Si (atom positions are marked by × in the figure). The overlaps of 3s and 3p wave functions are substantial at the midpoint between the two atoms (marked by an arrow). $P(r)=rR(r)$ (see caption to Fig. 2.1).

2.4 Motion of an electron in free space

The motion of an electron in free space, where the potential V is zero everywhere, can be described by the simplest form of the Schrödinger equation:

$$-\left(\frac{\hbar^2}{2m}\right)\nabla^2\psi(x,y,z) = -\left(\frac{\hbar^2}{2m}\right)\left(\frac{\partial^2}{\partial x^2}+\frac{\partial^2}{\partial y^2}+\frac{\partial^2}{\partial z^2}\right)\psi(x,y,z)=E\psi(x,y,z), \quad (2.1)$$

where $\hbar$, m, E, ψ are, respectively, the Planck constant divided by 2π, the mass of an electron, its energy eigenvalue and wave function. Equation (2.1) can be decomposed into three independent equations involving only a single variable x, y or z by setting $\psi(x,y,z)=X(x)Y(y)Z(z)$ and $E=E_x+E_y+E_z$:

2.4 Motion of an electron in free space

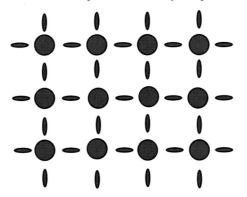

Figure 2.6. Schematic illustration of directional covalent bonding between neighboring atoms.

$$-\left(\frac{\hbar^2}{2m}\right)X''(x)=E_xX(x),$$

$$-\left(\frac{\hbar^2}{2m}\right)Y''(y)=E_yY(y), \qquad (2.2)$$

and

$$-\left(\frac{\hbar^2}{2m}\right)Z''(z)=E_zZ(z).$$

Here, superscript " denotes the second derivative. Equation (2.2) can be easily solved as

$$X(x)=A_1e^{ik_xx}+B_1e^{-ik_xx}, \; E_x=\frac{\hbar^2k_x^2}{2m},$$

$$Y(y)=A_2e^{ik_yy}+B_2e^{-ik_yy}, \; E_y=\frac{\hbar^2k_y^2}{2m}, \qquad (2.3)$$

and

$$Z(z)=A_3e^{ik_zz}+B_3e^{-ik_zz}, \; E_z=\frac{\hbar^2k_z^2}{2m}.$$

The total wave function $\psi(x, y, z) = X(x)Y(y)Z(z)$ is now expressed as a linear combination of eight different plane waves:

$$\psi(x,y,z)=\sum_{j=1}^{8}c_je^{i(\pm k_xx\pm k_yy\pm k_zz)}, \qquad (2.4)$$

where c_j ($j=1$ up to 8) is a numerical coefficient. Equation (2.4) represents a plane wave, which is characterized by wave numbers k_x, k_y and k_z corresponding to x, y and z components of the wave vector **k**.

The energy of an electron in three-dimensional free space is given by

$$E = \left(\frac{\hbar^2}{2m}\right)(k_x^2 + k_y^2 + k_z^2) = \frac{\hbar^2 k^2}{2m}, \tag{2.5}$$

where the wave vector **k** satisfies the relation

$$k^2 = k_x^2 + k_y^2 + k_z^2. \tag{2.6}$$

The wave number k is related to the wavelength λ through the equation

$$k = \frac{2\pi}{\lambda}. \tag{2.7}$$

The wave number is in units of the inverse of length. It is also clear from equation (2.5) that the energy of an electron is proportional to the square of the wave number, i.e., its wave number dependence is parabolic.

2.5 Free electron under the periodic boundary condition

An electron in a metal must be confined in a finite space. The effect of a finite size of a system on the motion of an electron must be taken into account. For the sake of simplicity, we set $y = z = 0$ in equation (2.4) and treat the problem as a one-dimensional system with x as a variable. The electron wave function $\psi(x)$ is assumed along a line with the length L. Let us impose now the following condition on it:

$$\psi(x + L) = \psi(x). \tag{2.8}$$

Equation (2.8) is obtained when both ends of the line are connected so as to form an endless ring. In this way, the finite size of a system can be taken into account while circumventing the difficulty associated with a singular end point. This is called the periodic boundary condition.

An insertion of equation (2.4) into equation (2.8) immediately leads to

$$c_1 e^{ik_x x}(e^{ik_x L} - 1) + c_2 e^{-ik_x x}(e^{-ik_x L} - 1) = 0.$$

This relation must hold for an arbitrary choice of c_1 and c_2. This is possible if the wave number k_x satisfies the relation:

$$k_x L = 2\pi n_x$$

or

$$k_x = \frac{2\pi}{L} n_x \quad (n_x = 0, \pm 1, \pm 2, \pm 3, \ldots). \tag{2.9}$$

2.5 Free electron under the periodic boundary condition

Equation (2.9) indicates that the wave number can take only a discrete set of values in units of $2\pi/L$, since n_x are integers including zero.[1] We have learned that a confinement of electrons to a system of a finite size (we selected a distance L along x) results in the quantization of the wave number.

An extension to three-dimensional space immediately leads to the following wave function:

$$\psi(x, y, z) = \sqrt{\frac{1}{V}} \exp(i\mathbf{k}\cdot\mathbf{r}) \tag{2.10}$$

and the wave vector

$$\mathbf{k} = \left(\frac{2\pi}{L}\right)(n_x\mathbf{i} + n_y\mathbf{j} + n_z\mathbf{k}), \tag{2.11}$$

where the wave vector $\mathbf{k}$ is expressed in the cartesian coordinate system with unit vectors $\mathbf{i}$, $\mathbf{j}$ and $\mathbf{k}$ and integers n_x, n_y and n_z including zero. Thus, the components k_x, k_y and k_z in the wave vector $\mathbf{k}$ are given by $k_i = (2\pi/L)n_i$ and take discrete sets of values. The quantity V in equation (2.10) represents the volume of a cube with the edge length L. The three-dimensional space encompassed by equation (2.11) is called reciprocal space or $\mathbf{k}$-space, since the wave vector $\mathbf{k}$ possesses the dimension reciprocal to the length L in the real space.

The periodic ionic potential is certainly present in a real metal. To a first approximation, however, the periodic potential may be replaced by an averaged constant value, which can be arbitrarily set equal to zero. This yields the Schrödinger equation (2.1) with the periodic boundary condition. This is the free-electron model in a metal. The energy of a free-electron subjected to the periodic boundary condition with the size L in x-, y- and z-directions can be written as

$$E = \left(\frac{\hbar^2}{2m}\right)\left(\frac{2\pi}{L}\right)^2 (n_x^2 + n_y^2 + n_z^2), \tag{2.12}$$

where n_x, n_y and n_z are integers including zero. The probability density of an electron at the position $\mathbf{r}$ with a wave vector $\mathbf{k}$ turns out to be constant:

$$|\psi_\mathbf{k}(\mathbf{r})|^2 = \psi_\mathbf{k}^*(\mathbf{r})\psi_\mathbf{k}(\mathbf{r}) = \frac{1}{V}. \tag{2.13}$$

This means that the wave function (2.10) of the free electron under the periodic boundary condition represents a travelling wave and that the probability density is uniform everywhere in a system.

[1] The function satisfying equation (2.8) is generally expanded in the Fourier series $\psi(x) = \sum_{n=-\infty}^{\infty} C_n e^{i(2\pi/L)nx}$ (see Table 4.1).

An operation of the momentum operator $\mathbf{p} = -i\hbar \nabla$ to the free-electron wave function immediately leads to

$$\mathbf{p}\psi = \left(\frac{-i\hbar}{\sqrt{V}}\right)\nabla \exp(i\mathbf{k}\cdot\mathbf{r}) = \hbar\mathbf{k}\psi. \tag{2.14}$$

Equation (2.14) gives us a very important relation

$$\mathbf{p} = \hbar\mathbf{k} \tag{2.15}$$

for the free electron.[2] Equation (2.15) means that the wave vector plays the same role as the momentum of an electron. In this sense, reciprocal space is sometimes referred to as momentum space.

2.6 Free electron in a box

Let us suppose that the potential $V(x, y, z)$ is zero everywhere inside a cube with edge length L but is infinite at each face. Then, the wave function $\psi(x, y, z)$ must be zero at the face. Here we use again a one-dimensional system with x as a variable. An application of the boundary condition $\psi(0) = 0$ and $\psi(L) = 0$ to equation (2.4) yields the relation $\sin(k_x L) = 0$. The k_x value satisfying this relation must be of the form:

$$k_x = \frac{\pi}{L} n_x \quad (n_x = 1, 2, 3, \ldots). \tag{2.16}$$

Equation (2.16) indicates that the value of k_x is discrete in units of π/L. The wave function after normalization is given by

$$\psi(x) = \sqrt{\left(\frac{2}{L}\right)} \sin\left(\frac{\pi n_x x}{L}\right). \tag{2.17}$$

Note that the wave function exists only in the range $0 \leq x \leq L$ and becomes strictly zero at both ends $x = 0$ and $x = L$. Therefore, equation (2.17) represents a stationary wave. The probability density $|\psi_k(x)|^2$ of electrons at x is no longer constant but changes as a function of x. Another important point to be noted is that, as opposed to equation (2.9), n_x in equation (2.16) takes only a positive integer. It is clear that the wave function with a negative n_x is the same as that with the corresponding positive one except for the reversal of a sign in the normalization factor and, hence, they are identical. In addition, the wave function with $n_x = 0$ is zero everywhere in the range $0 \leq x \leq L$. This must be excluded because of a physically meaningless solution.

---

[2] This relation fails for electrons in a periodic potential. We will learn in Section 5.3 that the wave vector $\mathbf{k}$ no longer represents solely the momentum of an electron.

The discussion above can be extended to a three-dimensional system without difficulty. The total wave function is written as

$$\psi(x, y, z) = \left(\frac{2}{L}\right)^{3/2} \sin\left(\frac{\pi n_x x}{L}\right) \sin\left(\frac{\pi n_y y}{L}\right) \sin\left(\frac{\pi n_z z}{L}\right). \tag{2.18}$$

The energy of free electron confined in a cube with edge length L is easily calculated as

$$E = \frac{\int \psi^* \left(\frac{-\hbar^2}{2m}\right) \nabla^2 \psi \, dV}{\int \psi^* \psi \, dV} = \left(\frac{\hbar^2}{2m}\right)\left(\frac{\pi}{L}\right)^2 (n_x^2 + n_y^2 + n_z^2). \tag{2.19}$$

As emphasized above, equation (2.10) represents a travelling wave whereas equation (2.18) represents a stationary wave. It must be kept in mind that the energy eigenvalue given in equation (2.19) is 4 times as large as that in equation (2.12). This difference is caused by the choice of different boundary conditions imposed on the free electrons. We will consider it again at the end of Section 2.7.

2.7 Construction of the Fermi sphere

As discussed in Section 2.2, the polar coordinate (r, θ, φ) representation is the most convenient to describe the motion of an electron revolving around a nucleus. Its stationary state can be specified in terms of four quantum numbers: principal quantum number n, azimuthal quantum number ℓ, magnetic quantum number m and spin quantum number s.[3] These four numbers, called good quantum numbers, comprise a set which describes the revolving motion of an inner electron. A unique quantum state (n, ℓ, m, s) is assigned to each inner electron according to the Pauli exclusion principle.

The motion of the free electron can be better described using cartesian coordinates. Hence, this means that a set of quantum numbers (n, ℓ, m, s) is no longer adequate. Instead, a set of (k_x, k_y, k_z, s) – three cartesian components of the wave vector $\mathbf{k}$ plus the spin quantum number – must be used as good quantum numbers to describe the motion of the free electron, as shown in equations (2.10) and (2.11).

---

[3] The magnetic quantum number m appears as a good quantum number in the z-component of the angular momentum L_z given by $L_z = -i\hbar \partial/\partial\phi$. Hence, the motion associated with the variable ϕ solely determines the value of m. The principal quantum number n appears in the energy eigenvalue of the Hamiltonian, which is expressed in terms of all three variables r, θ and ϕ. The azimuthal quantum number ℓ is related to the square of the angular momentum L^2, which involves two variables θ and ϕ. Accordingly, both the principal and azimuthal quantum numbers are determined from the electron motion involving more than two variables.

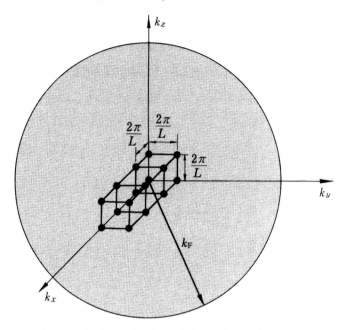

Figure 2.7. Construction of the Fermi sphere. The reciprocal space is quantized in units of $2\pi/L$ in the k_x-, k_y- and k_z-directions and is made up of cubes with edge length $2\pi/L$ as indicated in the figure. Electrons of up and down spins occupy the corner of each cube or integer set (n_x, n_y, n_z) in accordance with the Pauli exclusion principle while making $n_x^2+n_y^2+n_z^2$ as low as possible. The sphere with radius k_F represents the Fermi sphere.

There exist 6.02×10^{23} valence electrons per mole in a monovalent metal such as sodium discussed in Section 2.3. Suppose that the valence electrons in sodium metal are entirely free and that the molar shape of the metal piece is in the form of a cube with edge length L. As discussed in Section 2.5, reciprocal space is quantized in units of $2\pi/L$ in all three directions k_x, k_y and k_z, when the periodic boundary condition is employed. The Pauli exclusion principle should be applied to each electron; no two electrons can go into the same quantum state (k_x, k_y, k_z, s). In addition, we know from equation (2.12) that the energy E of the free electron is proportional to $n_x^2+n_y^2+n_z^2$. Keeping these two conditions in mind, we can construct the ground state for the assembly of free electrons in reciprocal space.

The reciprocal space is now filled with electrons so as to minimize the total energy or $n_x^2+n_y^2+n_z^2$ in accordance with the Pauli exclusion principle. First, two electrons with up and down spins can go into the lowest energy state ε_0 given by $n_x=n_y=n_z=0$ or $(0, 0, 0)$. As shown in Fig. 2.7, the origin in reciprocal space is filled by these two electrons. Next, twelve electrons can go to the next lowest energy states, which are given by the following six identical (n_x, n_y, n_z)

2.7 Construction of the Fermi sphere

values (1, 0, 0), (0, 1, 0), (0, 0, 1), (−1, 0, 0), (0, −1, 0) and (0, 0, −1). This process is continued until all electrons up to the Avogadro number of 6.02×10^{23} fill the reciprocal space. We end up with a sphere in reciprocal space, which is also illustrated in Fig. 2.7. The electron sphere thus obtained is called the Fermi sphere and its surface the Fermi surface. It should be emphasized that the Fermi sphere is constructed on the basis of the free-electron model with the periodic boundary condition described in Section 2.5. As will be discussed in Chapter 5, the deviation from the free-electron model becomes substantial and the distortion of the Fermi surface from a sphere occurs in many metals.

As discussed above, two electrons with up and down spins are accommodated in the volume $(2\pi/L)^3$ in reciprocal space. Let us suppose that the total number of free electrons per mole is equal to N_0 and that the Fermi sphere with the radius k_F is formed when N_0 electrons fill the reciprocal space. Then, we immediately obtain the following proportional relation:

$$\left(\frac{2\pi}{L}\right)^3 : 2 = \left(\frac{4\pi k_F^3}{3}\right) : N_0.$$

From this, we obtain the Fermi radius k_F given by

$$k_F = \left[3\pi^2\left(\frac{N_0}{V}\right)\right]^{1/3}, \qquad (2.20)$$

where V is the volume equal to $V = L^3$.

The radius k_F of the Fermi sphere for sodium metal is calculated in the following way. As shown in Table 2.1, a mole of sodium metal weighs 22.98 g with its density 0.97 g/cm³. Since it is a monovalent metal, N_0 in equation (2.20) is equal to the Avogadro number N_A. An insertion of numerical values $N_A = 6.02 \times 10^{23}$ and $V = 23.69 \times 10^{21}$ nm³ results in a Fermi radius $k_F = 9.1$ nm⁻¹ for sodium metal. Consider the mole of sodium metal to be a cube with edge length L. Then, L turns out to be 2.87 cm and the unit length $2\pi/L$ in reciprocal space to be of the order of 10^{-7} nm⁻¹. Hence, the condition $k_F \gg (2\pi/L)$ is well satisfied. This implies that the quantized points in units of $2\pi/L$ in reciprocal space are very densely distributed and, hence, the Fermi surface is very smooth and almost continuous.

The energy of a free electron with the Fermi radius k_F is calculated by inserting equation (2.20) into equation (2.5);

$$E_F = \frac{\hbar^2 k_F^2}{2m} = \left(\frac{\hbar^2}{2m}\right)\left[3\pi^2\left(\frac{N_0}{V}\right)\right]^{2/3}, \qquad (2.21)$$

where N_0 is obviously the total number of electrons in volume V and E_F is called the Fermi energy. As can be understood from the argument above, a

Table 2.1. Structures and fundamental properties for representative elements in the periodic table

atomic number	element	atomic weight (g)	crystal structure and lattice constants at 300 K	density (g/cm^3)	characteristic features
3	Li	6.941	bcc: $a=3.5$	0.534	lightest metal
6	C	12.011	hex.: $a=2.46$, $c=6.70$	2.25	graphite, semiconductor
			diamond: $a=3.567$	3.51	diamond, semiconductor
11	Na	22.98	bcc: $a=4.22$	0.97	the most free-electron-like metal
12	Mg	24.305	hcp: $a=3.2$, $c=5.21$	1.74	divalent light metal
13	Al	26.981	fcc: $a=4.04$	2.69	trivalent free-electron-like metal
14	Si	28.085	diamond: $a=5.43$	2.34	semiconductor
20	Ca	40.078	fcc: $a=5.582$	1.54	divalent fcc metal
22	Ti	47.867	hcp: $a=2.95$, $c=4.68$	4.54	one of the 3d-transition metals, non-magnetic
26	Fe	55.845	bcc: $a=2.86$	7.86	ferromagnetic metal, Curie temperature $T_C=1043$ K
27	Co	58.933	hcp: $a=2.56$, $c=4.07$	8.8	ferromagnetic metal, Curie temperature $T_C=1400$ K
28	Ni	58.693	fcc: $a=3.52$	8.85	ferromagnetic metal, Curie temperature $T_C=631$ K
29	Cu	63.546	fcc: $a=3.61$	8.93	noble metal, monovalent, electrically good conductor
30	Zn	65.39	hcp: $a=2.66$, $c=4.94$	7.12	divalent metal, easily cleaved in its c-plane
47	Ag	107.868	fcc: $a=4.085$	10.50	noble metal with the lowest resistivity
79	Au	196.966	fcc: $a=4.078$	19.3	noble metal
82	Pb	207.2	fcc: $a=4.95$	11.34	tetravalent, nearly free-electron-like metal, superconductor transition temperature $T_c=7.2$ K
83	Bi	208.980	rhomb: $a=4.54$, $c=11.86$	9.8	semimetal

Note:
bcc: body-centered cubic, hex: hexagonal, fcc: face-centered cubic, hcp: hexagonal close-packed, rhomb: rhombohedral. The lattice constants are in units of Å (1 Å = 0.1 nm).

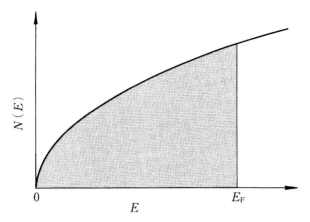

Figure 2.8. The parabolic density of states for free electrons. The states are filled with electrons up to the Fermi energy E_F. The filled area is shaded.

finite Fermi energy stems from the Pauli exclusion principle. Let us remove the suffix F in E_F and 0 in N_0 in equation (2.21) and assume that E and N are variables. Then, the variable E in equation (2.21) represents a maximum energy obtained when the N free electrons per volume V fill in the reciprocal space. The quantity dN/dE can be easily calculated from equation (2.21) and is given in the form of

$$N(E) = \frac{dN}{dE} = \left(\frac{V}{2\pi^2}\right)\left(\frac{2m}{\hbar^2}\right)^{3/2}\sqrt{E}, \qquad (2.22)$$

where $N(E)$ is called the electron density of states, since $N(E)\Delta E$ represents the number of electrons in an energy interval ΔE.

As is clear from equation (2.22), the density of states $N(E)$ exhibits a parabolic energy dependence in the free-electron model. This is shown schematically in Fig. 2.8. Note that the electrons fill energy levels from zero up to the Fermi energy. A total number N of free electrons per volume V is obtained by integrating equation (2.22) from zero to the Fermi energy E_F:

$$N = \int_0^{E_\mathrm{F}} N(E)dE. \qquad (2.23)$$

An average kinetic energy W_F per electron can be calculated:

$$W_F = \frac{\int_0^{E_F} E N(E) dE}{\int_0^{E_F} N(E) dE} = \frac{3}{5} E_F. \tag{2.24}$$

The magnitude of the Fermi energy for typical metals is now quantitatively evaluated on the basis of the free-electron model. First, numerical constants $\hbar = 1.05 \times 10^{-27}$ erg s and $m = 9.1 \times 10^{-28}$ g are inserted into equation (2.21). If we express the volume V and energy E in units of nm^3 and eV, respectively, we obtain

$$E_F = 36.46 \times 10^{-2} \left(\frac{N}{V}\right)^{2/3} \text{ eV}, \tag{2.25}$$

where N is the number of free electrons in volume V. It is often convenient to take the volume per atom, Ω, in place of V. Then, N becomes equal to the number of valence electrons per atom. This is often denoted as e/a. Let us take again sodium metal. It has the bcc structure with a lattice constant a of 0.422 nm. The volume per atom is then given as $\Omega = (0.422)^3/2 = 0.0376$ nm^3. Since sodium metal is monovalent, $e/a = 1$. The Fermi energy turns out to be 3.2 eV by inserting these values into equation (2.25).

Table 2.2 lists the Fermi energy E_F^{free} calculated from equation (2.25) in the free-electron model and the value of E_F^{band} from band calculations (see Chapter 8) for representative metals in the periodic table. It can be seen that the Fermi energy ranges from a few eV to above 10 eV and increases with increasing valency; 2–3 eV for monovalent alkali metals, 7 eV for divalent Mg, 11 eV for trivalent Al. It is to be noted that the Fermi energy is rather large for the noble metals Cu, Ag and Au, though they are also monovalent (see Exercise 2.4).

The Fermi energy is sometimes expressed in units of temperature through the relation $E_F = k_B T_F$. T_F is called the Fermi temperature. For instance, the Fermi temperature reaches about 60 000 K for a metal with $E_F = 5$ eV. This is higher than the temperature of the Sun. As already mentioned, the existence of such a high Fermi temperature for typical metals is the natural consequence of the Pauli exclusion principle. The Fermi wavelength is defined as $\lambda_F = 2\pi/k_F$ from equation (2.7). It is easily checked that the value of the Fermi wavelength is a few tenths nm for metals like sodium and turns out to be comparable to the lattice constant. Electrons deep below the Fermi surface possess lower energies and, hence, longer wavelengths. It is also worthwhile mentioning that in Table 2.2 E_F^{free} does not always agree well with E_F^{band} but the disagreement is generally not too serious in many metals, indicating that the free-electron model is not

Table 2.2. *Fermi energies in representative metals*

element	e/a	$\Omega(\text{Å})^3$	E_F^{free} (eV)	E_F^{band} (eV)	E_F^{free}/E_F^{band}
Cu	1.0	11.81	7.03	9.09	0.77
Ag	1.0	17.06	5.50	7.5	0.73
Au	1.0	16.96	5.52	9.4	0.58
Zn	2.0	15.24	9.42	10.8	0.87
Cd	2.0	21.58	7.59	8.85	0.86
Be	2.0	8.13	14.31	11.9	1.20
Mg	2.0	23.23	7.09	7.1	1.0
Al	3.0	16.60	11.65	11.3	1.0
K	1.0	71.32	2.12	2.24	0.94
Na	1.0	37.71	3.24	3.30	0.98
β-Cu$_{50}$Zn$_{50}$	1.50	12.76	8.75	9.93	0.88

Source:
T. B. Massalski and U. Mizutani, *Prog. Mat. Sci.* **22** (1978) 151

too bad. The electron theory of metals beyond the free-electron model will be discussed in Chapter 5 and subsequent chapters.

We end this section by considering the Fermi sphere when the free electrons are confined in a cubical box, as described in Section 2.6. We learned that the scale of the quantization for the wave vector **k** is different, depending on the choice of the boundary conditions. The value of n_x in equation (2.16) takes only a positive integer and the interval π/L is one-half that in equation (2.9). On the other hand, the energy eigenvalue given by equation (2.19) is one-quarter that given by equation (2.12). Physical quantities like the Fermi energy and the Fermi radius should be independent of the boundary condition imposed. Remember that the Fermi sphere, when being confined in a box, is defined only in the positive octant $k_x > 0$, $k_y > 0$ and $k_z > 0$ in reciprocal space with the interval π/L. Therefore, the Fermi radius for N_0 electrons per volume V is calculated from the proportional relation;

$$\left(\frac{\pi}{L}\right)^3 : 2 = \frac{1}{8}\left(\frac{4\pi k_F^3}{3}\right) : N_0. \qquad (2.26)$$

This leads to the same formula as equation (2.20) obtained under the periodic boundary condition. In this way, we could prove that physical quantities like the Fermi energy and the Fermi radius are indeed independent of the boundary conditions. Unless otherwise stated, the reciprocal space defined by the periodic boundary condition will be employed in the remaining chapters.

Exercises

2.1 We found the expectation value of the momentum of the travelling wave to be proportional to the wave vector (see equation (2.15)). Show that the expectation value for the momentum becomes zero when the electron wave function is given by equation (2.17). Remember that equation (2.17) is obtained by the superposition of two travelling waves with wave numbers k and $-k$ and represents a stationary wave.

2.2 An equilibrium position of atoms in a metal is slightly shifted from the value of r_{min} corresponding to the minimum of the eigenvalue ε_0 for the wave function with the wave number $k=0$. Because of the Pauli exclusion principle, electrons are distributed over the energy range from zero to the Fermi energy. Use equation (2.24) and show that the average kinetic energy W_F is expressed as

$$W_F = \frac{3}{10}\frac{\hbar^2}{m}\left(\frac{9\pi}{4}\right)^{2/3}\frac{1}{r^2}. \tag{2Q.1}$$

The r dependence of W_F is drawn in Fig. 2.3. (The r dependence of the lowest state energy ε_0 is derived from the Wigner–Seitz method. See details in reference 1 or 2.)

2.3 The wave vector is quantized in units of $2\pi/L$ by applying the periodic boundary condition to a cube with edge length L. Suppose that we have a sodium thin film with dimensions $L_x = L_y = 1$ cm and $L_z = 10^{-6}$ cm $= 10$ nm and apply the periodic boundary condition to this system. Show how reciprocal space is quantized in this two-dimensional system and calculate the Fermi energy. Calculate also the density of states and compare the results with the corresponding three-dimensional system.

2.4 Consider why the Fermi energy in noble metals like Cu, Ag and Au is much higher than that in the alkali metals, despite the fact that they are all monovalent.

Chapter Three

Electrons in a metal at finite temperatures

3.1 Prologue

In Chapter 2, we constructed the Fermi sphere of free-electrons with the radius k_F in reciprocal space. It represents the distribution of the quantized electronic states at absolute zero, in which the states in $\mathbf{k} \leq \mathbf{k}_F$ are all occupied but those in $\mathbf{k} > \mathbf{k}_F$ are vacant. At finite temperatures, thermal energy would excite some electrons in the range $\mathbf{k} \leq \mathbf{k}_F$ into the range $\mathbf{k} > \mathbf{k}_F$. The redistribution of electrons will occur so as not to violate the Pauli exclusion principle. As noted in Section 2.7, the Fermi energy in typical metals is of the order of several eV and is equivalent to $\sim 10000\,\text{K}$ on the temperature scale. Hence, only electrons near the Fermi surface can be excited at temperatures below $\sim 1000\,\text{K}$. The aim of the present chapter is to formulate first the Fermi–Dirac distribution function, which determines the distribution of electronic states or the Fermi surface at finite temperatures, and then to deduce the temperature dependence of various physical properties due to conduction electrons by calculating relevant quantities involving the Fermi–Dirac distribution function.

3.2 Fermi–Dirac distribution function (I)

We know that the velocity of dilute gas molecules obeys the Maxwell–Boltzmann distribution law. Unfortunately, however, classical statistics cannot be applied to the conduction electron system in metals because of an extremely high electron density of the order of 10^{28}–$10^{29}/\text{m}^3$. As emphasized in the preceding chapter, an electron carries a spin of $\frac{1}{2}$ and particles with a half-integer spin should obey the Pauli exclusion principle. In addition, they are indistinguishable from each other. Our first objective in this section is to deduce the statistical distribution function under these two conditions imposed by quantum mechanics.

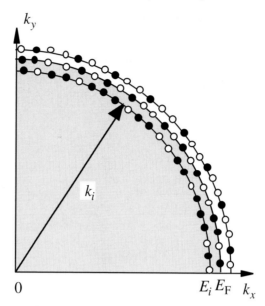

Figure 3.1. A quarter of the cross-section of the Fermi sphere at a finite temperature. The cross-section is cut through its origin. Equi-energy surfaces marked as E_i consist of a series of concentric circles. E_F refers to the Fermi energy. The occupied and unoccupied seats are marked by solid and open circles, respectively.

Electrons occupy discrete states in intervals of $2\pi/L$ in **k**-space so as to conform with the Pauli exclusion principle. For the sake of simplicity, we assume that these allowed states are seats for electrons to be taken. At absolute zero, seats below the Fermi energy E_F are occupied and those above E_F are vacant. At finite temperatures, some electrons below E_F would be excited to higher energies, thus leaving some seats below it vacant and some seats above it occupied. We now try to derive mathematical expressions for the electron distribution as functions of energy and temperature.

Figure 3.1 shows a quarter of the cross-section of the Fermi sphere cut through its origin. The **k**-space is sliced into an assembly of a number of concentric spheres centered at its origin. Each sphere is assumed to be equally spaced in energy. A constant energy $E_i = \hbar^2 k_i^2/2m$ is assigned to the i-th sphere. As emphasized in Section 2.7, the **k**-space is quantized with an interval of $2\pi/L$. Hence, electrons on the concentric sphere with energy E_i can be seated only on $(2\pi/L)$-spaced allowed seats. Z_i, the total number of seats available for electrons with energy E_i is proportional to the surface area of the sphere with the radius k_i. As illustrated in Fig. 3.1, at finite temperatures some electrons take seats above E_F, leaving some unoccupied seats below E_F.

Let us suppose that N_i electrons are seated on a spherical surface with energy E_i. Here $N_i \leq Z_i$ holds. We can calculate in a statistical manner how many ways

3.2 Fermi–Dirac distribution function (I)

there are of distributing N_i electrons over Z_i seats. Remember that N_i electrons are indistinguishable from each other and that the number of electrons taken on each seat is at most one because of the Pauli exclusion principle. Let us name the Z_i seats as $\alpha_1, \alpha_2, \alpha_3, \alpha_4, \ldots$ and let the number of electrons on a seat be 0, 1, 1, 0, 1, 0, 1, ... In this case, the N_i occupied seats are $\alpha_2, \alpha_3, \alpha_5, \alpha_7, \ldots$, whereas the $(Z_i - N_i)$ vacant seats are $\alpha_1, \alpha_4, \alpha_6, \ldots$. In this example, the first electron was seated on α_2. But there are Z_i ways for the first electron to choose a seat. The second electron has $(Z_i - 1)$ ways for its choice. There are in total $Z_i!$ ways of choosing seats, since vacant seats must also be counted. Among them, however, the configuration $\alpha_2, \alpha_3, \ldots$ cannot be distinguished from $\alpha_3, \alpha_2, \ldots$. Hence, the $N_i!$ ways for the occupied seats and $(Z_i - N_i)!$ ways for the unoccupied seats must be excluded. Thus, the number of distinguishable ways ω_i for distributing N_i electrons over Z_i seats must satisfy the relation:

$$\omega_i N_i!(Z_i - N_i)! = Z_i!. \tag{3.1}$$

Equation (3.1) holds for any energy E_i. Hence, the total number of distinguishable ways W is given by the product of ω_i over all possible states:

$$W = \prod_{i=1}^{N} \omega_i = \prod_{i=1}^{N} \frac{Z_i!}{[N_i!(Z_i - N_i)!]} \tag{3.2}$$

According to statistical mechanics, an average value of any macroscopically observed physical quantity is given by the most probable distribution of microscopically possible states [1]. This is equivalent to maximizing the value of W under the condition that the total number of electrons $N = \sum_i N_i$ and the total energy $E = \sum_i N_i E_i$ are kept constant. For this particular purpose, it is convenient to use the method of Lagrangian multipliers and to find the maximum of $\ln W$ in place of W itself. This is reduced to solving a set of equations involving two unknown coefficients α and β:

$$\frac{\partial}{\partial N_j}\left[\ln W + \alpha\left(N - \sum_{i=1}^{N} N_i\right) + \beta\left(E - \sum_{i=1}^{N} N_i E_i\right)\right] = 0. \tag{3.3}$$

By taking the logarithm of equation (3.2), we obtain

$$\ln W = \sum_{i=1}^{N} [\ln Z_i! - \ln(Z_i - N_i)! - \ln N_i!]. \tag{3.4}$$

Equation (3.4) can be rewritten by using the Stirling formula $\ln N! \approx N \ln N - N$, which holds well when $N \gg 1$:

$$\ln W = \sum_{i=1} [Z_i \ln Z_i - (Z_i - N_i)\ln(Z_i - N_i) - N_i \ln N_i]. \tag{3.5}$$

This is valid, since $Z_i \gg 1$ and $N_i \gg 1$. Now the calculation of equation (3.3) is straightforward and leads to the following expression:

$$\ln\left(\frac{Z_j - N_j}{N_j}\right) = \alpha + \beta E_j$$

or

$$\frac{N_j}{Z_j} = \frac{1}{1 + \exp(\alpha + \beta E_j)}, \tag{3.6}$$

where the ratio N_j/Z_j represents the probability of occupying states of energy E_j. From now on, we remove the suffix j and continue our discussion by assuming equation (3.6) to hold as a continuous function of energy E.

The coefficients α and β in equation (3.6) must be determined. Let us consider first the coefficient β. Suppose that the temperature of the electron system is raised so high that the density of the electron gas becomes dilute, i.e., $N \ll Z$. Under such a high-temperature limit, the denominator in equation (3.6) must become very large and, hence, $\exp(\alpha + \beta E) \gg 1$ holds. Then, equation (3.6) can be approximated as

$$\frac{N}{Z} \approx \exp(-\alpha - \beta E) \tag{3.7}$$

at high temperatures. The statistical distribution (3.7) obtained at the high-temperature limit should approach the Boltzmann distribution function, from which we can deduce the relation $\beta = 1/k_B T$.

The other Lagrangian multiplier α is determined in relation to the total number of electrons $N = \sum_i N_i$, which must be conserved at any temperature T. As discussed in Section 2.7, the total number of electrons at absolute zero is directly linked with the Fermi energy E_F through equation (2.21). We learned above that the Fermi surface is blurred at high temperatures. Nevertheless, we assume the Fermi energy at temperature T to be still well defined and denote it as $E_F(T)$. We will learn below that a proper choice of α must be $-E_F(T)/k_B T$.[1]

An insertion of α and β thus obtained into equation (3.6) leads to

$$f(E,T) = \frac{N}{Z} = \frac{1}{1 + \exp\left(\dfrac{E - E_F(T)}{k_B T}\right)}. \tag{3.8}$$

---

[1] An alternative derivation of the coefficient α and β will be explained in Section 4.7, where the Bose–Einstein distribution function is discussed.

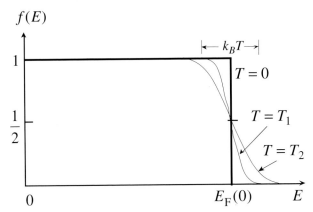

Figure 3.2. Fermi–Dirac distribution function. The smeared region $0 \leq f(E) \leq 1$ extends over $k_B T$ about the Fermi energy E_F at finite temperatures. It increases with increasing temperature ($T_2 > T_1$). A change in the Fermi energy is ignored here.

The function $f(E,T)$ or $f(E)$ is called the Fermi–Dirac distribution function or simply the Fermi distribution function. Particles like electrons, obeying the Fermi–Dirac statistics, are called fermions. The energy dependence of the Fermi–Dirac distribution function $f(E,T)$ is shown in Fig. 3.2. One can easily check that $f(E,T)$ in the limit $T \to 0$ becomes a step function: $f(E) = 1$ meaning all seats are filled for $E \leq E_F(0)$ and $f(E) = 0$ meaning all seats are vacant for $E > E_F(0)$. This reproduces well the Fermi sphere at absolute zero discussed in Section 2.7. At finite temperatures, equation (3.8) yields $0 < f(E) < 1$ only in the vicinity of $E = E_F(T)$. This can also reproduce well the Fermi sphere at finite temperatures, where both occupied and unoccupied states coexist only around $E = E_F(T)$. The distribution functions at two different temperatures T_1 and T_2 are shown in Fig. 3.2. In this way, we have proved a proper choice of α and β and obtained the mathematical formula to describe the electron distribution at finite temperatures.

The Fermi energy $E_F(T)$ at a finite temperature is sometimes called the Fermi level. Since the value is determined from the total number N of particles, it is also called the chemical potential and denoted as the Greek letter ζ, pronounced "zeta".[2] The region in which $0 < f(E) < 1$ holds extends further about $E_F(T)$ as the temperature rises. It can be easily checked from equation (3.8) that the smeared region is of the order of $k_B T$. It is also noted that $f(E_F) = \frac{1}{2}$ holds at any temperature.

As mentioned earlier, the Fermi energy in typical metals is of the order of several eV, which is far larger than the thermal energy at room temperature

---

[2] The Helmholtz free energy F for a system consisting of N identical particles is expressed as $dF = SdT - pdV + \zeta dN$. The chemical potential ζ is defined as $\zeta = (\partial F/\partial N)_{T,V}$. See also Section 4.7.

($k_B T = 0.025$ eV). Hence, the degree of smearing about the Fermi energy is very small at ordinary temperatures. This means that only electrons in the very vicinity of the Fermi energy can possess the freedom to be excited and all other electrons below it are essentially frozen. Electrons in the frozen state are often referred to as being degenerate. The tail of the Fermi–Dirac distribution function is extended to a higher energy with increasing temperature, as is seen in Fig. 3.2. The energy distribution of the tail can be approximated by the Boltzmann distribution function given by equation (3.7).

We learned that equation (3.8), with α and β chosen in the way discussed above, can describe well the energy distribution of electrons at finite temperatures. We still need to know how the Fermi energy $E_F(T)$ at a finite temperature introduced in connection with the coefficient α is related to the Fermi energy $E_F(0)$ at absolute zero defined by equation (2.21).

3.3 Fermi–Dirac distribution function (II)

In this section, we discuss the series expansion of the integral involving the Fermi–Dirac distribution function at ordinary temperatures, where the relation $k_B T \ll E_F(T)$ holds. The integral we consider is expressed in the form

$$I = \int_0^\infty f(E,T) \left(\frac{dF(E)}{dE} \right) dE, \qquad (3.9)$$

where $f(E,T)$ is the Fermi–Dirac distribution function and $F(E)$ is a physically meaningful arbitrary function, which vanishes at $E=0$. Examples of $F(E)$ will be described later. Integrating equation (3.9) by parts leads to

$$I = [f(E)F(E)]_0^\infty - \int_0^\infty F(E) \frac{df(E)}{dE} dE. \qquad (3.10)$$

The first term vanishes, since $f(\infty) = 0$ and $F(0) = 0$. One can easily check from equation (3.8) that the derivative $-df(E)/dE$ is finite only in the vicinity of $E_F(T)$. Hence, it is legitimate to expand the function $F(E)$ about $E_F(T)$ by using the Taylor theorem:

$$F(E) = F(E_F(T)) + [E - E_F(T)] \left[\frac{dF(E)}{dE} \right]_{E=E_F(T)}$$

$$+ \left(\frac{1}{2} \right)[E - E_F(T)]^2 \left[\frac{d^2 F(E)}{dE^2} \right]_{E=E_F(T)} + \cdots. \qquad (3.11)$$

3.3 Fermi–Dirac distribution function (II)

The integral I is rewritten by inserting equation (3.11) into equation (3.10):

$$I = -F(E_F) \int_0^\infty \frac{df(E)}{dE} dE - \left[\frac{dF(E)}{dE}\right]_{E=E_F(T)} \int_0^\infty (E-E_F)\left[\frac{df(E)}{dE}\right] dE$$

$$-\left(\frac{1}{2}\right)\left[\frac{d^2F(E)}{dE^2}\right]_{E_F(T)} \int_0^\infty (E-E_F)^2 \left[\frac{df(E)}{dE}\right] dE - \cdots. \quad (3.12)$$

Equation (3.12) is further rewritten by using the following relations:

$$-\int_0^\infty \frac{df(E)}{dE} dE = f(0) - f(\infty) = 1 \quad (3.13)$$

and

$$\left(\frac{1}{n!}\right)\int_0^\infty (E-E_F)^n \left[-\frac{df(E)}{dE}\right] dE$$

$$= \frac{(k_B T)^n}{n!} \int_0^\infty \frac{\left(\frac{E-E_F}{k_B T}\right)^n \exp\left(\frac{E-E_F}{k_B T}\right)}{\left[\exp\left(\frac{E-E_F}{k_B T}\right)+1\right]^2} d\left(\frac{E-E_F}{k_B T}\right)$$

$$= \frac{(k_B T)^n}{n!} \int_{-\frac{E_F}{k_B T}}^\infty \frac{z^n dz}{(e^z+1)(e^{-z}+1)} \approx \frac{(k_B T)^n}{n!} \int_{-\infty}^\infty \frac{z^n dz}{(e^z+1)(e^{-z}+1)}$$

$$= 2c_n (k_B T)^n \quad (n\text{: even integer})$$
$$= 0 \quad (n\text{: odd integer}). \quad (3.14)$$

Here the lower limit $-E_F(T)/k_B T$ in the integral was replaced by $-\infty$ in the last expression. The coefficient $2c_n$ is calculated as $2c_2 = \frac{\pi^2}{6}$, $2c_4 = \frac{7\pi^4}{360}$, $2c_6 = \frac{31\pi^6}{15120}$, $\cdots$.

Equation (3.12) is finally reduced to the expansion formula:

$$I = F[E_F(T)] + \left(\frac{\pi^2}{6}\right)(k_B T)^2 \left[\frac{d^2 F(E)}{dE^2}\right]_{E=E_F(T)} + \cdots. \quad (3.15)$$

We can calculate various electronic properties at a finite temperature by using equation (3.15).

As one of its applications, we now show how the Fermi energy at a finite temperature $E_F(T)$ is related to that at absolute zero $E_F(0)$. An arbitrary function

$F(E)$ is taken as $F(E) = \int_0^E N(E)dE$, where $N(E)$ is the density of states of conduction electrons defined by equation (2.22). Since $dF(E)/dE = N(E)$, equation (3.9) is reduced to $I = \int_0^\infty N(E)f(E)dE$, which obviously represents the total number of electrons N. Now, equation (3.15) is explicitly written as

$$N = \int_0^\infty N(E)f(E)dE = \int_0^{E_F(T)} N(E)dE + \left(\frac{\pi^2}{6}\right)(k_B T)^2 \left[\frac{dN(E)}{dE}\right]_{E=E_F(T)} + \cdots. \quad (3.16)$$

The left-hand side can be replaced by $N = \int_0^{E_F(0)} N(E)dE$, since N is independent of temperature. By equating this with equation (3.16), we obtain

$$\int_0^{E_F(0)} N(E)dE = \int_0^{E_F(T)} N(E)dE + \left(\frac{\pi^2}{6}\right)(k_B T)^2 \left[\frac{dN(E)}{dE}\right]_{E=E_F(T)} + \cdots,$$

which is approximated as follows by ignoring higher-order terms:

$$\int_0^{E_F(T)} N(E)dE - \int_0^{E_F(0)} N(E)dE + \left(\frac{\pi^2}{6}\right)(k_B T)^2 \left[\frac{dN(E)}{dE}\right]_{E=E_F(T)} \approx 0. \quad (3.17)$$

The difference between $E_F(T)$ and $E_F(0)$ is so small that the integrand $N(E)$ may be treated as a constant in the interval between them. Equation (3.17) is then simplified as

$$[E_F(T) - E_F(0)]N(E_F(0)) + \left(\frac{\pi^2}{6}\right)(k_B T)^2 \left[\frac{dN(E)}{dE}\right]_{E=E_F(0)} = 0$$

or

$$E_F(T) = E_F(0) - \left(\frac{\pi^2}{6}\right)(k_B T)^2 \left[\frac{\left[\frac{dN(E)}{dE}\right]_{E=E_F(0)}}{N(E_F(0))}\right]. \quad (3.18)$$

We have derived in this way the temperature dependence of the Fermi energy.

Suppose that the density of states is given by the free-electron model and is written as $N(E) = C\sqrt{E}$ from equation (2.22). We have $N'(E_F)/N(E_F) = E_F/2$, since $N'(E) = dN(E)/dE = C/2\sqrt{E}$. Its insertion into equation (3.18) results in

3.4 Electronic specific heat

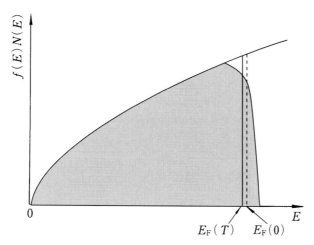

Figure 3.3. Free-electron-like density of states curve $N(E)$ multiplied by the Fermi–Dirac distribution function $f(E)$. The shaded area represents the density of states occupied by electrons at a finite temperature. The Fermi energy at a finite temperature changes by an amount given by equation (3.19) relative to that at 0 K.

$$E_F(T) = E_F(0)\left[1 - \left(\frac{\pi^2}{12}\right)\left(\frac{k_B T}{E_F(0)}\right)^2\right] \quad (3.19)$$

in the free-electron model. Equation (3.19) tells us that the Fermi energy decreases with increasing temperature. Its magnitude is of the order of $[k_B T/E_F(0)]^2$ and amounts to only 1/10 000 at room temperature. The change in the Fermi energy in the free-electron model is illustrated schematically in Fig. 3.3. Note that equation (3.19) holds only for the free-electron model. Indeed, an increase or decrease in the Fermi energy with increasing temperature is decided by the sign of $dN(E)/dE$ at the Fermi level.

3.4 Electronic specific heat

Various electronic properties can be calculated by making use of equation (3.15). The electronic specific heat is considered first. When a given heat ΔQ is fed into a system at temperature T from outside under adiabatic conditions, the temperature of the system will increase by ΔT. The ratio $\Delta Q/\Delta T$ after normalization to a molar quantity defines the specific heat at temperature T. The specific heat of the conduction electron system, which is called the electronic specific heat, can be easily calculated by using equation (3.15).

The internal energy U_{el} of the electron system is expressed as

$$U_{el}(T) = \int_0^\infty EN(E)f(E,T)dE, \qquad (3.20)$$

where the subscript "el" refers to conduction electrons. $F(E)$ in this case is chosen as $F(E) = \int_0^E EN(E)dE$ and is inserted into equation (3.15). We obtain

$$U_{el}(T) = \int_0^{E_F(T)} EN(E)dE + \left(\frac{\pi^2}{6}\right)(k_B T)^2 \left[\frac{d(EN(E))}{dE}\right]_{E=E_F(T)} + \cdots$$

$$= \int_0^{E_F(0)} EN(E)dE + \int_{E_F(0)}^{E_F(T)} EN(E)dE + \left(\frac{\pi^2}{6}\right)(k_B T)^2 N(E_F(0))$$

$$+ \left(\frac{\pi^2}{6}\right)(k_B T)^2 E_F(0) \left[\frac{d(N(E))}{dE}\right]_{E=E_F(0)} + \cdots$$

$$= U_0 - \left(\frac{\pi^2}{6}\right)(k_B T)^2 E_F(0) \left[\frac{d(N(E))}{dE}\right]_{E=E_F(0)}$$

$$+ \left(\frac{\pi^2}{6}\right)(k_B T)^2 E_F(0) \left[\frac{d(N(E))}{dE}\right]_{E=E_F(0)}$$

$$+ \left(\frac{\pi^2}{6}\right)(k_B T)^2 N(E_F(0)) + \cdots$$

$$= U_0 + \left(\frac{\pi^2}{6}\right)(k_B T)^2 N(E_F(0)) + \cdots. \qquad (3.21)$$

Here U_0 represents the internal energy of the electron system at absolute zero. The electronic specific heat C_{el} is calculated by differentiating equation (3.21) with respect to temperature:

$$C_{el}(T) = \left(\frac{\partial U_{el}(T)}{\partial T}\right)_V = \left(\frac{\pi^2}{3}\right) k_B^2 N(E_F(0)) T, \qquad (3.22)$$

where the subscript "V" is attached to emphasize that the temperature derivative is taken under constant volume. Equation (3.22) is known as the formula for the electronic specific heat. It is proportional to the absolute temperature and the density of states at the Fermi level at 0 K. The linearly temperature

3.4 Electronic specific heat

dependent coefficient in equation (3.22) is called the electronic specific heat coefficient and is frequently expressed in the units of mJ/mol.K² as

$$\gamma = \left(\frac{\pi^2}{3}\right) k_B^2 N(E_F(0)) = 2.358 N(E_F(0)), \quad (3.23)$$

where $N(E_F(0))$ is in units of states/eV. atom. Note here that equation (3.22) is derived without reference to the free-electron model. This formula can be applied to any electron system, where the free-electron model does not necessarily hold.

Figure 3.3 shows the parabolic free-electron-like density of states multiplied by the Fermi–Dirac distribution function at temperature T. The shaded area indicates the region occupied by electrons. As emphasized earlier, both occupied and vacant states coexist in a region $k_B T$ about $E_F(T)$. Only electrons in this energy region can be excited to higher energy states by absorbing energy from outside. The number of electrons involved is roughly given by $N(E_F(0)) \cdot k_B T$. The magnitude of energy that these electrons can receive must be of the order of $k_B T$ at temperature T and, hence, a change in the internal energy ΔU is approximately given by $N(E_F(0))(k_B T)^2$. The electronic specific heat is derived by differentiating this with respect to temperature. This gives rise to the electronic specific heat C_{el} proportional to $N(E_F(0)) k_B^2 T$. The present argument clearly explains why the Fermi–Dirac statistics result in the electronic specific heat in the form of equation (3.22).

We discuss as a next step the magnitude of the electronic specific heat. For the sake of simplicity, the free-electron model is assumed. Equation (2.22) is easily rewritten as

$$N(E_F(0)) = \left(\frac{V}{2\pi^2}\right) \left(\frac{2m}{\hbar^2}\right)^{3/2} [E_F(0)]^{1/2} = \frac{3N}{2E_F(0)} = \frac{3N}{2k_B T_F}, \quad (3.24)$$

where N is the total number of electrons and T_F is the Fermi temperature defined in Section 2.7. Therefore, the electronic specific heat is reduced to

$$C_{el} = \left(\frac{\pi^2 k_B^2 T}{3}\right) \left(\frac{3N}{2k_B T_F}\right) = \left(\frac{\pi^2 k_B T}{2 T_F}\right) N$$

in the free-electron model. The total number of electrons per mole is given by the product of the Avogadro number N_A and the valency n_0 of the constituent atom, so $k_B N = n_0 k_B N_A = n_0 R$, where R is the gas constant. Thus the electronic specific heat can be expressed in the form

$$C_{el}^{\text{Fermi–Dirac}} = \left(\frac{\pi^2 n_0 R}{2}\right) \left(\frac{T}{T_F}\right), \quad (3.25)$$

where the superscript "Fermi–Dirac" emphasizes that equation (3.25) is derived by applying the Fermi–Dirac statistics to the electron system.

It is interesting, at this stage, to compare the results above with the case where electrons are treated as classical particles. According to the Boltzmann equipartition law, the energy $k_B T/2$ is furnished to each degree of freedom. Therefore, the specific heat of $n_0 N_A$ electrons obeying classical statistics is reduced to

$$C_{el}^{classical} = \frac{\partial(3n_0 N_A k_B T/2)}{\partial T} = \frac{3n_0 R}{2}. \quad (3.26)$$

A ratio of equation (3.25) over (3.26) leads to

$$\frac{C_{el}^{Fermi-Dirac}}{C_{el}^{classical}} = \left(\frac{\left(\frac{\pi^2 n_0 R}{2}\right)\frac{T}{T_F}}{\left(\frac{3n_0 R}{2}\right)}\right) = \left(\frac{\pi^2}{3}\right)\left(\frac{T}{T_F}\right). \quad (3.27)$$

The ratio turns out to be about 1/60 at room temperature, if the Fermi temperature T_F of 5×10^4 K is inserted. This means that the electronic specific heat at room temperature is merely 1/60 that derived in the classical statistics.

The well-known Dulong–Petit law states that the specific heat near room temperature is almost $3R$, regardless of whether the substances are metals or insulators. The Boltzmann equipartition law yields the electronic specific heat of $3/2 n_0 R$ in addition to the lattice specific heat of $3R$. The reason why the Dulong–Petit law holds even for metals used to be a puzzle. However, this dilemma was resolved when the electron theory based on the Fermi–Dirac statistics was established by Sommerfeld, as described above (see also Section 1.2).

3.5 Low-temperature specific heat measurement

We discuss in this section how we measure experimentally the electronic specific heat of a given sample. Generally speaking, metals are characterized by possessing a finite electronic specific heat coefficient.[3] As far as non-magnetic metals are concerned, their total specific heat is well expressed as a sum of the electronic specific heat and the lattice specific heat (see the definition of non-magnetic metals in Section 3.6). As will be described in Section 4.8, the lattice specific heat below about 10 K can be well approximated by the Debye model and is expressed as a function of temperature in the form

---

[3] A material is definitely an insulator if $N(E_F(0))$ is zero. However, localized states can be formed at the Fermi level in a disordered system. Hence, some insulators possess a finite $N(E_F(0))$. See more details in Sections 15.8.5 and 15.14.

3.5 Low-temperature specific heat measurement

$$C_{\text{lattice}}(T) = \alpha T^3, \tag{3.28}$$

where α is the lattice specific heat coefficient. As emphasized in the preceding section, the electronic specific heat at room temperature amounts to only 1/100 time the lattice specific heat of approximately $3R$ and, hence, is negligibly small.[4] However, the lattice specific heat decreases in proportion to T^3 at low temperatures, whereas the electronic specific heat decreases only in proportion to T. Therefore, the electronic specific heat becomes larger than the lattice specific heat below a certain temperature. This generally occurs below about 10 K in ordinary metals. When the specific heat measurement is carried out in this temperature range, the electronic specific heat can be easily separated from the lattice specific heat by following the prescription described below.

The specific heat below about 10 K in non-magnetic metals is given by the sum of the electronic specific heat and the T^3-dependent lattice specific heat:

$$C(T) = \gamma T + \alpha T^3, \tag{3.29}$$

where the lattice specific heat coefficient α, as will be discussed in Section 4.8, is related to the Debye temperature Θ_D through the relation

$$\Theta_D = \left(\frac{12\pi^4 R}{5\alpha}\right)^{1/3}. \tag{3.30}$$

By dividing both sides of equation (3.29) by T, we obtain

$$\frac{C}{T} = \gamma + \alpha T^2. \tag{3.31}$$

The measured specific heat C at temperature T is divided by T and the resulting C/T values are plotted against T^2. The data would fall on a straight line with a slope α and an intercept γ, provided that equation (3.31) holds well.

Figure 3.4 shows the data for pure Zn in the form of C/T versus T^2. It is seen that the data are well fitted to a straight line except for a small gradual upward deviation at high temperatures and also a large sharp deviation of a few data points in the lowest temperature range. An upward deviation at higher temperatures is due to the deviation of the lattice specific heat from the Debye model. The deviation below 1 K is caused by the superconducting transition of pure Zn.[5] Both the electronic and lattice specific heat coefficients can be determined from the intercept and slope, respectively, of the data shown in Fig. 3.4. Table 3.1 lists the electronic specific heat coefficient and the Debye temperature

---

[4] Note that the ratio $C_{\text{el}}^{\text{Fermi-Dirac}}/C_{\text{lattice}}$ is $n_0/2$ times that in equation (3.27), when $C_{\text{lattice}} = 3R$. A ratio of 1/120 is roughly obtained when $n_0 = 1$.

[5] The specific heat in the superconducting state is discussed in Section 12.7. The superconducting transition temperature of pure Zn is 0.85 K (see Table 12.1).

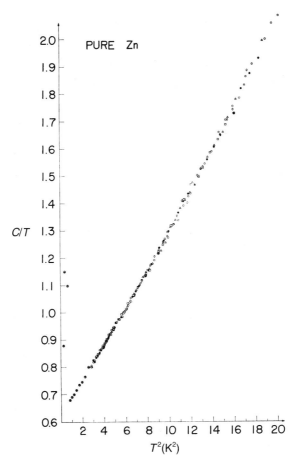

Figure 3.4. Low-temperature specific heat of pure Zn. The electronic specific heat coefficient can be determined from the intercept by extrapolating the data to 0 K. The lattice specific heat coefficient is derived from its slope. [U. Mizutani, *Japan. J. Appl. Phys.* **10** (1971) 367]

deduced from low-temperature specific heat measurements for representative metals. Included are the electronic specific heat coefficient derived from band calculations (see Chapter 8) and also from equation (3.25) in the free-electron model.

As can be seen from Table 3.1, the experimentally derived electronic specific heat coefficient γ_{exp} generally deviates from the corresponding free-electron value γ_F, which is calculated from equations (3.23) and (3.24) as

$$\gamma_F = 0.136(A/d)^{2/3}(e/a)^{1/3} \text{ mJ/mol·K}^2, \tag{3.32}$$

by inserting the atomic weight A in g, the density d in g/cm^3 and the number of

Table 3.1. *Electronic specific heat coefficient and Debye temperature in metals*

element	e/a	γ_{exp}(mJ/mol·K^2)	γ_F(mJ/mol·K^2)	γ_{band}(mJ/mol·K^2)	γ_{exp}/γ_F	Θ_D (K)
Na	1	1.38	1.13	1.06	1.22	157
K	1	2.08	1.67	1.72	1.24	91
Cu	1	0.690	0.503	0.68	1.37	342
Ag	1	0.641	0.642	0.636	0.99	223
Au	1	0.725	0.640		1.14	163
Mg	2	1.30	0.995	1.06	1.30	396
Zn	2	0.638	0.750	0.70	0.85	319
Cd	2	0.688	0.947	0.85	0.73	209
Al	3	1.348	0.910	0.966	1.48	428
Pb	4	3.04	1.49		2.04	106

Note:
e/a: number of electrons per atom
γ_{exp}: measured electronic specific heat coefficient
γ_F: electronic specific heat coefficient in the free-electron model
γ_{band}: electronic specific heat coefficient derived from band calculations
Θ_D: Debye temperature derived from low-temperature specific heat measurements

electrons per atom e/a introduced in Section 2.7. According to equation (3.24), the density of states at the Fermi level is inversely proportional to $E_F(0)$ ($=\hbar^2 k_F^2/2m$) in the free-electron model. Hence, the measured electronic specific heat may be linearly scaled in terms of the mass of the electron. The ratio γ_{exp}/γ_F is often employed as a parameter to judge the deviation from the free-electron model:

$$\frac{\gamma_{exp}}{\gamma_F} = \frac{m^*}{m} \equiv m^*_{th}, \qquad (3.33)$$

where the dimensionless parameter m^*_{th} is referred to as the thermal effective mass. The density of states at the Fermi level in a real metal certainly deviates from the free-electron model. The value obtained from band calculations is listed as γ_{band} in Table 3.1. The value of γ_{band} is sometimes lower but, in other cases, higher than γ_F, depending on the band structure of a given substance. In addition to the band structure effect, many-body effects including electron–phonon and electron–electron interactions are known to affect the value of m^*_{th}. In particular, the electron–phonon interaction is theoretically predicted to enhance m^*_{th} and is believed to be responsible for the reason why the value of γ_{exp}, even for a free-electron-like metal such as sodium, is always 10–40% higher than the value of γ_{band} [2].

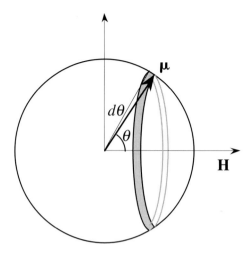

Figure 3.5. The magnetic moment **μ** makes an angle θ with the applied field **H**.

3.6 Pauli paramagnetism

As mentioned in Section 2.6, each electron possesses a freedom associated with its spin and carries a magnetic moment arising from its angular momentum. In this section, we dicuss first the magnetism in the electron system by treating electrons as classical particles. In Fig. 3.5, the magnetic moment carried by a classical particle is denoted by **μ**. Magnetic moments in a unit volume are randomly oriented at high temperatures because of thermal agitation. The magnetic field **H** is applied to this system. An energy $-\mu H\cos\theta$ is gained, when the magnetic moment **μ** makes an angle θ with the applied field **H**. In classical statistics, a component of the magnetic moment along the magnetic field, $\mu_z = \mu\cos\theta$, can take arbitrary values in the range $-\mu \leq \mu_z \leq \mu$.

The probability of aligning the magnetic moment at angle θ must be proportional to the Boltzmann factor $\exp(\mu H\cos\theta/k_B T)$. In addition, the number of magnetic moments aligned at angles between θ and $\theta + d\theta$ is proportional to the shaded area shown in Fig. 3.5. Hence, the probability $p(\theta)d\theta$ of finding the magnetic moment at angles between θ and $\theta + d\theta$ is given by

$$p(\theta)d\theta = \frac{\exp\left(\dfrac{\mu H\cos\theta}{k_B T}\right)\sin\theta d\theta}{\displaystyle\int_0^\pi \exp\left(\dfrac{\mu H\cos\theta}{k_B T}\right)\sin\theta d\theta}. \quad (3.34)$$

The magnetization M is defined as the sum of the magnetic moments in a unit volume. Hence, its component parallel to **H** is calculated as follows:

3.6 Pauli paramagnetism

$$M(T) = N\mu \langle \cos\theta \rangle$$

$$= N\mu \int_0^\pi \cos\theta\, p(\theta) d\theta$$

$$= N\mu \frac{\int_0^\pi \exp\left(\frac{\mu H \cos\theta}{k_B T}\right) \cos\theta \sin\theta\, d\theta}{\int_0^\pi \exp\left(\frac{\mu H \cos\theta}{k_B T}\right) \sin\theta\, d\theta}, \qquad (3.35)$$

where N is the number of electrons in a unit volume. Equation (3.35) is rewritten using new variables α and x defined as $\alpha = \mu H / k_B T$ and $x = \cos\theta$:

$$M(T) = N\mu \frac{\int_{-1}^{1} \exp(\alpha x) x\, dx}{\int_{-1}^{1} \exp(\alpha x)\, dx}$$

$$= N\mu \left[\left(\frac{e^\alpha + e^{-\alpha}}{e^\alpha - e^{-\alpha}}\right) - \frac{2}{\alpha}\right]$$

$$= N\mu \left(\coth\alpha - \frac{1}{\alpha}\right). \qquad (3.36)$$

The function $L(\alpha) = \coth\alpha - 1/\alpha$ appearing in equation (3.36) is often called the Langevin function. $L(\alpha) \to 1$ is obtained in the limit $\alpha \to \infty$. In other words, the magnetization M approaches $N\mu$ and all magnetic moments are aligned along the direction of the magnetic field, when the magnetic field is increased. As its opposite limit, we consider the case $\alpha \ll 1$, which corresponds to a very weak magnetic field or high temperatures. Since α is very small, the Langevin function $L(\alpha)$ can be expanded in a series:

$$L(\alpha) = \frac{\alpha}{3} - \frac{\alpha^3}{45} + \cdots. \qquad (3.37)$$

If only the first term in the series is retained, we obtain

$$M(T) = \frac{N\mu\alpha}{3} = \left(\frac{N\mu^2}{3k_B T}\right) H. \qquad (3.38)$$

Equation (3.38) indicates that the magnetization is proportional to the magnetic field and the magnetic susceptibility, defined as the ratio of M over H, is reduced to the form:

$$\chi(T) \equiv \frac{M(T)}{H} = \frac{\mu^2 N}{3k_B T}. \tag{3.39}$$

Thus, the magnetic susceptibility in a system consisting of N particles, each carrying magnetic moment μ, is found to be inversely proportional to the absolute temperature. This is known as the Curie law.

We find the Curie law to hold when electrons are treated as classical particles bearing a magnetic moment. But this is certainly not true for electron systems obeying the Pauli exclusion principle. We have to treat them using quantum statistics. The component of the magnetic moment along the direction of the magnetic field is quantized.[6] In Fig. 2.7, we assigned both a spin-up electron with $m_s = \frac{1}{2}$ and a spin-down electron with $m_s = -\frac{1}{2}$ to a given quantized state in **k**-space without differentiating between them. But now they have to be treated separately in the magnetic field, since a spin-up electron parallel to the field raises its energy by $\mu_B H$, whereas a spin-down electron antiparallel to the field lowers its energy by the same amount.[7] As a result, the density of states curves for spin-up and spin-down electrons are shifted relative to each other, as shown in Fig. 3.6.

We have just stated that the energy difference $2\mu_B H$ arises between spin-up and spin-down electrons when a magnetic field is applied. Alternatively, we may say that the energy $2\mu_B H$ is gained by flipping an electron spin along the energetically favorable direction. Because of the Pauli exclusion principle, only electrons in the range $k_B T$ in the neighborhood of the Fermi level are involved. The number of relevant electrons is roughly given by $N(E_F(0)) \cdot k_B T$ and, hence, N in equation (3.38) should be replaced by $N(E_F(0)) \cdot k_B T$. Equation (3.39) is thus rewritten as

$$\chi = \frac{\mu_B^2 N(E_F(0)) k_B T}{3 k_B T} = \frac{\mu_B^2 N(E_F(0))}{3} \tag{3.40}$$

---

[6] In quantum mechanics, a component of the magnetic moment along the direction of the magnetic field is expressed as $\mu_z = -g m_s \mu_B$, where m_s is the spin quantum number equal to $\frac{1}{2}$ or $-\frac{1}{2}$, the g-factor is equal to 2.0023 and μ_B is the Bohr magneton. Note that a minus sign arises from the fact that the electronic charge $(-e)$ is negative. The Bohr magneton $\mu_B = \mu_0 e\hbar/2m = 0.927 \times 10^{-20}$ erg·gauss^{-1} = 1.165×10^{-29} Wb·m refers to the magnetic moment on the atomic scale. The value of μ_z takes only two possible values $\mu_z = \mp \mu_B$ for $m_s = \pm \frac{1}{2}$, respectively. The interaction energy $U = -\boldsymbol{\mu} \cdot \mathbf{H}$ of the electron spin in the presence of a magnetic field is accordingly given by $U = \pm \mu_B H$.

[7] The names "spin-up" and "spin-down" are assigned to the positive and negative sign of the spin quantum number m_s, respectively.

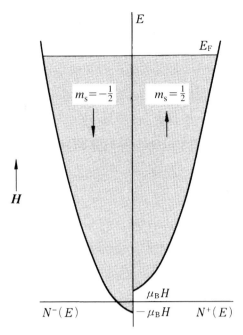

Figure 3.6. The free-electron density of states in the presence of a magnetic field, the direction of which is shown in the figure. The arrow in the bands indicates the direction of the electron spin. Note that the direction of the magnetic moment is opposite to it, as mentioned in footnote 6. The spin-up electrons with $m_s = \tfrac{1}{2}$ raise their energies by $\mu_B H$, whereas the spin-down electrons with $m_s = -\tfrac{1}{2}$ lower their energy by the same amount.

for conduction electrons subjected to quantum statistics. Equation (3.40) indicates that the magnetic susceptibility associated with spins of conduction electrons is temperature independent and proportional to the density of states at the Fermi level.

Equation (3.15) is now applied to calculate more rigorously the magnetic susceptibility at a finite temperature. As is clear from Fig. 3.6, the magnetization at an arbitrary temperature arises as a result of the shift of the spin-up band relative to the spin-down band and is expressed as

$$M(T) = \left(\frac{\mu_B}{2}\right) \int_0^\infty [N(E + \mu_B H) - N(E - \mu_B H)] f(E,T) dE, \qquad (3.41)$$

where the coefficient $\left(\tfrac{1}{2}\right)$ is introduced to reduce the density of states shown in Fig. 2.8 to one-half so as to differentiate spin-up and spin-down electrons. If the magnetic field is sufficiently small, the quantity in the square brackets can be expanded in a series:

$$M(T) = \left(\frac{\mu_B}{2}\right) \int_0^\infty \left\{ \left[N(E) + \mu_B H \frac{dN(E)}{dE} + \cdots \right] \right.$$

$$\left. - \left[N(E) - \mu_B H \frac{dN(E)}{dE} + \cdots \right] \right\} f(E,T) dE \cong \mu_B^2 H \int_0^\infty \left[\frac{dN(E)}{dE} \right] f(E,T) dE. \quad (3.42)$$

By taking $F(E) = \int_0^E \left[\frac{dN(E)}{dE} \right] dE$ in equation (3.42), we can write equation (3.15) as

$$I = \int_0^\infty \left[\frac{dN(E)}{dE} \right] f(E,T) dE$$

$$= N(E_F(T)) + \left(\frac{\pi^2}{6}\right)(k_B T)^2 \left[\frac{d^2 N(E)}{dE^2} \right]_{E=E_F(T)} + \cdots. \quad (3.43)$$

The magnetic susceptibility is then deduced to be

$$\chi(T) = \mu_B^2 N(E_F(0)) + O(T^2). \quad (3.44)$$

Note here that $N(E_F(T))$ is replaced by $N(E_F(0))$, since the difference between them is of the second-order correction proportional to T^2 and is included in $O(T^2)$ together with the second term of equation (3.43) (see Exercise 3.1). Equation (3.44) expresses the paramagnetic susceptibility due to spins of conduction electrons. This is known as the Pauli paramagnetism, since it was first derived by Pauli, as introduced in Section 1.2. It is proportional to the density of states at the Fermi level at absolute zero, $N(E_F(0))$, in the same manner as the electronic specific heat coefficient discussed in the preceding section.

The density of states at the Fermi level reduces to $N(E_F(0)) = 3N/2k_B T_F$ in the free-electron model. Hence the magnetic susceptibility for the free-electron system can be expressed as

$$\chi = \frac{3N\mu_B^2}{2k_B T_F}, \quad (3.45)$$

where T_F is the Fermi temperature. Equation (3.45) is now compared with equation (3.39) derived from classical statistics. Its ratio turns out to be

$$\frac{\chi_{el}^{\text{Fermi-Dirac}}}{\chi_{el}^{\text{classical}}} = \frac{\left(\frac{3\mu_B^2 N}{2k_B T_F}\right)}{\left(\frac{\mu_B^2 N}{3k_B T}\right)} = \frac{9}{2}\left(\frac{T}{T_F}\right). \quad (3.46)$$

3.6 Pauli paramagnetism

Here the value of μ in equation (3.39) is replaced by $\pm \mu_B$ to allow a direct comparison with equation (3.45). The ratio is calculated to be about 3/100 at room temperature, if $T_F = 5 \times 10^4$ K is inserted.

There are contributions to magnetic susceptibility other than the Pauli paramagnetism in metals. First, we consider the motion of conduction electrons in the presence of a magnetic field. They are subjected to a spiral motion due to the Lorentz force in the magnetic field (see Section 7.2). Suppose that the magnetic field is applied in the z-direction. The electron orbit is projected to a circle in the xy-plane and gives rise to a circular current. This yields a magnetic field opposite to the direction of the applied field. Magnetism induced in the direction opposite to the applied field is called diamagnetism. The magnetism associated with spiral motion caused by the Lorentz force is rigorously treated via quantum mechanics by Landau [3] and is called the Landau diamagnetism. The absolute value of the Landau diamagnetism becomes equal to one-third that of the Pauli paramagnetism in the free-electron model. The net magnetic susceptibility of the free-electron system is given by the sum of the Pauli paramagnetism and the Landau diamagnetism and is explicitly written in units of e.m.u./mol in the following form:

$$\chi_{\text{free}} = \left(\frac{2}{3}\right) \mu_B^2 N(E_F)$$
$$= 1.243 \times 10^{-6} (A/d)^{2/3} (e/a)^{1/3}, \qquad (3.47)$$

where A is the atomic weight in g, d is the density in (g/cm^3) and e/a is the number of electrons per atom.

Each inner core electron also contributes to the magnetic susceptibility. The applied magnetic field penetrates the orbit of each core electron and induces a current in the direction opposite to the field. This leads again to diamagnetism. The diamagnetism of ions is shown to be independent of temperature and its magnitude is comparable to that of the Pauli paramagnetism [4]. Therefore, there exist three temperature independent contributions to the magnetic susceptibility: Pauli paramagnetism, Landau diamagnetism and diamagnetism due to ions. Hence, their separation by utilizing the temperature dependence is not feasible in contrast to the electronic specific heat discussed in the preceding section.

We have so far studied the temperature independent para- and diamagnetism. Substances characterized by a sum of these temperature independent magnetic susceptibilities are called non-magnetic. They are treated separately from magnetic substances characterized by ferromagnetism, antiferromagnetism, paramagnetism obeying the Curie–Weiss law and other complicated magnetisms like spin-glass. More details concerning magnetic metals will be

Table 3.2. *Magnetic susceptibility in non-magnetic metals*

element	e/a	d (g/cm³)	A (g)	χ_{exp} (10⁻⁶/mol)	χ_{ion} (10⁻⁶/mol)	χ_{free} (10⁻⁶/mol)
Na	1.0	0.966	22.989	13.8	−3.7	10.3
K	1.0	0.909	39.102	17.9	−13.1	15.2
Cu	1.0	8.932	63.54	−5.46	−19.3	4.60
Ag	1.0	10.50	107.87	−19.52	−42	5.87
Au	1.0	19.281	196.967	−27.9	−58.4	5.85
Mg	2.0	1.737	24.305	6.07	−2.9	9.09
Zn	2.0	7.134	65.37	−9.15	−15.5	6.85
Cd	2.0	8.647	112.4	−19.67	−33.9	8.65
Al	3.0	2.698	26.981	16.19	−2.3	8.32

Note:
d: mass density, A: atomic weight, χ_{exp}: experimentally observed magnetic susceptibility, χ_{ion}: calculated magnetic susceptibility due to ions and χ_{free}: free-electron magnetic susceptibility calculated from equation (3.45).

discussed in Chapter 13. The magnetic susceptibility data for typical non-magnetic metals are summarized in Table 3.2.

3.7 Thermionic emission

We have studied, in the previous two sections, physical properties that conduction electrons exhibit and learned how the Fermi–Dirac distribution function determines their quantities and temperature dependence. In this section, we discuss the phenomenon of conduction electrons at high temperatures such that the Fermi–Dirac distribution function is replaced by the Maxwell–Boltzmann distribution function. Our aim is to determine under what conditions electrons confined in a metal can escape from it. It is more convenient to treat electrons as particles possessing a velocity **v** and a momentum **p**. As is clear from Fig. 3.7, a minimum energy ϕ required to remove an electron having the Fermi energy E_F from a metal is given by $\phi = E_0 - E_F$. Here E_0 is the work needed to remove to infinity an electron at the lowest energy state in the valence band and ϕ is called the work function. Obviously, an electron in a metal can be excited into a vacuum if its energy E is higher than $E_F + \phi$.

Let us take the yz-plane as the surface of a metal and consider the electron whose x-component of the velocity is v_x. The condition for the electron to escape from a metal is given by

$$\frac{mv_x^2}{2} = \frac{p_x^2}{2m} \geq E_F + \phi. \tag{3.48}$$

3.7 Thermionic emission

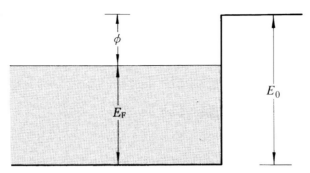

Figure 3.7. Work function ϕ in a metal. The shaded area represents the valence band. The energy E_0 is needed to excite an electron at the bottom of the valence band to the vacuum level.

For example, heating a metal in a vacuum causes some electrons to escape from the metal, since the tail of the Fermi–Dirac distribution function is extended to such high energies that there appear electrons satisfying equation (3.48). This phenomenon is called thermionic emission.[8]

The emission current density J_x per unit area of metal surface is expressed as

$$J_x = ne\langle v_x \rangle = \frac{ne\int_{v_{x_0}}^{\infty} v_x dN}{N} = \left(\frac{eN}{V}\right)\left(\frac{\int_{v_{x_0}}^{\infty} v_x dN}{N}\right), \qquad (3.49)$$

where $\langle v_x \rangle$ is an average velocity component v_x per electron, n is the number of electrons per unit volume $n = N/V$. A lower limit v_{x_0} of the integral in equation (3.49) is determined from the relation (3.48):

$$\frac{mv_{x_0}^2}{2} = \frac{p_{x_0}^2}{2m} = E_F + \phi.$$

The total number of electrons in a metal at temperature T can be written as

$$N = \left(\frac{2V}{8\pi^3}\right) \int_{-\infty}^{\infty}\int_{-\infty}^{\infty}\int_{-\infty}^{\infty} f(\mathbf{k},T) d\mathbf{k} = \left(\frac{2V}{8\pi^3}\right) \int_{-\infty}^{\infty}\int_{-\infty}^{\infty}\int_{-\infty}^{\infty} f(\mathbf{p},T) \frac{d\mathbf{p}}{\hbar^3}. \qquad (3.50)$$

Thus, one can easily derive the relation $dN = \left(\frac{2V}{8\pi^3\hbar^3}\right) f(\mathbf{p},T) d\mathbf{p}$ by taking its

---

[8] The entities ejected from a metal by heating it in a vacuum were originally thought to be ions and the term "thermionic" was adopted. Strictly, it should be called "thermoelectronic emission".

Table 3.3. *Thermionic properties of metals*

element	A (amp/cm$^2 \cdot$ K^2)	ϕ (eV)
W	75	4.5
Ta	55	4.2
Ni	30	4.6
Cr	48	4.6
Ca	60	3.2
Cs	160	1.8
Mo	60	4.3

Note:
A is defined by equation (3.54)

derivative. Equation (3.49) is now reduced to the form of

$$J_x = \left(\frac{e}{V}\right)\left(\frac{2V}{8\pi^3\hbar^3}\right) \int_{p_{x0}}^{\infty} \left(\frac{p_x}{m}\right)\left(\int_{-\infty}^{\infty}\int_{-\infty}^{\infty} f(p_x,p_y,p_z,T)dp_y\,dp_z\right)dp_x. \quad (3.51)$$

The work function ϕ in ordinary metals is of the order of a few eV, as listed in Table 3.3, and is far larger than the thermal energy $k_B T$. Hence, the relation $E - E_F > \phi \gg k_B T$ holds. This means that the Fermi–Dirac distribution function $f(\mathbf{p},T)$ can be well approximated by the Maxwell–Boltzmann distribution function:

$$f(\mathbf{p},T) \cong \exp\left[\frac{-(E-E_F)}{k_B T}\right] = \exp\frac{E_F}{k_B T}\exp\left[-\frac{(p_x^2+p_y^2+p_z^2)}{2mk_B T}\right]. \quad (3.52)$$

Equation (3.52) is inserted into equation (3.51). We can easily calculate integrals involved in equation (3.51) as follows:

$$\int_{-\infty}^{\infty}\exp\left(-\frac{p_y^2}{2mk_B T}\right)dp_y = \sqrt{2mk_B T}\int_{-\infty}^{\infty}e^{-y^2}dy = \sqrt{2m\pi k_B T}$$

and

$$\int_{-\infty}^{\infty}\exp\left(-\frac{p_z^2}{2mk_B T}\right)dp_z = \sqrt{2\pi mk_B T},$$

where $\int_{-\infty}^{\infty} e^{-y^2} dy = \sqrt{\pi}$ is used. The integral involving the x-component is different from the others and is calculated as

$$\int_{p_{x0}}^{\infty} \left(\frac{p_x}{m}\right) \exp\left(-\frac{p_x^2}{2mk_BT}\right) dp_x = 2k_BT \int_{x_0}^{\infty} x\exp(-x^2)\, dx = k_BT\exp(-x_0^2)$$

$$= k_BT\exp\left(-\frac{p_{x0}^2}{2mk_BT}\right) = k_BT\exp\left(-\frac{\phi + E_F}{k_BT}\right),$$

where we put $p_x = (2mk_BT)^{1/2}x$. A substitution of these results back into equation (3.51) leads to the well-known formula:

$$J_x = AT^2\exp\left(-\frac{\phi}{k_BT}\right), \tag{3.53}$$

where A is given by

$$A = \frac{4\pi m e k_B^2}{h^3} \approx 120\ \frac{\text{amp}}{\text{cm}^2\cdot\text{K}^2}. \tag{3.54}$$

Equation (3.53) is known as the Richardson–Dushman equation [5]. The emission current density J_x is measured as a function of temperature. The data fall on a straight line when $\log(J_x/T^2)$ is plotted against $1/T$. This implies that the use of Maxwell–Boltzmann statistics even for the conduction electron system is valid at high temperatures. The slope of the line determines the work function ϕ and the intercept the value of A. Some data for typical metals are listed in Table 3.3. The emission current is known to be very sensitive to the degree of oxidation of the metal surface. This is the main reason why the measured value often deviates from the theoretical value of 120 amp/cm²·K².

Exercise

3.1 The Pauli paramagnetism is given by equation (3.44). Calculate its second-order correction term $O(T^2)$, proportional to T^2. Estimate its contribution (%) relative to the first term for pure Cu in the free-electron model.

Chapter Four
Periodic lattice, and lattice vibrations in crystals

4.1 Prologue

In Chapter 3, we discussed fundamental properties manifested by conduction electrons without explicitly considering possible effects of a periodic ion potential. In other words, our discussion was limited to the context of the free-electron model. However, we will soon realize that the free-electron model is too simple and that the behavior of the conduction electrons in a real metal can be understood only if the effect of the periodic ion potential is properly taken into account. Prior to our discussion, we need to study lattice properties associated with the periodic array of ions. We learned in Chapters 2 and 3 that reciprocal space is convenient to describe the electronic state of a conduction electron subject to the Pauli exclusion principle. We show in this chapter that both static and dynamical properties of the periodic lattice are also conveniently described in terms of reciprocal space.

4.2 Periodic structure and reciprocal lattice vectors

An arbitrary perodic function $f(\mathbf{r})$ having the translational symmetry of period $\mathbf{l}$ is expressed as

$$f(\mathbf{r}+\mathbf{l})=f(\mathbf{r}), \tag{4.1}$$

where $\mathbf{r}$ is a position vector in three-dimensional real space. For the sake of simplicity, we consider a one-dimensional system, where equation (4.1) is reduced to $f(x+l)=f(x)$. Note that equation (4.1) is mathematically equivalent to equation (2.8), where the size L of a crystal is simply replaced by l. A crystal has another characteristic periodicity arising from a periodic arrangement of its atoms. This is the lattice constant a in a crystal, which is now taken as the period l in equation (4.1).

4.2 Periodic structure and reciprocal lattice vectors

Any periodic function with a lattice constant a can be expanded into the Fourier series:

$$f(x) = \sum_{n=-\infty}^{\infty} A_n e^{i\frac{2\pi}{a}nx}, \qquad (4.2)$$

where n represents integers ranging over $-\infty$ to ∞. A new variable g_n is assigned to the variable $(2\pi/a)n$ appearing in the exponent. Note that g_n is a discrete variable in reciprocal space. Equation (4.2) is then rewritten as

$$f(x) = \sum_{g_n=-\infty}^{\infty} A_{g_n} e^{ig_n x}. \qquad (4.3)$$

Since any lattice vector l in a one-dimensional crystal is expressed as $l_m = ma$ with an arbitrary integer m, we have the relation $e^{ig_n \cdot l_m} = 1$ for any $g_n = (2\pi/a)n$ with an arbitrary integer n. Indeed, this relation assures $f(x)$ to be a periodic function of a. The Fourier coefficient A_{g_n} in equation (4.3) is easily deduced to be

$$A_{g_n} = \left(\frac{1}{a}\right) \int_{\text{cell}} f(x) e^{-ig_n x} dx. \qquad (4.4)$$

It is important to realize that the coefficient A_{g_n} is determined solely from information about $f(x)$ in the unit cell with the lattice constant a.

The discussion above may be easily extended to a three-dimensional crystal characterized by the lattice constant a in x-, y- and z-directions. The lattice vector $\mathbf{l}$ is written as $a(l_x, l_y, l_z)$, where l_x, l_y and l_z are integers. Equation (4.3) is then generalized to

$$f(\mathbf{r}) = \sum_{g_{n_x n_y n_z}} A_{g_{n_x n_y n_z}} e^{i g_{n_x n_y n_z} \cdot \mathbf{r}}, \qquad (4.5)$$

where the vector $\mathbf{g}_{n_x n_y n_z}$ is defined as $(2\pi/a)(n_x, n_y, n_z)$ with integers n_x, n_y and n_z. An inner product of the vectors $\mathbf{g}_{n_x n_y n_z}$ and $\mathbf{l}_{l_x l_y l_z}$ results in

$$\mathbf{g}_{n_x n_y n_z} \cdot \mathbf{l}_{l_x l_y l_z} = \left(\frac{2\pi}{a}\right) n_x l_x a + \left(\frac{2\pi}{a}\right) n_y l_y a + \left(\frac{2\pi}{a}\right) n_z l_z a$$

$$= 2\pi \times (n_x l_x + n_y l_y + n_z l_z), \qquad (4.6)$$

justifying again the relation $\exp(i\mathbf{g}_{n_x n_y n_z} \cdot \mathbf{l}_{l_x l_y l_z}) = 1$.

Equation (4.5) can be further extended to more general three-dimensional periodic structures. The lattice vector $\mathbf{l}_{l_x l_y l_z}$ or simply $\mathbf{l}$ is written in the form of

$$\mathbf{l} = l_x \mathbf{a}_x + l_y \mathbf{a}_y + l_z \mathbf{a}_z, \qquad (4.7)$$

where $\mathbf{a}_x, \mathbf{a}_y$ and $\mathbf{a}_z$ are basic vectors (also called primitive translation vectors) in real space and l_x, l_y and l_z are integers to specify a particular lattice site. As its complementary quantity, we introduce the reciprocal lattice vector $\mathbf{g}_{n_x n_y n_z}$ or $\mathbf{g}$ defined as

$$\mathbf{g} = 2\pi(n_x \mathbf{b}_x + n_y \mathbf{b}_y + n_z \mathbf{b}_z), \tag{4.8}$$

where each basic vector $\mathbf{b}_i$ ($i=x, y$ and z) is chosen perpendicular to the plane formed by the two basic vectors $\mathbf{a}_j$ ($j=y, z$ and x) and $\mathbf{a}_k$ ($k=z, x$ and y) in real space. In other words, the relation $(\mathbf{a}_i \cdot \mathbf{b}_j) = 0$ holds in the case of $i \neq j$. In addition, the vectors are normalized so as to satisfy the relation $(\mathbf{a}_i \cdot \mathbf{b}_i) = 1$. The three basic vectors $\mathbf{b}_x, \mathbf{b}_y$ and $\mathbf{b}_z$ satisfying these conditions can be expressed[1] as

$$\mathbf{b}_x = \frac{[\mathbf{a}_y \mathbf{a}_z]}{(\mathbf{a}_x[\mathbf{a}_y \mathbf{a}_z])}, \quad \mathbf{b}_y = \frac{[\mathbf{a}_z \mathbf{a}_x]}{(\mathbf{a}_x[\mathbf{a}_y \mathbf{a}_z])} \text{ and } \mathbf{b}_z = \frac{[\mathbf{a}_x \mathbf{a}_y]}{(\mathbf{a}_x[\mathbf{a}_y \mathbf{a}_z])}. \tag{4.9}$$

An inner product of the two complementary vectors $\mathbf{l}$ and $\mathbf{g}$ results in multiples of 2π and, hence, the relation $e^{i\mathbf{g}\cdot\mathbf{l}} = 1$ holds again. The space encompassed by the basic vectors given by equation (4.9) or the $2\pi\mathbf{b}_x, 2\pi\mathbf{b}_y, 2\pi\mathbf{b}_z$ forms the reciprocal space because of the possession of a dimension reciprocal to the lattice vector $\mathbf{l}$ in real space.

The reciprocal lattice vector $\mathbf{g}$, say, for a simple cubic lattice is quantized in intervals of $2\pi/a$ and possesses the same dimension as the wave vector of conduction electrons defined in Section 2.5. Indeed, as mentioned above, equation (4.1) is of the same form as the periodic boundary condition of equation (2.8). The period in equation (2.8) is the system size or the edge length L of a metal cube. Instead, the lattice constant a in a crystal is taken as the period in equation (4.1). This difference is reflected in the interval of the quantization in reciprocal space. An interval of the quantized wave vector in equation (2.9) is $2\pi/L$, whereas it is now $2\pi/a$. The ratio L/a is of the order of 10^8 for a metal in a molar quantity, since the edge length L of the metal is about 1 cm while the lattice constant a is a few-tenths nm. It can be easily shown that the magnitude of the shortest reciprocal lattice vector appearing in equation (4.8) is $2\pi/a$, which is comparable to the Fermi diameter $2k_F$ obtained from equation (2.20).

Various band calculation techniques are discussed in Chapter 8, where we repeatedly use the series expansion like equation (4.5) in terms of the reciprocal lattice vector $\mathbf{g}_{n_x n_y n_z}$ ($-\infty < n_i < \infty$; $i=x, y$ and z) for any periodic function with the lattice constant a.

---

[1] It is easily checked that the relation $\mathbf{a}_x \cdot (\mathbf{a}_y \times \mathbf{a}_z) = \mathbf{a}_y \cdot (\mathbf{a}_z \times \mathbf{a}_x) = \mathbf{a}_z \cdot (\mathbf{a}_x \times \mathbf{a}_y) = V$ holds, where V represents the volume of the cell formed by the basis vectors.

4.3 Periodic lattice in real space and in reciprocal space

As is clear from the argument above, a three-dimensional arrangement of points satisfying the translational symmetry in the form of equation (4.7) forms a lattice and its space is called a space lattice. A crystal is made up of structurally identical units or bases (singular *basis*) consisting of either a single atom, a pair of atoms or even a molecule located at every lattice point in a Bravais lattice.[2] In the preceding section, we pointed out that the three basic vectors $\mathbf{a}_x$, $\mathbf{a}_y$ and $\mathbf{a}_z$ are defined in real space such that all lattice points can be mapped by the lattice vector given by equation (4.7). Likewise, the three basic vectors $\mathbf{b}_x$, $\mathbf{b}_y$ and $\mathbf{b}_z$ are defined in reciprocal space such that all reciprocal points can be mapped by the reciprocal lattice vector given by equation (4.8).

A simple cubic lattice with the lattice constant a is obtained, if the basic vectors $\mathbf{a}_x$, $\mathbf{a}_y$ and $\mathbf{a}_z$ are perpendicular to each other and their magnitudes are all equal to a. The corresponding reciprocal space is constructed by the set of the basic vectors $\mathbf{b}_x$, $\mathbf{b}_y$ and $\mathbf{b}_z$ being perpendicular to each other with their magnitude equal to $1/a$. It can be easily checked that each basic vector $\mathbf{b}_i$ is parallel to $\mathbf{a}_i$. Obviously, a simple cubic lattice is formed in reciprocal space and the reciprocal lattice vector $\mathbf{g}_{n_x n_y n_z}$ is expressed as $(2\pi/a)(n_x\, n_y\, n_z)$.

Let us take in a given space lattice three arbitrary lattice points, which do not happen to fall simultaneously on a single straight line. Then, we can define a plane on which these three lattice points are included. We can construct an infinite number of equivalent parallel planes in the space lattice. They are called a set of the lattice planes or crystal planes.

Figure 4.1 shows some important lattice planes in a cubic crystal. A set of the lattice planes can be specified by using the Miller indices. First, we find the intercept of a given plane with the crystal axes defined by the three basic vectors $\mathbf{a}_x$, $\mathbf{a}_y$ and $\mathbf{a}_z$. Then, the reciprocals of these three numbers are reduced to a set of the smallest integers by multiplying some common integer in the case of fractional numbers. The set of integers h, k and l thus obtained is enclosed in parentheses and is expressed as (hkl). These are called the Miller indices. Note here that the construction of the Miller indices involves the process of taking reciprocals and, hence, produces a quantity in reciprocal space. For example, let us consider the plane intersecting with the x- and y-axes at the coordinates $(1,0,0)$ and $(0,1,0)$ but parallel to the z-axis in the cartesian coordinates. The reciprocal of the intercept is given by $\left(\frac{1}{1}, \frac{1}{1}, \frac{1}{\infty}\right)$ and the Miller indices are denoted as (110).

---

[2] The arrangement of a unit assembly or *basis* in a crystal remains invariant under translation and symmetry operations [1]. The symmetry elements include the one-, two-, three-, four- and six-fold rotation axes about a lattice point, mirror plane, inversion center and rotation–inversion axes. In three-dimensional crystals, the thirty-two permissible point groups give rise to 14 different Bravais or space lattices.

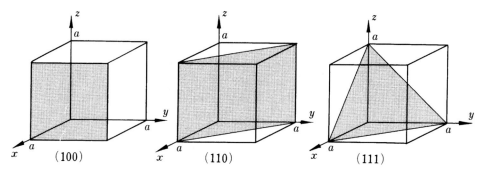

Figure 4.1. (100), (110) and (111) lattice planes in a cubic crystal.

We consider the lattice plane specified by the Miller indices (hkl) in a cubic crystal. By definition, the three coordinates, where one of the planes intersects with the x-, y- and z-axes, are given as $\left(\frac{N}{h},0,0\right)$, $\left(0,\frac{N}{k},0\right)$ and $\left(0,0,\frac{N}{l}\right)$, respectively, where N is a common multiple of h, k and l. Note that the number of possible N is infinite so that the Miller indices (hkl) refer to the set of an infinite number of parallel planes.

Let us consider a plane in the set of the (hkl) planes. Its equation is obviously expressed as

$$hx + ky + lz = N, \tag{4.10}$$

where N is an integer. Take a lattice point (x_o, y_o, z_o) in this plane. The lattice vector pointing to the coordinate (x_o, y_o, z_o) is given by $\mathbf{l}_{x_o y_o z_o} = x_o \mathbf{a}_x + y_o \mathbf{a}_y + z_o \mathbf{a}_z$. Consider the reciprocal lattice vector $\mathbf{g}_{hkl} = 2\pi(h\mathbf{b}_x + k\mathbf{b}_y + l\mathbf{b}_z)$, in which the three integers h, k and l refer to the Miller indices (hkl) of this plane. An inner product of these two vectors immediately results in

$$\mathbf{g} \cdot \mathbf{l} = 2\pi(hx_o + ky_o + lz_o) = 2\pi N \tag{4.11}$$

by using equations (4.9) and (4.10).

The left-hand side of equation (4.11) is rewritten as $|\mathbf{g}| \cdot |\mathbf{l}| \cos \theta = |\mathbf{g}| d_N$, where θ is the angle between $\mathbf{g}$ and $\mathbf{l}$ and $d_N \equiv |\mathbf{l}| \cos \theta$ is the length of the vector $\mathbf{l}$ projected on the vector $\mathbf{g}$, as shown in Fig. 4.2. Now equation (4.11) is reduced to

$$d_N = \frac{2\pi N}{|\mathbf{g}|}, \tag{4.12}$$

representing the distance from the origin to this plane. This implies that the reciprocal lattice vector $\mathbf{g}_{hkl}$ is perpendicular to the set of (hkl) lattice planes. It is also noted that the value of $d = d_N/N$ corresponds to the distance between the adjacent planes and is given by $d = 2\pi/|\mathbf{g}|$. To summarize, a single reciprocal

4.3 Periodic lattice in real space and in reciprocal space

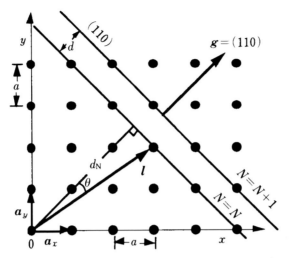

Figure 4.2. Cross-section cut across the xy-plane for a simple cubic crystal with lattice constant a. The z-axis is perpendicular to the page. The cross-section of the family of (110) lattice planes and the corresponding (110) reciprocal lattice vector are shown. They are perpendicular to each other.

lattice vector $\mathbf{g}_{hkl}$ is uniquely assigned to a set of (hkl) lattice planes and is perpendicular to it. This is illustrated schematically in Fig. 4.2.

The distance between adjacent planes in the set of the (hkl) planes in the simple cubic lattice is calculated from the relation $d = 2\pi/|\mathbf{g}|$:

$$d = \frac{a}{\sqrt{(h^2 + k^2 + l^2)}}, \qquad (4.13)$$

where a is the lattice constant. This means that the set of the lattice planes consisting of smaller Miller indices gives rise to a wider value of d. Since the atom density in a crystal is everywhere uniform, its density on a given plane becomes denser with increasing interplanar distance.[3] For example, Fig. 4.3 shows the (100) and (310) planes projected onto the xy-plane in the simple cubic lattice. It is seen that the atoms are more densely distributed on the (100) planes than on the (310) planes.

We now study the important role of the reciprocal lattice vector in relation to x-ray diffraction phenomena. Suppose that an incident x-ray with wavelength λ falls at a glancing angle θ on the crystal plane. We consider this situation in reciprocal space. As shown schematically in Fig. 4.4(a), the center of the

---

[3] The parellelepiped defined by the three basis vectors $\mathbf{a}_x$, $\mathbf{a}_y$ and $\mathbf{a}_z$ is called a primitive cell. There is always one lattice point per primitive cell. The argument in this paragraph holds true, as long as a primitive cell is concerned. But it is no longer valid when the Bravais lattice is employed. Note that the Bravais lattices for the fcc and bcc structures contain 4 and 2 atoms in their unit cells, respectively. The primitive cells for the bcc and fcc lattices are shown in Figs. 5.12 and 5.14, respectively.

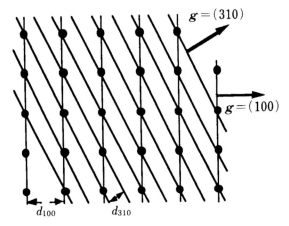

Figure 4.3. (100) and (310) lattice planes with the corresponding reciprocal lattice vectors.

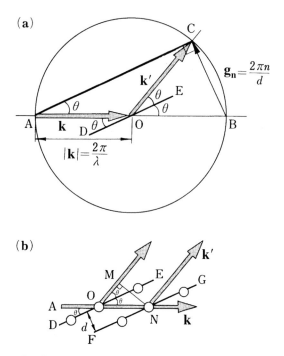

Figure 4.4. Bragg reflections in (a) reciprocal space and (b) real space. Open circles in (b) represent atoms to emphasize the real space.

4.3 Periodic lattice in real space and in reciprocal space

circle with the radius $2\pi/\lambda$ is positioned at a point O, where the incident x-ray intersects with the lattice plane DE. The line AO refers to the direction of the incident x-ray beam. The point B, at which the extrapolated line AO intersects with the circle, is chosen as the origin in reciprocal space. The point C represents the intersection of the reflected x-ray beam with the circle. A simple geometric consideration immediately leads to the relation AC // DE and, hence, the plane AC would possibly belong to the same set as the lattice plane DE. In addition, we see that the vector **BC** is always perpendicular to the set of the planes and, thus, it coincides with the direction of the reciprocal lattice vector. If the point C coincides with the reciprocal lattice point, then the vector **BC** obviously represents the reciprocal lattice vector $\mathbf{g_n}$ associated with the set of the lattice planes (DE and AC). Equation (4.12) tells us that the magnitude of the vector **BC** must be equal to multiples of $2\pi/d$ or $2\pi n/d$, where n is a positive integer, and the set of the lattice planes (DE and AC) is characterized by the interplanar distance d.

As a next step, we discuss how the geometry shown in Fig. 4.4(a) is related to x-ray diffraction phenomena. The circle in Fig. 4.4(a) represents the cross-section of a sphere in three-dimensional space. The sphere is called the Ewald sphere. The vector **AO** corresponds to the wave vector **k** of the incident x-ray beam, since its magnitude is equal to the radius $2\pi/\lambda$ of the Ewald sphere. The wave vector **k'** of the reflected x-ray beam makes an angle 2θ with the incident beam but its magnitude OC is still $2\pi/\lambda$. This indicates that the scattering of the x-ray beam due to the lattice planes is elastic. We see, therefore, that the two relations $\mathbf{k'} = \mathbf{k} + \mathbf{g_n}$ and $|\mathbf{k}| = |\mathbf{k'}|$ are simultaneously satisfied. This is called the Laue condition. The simple trigonometry in Fig. 4.4(a) easily proves the Laue condition to be equivalent to the Bragg law discussed below.

Figure 4.4(b) illustrates the situation where the incident x-ray beam is reflected from two successive crystal planes DE and FG in real space. It can be seen that the waves reflected at N in the plane FG take a path longer by the distance (ON–OM) than the waves reflected at O in the plane DE. A constructive interference between these two waves will occur when the path difference (ON–OM) equals a multiple of the wavelength or $n\lambda$. This is the basic idea for diffraction phenomena to occur and leads to the famous Bragg law $2d\sin\theta = n\lambda$, where d is the interplanar distance. The integer n is called the order of reflection. Reflections with $n=1$ and $n \geq 2$ are often referred to as the first- and higher-order reflections, respectively.

The interrelationship between Fig. 4.4(a) and (b) becomes clearer if we understand the physical meaning of the integer n appearing in the reciprocal lattice vector $\mathbf{g_n}$ in Fig. 4.4(a). The integer n is different from the variable N in

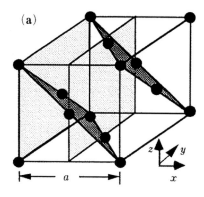
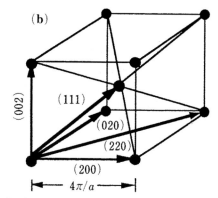

Figure 4.5. (a) (200) and (111) lattice planes in fcc unit cell. (b) Corresponding bcc unit cell in reciprocal space.

equation (4.12).[4] Let us denote the Miller indices for the set of lattice planes shown in Fig. 4.4(b) as (hkl) and their interplanar distance as d. The Bragg condition leads to $2d\sin\theta = \lambda$ and the magnitude of the corresponding reciprocal lattice vector is $2\pi/d$. Consider the reciprocal lattice vector $\mathbf{g}_{nh\,nk\,nl}$. Its magnitude is n times as large as that of the reciprocal lattice vector $\mathbf{g}_{hkl}$ and, hence, is equal to $2\pi/d' = 2\pi n/d$, which may be the case in Fig. 4.4(a). The corresponding Bragg condition $2d'\sin\theta = \lambda$ is alternatively written as $2d\sin\theta = n\lambda$. The diffracted x-ray is, therefore, considered as either a first-order reflection from the set of $(nh\,nk\,nl)$ planes with a spacing $d' (=d/n)$ or the n-th order reflection from the set of (hkl) lattice planes with the spacing d. For example, let us consider the set of (220) planes in the simple cubic lattice. The length of the reciprocal lattice vector (220) is twice as large as that of the reciprocal lattice vector (110). Note that the (220) planes in the simple cubic lattice do not contain atoms but the reciprocal lattice vector (220) is meaningful and refers to second-order reflections from the (110) planes.

As a practical example, let us take the fcc lattice with lattice constant a, as shown in Fig. 4.5(a). The lattice plane which intersects with the x-, y- and z-axes at $\left(\frac{a}{2},0,0\right)$, $(0,\infty,0)$ $(0,0,\infty)$, respectively, is expressed as (200) in the Miller indices. Planes which are parallel to this plane and separated by multiples of the interplanar distance $a/2$, are all equivalent and constitute the set of (200) planes. The corresponding reciprocal lattice vector is denoted as $\mathbf{g} =$

---

[4] The integer N in equation (4.12) is used to assign a particular lattice plane among an infinite number of equivalent lattice planes. On the other hand, the integer n appearing in Fig. 4.4(a) represents a higher-order reflection of diffracted waves. Since $d_N = Nd$ and $\mathbf{g}_n \equiv n\mathbf{g}$ hold, N and n refer to an integer in real and reciprocal spaces, respectively.

4.3 Periodic lattice in real space and in reciprocal space

$(2\pi/a)(200)$ or simply (200).[5] The (200) reciprocal lattice vector is directed along the k_x-axis in reciprocal space with a magnitude $|\mathbf{g}| = (2\pi/a) \cdot 2 = 4\pi/a$. The $(2n\,00)$ reciprocal lattice vector corresponds to the n-th-order reflections from the (200) planes. This is illustrated in Fig. 4.5(b).

The waves diffracted by the set of (100) planes in the fcc lattice cancel out as a result of the phase difference π because of the presence of an intervening plane (see the (200) plane in Fig. 4.5(a)) between two adjacent (100) planes and, hence, their intensity is reduced to zero. This is known as the extinction rule.[6] Hence, the (100) reciprocal lattice vector does not appear in the case of the fcc lattice. The reciprocal lattice vector $\mathbf{g}_{111}$ with a magnitude $2\pi\sqrt{3}/a$ is assigned to the set of the (111) planes. In a similar manner, we see that the lattice plane passing through the atoms at $(a/2, 0, a/2)$ and $(0, a/2, a/2)$ and parallel to the z-axis constitutes the set of the (220) planes. The corresponding $\mathbf{g}_{220}$ reciprocal lattice vector points to the corner of the cube in reciprocal space, as shown in Fig. 4.5(b).

The non-vanishing reciprocal lattice vectors associated with (111), (200), (020), (002), (220), (202), (022) and (222) planes obtained from the fcc lattice form the unit of the bcc lattice in reciprocal space, as shown in Fig. 4.5(b). The space outside the unit cell is filled by the reciprocal lattice vector of the higher-order reflections. For example, the reciprocal lattice vectors (400) and (600) represent the second- and third-order reflections from the set of the (200) planes. Thus, we see that the fcc structure in real space yields the bcc structure in reciprocal space. We can also easily check that the bcc structure in real space yields the fcc structure in reciprocal space.

To summarize, an infinite number of equivalent lattice planes in real space is reduced to a single reciprocal lattice vector in reciprocal space. It is not convenient to treat an infinite number of periodic lattice planes in real space. Instead, we can treat them all together as a single point in reciprocal space. This is made possible because of the periodic nature of the lattice structure in real space. We will study more about this unique feature in Chapter 5 in connection with the behavior of conduction electrons in a periodic potential field.

---

[5] Since the Miller indices are defined as a set of three mutually prime integers, the (nh, nk, nl) can be reduced to (hkl). For example, the (200) planes in the fcc lattice are reduced to (100) planes. This definition has been conventionally used in discussing the lattice planes in real space such as the slip planes in a crystal. However, the discussion in reciprocal space needs to differentiate the (200) planes from the (100) planes, as is explained here.

[6] The extinction rule for the fcc and bcc lattices is stated as follows. The diffraction intensity in the fcc lattice vanishes when the Miller indices involve both even and odd integers, whereas in the bcc lattice it vanishes when the sum of indices $h+k+l$ is odd.

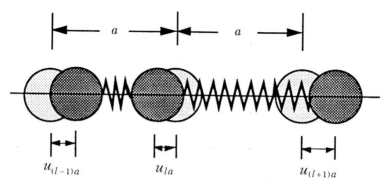

Figure 4.6. Lattice vibrations in a one-dimensional monatomic lattice with lattice constant a. The cases $u_{(l-1)a}>0$, $u_{la}<0$ and $u_{(l+1)a}>0$, are shown.

4.4 Lattice vibrations in one-dimensional monatomic lattice

We have so far treated the assembly of atoms as being fixed at their equilibrium positions in a perfect periodic lattice. This is the argument at absolute zero. But the atoms are vibrating about their equilibrium positions at finite temperatures, resulting in disruption of the periodicity of the lattice. The aim in this section is to study how disruption of the lattice periodicity due to thermal virbations can be treated in reciprocal space.

We consider a one-dimensional model, in which an infinite number of identical atoms of mass M are spaced periodically with the lattice constant a, as shown in Fig. 4.6. Each atom is connected to its neighboring atoms through a spring with a force constant β. This is called the linear chain model. A displacement of the l-th atom or the atom at the position la is denoted as u_{la}, where l is an integer. The equation of motion of the l-th atom is then expressed as

$$M\frac{d^2 u_{la}}{dt^2} = \beta(u_{(l+1)a} + u_{(l-1)a} - 2u_{la}). \tag{4.14}$$

Such an equation can be written for every atom in a linear chain. Hence, we have an infinite number of similar equations of motion in real space. To avoid handling an infinite number of equations, we transform equation (4.14) into reciprocal space by assuming its solution in the form

$$u_{la} = \xi(t)e^{iqla}. \tag{4.15}$$

An insertion of equation (4.15) into equation (4.14) yields

$$M\frac{d^2\xi}{dt^2} = \beta(e^{iqa} + e^{-iqa} - 2)\xi$$

4.4 Lattice vibrations in one-dimensional monatomic lattice

$$= \left[-4\beta\sin^2\left(\frac{qa}{2}\right)\right]\xi. \qquad (4.16)$$

Note here that this transformation can eliminate the variable l associated with a position of the atom and gives rise to the equation of motion for a simple-harmonic oscillator. Indeed, an infinite number of equations in real space are replaced by a single equation for the amplitude of the wave with the wave number q, which describes the collective motion of atoms over a linear chain. This is called a lattice wave, characterized by a wave number in reciprocal space.

As long as the stationary state is concerned, the time-dependent amplitude $\xi(t)$ must take the form of $\xi(t)=\xi_0\exp(i\omega t)$. Then, we easily obtain a solution of equation (4.16) as

$$\omega = \sqrt{\frac{4\beta}{M}}\left|\sin\left(\frac{qa}{2}\right)\right|. \qquad (4.17)$$

Equation (4.17) gives the relation between the angular frequency ω and wave number q and is called the dispersion relation for the lattice wave. As shown in Fig. 4.7, the frequency ω takes it maximum value of $\omega_{max} = \sqrt{(4\beta/M)}$ at $q_{max} = \pm\pi/a$.

As is clear from equation (4.15), motions of individual atoms are transformed into lattice waves propagating throughout the lattice and are best described in terms of the wave number in reciprocal space. But, the entities responsible for the displacement are periodically arranged atoms with a lattice constant a. This feature must be reflected in reciprocal space. The lattice vector in one-dimensional (real) space is expressed as $l_m = ma$ (m: integer), whereas the corresponding reciprocal lattice vector is expressed as $g_n = (2\pi/a)n$ (n: integer) where l and g are not shown in bold type in one-dimensional (reciprocal) space. Hence, the relation $e^{ig\cdot l}=1$ always holds between them. Let us choose an arbitrary wave vector[7] q outside the region $-\pi/a < q \leq \pi/a$ and add the reciprocal lattice vector $g_n = (2\pi/a)n$ so as to bring $q+g_n$ into the region $-\pi/a < q \leq \pi/a$. According to equation (4.15), the displacement of the m-th atom is given by $u_{ma} = \xi e^{i(q+g_n)ma}$ but is reduced to $u_{ma} = \xi e^{iqma}$ because $e^{i[(2\pi/a)n]\cdot ma}=1$. This holds for an arbitrary atom in the chain. Therefore, we confirm that lattice waves with wave vectors q and $q+g_n$ give rise to the same atom displacements in the linear chain. This proves that a given normal mode of lattice vibrations in real space can be equally described by any lattice waves characterized by the wave vectors differing by the reciprocal lattice vector (see Exercise 4.1 and Fig. 4A.1). Therefore,

---

[7] The words "wave vector" are intentionally used in place of "wave number" in spite of this being in one-dimensional space.

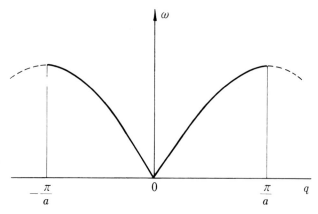

Figure 4.7. Dispersion relation of the lattice vibrations for a one-dimensional monatomic lattice with lattice constant a.

we need to consider wave vectors only in the range $-\pi/a < q \leq \pi/a$ in the description of lattice vibrations. This region in reciprocal space is called the first Brillouin zone and is obtained by bisecting the shortest reciprocal lattice vectors $g_1 = 2\pi/a$ and $g_{-1} = -2\pi/a$.

Equation (4.17) can be approximated as $\omega = (\beta/M)^{1/2} aq$, when the wave number q is small ($aq \ll 1$). A linearly q-dependent frequency reminds us of the well-known formula $\omega = sq = (c/\rho)^{1/2} q$ for vibrations of an elastic medium like a rubber [1]. Here s is the sound velocity in cm/s, c is the elastic stiffness constant in dyne/cm² and ρ is the density in g/cm³. One can easily check that $\beta = ca$ holds, since β is in units of dyne/cm and $\rho = M/a^3$ in a three-dimensional system. Though atoms are assumed to be periodically spaced with the lattice constant a in the linear chain model, we see that, in the long wavelength limit where $aq \ll 1$ holds, the system can be treated as an elastic continuum. This is indeed the region where the wavelength of the lattice wave is much longer than the lattice constant. It must be noted that the transformation (4.15) is based on the periodicity of the lattice with a period a. This yields the periodic structure with the period of $2\pi/a$ in reciprocal space, as discussed in the preceding section. Therefore, the dispersion relation in the range $-\pi/a < q \leq \pi/a$ shown in Fig. 4.7 can be extended periodically outside this range in reciprocal space.

4.5 Lattice vibrations in a crystal

We saw in Section 2.5 that the wave vector of the conduction electron is quantized by imposing the periodic boundary condition to its wave function. The same boundary condition must be applied to lattice vibrations because of the finite size of a crystal. A one-dimensional crystal containing N atoms with the

4.5 Lattice vibrations in a crystal

lattice constant a is first considered. Its total length is obviously $L = Na$.[8] The periodic boundary condition is applied in such a way that the atom at the position $l' = l$ coincides with that at the position $l' = l + L$ by connecting both ends of a linear chain. Now the displacement of the atom at the position l should be the same as that of the atom at the position $(l+L)$:

$$u(l) = u(l+L). \tag{4.18}$$

The periodic boundary condition (4.18) is equivalent to that given by equation (2.8) for the conduction electron and equation (4.1) for any periodic function. By inserting equation (4.15) into equation (4.18), we immediately deduce the relation $e^{iqL} = 1$, which results in[9]

$$q = \frac{2\pi}{L} n \left(n = \pm 1, \pm 2, \ldots, \pm \frac{N}{2} \right). \tag{4.19}$$

Equation (4.19) indicates that the wave number of the lattice wave is quantized and can take only a discrete set of values. One has to recognize an essential difference between equations (4.19) and (2.9). In the case of the conduction electron system, there exists no limitation to the quantized wave number given by equation (2.9). On the other hand, a one-dimensional linear chain consists of N atoms and, hence, the degrees of freedom are limited to N. The degrees of freedom must be conserved upon the transformation from real space to reciprocal space. This implies that the number of allowed integers n in equation (4.19) must be N, which results in $n = N/2$ and $n = -N/2$ as the maximum and minimum in n, respectively. The corresponding maximum and minimum wave numbers are immediately reduced to $q_{max} = \pi/a$ and $q_{min} = -\pi/a$ from equation (4.19). This agrees with an upper and a lower limit of the first Brillouin zone discussed in Section 4.4. To summarize, lattice vibrations in a one-dimensional periodic lattice consisting of N atoms with the lattice constant a are described as N independent lattice waves, which are often referred to as N modes of lattice waves. They are discretely distributed in intervals of $2\pi/L$ in the range $-\pi/a < q \leq \pi/a$.

The discussion above can be easily extended to a three-dimensional system. We consider a rectangular crystal with edge length L_x, L_y and L_z, which crystallizes in a simple cubic structure with the lattice constant a. Since $L_i = N_i a$ ($i = x, y, z$) holds along the three crystal axes, the total degrees of freedom are

---

[8] In Section 2.6, a uniform potential was assumed in a box with edge length L. Here, in the discussion of lattice vibrations, N atoms are periodically spaced with a lattice constant a in a box with edge length $L = Na$.

[9] Equation (4.18) assures the translational symmetry with the period L. Hence, the function $u(l)$ can be expanded in the Fourier series as $u(l) = \sum_{n=-N/2}^{N/2} B_n e^{i(2\pi/L)nl}$ (see Table 4.1).

obviously equal to $3N_xN_yN_z$ or simply $3N$, where $N=N_xN_yN_z$ is the total number of atoms in the crystal. The corresponding lattice waves should also possess $3N$ degrees of freedom. An extension of the argument for the one-dimensional linear chain immediately leads to the conclusion that x-, y- and z-components of the wave vector $\mathbf{q}$ are quantized in the form of $[(2\pi/L_x)n_x, (2\pi/L_y)n_y, (2p/L_z)n_z]$, when the periodic boundary condition (4.18) is applied to the respective directions in real space. Obviously, the integer n_i is confirmed in the region $-N_i/2 < n_i \leq N_i/2$ ($i=x,y,z$) and, hence, the component of the wave vector is confined in the region $-\pi/a < q_i \leq \pi/a$ ($i=x,y,z$).

There are six shortest equivalent reciprocal lattice vectors in a simple cubic lattice with the lattice constant a. They are denoted as $\mathbf{g}=(2\pi/a)(\pm 100)$, $\mathbf{g}=(2\pi/a)(0\pm 10)$ and $\mathbf{g}=(2\pi/a)(00\pm 1)$. The cube bounded by $\pi/a < q_i \leq \pi/a$ ($i=x,y,z$) in reciprocal space is constructed by bisecting perpendicularly these six equivalent reciprocal lattice vectors.[10] The cube thus obtained with edge length $2\pi/a$ is called the first Brillouin zone for the simple cubic lattice. As is clear from the argument above, N independent wave vectors are accommodated in the first Brillouin zone. However, this is only one-third the degrees of freedom in the three-dimensional lattice consisting of N atoms, since each atom has three degrees of freedom. Hence, the lattice wave for a three-dimensional crystal cannot be uniquely specified even when N independent wave vectors are designated. The remaining degrees of freedom can be specified in relation to the freedom associated with the direction of the displacement of atoms in a crystal.

Two transverse and one longitudinal wave modes can exist for each wave vector. The transverse wave mode refers to the wave where the displacement of atoms is perpendicular to the propagation direction or the direction of the wave vector. There are two transverse modes corresponding to two degrees of freedom associated with the in-plane motion of atoms. On the other hand, the longitudinal wave mode refers to the wave where the displacement of atoms is parallel to the propagation direction and, thus, there is only one degree of freedom in the longitudinal wave. In this way, the lattice wave in the three-dimensional lattice can be uniquely assigned by specifying the wave vector $\mathbf{q}$ and the direction of the displacement of atoms, the latter of which is differentiated by the type of polarization p_i ($i=1, 2, 3$). To summarize, lattice vibrations in a three-dimensional crystal consisting of N atoms are described by $3N$ independent lattice wave modes in the first Brillouin zone.

---

[10] The states $q_i = \pi/a$ and $q_i = -\pi/a$ are shown to be identical and, hence, the first Brillouin zone is defined in the range $-\pi/a < q_i \leq \pi/a$.

4.6 Lattice waves and phonons

As mentioned in Section 1.2, Planck explained the wavelength dependence of the intensity of radiation emitted from a black-body at high temperatures by assuming that a light wave is composed of a set of oscillators and that the energy of an oscillator with the frequency v is quantized in discrete units of hv in the form

$$E = nhv \quad (n = 0, 1, 2, 3, \ldots), \tag{4.20}$$

where h is the Planck constant. This is indeed a revolutionary idea, since an oscillator takes an arbitrary energy in classical theory.

In 1905, Einstein explained the photoelectric effect by assuming that light is made up of photons having discrete energies given by equation (4.20). Two years later, Einstein further extended the concept of the particle-like nature of a light wave to lattice vibrations and calculated the temperature dependence of the lattice specific heat by assuming lattice vibrations to be described as an assembly of harmonic oscillators possessing a common frequency v_0. The Einstein model could successfully account for a rapid decrease in the lattice specific heat from $3R$ with decreasing temperature below room temperature (see Fig. 4.10).

According to equation (4.16), the motion of the lattice wave in each mode can be described by that of a harmonic oscillator. In quantum mechanics, the energy eigenvalue of a harmonic oscillator with wave vector $\mathbf{q}$ and angular frequency $\omega_{\mathbf{q}}$ can be expressed as

$$E_{\mathbf{q},p_i} = \left(n_{\mathbf{q},p_i} + \tfrac{1}{2}\right) \hbar \omega_{\mathbf{q},p_i} \quad (n_{\mathbf{q}} = 0, 1, 2, 3, \ldots), \tag{4.21}$$

where $\hbar = h/2\pi$, $\omega = 2\pi v$ and p_i ($i = 1, 2, 3$) is the type of polarization [1]. For brevity, the suffix p_i is hereafter dropped, unless otherwise stated. We say from equation (4.21) that the lattice vibration in the mode $\mathbf{q}$ gives rise to an energy state, which is gained by creating $n_{\mathbf{q}}$ particles having energy $\hbar \omega_{\mathbf{q}}$. The quantum of energy for excitations of the lattice wave is called a phonon in analogy with the photon in the electromagnetic wave. It is stated that a given lattice mode is excited to $n_{\mathbf{q}}$ phonons or the mode is occupied by $n_{\mathbf{q}}$ phonons. As temperature increases, lattice waves at shorter wavelengths or larger wave vectors are more excited. In the next section, the number of phonons at a finite temperature T is calculated.

4.7 Bose–Einstein distribution function

The Pauli exclusion principle acts on particles with a half-integer spin and forces only one particle to be assigned to a given quantum state. On this basis,

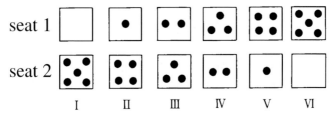

Figure 4.8. Ways of distribution in the case of $Z=2$ and $N=5$ in Bose–Einstein statistics.

in Section 3.2, the Fermi–Dirac distribution function was deduced. In contrast, the Pauli exclusion principle is no longer applicable to particles with integer spin including zero. Both photons and phonons are in this family, since their spin is zero. Thus, any number of particles can be assigned to a given quantum state. Quantum statistics in this family is described by the Bose–Einstein distribution function, the derivation of which is shown below.

For the sake of simplicity, we show in Fig. 4.8 all possible ways in which five indistinguishable particles are distributed over two seats. It is possible to put all five particles in seat '1' or 'conversely' to put them in seat '2'. There are, in total, six different ways as shown in Fig. 4.8. These are the ways in which the particles obeying the Bose–Einstein distribution function are distributed.

The argument above can be easily extended to the calculation for the ways of distributing N_i particles over Z_i seats at energy E_i. Let us remind ourselves how we deduced the ways of distribution in the case of $N=5$ and $Z=2$. One realizes that the illustration shown in Fig. 4.8 is equivalent to calculating the coefficient of x^5 in the expansion of the binomial $(1+x+x^2+\cdots)^2$. In general, the ways of distribution of N_i particles over Z_i seats can be obtained by calculating the coefficient x^{N_i} in the binomial $(1+x+x^2+\cdots)^{Z_i}$. The binomial can be expanded as

$$(1+x+x^2+\cdots)^{Z_i}$$
$$=(1-x)^{-Z_i}$$
$$=1+Z_i x+\frac{Z_i(Z_i+1)}{2!}x^2+\cdots+\frac{Z_i(Z_i+1)\cdots(Z_i+N_i-1)}{N_i!}x^{N_i}+\cdots. \quad (4.22)$$

Hence, the coefficient of x^{N_i} is obtained as

$$\frac{Z_i(Z_i+1)\cdots(Z_i+N_i-1)}{N_i!}=\frac{(Z_i+N_i-1)!}{N_i!(Z_i-1)!}. \quad (4.23)$$

The total ways of distributing particles with integer spin are given by a product of the terms like equation (4.23) over all possible energy states E_i:

4.7 Bose–Einstein distribution function

$$W = \prod_i \frac{(Z_i + N_i - 1)!}{N_i!(Z_i - 1)!}. \tag{4.24}$$

As discussed in Section 3.2, we must find a maximum in $\ln W$ in place of W under the conditions that the total energy $E = \sum_i E_i N_i$ and the total number of particles $N = \sum_i N_i$ are conserved. The method of Lagrangian multipliers is again employed. The resulting equation is explicitly written as

$$\frac{\partial}{\partial N_j}\left[\ln W + \alpha\left(N - \sum_{i=1}^{\infty} N_i\right) + \beta\left(E - \sum_{i=1}^{\infty} N_i E_i\right)\right] = 0. \tag{4.25}$$

A straightforward calculation by inserting equation (4.24) into (4.25) leads to

$$\frac{N_i}{Z_i} = \frac{1}{e^{\alpha + \beta E_i} - 1}. \tag{4.26}$$

In this section, we try to derive the coefficients α and β in a way different from that described in Section 3.2. Equation (4.25) may be expressed in the form of a total derivative:

$$\delta(\ln W - \alpha N - \beta E) = 0. \tag{4.27}$$

The $\ln W$ in equation (4.27) is related to entropy S through the Boltzmann relation $S = k_B \ln W$. Now equation (4.27) is rewritten as

$$\delta\left(\frac{S}{k_B} - \alpha N - \beta U\right) = 0, \tag{4.28}$$

where the energy E is replaced by the symbol U to emphasize the internal energy of a system. The coefficients α and β are obviously derived from equation (4.28) as $\alpha = (1/k_B)(\partial S/\partial N)_{UV}$ and $\beta = (1/k_B)(\partial S/\partial U)_{NV}$. According to the thermodynamics, we have the relation

$$TdS = dU + pdV - \zeta dN, \tag{4.29}$$

where ζ is the chemical potential and p is pressure. A comparison of equations (4.28) with (4.29) under the condition $dV = 0$ immediately leads to $\alpha = -\zeta/k_B T$ and $\beta = 1/k_B T$. In this way, we arrive at the final form of equation (4.26):

$$n(E, T) = \frac{1}{\exp[(E - \zeta)/k_B T] - 1}. \tag{4.30}$$

This is the Bose–Einstein distribution function at temperature T.

The total number of particles is not conserved in a system such as phonons, since it is zero at absolute zero but increases with increasing temperature.

Hence, the condition $N=\Sigma_i N_i$ must be omitted and the coefficient α disappears from equation (4.27). The resulting distribution function turns out to be

$$n(E, T) = \frac{1}{\exp(E/k_B T) - 1}. \quad (4.31)$$

This is often called the Planck distribution function. The number of phonons excited at a finite temperature T should obey the Planck distribution function. Since the energy of phonons is given by equation (4.21),[11] its insertion into equation (4.31) yields the Planck distribution function for phonons:

$$n_q(T) = \frac{1}{\exp(\hbar\omega_q/k_B T) - 1}. \quad (4.32)$$

By using equation (4.32), we can express the internal energy due to lattice vibrations at temperature T as follows:

$$U_{\text{lattice}}(T) = \sum_{p_i}\sum_q n_{q,p_i}(T)\hbar\omega_{q,p_i} = \sum_{p_i}\sum_q \hbar\omega_{q,p_i} \Big/ \left[\frac{1}{\exp(\hbar\omega_{q,p_i}/k_B T) - 1}\right], \quad (4.33)$$

where the sum is taken over all permissible wave vectors $\mathbf{q}$ and types of polarization p_i. If the thermal energy $k_B T$ exceeds the maximum possible phonon energy $\hbar\omega_q$ at high temperatures, the exponential function can be expanded and the lattice energy for a molar quantity is reduced to $3k_B N_A T$ or $3RT$, consistent with the equipartition law derived from the classical theory. However, when the temperature is lowered, phonons preferentially occupy low energy states in accordance with the Planck distribution function. We will discuss this situation in more detail in the next section.

4.8 Lattice specific heat

We discussed in Section 3.4 the specific heat of conduction electrons obeying Fermi–Dirac statistics and compared it with that derived from classical statistics. According to the equipartition law, the energy $k_B T/2$ is partitioned to each degree of freedom. In the case of lattice vibrations, the total energy per atom consists of its kinetic energy and potential energy, thus resulting in an energy of $3k_B T$ per atom in a three-dimensional system.[12] This leads to the lattice

---

[11] Note that the zero-point energy appearing in equation (4.21) is omitted.
[12] We consider a system in which the Hamiltonian H is expressed as $H = \Sigma_i(\alpha_i p_i^2 + \beta_i q_i^2)$ in terms of generalized momenta $p_1, p_2, p_3, \ldots$ and generalized coordinates $q_1, q_2, q_3, \ldots$. In thermal equilibrium, $\langle\alpha_i p_i^2\rangle = k_B T/2$ and $\langle\beta_i q_i^2\rangle = k_B T/2$ hold, where $\langle\,\rangle$ means a thermal average. This is known as the equipartition law of energy.

4.8 Lattice specific heat

specific heat per mole equal to $3k_B N_A = 3R$ and explains the Dulong–Petit law in the framework of the classical equipartition law.

The lattice specific heat can be calculated over a whole temperature range by utilizing the concept of phonons and the Planck distribution function. For this purpose, we need to calculate the internal energy given by equation (4.33) for lattice vibrations. Equation (4.33) is, however, not convenient because of the summation over wave vectors. In the case of conduction electrons, the internal energy was calculated by employing the density of states defined by equation (3.20). In the same spirit, we replace the summation over a vector quantity by an integral over a scalar quantity. We introduce the phonon density of states or the frequency spectrum $D(\omega)d\omega$, which represents the number of lattice modes in a frequency interval between ω and $\omega + d\omega$. By using the frequency spectrum, we can rewrite equation (4.33) as

$$U_{\text{lattice}}(T) = \int_0^{\omega_D} \hbar\omega D(\omega) n(\omega, T) d\omega, \tag{4.34}$$

where $\hbar\omega$ is the energy of phonons and $n(\omega, T)$ is the Planck distribution function given by equation (4.32). An upper limit ω_D in the integral represents the maximum frequency available in the phonon system. Its derivation will be shown below. Equation (4.34) has the same form as equation (3.20) for the internal energy of conduction electrons.

The frequency spectrum must be calculated for a given solid. As mentioned earlier, Einstein regarded lattice vibrations of N atoms as an assembly of $3N$ independent harmonic oscillators with a given frequency ω_0. Specifically, the Einstein model assumes the delta-function at $\omega = \omega_0$ in the frequency spectrum. However, as mentioned in Section 3.5, the Debye model is more appropriate to describe the lattice specific heat over a wide temperature range.

In the Debye model, the dispersion relation $\omega = sq$ is assumed over the frequency range where the integral in equation (4.34) is carried out. Here a proportional constant s represents the sound velocity propagating through a substance. Strictly speaking, the sound velocity depends not only on the type of polarization but also on the direction along which it propagates (see Fig. 4.14). The Debye model ignores such complexities and assumes an averaged velocity s, irrespective of the polarization modes as well as the directions of the sound propagation. For example, the dispersion relation shown in Fig. 4.7 obtained from the linear chain model conforms well with the Debye model $\omega = sq$ at small wave numbers.

The wave vector in reciprocal space is quantized in intervals of $2\pi/L$. Since three degrees of freedom are assigned to each wave vector in a three-dimensional system, the number of independent lattice modes, $N(q)$, enclosed by a sphere with a radius q is calculated from the following proportional relation:

$$\frac{4\pi q^3}{3} : N(q) = \left(\frac{2\pi}{L}\right)^3 : 3 \tag{4.35}$$

or

$$N(q) = \frac{V}{2\pi^2} q^3. \tag{4.36}$$

The total number of independent lattice modes must be equal to $3N$ for a system containing N atoms. Since this gives a maximum in $N(q)$, the corresponding maximum wave number q_D, which is called the Debye radius, is obtained as

$$q_D = \left(\frac{6\pi^2 N}{V}\right)^{1/3}. \tag{4.37}$$

The sphere with the radius q_D is called the Debye sphere and its volume is equal to that of the first Brillouin zone in reciprocal space. It can be easily checked that $1/q_D$ is of the order of an atomic distance. The maximum or cut-off frequency ω_D corresponding to the Debye radius is therefore given by

$$\omega_D = s\left(6\pi^2 \frac{N}{V}\right)^{1/3}, \tag{4.38}$$

which is employed as an upper limit of the integral in equation (4.34).

Since the frequency spectrum $D(\omega)d\omega$ represents the number of lattice modes in the interval ω and $\omega + d\omega$ per unit volume, it is easily calculated in the Debye model as follows:

$$D(\omega)d\omega = \left(\frac{dN(q)}{dq}\right)\left(\frac{dq}{d\omega}\right)d\omega = \frac{d}{dq}\left(\frac{q^3}{2\pi^2}\right)\left(\frac{dq}{d\omega}\right)d\omega$$

$$= \left(\frac{3q^2}{2\pi^2}\right)\left(\frac{1}{s}\right)d\omega = \left(\frac{3\omega^2}{2\pi^2}\right)\left(\frac{1}{s^3}\right)d\omega, \tag{4.39}$$

where the dispersion relation $\omega = sq$ is inserted. The frequency spectrum given by equation (4.39) is depicted in Fig. 4.9. It is proportional to ω^2 and is cut off at the Debye frequency ω_D.

The Debye temperature is defined as the characteristic temperature corresponding to the cut-off frequency through the relation $\hbar\omega_D = k_B \Theta_D$ and is explicitly written as

4.8 Lattice specific heat

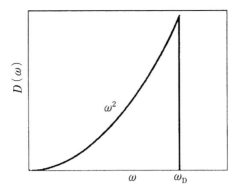

Figure 4.9. Frequency spectrum $D(\omega)$ in the Debye model. $D(\omega)$ is also called the phonon density of states. The cut-off frequency ω_D is determined from the equation

$$3N = \int_0^{\omega_D} D(\omega) d\omega.$$

$$\Theta_D = \left(\frac{\hbar s}{k_B}\right)\left(6\pi^2 \frac{N}{V}\right)^{1/3}. \quad (4.40)$$

The Debye temperature depends linearly on the sound velocity s propagating in a substance.

The internal energy of lattice vibrations is now explicitly written by inserting equations (4.32) and (4.39) into equation (4.34):

$$U_{\text{lattice}} = \left(\frac{3k_B^4 T^4}{2\pi^2 \hbar^3 s^3}\right)\int_0^{x_D} \frac{x^3 dx}{e^x - 1} \quad (4.41)$$

where $x = \hbar\omega/k_B T$ and

$$x_D = \frac{\hbar\omega_D}{k_B T} = \frac{\hbar s q_D}{k_B T} = \frac{\hbar s}{k_B T}\left(6\pi^2 \frac{N}{V}\right)^{1/3} = \frac{\Theta_D}{T}. \quad (4.42)$$

The lattice specific heat in the Debye model is obtained by differentiating equation (4.41) with respect to temperature:

$$C_{\text{lattice}} = 9Nk_B \left(\frac{T}{\Theta_D}\right)^3 \int_0^{x_D} \frac{e^x x^4 dx}{(e^x - 1)^2}. \quad (4.43)$$

Equation (4.43) is called the Debye formula for the lattice specific heat, indicating that the lattice specific heat can be calculated as a function of temperature, once the Debye temperature is given.

It is interesting at this stage to examine the Debye formula at high and low temperatures. At high temperatures satisfying $T \gg \Theta_D$, the upper limit x_D in the integral becomes small and, hence, the variable x must be small. Thus, the exponential function can be expanded into a series. In this limit, we can easily see that the lattice specific heat approaches $3R$, in good agreement with the value expected from the classical equipartition law discussed earlier. In contrast, the upper limit in the integral may be replaced by infinity at low temperatures $T \ll \Theta_D$. Now the integral in equation (4.41) is reduced to

$$\int_0^\infty \frac{x^3 dx}{e^x - 1} = 6\zeta(4) = 6\sum_{n=1}^\infty \frac{1}{n^4} = \frac{\pi^4}{15}, \tag{4.44}$$

where $\zeta(n)$ is called the Riemann zeta function. Thus, the internal energy at low temperatures $T \ll \Theta_D$ is calculated as

$$U_{\text{lattice}} = \frac{3\pi^4 N k_B T^4}{5\Theta_D^3}. \tag{4.45}$$

The lattice specific heat at low temperatures is obtained by differentiating equation (4.45) with respect to temperature:

$$C_{\text{lattice}} = \left(\frac{12\pi^4 N k_B}{5}\right)\left(\frac{T}{\Theta_D}\right)^3 = 234 N k_B \left(\frac{T}{\Theta_D}\right)^3. \tag{4.46}$$

As is clear from equation (4.46), the lattice specific heat at low temperatures decreases in proportion to the cube of absolute temperature but is inversely proportional to the cube of the Debye temperature. It is experimentally well confirmed that the T^3-law holds well in the temperature range $T/\Theta_D < 0.1$ for various solids as shown below.

Equation (4.43) indicates that the lattice specific heat is scaled in terms of the reduced temperature T/Θ_D. Figure 4.10 shows the temperature dependence of the lattice specific heat calculated from equation (4.43), together with experimentally derived specific heats for various substances, which include Ag and Al as metals, graphite as a semimetal, alumina (Al_2O_3) as an oxide and KCl as an ionic compound. It is clear that the lattice specific heat for all kinds of substances, when plotted against the reduced temperature T/Θ_D, falls well on the calculated curve, indicating the validity of the Debye model. The lattice specific heat approaches the value of $3R$ at temperatures $T/\Theta_D > 1$, being well consistent with the Dulong–Petit law at high temperatures. At low temperatures $T \ll \Theta_D$, however, it rapidly decreases and approaches zero in accordance with the T^3-law.

4.9 Acoustic phonons and optical phonons 77

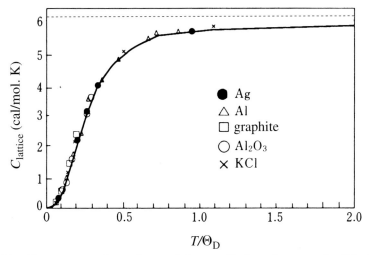

Figure 4.10. Temperature dependence of the specific heat for several solids. The temperature is normalized with respect to the Debye temperature. For example, the Debye temperatures for Ag and Al are 220 and 430 K, respectively. The theoretical curve derived from the Debye model is shown as a solid curve. The value of $3R$ is shown as a dashed line. [F. Seitz, *The Modern Theory of Solids* (McGraw-Hill, New York, 1940)]

4.9 Acoustic phonons and optical phonons

We have so far discussed lattice vibrations in a monatomic system. Different modes appear in lattice vibrations when two or more different atoms are periodically arranged. As its simplest form, we consider the diatomic linear chain model, in which two different atoms possess masses M and m ($M>m$) and they are alternatively arranged in a periodic lattice with a lattice constant a, as shown in Fig. 4.11. The unlike atoms are coupled through a force constant β. Atoms with mass m are located at even-numbered lattice points $2l$, $2l+2$, ..., while atoms with mass M are located at at odd-numbered lattice points $2l-1$, $2l+1$, ...

Equations of motion for atoms at even- and odd-numbered lattice points are written as

$$m\frac{d^2u_{2la}}{dt^2} = \beta(u_{(2l+1)a} + u_{(2l-1)a} - 2u_{2la})$$

$$M\frac{d^2u_{(2l+1)a}}{dt^2} = \beta(u_{(2l+2)a} + u_{2la} - 2u_{(2l+1)a}). \qquad (4.47)$$

Solutions of these two equations may be found by assuming the displacements of atoms in the form of

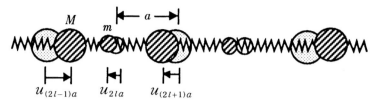

Figure 4.11. Lattice vibrations for one-dimensional diatomic lattice with lattice constant a. A larger atom with mass M is placed at odd-numbered sites and a smaller atom with mass m at even-numbered sites.

$$u_{2la} = \xi_0 \exp[i(\omega t + 2laq)]$$
$$u_{(2l+1)a} = \eta_0 \exp\{i[\omega t + (2l+1)aq]\}. \quad (4.48)$$

Equation (4.47) is then reduced to

$$-\omega^2 m \xi_0 = \beta \eta_0 (e^{iqa} + e^{-iqa}) - 2\beta \xi_0$$
$$-\omega^2 M \eta_0 = \beta \xi_0 (e^{iqa} + e^{-iqa}) - 2\beta \eta_0. \quad (4.49)$$

The set of homogeneous equations has a non-trivial solution only when the determinant of the coefficients of the two variables ξ_0 and η_0 is equal to zero:

$$\begin{vmatrix} 2\beta - m\omega^2 & -2\beta \cos qa \\ -2\beta \cos qa & 2\beta - M\omega^2 \end{vmatrix} = 0.$$

This yields the dispersion relation between the frequency ω and wave number q in the following form:

$$\omega^2 = \beta\left(\frac{1}{m} + \frac{1}{M}\right) \pm \beta\left[\left(\frac{1}{m} + \frac{1}{M}\right)^2 - \left(\frac{4\sin^2 qa}{Mm}\right)\right]^{1/2}. \quad (4.50)$$

Figure 4.12 shows the dispersion relation in the case of $M > m$. In the region where q is small, the two solutions in equation (4.50) can be simplified as

$$\omega = \left(\sqrt{\frac{2\beta}{M+m}}\right) qa \quad (4.51)$$

and

$$\omega = \sqrt{2\beta\left(\frac{1}{m} + \frac{1}{M}\right)}. \quad (4.52)$$

Equation (4.51) arises from a negative sign in equation (4.50) and indicates that the frequency ω is proportional to the wave number q (note that $\sin q \simeq q$). This agrees with equation (4.17) at small q obtained from the monatomic linear chain model. The dispersion relation (4.50) with a negative sign yields the curve appearing at a lower frequency in Fig. 4.12. This is referred to as the acoustic

4.9 Acoustic phonons and optical phonons

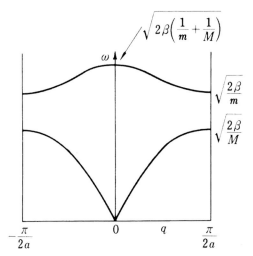

Figure 4.12. Dispersion relation of the lattice vibrations for a one-dimensional diatomic lattice with lattice constant a.

branch. Phonons associated with this branch are called acoustic phonons. The frequency spectrum in the acoustic branch in the low-q region is well approximated by the Debye model.

Equation (4.52), which arises from a positive sign in equation (4.50), indicates that the frequency ω is independent of the wave number q. Indeed, the dispersion relation (4.50) with a positive sign yields the curve appearing at a high-frequency region, as shown in Fig. 4.12. This is called the optical branch. Phonons in this branch are called optical phonons. The frequency spectrum in the low-q region is characterized by a delta-function-like peak at the frequency given by equation (4.52). Thus, the Einstein model is appropriate to evaluate the contribution of optical phonons to the specific heat.[13]

The difference in the characteristic features of acoustic and optical modes may be explained by using a diatomic linear chain model. One can easily check from equations (4.49) and (4.52) that the ratio of the amplitudes for small q in the optical branch results in $\xi_0/\eta_0 = -M/m$. This means that the two atoms vibrate against each other, as shown in Fig. 4.13(b). Suppose that the two types of atoms are of opposite electric charge as in ionic crystals such as KF and NaCl. Then, the vibration of positive and negative ions in opposition to one another will be excited by an electric fields or an electromagnetic wave. This is the reason why it is called the optical branch.

---

[13] The specific heat in the Einstein model approaches $3R$ at high temperatures in good agreement with the Dulong–Petit law. At low temperatures, the exponential temperature dependence appears instead of the T^3-law (see Exercise 4.2). For example, the alkali metal–graphite intercalation compounds like C_8Rb and C_8Cs consist of light carbon atoms and heavy alkali atoms. The optical modes appear in their dispersion relation. The low-temperature specific heat exhibits an exponential temperature dependence and can be well analyzed using the Einstein model [2].

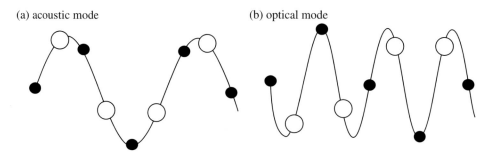

Figure 4.13. (a) Acoustic mode, and (b) optical mode in a one-dimensional diatomic lattice consisting of two different atoms, masses M and m. Open and solid circles represent heavy and light atoms, respectively.

4.10 Lattice vibration spectrum and Debye temperature

The dispersion relation for lattice vibrations can be experimentally derived from the measurement of neutron inelastic scattering [1]. Fig. 4.14 shows the dispersion relation measured for pure Cu where the ordinate v is the frequency of lattice vibrations. It can be seen that the slope of the longitudinal wave, L, is steeper than those of the two transverse waves, T_1 and T_2, indicating that the sound velocity of the longitudinal wave is faster than that of the transverse waves.

The frequency spectrum for pure Cu can be calculated from the dispersion relation shown in Fig. 4.14. The results are shown in Fig. 4.15(a), along with the Planck distribution function at three different temperatures in (b). Note that the abscissas for both data are shown on the same scale. We can see that the spectrum is parabolic at low frequencies in good agreement with equation (4.39), expected from the Debye model. By comparing (a) with (b), we can immediately see how an increase in temperature contributes to the excitation of lattice waves having high frequencies. For example, only lattice waves consistent with the Debye model are excited at low temperatures but the deviation from the Debye model is obviously significant at high temperatures.

The Debye temperature is an important parameter in characterizing lattice vibrations of a solid. As will be discussed in Chapter 10, the Debye temperature plays a key role in the discussion of the temperature dependence of the electrical resistivity in metals. The Debye temperature defined as equation (4.40) is proportional to the sound velocity. As mentioned in Section 4.4, the sound velocity in a continuum is expressed as $s=\sqrt{c/\rho}$, whereas the linear chain model gives rise to the expression $s=\sqrt{(\beta/M)}a$. Therefore, the Debye temperature is increased when the interatomic force constant β or the elastic

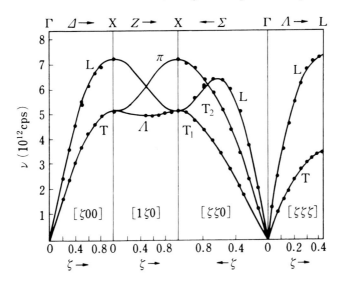

Figure 4.14. Dispersion relation of pure Cu derived from an inelastic neutron scattering experiment. The measurements were done using different orientations of a single crystal. T_1, T_2 and L refer to the two transverse and one longitudinal wave modes, respectively. [R. M. Nicklow et al., Phys. Rev. **164** (1967) 922]

stiffness constant c is increased. For instance, the Debye temperature of diamond reaches a value as high as 2000 K, since its mass is low and its interatomic force constant is high because of the prevailing covalent bonding. On the other hand, the Debye temperature of lead is as low as only 106 K, as listed in Table 3.1. This is because its mass is heavy and the interatomic force constant is low.

Phenomena concerning lattice vibrations are often universally scaled in terms of the dimensionless temperature T/Θ_D. The temperature dependence of the specific heat shown in Fig. 4.10 is one of the examples. Similarly, the data for the temperature dependence of the electrical resistivity, when normalized with respect to the Debye temperature, fall on a universal curve, regardless of the metals involved (see Fig. 10.9). The Debye temperature is experimentally determined from measurements of the specific heat and the elastic stiffness constant [3]. Values for representative metals deduced from specific heat measurements are listed in Table 3.1.

4.11 Conduction electrons, set of lattice planes and phonons

In the free-electron system, we pointed out that the wave vector is equivalent to the momentum of the electron and is quantized in intervals of $2\pi/L$. Owing to the requirement of the Pauli exclusion principle, the ground state for the

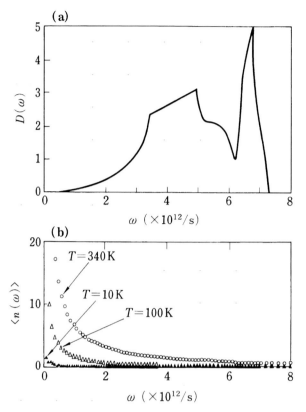

Figure 4.15. (a) Frequency spectrum of pure Cu. The spectrum is deduced from the dispersion relation shown in Fig. 4.14. (b) Planck distribution function at 10, 100 and 340 K. Note that the Debye temperature of pure Cu is 340 K. A deviation from the Debye model at high energies is due to the proximity of the Debye sphere to the Brillouin zone.

assembly of conduction electrons forms a well-defined Fermi sphere in reciprocal space. Sets of an infinite number of equivalent lattice planes are formed in three-dimensional periodic arrays of atoms in a crystal. Each set of lattice planes denoted by the Miller indices (hkl) gives rise to a corresponding reciprocal lattice vector $\mathbf{g}_{hkl}$ in reciprocal space. A deviation from the equilibrium position of atoms due to thermal vibrations can be described in terms of lattice waves propagating through a crystal. We showed that the wave vector characterizing the lattice wave is quantized in intervals of $2\pi/L$ but is confined within the first Brillouin zone bounded by planes bisecting perpendicularly the shortest reciprocal lattice vectors. We learned that states of both conduction electrons and phonons are described in terms of the wave vector in reciprocal space. Their characteristic features are summarized in Table 4.1 (pp. 84–5).

Exercises

4.1 Consider a one-dimensional ring onto which 100 identical atoms are periodically positioned with a spacing $a=2$. Thus, its circumference L is 200. Suppose that a transverse lattice wave with wave vector $q=(2\pi/L)n=(2\pi/200)40$ is excited. Draw this lattice wave and mark the vertical displacement of atoms. Add the reciprocal lattice vector $g=(2\pi/a)=(2\pi/2)$ to the wave vector above and draw the lattice wave thus obtained. Check that the displacement of atoms remains unchanged when the wave vector is shifted by the reciprocal lattice vector.

4.2(a) Derive the internal energy and specific heat due to lattice vibrations in the Einstein model with the characteristic frequency ω_0.

(b) Show that the lattice specific heat at low temperatures $T \ll \Theta_E$ is approximated as

$$C_{\text{lattice}} = 3R \left(\frac{\Theta_E}{T}\right)^2 \exp\left(-\frac{\Theta_E}{T}\right) + \cdots \quad (4Q.1)$$

and that at high temperatures $T > \Theta_E$, as

$$C_{\text{lattice}} = 3R \left[1 - \frac{1}{12}\left(\frac{\Theta_E}{T}\right)^2 + \cdots \right], \quad (4Q.2)$$

where the Einstein temperature Θ_E is defined as $\Theta_E = \dfrac{\hbar \omega_0}{k_B}$.

Table 4.1. *Conduction electrons, set of lattice planes and lattice vibrations in a crystal metal*
[Numbers in parentheses, e.g. (2.8), refer to equations in text.]

	conduction electrons	set of lattice planes	lattice vibrations (phonons)
periodic functions in real space (one-dimensional system)	periodic boundary condition $\psi(x+L) = \psi(x)$ (2.8) $\psi(x)$: wave function of electrons period: crystal size L $L \approx 10^8$ Å	lattice periodicity $f(x+a) = f(x)$ (4.1) $f(x)$: a function having lattice periodicity period: lattice constant a $a \approx 3$–4 Å	periodic boundary condition plus lattice periodicity $u(l+L) = u(l)$ (4.18) $f(x+a) = f(x)$ (4.1) $u(l)$: displacement of atoms period: crystal size L and lattice constant a
quantization in reciprocal space (one-dimensional system)	$\psi(x) = \sum\limits_{n=-\infty}^{\infty} C_n e^{i\frac{2\pi}{L}nx}$ quantization of wave vector $k_n = \dfrac{2\pi}{L} n \; (n = 0, \pm 1, \pm 2, \ldots)$ $(-\infty < k_n < \infty)$	$f(x) = \sum\limits_{n=-\infty}^{\infty} A_n e^{i\frac{2\pi}{a}nx}$ quantization of reciprocal lattice vector $g_n = \dfrac{2\pi}{a} n \; (n = \pm 1, \pm 2, \ldots)$ $(-\infty < g_n < \infty)$	$u(x) = \sum\limits_{n=-N/2}^{N/2} B_n e^{i\frac{2\pi}{L}nx}$ quantization of wave vector $q_n = \dfrac{2\pi}{L} n \; \left(n = \pm 1, \pm 2, \ldots, \pm \dfrac{N}{2}\right)$ in the first Brillouin zone $-\dfrac{\pi}{a} < q_n \leq \dfrac{\pi}{a}$
Fermi sphere and Debye sphere (three-dimensional system)	Fermi radius $k_\mathrm{F} = \left[3\pi^2\left(\dfrac{N}{V}\right)\right]^{1/3}$ (2.20)		Debye radius $q_\mathrm{D} = \left[6\pi^2\left(\dfrac{N}{V}\right)\right]^{1/3}$ (4.37)

	N: number of electrons in volume V	N: number of atoms in volume V
characteristic features in reciprocal space	conduction electron in periodic lattice (Bloch wave) $\rightarrow$ Chapter 5	lattice vibration with wave vector $\mathbf{q} + \mathbf{g}$ is identical to that with wave vector $\mathbf{q}$.
	Bloch wave function $\psi_k(x+l) = \exp(ikl)\psi_k(x)$ (5.14) $\rightarrow$ another form of a periodic function	
	electronic state with wave factor $\mathbf{k} + \mathbf{g}$ is identical to that with wave vector $\mathbf{k}$.	
	reduction to the first Brillouin zone $\rightarrow$ see Section 5.3	3N independent modes in the first Brillouin zone
	2 electrons per atom are accommodated in the first Brillouin zone.	
quantum statistics	$f(E, T)$: Fermi–Dirac distribution function (3.8)	$n(\omega, T)$: Planck distribution function (4.32)
dispersion relation	In the free-electron model, $E = \hbar^2 k^2 / 2m$ (2.5)	In the Debye model, $\omega = sq$
density of states	$N(E) = C\sqrt{E}$ (2.22)	$D(\omega) = C\omega^2$ (4.39)
internal energy	$U_{el} = \int_0^{\infty} f(E,T)N(E)EdE$ (3.20)	$U_{lattice} = \int_0^{\omega_D} n(\omega,T)D(\omega)\hbar\omega d\omega$ (4.34)
specific heat	$C_{el} = \gamma T$ (3.22)	$C_{lattice} = \alpha T^3$ ($T \ll \Theta_D$) (4.46)

Chapter Five
Conduction electrons in a periodic potential

5.1 Prologue

In the present chapter, we study first the Bloch theorem, which plays a key role in describing the motion of the conduction electron in a periodic potential, and discuss how the free-electron E–**k** relation is perturbed by the periodic potential. The reciprocal space is partitioned into polyhedra bounded by planes normal to the reciprocal lattice vectors at their midpoints, and the energy gap, the magnitude of which depends on the Fourier component of the ionic potential, appears across the plane of each polyhedron. This is the Brillouin zone introduced in Chapter 4. The Fermi surface representing the momentum distribution of conduction electrons is also constructed in reciprocal space. The effect of the periodic potential on the conduction electron can be treated in reciprocal space in terms of the interaction of the Fermi surface with the Brillouin zone. We will show that the band structure unique to a given material emerges as a result of distortion of the Fermi surface upon its approach to the zone planes, its subsequent contacts with and overlaps across them.

5.2 Cosine-type periodic potential

We consider first the motion of the conduction electron in a one-dimensional cosine-type periodic potential. Its potential is expressed as

$$V(x) = -A \cos\left(\frac{2\pi x}{a}\right), \tag{5.1}$$

where $2A$ is the amplitude of the potential and a is the lattice constant. The cosine-type potential is illustrated in Fig. 5.1. The motion of the conduction electron in the cosine-type potential (5.1) can be described by the Schrödinger equation:

5.2 Cosine-type periodic potential

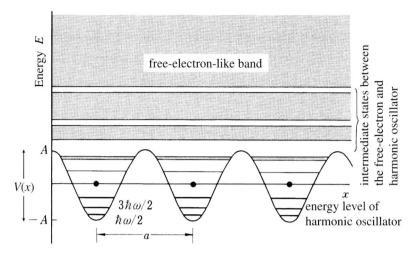

Figure 5.1. Energy of conduction electron in cosine-type periodic potential.

$$\left(\frac{-\hbar^2}{2m}\right)\frac{d^2\psi(x)}{dx^2} - A\cos\left(\frac{2\pi x}{a}\right)\psi(x) = E\psi(x), \qquad (5.2)$$

where E is the energy eigenvalue of the conduction electron. Let us rewrite equation (5.2) by using dimensionless variables defined as $\xi = (\pi x/a)$, $\varepsilon = (8mEa^2/h^2)$ and $\eta = (8mAa^2/h^2)$. The Schrödinger equation is now simplified as

$$\frac{d^2\psi}{d\xi^2} + (\varepsilon + \eta\cos 2\xi)\psi = 0. \qquad (5.3)$$

We solve equation (5.3) under the two extreme conditions.

(1) $\varepsilon \gg \eta$ or $E \gg A$

Since the energy E of the conduction electron is much higher than the amplitude of the potential A, the term $\eta\cos 2\xi$ in equation (5.3) can be ignored. The Schrödinger equation is reduced to

$$\frac{d^2\psi}{d\xi^2} + \varepsilon\psi = 0.$$

This is similar to equation (2.1) for the free electron and, hence, the wave function can be expressed as a linear combination of two plane waves:

$$\psi(\xi) = c_1 e^{ik_\xi \xi} + c_2 e^{-ik_\xi \xi}.$$

(2) $\varepsilon \ll \eta$ or $E \ll A$

In contrast to case (1), the energy E of the conduction electron is much smaller than the amplitude of the potential A and, thus, the electron must be confined near the bottom of the cosine-type potential. This means that only the motion near $\xi = 0$ needs to be considered. The potential in equation (5.3) is then approximated as

$$\eta \cos 2\xi \approx \eta(1 - 2\xi^2),$$

since ξ is small. The resulting Schrödinger equation is reduced to

$$\frac{d^2\psi}{d\xi^2} + (\varepsilon' - 2\eta\xi^2)\psi = 0.$$

The potential in this equation is proportional to the square of the coordinate ξ, thereby representing the motion of a harmonic oscillator. As discussed in Section 4.6, its energy eigenvalue is given by

$$\varepsilon' = \left(n + \frac{1}{2}\right)\hbar\omega \quad (n = 0, 1, 2, 3, \ldots),$$

where ω represents the angular frequency characterizing the oscillation of the electron at the bottom of the potential. The energy eigenvalue obtained under the two extreme cases is illustrated schematically in Fig. 5.1. In case (1), the electron can propagate freely in space, since the amplitude of the potential is negligibly small relative to the kinetic energy of the electron. The energy eigenvalue is given by equation (2.5) and forms a continuous band, as discussed in Chapter 2. In contrast, the electron is captured in the potential well in case (2). The discrete energy level is given by that of a harmonic oscillator. In this chapter, we deal with the situation where the energy of the electron is comparable to the amplitude of the potential or $E \approx A$. We will learn that energy gaps open up within the continuous band, as schematically illustrated in Fig. 5.1.

5.3 Bloch theorem

We study in this section the Bloch theorem and prove it by using a one-dimensional periodic lattice consisting of N monatomic ions with lattice constant a. The ionic potential located at the origin $x = 0$ is defined in the range $-a/2 < x \leq a/2$ and denoted as $V(x)$. The ionic potential located at its nearest neighbor position $x = a$ is then expressed as $V(x + a)$ in the range $a/2 < x \leq 3/2a$. Because of its identical nature, $V(x) = V(x + a)$ holds. In the same manner, we obtain

$$V(x) = V(x + a) = V(x + 2a) = \cdots = V(x + (N-1)a), \tag{5.4}$$

5.3 Bloch theorem

where $V(x+ma)$ is defined in the range $(-a/2)+ma < x \le (a/2)+ma$ ($m = 0, 1, 2, \ldots, N-1$).

The periodic boundary condition is imposed in such a way that the ion at $x = 0$ coincides with that at the position $x = Na$. This forms a ring of length Na, onto which N lattice points are evenly distributed with the lattice constant a. Now we have the relation

$$V(x) \equiv V(x+Na), \tag{5.5}$$

where the symbol $\equiv$ emphasizes that both $V(x)$ and $V(x+Na)$ refer to the same potential. The Schrödinger equation in each unit cell can be expressed as

$$\left(\frac{-\hbar^2}{2m}\right)\frac{d^2\psi(x)}{dx^2} + V(x)\psi(x) = E\psi(x)$$

$$\left(\frac{-\hbar^2}{2m}\right)\frac{d^2\psi(x+a)}{dx^2} + V(x+a)\psi(x+a) = E\psi(x+a)$$

$$\vdots$$

$$\left(\frac{-\hbar^2}{2m}\right)\frac{d^2\psi(x+(N-1)a)}{dx^2} + V(x+(N-1)a)\psi(x+(N-1)a) = E\psi(x+(N-1)a). \tag{5.6}$$

Note that ionic potentials periodically arranged with the lattice constant a are identical. For example, the identity of the wave functions $\psi(x)$ with $\psi(x+a)$ means that they should possess the same energy eigenvalue E but that $\psi(x)$ can differ from $\psi(x+a)$ by a phase factor. They are therefore written as

$$\psi(x+a) = \lambda\psi(x) \quad |\lambda| = 1. \tag{5.7}$$

By repeating this process N times to reach the N-th unit cell, we finally obtain the relation

$$\psi(x+Na) = \lambda^N \psi(x).$$

However, the N-th one is nothing but the cell at $x = 0$ and, hence, $\psi(x+Na) = \psi(x)$ holds. This results in

$$\lambda^N = 1,$$

which is solved as

$$\lambda = \exp\left(\frac{2\pi n i}{N}\right),$$

where n is an integer in the range of 0 up to $N-1$.

The Bloch theorem in a one-dimensional lattice is stated as follows. The wave function $\psi(x)$ for an electron propagating in the periodic potential with the period a can be expressed as

$$\psi(x) = \exp\left(\frac{2\pi i n x}{Na}\right) u(x), \tag{5.8}$$

where an arbitrary function $u(x)$ is a periodic function of a and satisfies the relation

$$u(x + ma) = u(x), \tag{5.9}$$

with a positive integer m. In order to prove the Bloch theorem, we first assume equation (5.8) to hold. Then, we can prove below that the function $u(x)$ must satisfy equation (5.9). Let the variable x in equation (5.8) to be replaced by $x + ma$. We have

$$\psi(x + ma) = \exp\left[\frac{2\pi n i (x + ma)}{Na}\right] u(x + ma)$$

$$= \exp\left(\frac{2\pi m n i}{N}\right) \exp\left(\frac{2\pi n i x}{Na}\right) u(x + ma)$$

$$= \lambda^m \exp\left(\frac{2\pi n i x}{Na}\right) u(x + ma). \tag{5.10}$$

The relation $\lambda = \exp(2\pi n i / N)$ obtained above is inserted to reach the last line. If we apply equation (5.7) m times to $\psi(x)$, then we get the relation $\psi(x + ma) = \lambda^m \psi(x)$. By inserting equation (5.8) into it, we have

$$\psi(x + ma) = \lambda^m \psi(x) = \lambda^m \exp\left(\frac{2\pi n i x}{Na}\right) u(x). \tag{5.11}$$

A comparison of equations (5.10) and (5.11) immediately leads us to conclude that an arbitrary function $\psi(x)$ must satisfy equation (5.9).

A quantity of $(2\pi/Na)n$ or $(2\pi/L)n$ in equation (5.8) may be replaced by a new variable k, since it is of the same form as the wave number defined by equation (2.9) for free electrons. By this replacement, the wave function (5.8) is simplified to $\psi(x) = \exp(ikx)u(x)$, allowing us to envisage $\psi(x)$ as the plane wave $\exp(ikx)$ modulated by the periodic function $u(x)$. Here it is important to keep in mind that the variable k of the free electron was originally introduced as the wave number of the plane wave in free space, whereas the new variable above appeared in relation to the periodicity of the lattice. Before discussing its

5.3 Bloch theorem

unique nature, we extend the Bloch theorem to a three-dimensional periodic lattice.

Let us assume the periodic potential in a crystal where the position of each ion is specified by the lattice vector $\mathbf{l} = l_x \mathbf{a}_x + l_y \mathbf{a}_y + l_z \mathbf{a}_z$ (l_x, l_y, l_z = integers) in equation (4.7). The wave function $\psi(\mathbf{r})$ of the electron in the periodic potential can be expressed in the form

$$\psi_\mathbf{k}(\mathbf{r}) = \exp(i\mathbf{k}\cdot\mathbf{r})u_\mathbf{k}(\mathbf{r}), \tag{5.12}$$

where $u_\mathbf{k}(\mathbf{r})$ satisfies the relation

$$u_\mathbf{k}(\mathbf{r}+\mathbf{l}) = u_\mathbf{k}(\mathbf{r}). \tag{5.13}$$

Here the vector $\mathbf{k}$ is of the same form as the wave vector in equation (2.11) for the free electron. This is called the Bloch theorem. The wave function expressed by equation (5.12) is called the Bloch wave or Bloch state.

The Bloch theorem is very important. By applying this theorem, the wave function in a macroscopic crystal containing as many atoms as the Avogadro number can be determined by solving the Schrödinger equation into which information from just one unit cell is inserted. This unique advantage stems from the fact that the wave function everywhere in a crystal is automatically decided, once $u_\mathbf{k}(\mathbf{r})$ in the unit cell, say, at $\mathbf{l}=0$ is specified. Therefore, the Bloch theorem is responsible for the successful development of band structure calculations for a 'macroscopic' crystal, which we will study in Chapter 8. We show, in Fig. 5.2, an example of the Bloch wave in a one-dimensional system, where $u_\mathbf{k}(x)$ is positioned at the center of the unit cell. Once the function $u_\mathbf{k}(\mathbf{r})$ in the unit cell is given, the wave function extending over the crystal is completely decided by the product of the plane wave $e^{i\mathbf{k}\cdot\mathbf{r}}$ and the periodic function $u_\mathbf{k}(\mathbf{r})$, as shown in Fig. 5.2(b).

It is of great importance for the reader to recognize how physical quantities associated with the wave vector $\mathbf{k}$ in the Bloch wave differ from those derived from the free-electron model. For example, $\hbar\mathbf{k}$ is found to be the eigenvalue of the momentum operator $-i\hbar\nabla$ in the free-electron model (see equation (2.14)). If it is operated to the Bloch wave function (5.12), one can easily find that $\hbar\mathbf{k}$ is no longer its eigenvalue. This is because the ionic potential exerts a force on the electron through the function $u_\mathbf{k}(\mathbf{r})$.

To study further the characteristic features of the Bloch wave function, we can rewrite equation (5.12) in the following form:

$$\psi_\mathbf{k}(\mathbf{r}+\mathbf{l}) = \exp(i\mathbf{k}\cdot\mathbf{l})\psi_\mathbf{k}(\mathbf{r}). \tag{5.14}$$

As discussed in Section 4.2, the reciprocal lattice vector $\mathbf{g}$ is defined so as to satisfy the relation $\exp(\pm i\mathbf{g}\cdot\mathbf{l}) = 1$, where $\mathbf{l}$ is the lattice vector defined by

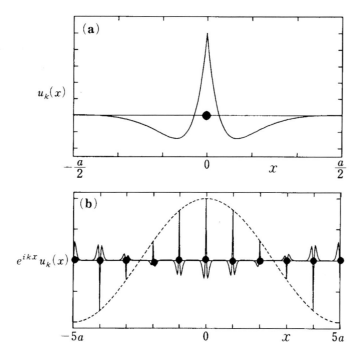

Figure 5.2. (a) The periodic function $u_k(x)$ centered at the origin of the unit cell in the range $-a/2 < x \leq a/2$. (b) The Bloch wave constructed by using the function shown in (a). Only the real part is shown. The function $u_k(x)$ is placed at every origin of the unit cell and modulated by the plane wave $\exp(ikx)$. A solid circle represents the ion at the center of each unit cell.

equation (4.7). Keeping this in mind, we can replace the wave vector $\mathbf{k}$ of the Bloch wave by the wave vector $\mathbf{k} = \mathbf{k}' \pm \mathbf{g}$:

$$\psi_\mathbf{k}(\mathbf{r}+\mathbf{l}) = \exp(i\mathbf{k}\cdot\mathbf{l})\psi_\mathbf{k}(\mathbf{r})$$
$$= \exp(\pm i\mathbf{g}\cdot\mathbf{l}) \exp(i\mathbf{k}'\cdot\mathbf{l})\psi_\mathbf{k}(\mathbf{r})$$
$$= \exp(i\mathbf{k}'\cdot\mathbf{l})\psi_\mathbf{k}(\mathbf{r}). \qquad (5.15)$$

A comparison of equations (5.14) and (5.15) tells us that the Bloch state of the wave vector $\mathbf{k}$ is equally describable in terms of the wave vector $\mathbf{k}'$ different from it by the reciprocal lattice vector $\mathbf{g}$.

This is a property unique to the Bloch wave or Bloch electron. Multiply both sides of $\mathbf{k} = \mathbf{k}' \pm \mathbf{g}$ by $\hbar$. Then, it is viewed as representing the momentum conservation law of the Bloch electron, indicating that the Bloch electron exchanges its momentum with the lattice by the amount $\pm \hbar \mathbf{g}$. What does $\pm \hbar \mathbf{g}$ mean? It is assigned to an infinite array of identical lattice planes specified by the reciprocal lattice vector **g** and has nothing to do with phonons. It may merely refer to the motion of the lattice as a whole. Thus, the momentum $\hbar\mathbf{k}$ of the Bloch wave cannot be uniquely determined as the momentum inherent to an electron but involves arbitrariness associated with a whole motion of the lattice. This is the reason why the momentum $\hbar\mathbf{k}$ is often called the crystal momentum of the Bloch wave.

Let us consider a special case where the magnitude of the periodic potential is reduced infinitesimally small. We call it the periodic empty-lattice, under which the electron should resume the free-electron band structure but the periodicity of the lattice and, hence, the concept of the Bloch wave remains valid. This is a hypothetical model but helps the reader to gain further insight into the role of the periodic potential. The free-electron wave function $\psi_\mathbf{k}(\mathbf{r})=\exp(i\mathbf{k}\cdot\mathbf{r})$ must be its eigenfunction but still obeys the Bloch theorem. The wave function may be rewritten as

$$\psi_\mathbf{k}(\mathbf{r}) = \exp[i(\mathbf{k}\pm\mathbf{g})\cdot\mathbf{r}]\exp(\mp i\mathbf{g}\cdot\mathbf{r})$$

$$= \exp(i\mathbf{k}'\cdot\mathbf{r})u_{\mp\mathbf{g}}(\mathbf{r}), \qquad (5.16)$$

where $\mathbf{k}'=\mathbf{k}\pm\mathbf{g}$ and $u_{\mp\mathbf{g}}(\mathbf{r})=\exp(\mp i\mathbf{g}\cdot\mathbf{r})$. Equation (5.16) satisfies the Bloch theorem, since $u_{\mp\mathbf{g}}(\mathbf{r}+\mathbf{l})=\exp[\mp i\mathbf{g}\cdot(\mathbf{r}+\mathbf{l})]=\exp(\mp i\mathbf{g}\cdot\mathbf{r})=u_{\mp\mathbf{g}}(\mathbf{r})$. It is now interesting to examine the E–k relation of the Bloch electron in the periodic empty-lattice potential. By reflecting the periodic nature of the lattice, the Bloch state of the wave vector **k** should be identical to that of the wave vector $\mathbf{k}\pm\mathbf{g}$ but yet the energy eigenvalue is given by the free-electron value (see Exercise 5.2).

A one-dimensional monatomic lattice with lattice constant a is assumed. In this particular case, the reciprocal lattice vector becomes multiples of $2\pi/a$. Since the Bloch states **k** and $\mathbf{k}\pm\mathbf{g}$ possess the same eigenstate, we can always transfer the Bloch state of any wave vector into the region $-\pi/a < k_x \leq \pi/a$. This is called the reduction to the first Brillouin zone. We will learn more about the operation of the reduction in Section 5.11. This unique property in reciprocal space is caused by the periodic array of ions in a crystal and has already been discussed in relation to lattice vibrations in Section 4.4.

5.4 Kronig–Penney model

By making full use of the Bloch theorem, we can study the effect of the periodic potential on the E–k relation of the conduction electron. For this purpose, the Kronig–Penney model is known to be quite instructive. The model assumes a periodic square-well potential in one-dimensional space, as indicated in Fig. 5.3. The Schrödinger equation in one-dimensional space is generally written as

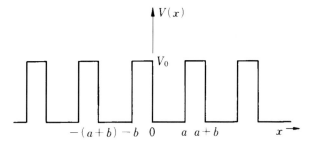

Figure 5.3. Periodic potential in the Kronig–Penney model.

$$\left(\frac{-\hbar^2}{2m}\right)\left(\frac{d^2\psi(x)}{dx^2}\right) + V(x)\psi(x) = E\psi(x). \tag{5.17}$$

The potential $V(x)$ in Fig. 5.3 is zero in the range $0 < x \le a$ and, hence, the free-electron wave function given by equation (2.3a) is obviously its solution:

$$\psi(x) = Ae^{i\alpha x} + Be^{-i\alpha x}, \quad E = \frac{\hbar^2 \alpha^2}{2m} \tag{5.18}$$

i.e., $\hbar\alpha = \sqrt{2mE}$. The solution in the range $-b < x \le 0$ depends on the energy of the conduction electron. Let us assume that the potential height V_0 is higher than the energy of the electron, i.e., $V_0 > E$. The wave function is then written as

$$\psi(x) = Ce^{\beta x} + De^{-\beta x}, \tag{5.19}$$

where $\hbar\beta = \sqrt{2m(V_0 - E)}$. Similarly, the wave function in the range $a < x \le a+b$ can be expressed as

$$\psi(x) = C'e^{\beta[x-(a+b)]} + D'e^{-\beta[x-(a+b)]}, \tag{5.20}$$

where the coefficients C' and D' are no longer independent of the coefficients C and D in equation (5.19), as is discussed below.

Remember that the square-well potential is periodic and, hence, the wave function must satisfy the Bloch theorem. We emphasized in connection with equation (5.14) that the wave function at any lattice site l is decided, once the wave function, say at the origin, is given. The two wave functions separated by the interatomic distance a are related to each other through $\psi(x+a) = \exp(ika)\psi(x)$. As is clear from the argument above, the wave number k of the Bloch wave serves to connect the wave functions at different lattice sites over a whole crystal. In the present case, the wave function in the region $a < x \le a+b$ should differ from that in the region $-b < x \le 0$ by the phase $\exp[ik(a+b)]$, since its period is $(a+b)$. Therefore, the Bloch theorem results in

5.4 Kronig–Penney model

$$C' = C\exp[ik(a+b)] \tag{5.21}$$

and

$$D' = D\exp[ik(a+b)]. \tag{5.22}$$

In other words, we say that the Bloch theorem introduces the wave number k as a variable to let the wave function extend over a whole system.

The wave functions given by equations (5.18), (5.19) and (5.20) must be smoothly connected across two boundaries $x=0$ and $x=a$. This is done by causing both the wave function and its derivative $d\psi(x)/dx$ to be continuous across the boundaries. Thus, we obtain four linear homogeneous equations from the boundary conditions. The non-trivial solutions can be derived only if the determinant of the coefficients vanishes. The determinantal equation yields

$$\frac{\beta^2 - \alpha^2}{2\alpha\beta}\sinh\beta b\sin\alpha a + \cosh\beta b\cos\alpha a = \cos k(a+b). \tag{5.23}$$

Equation (5.23) is too complex to conceive its physical meaning. The periodic square-well potential may be replaced by a periodic delta-function by taking the limits $b \to 0$ and $V_0 \to \infty$ while keeping $\beta^2 b$ finite. By introducing a new parameter defined as

$$\lim_{\substack{b \to 0 \\ \beta \to \infty}} \left(\frac{\beta^2 ab}{2}\right) = P, \tag{5.24}$$

we can reduce equation (5.23) to

$$P\frac{\sin \alpha a}{\alpha a} + \cos \alpha a = \cos ka. \tag{5.25}$$

Equation (5.25) represents the E–k relation of the conduction electron, since α is a function of E through the relation $\hbar\alpha = \sqrt{2mE}$.

Unfortunately, equation (5.25) is a transcendental equation and cannot be solved analytically. Its graphical analysis allows us to extract essential features in the Kronig–Penney model. The parameter P in equation (5.24) is set equal to the arbitrary value $P = 3\pi/2$. The left-hand side of equation (5.25) is plotted in Fig. 5.4 as a function of αa. Since the right-hand side of equation (5.25) is a cosine function, the value in the left-hand side must fall in the range -1 to 1. In other words, the allowed solution αa of equation (5.25) is found only in the range marked by a thick line in Fig. 5.4.

The minima and maxima of the allowed αa values are derived from the condition $\cos ka = \pm 1$, which results in $k = n\pi/a$, $(n = \pm 1, \pm 2, \ldots)$. The dispersion

96   5 Conduction electrons in a periodic potential

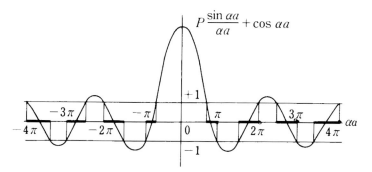

Figure 5.4. The function appearing in the left-hand side of equation (5.25) in the Kronig–Penney model. Its allowed region is limited from -1 to $+1$. Hence, the value of αa is allowed only in regions marked by thick lines.

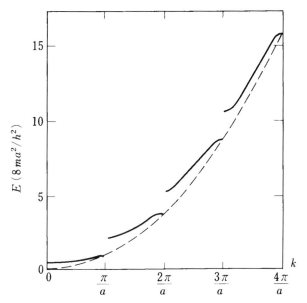

Figure 5.5. E–k relation obtained from the Kronig–Penney model. The dashed curve represents the free-electron parabolic band. [A. Sommerfeld and H. Bethe, *Elektronentheorie der Metalle*, (Springer-Verlag, 1967)]

relation can be numerically calculated and the results are shown in Fig. 5.5. We see that the E–k relation deviates from the free-electron parabola given by equation (2.5) and an energy discontinuity appears at every $k = n\pi/a$, $(n = \pm 1, \pm 2, \ldots)$. Obviously, there exist energy ranges which electrons are not allowed to occupy. Each of these energy ranges is called a forbidden energy band. Notice that forbidden energy band appears as a result of the interaction of the conduction electron with the periodic lattice potential.

What happens when the parameter P and, hence, the magnitude of the periodic potential is reduced to zero? The allowed αa range expands while the forbidden band disappears. As a result, the free-electron-like parabolic band is resumed. On the other hand, the αa range is converged into points $n\pi$, $(n = \pm 1, \pm 2, \ldots)$, when P is increased to infinity. The relation $\alpha a = n\pi$ immediately leads to

$$E = \frac{\hbar^2}{2m}\left(\frac{n\pi}{a}\right)^2. \tag{5.26}$$

We see that equation (5.26) agrees with the energy eigenvalue given by equation (2.19), for an electron confined in a one-dimensional box with length a.

5.5 Nearly-free-electron model

The Schrödinger equation for the conduction electron in a periodic potential can be solved in a more general way than that in the Kronig–Penney model. In a three-dimensional periodic lattice, the Schrödinger equation is expressed as

$$\left(\frac{-\hbar^2}{2m}\right)\nabla^2 \psi(\mathbf{r}) + V(\mathbf{r})\psi(\mathbf{r}) = E\psi(\mathbf{r}), \tag{5.27}$$

where $V(\mathbf{r})$ is a periodic potential and $\psi(\mathbf{r})$ is the Bloch wave function given by equation (5.12). For simplicity, we assume a simple cubic lattice with lattice constant a. The ionic potential $V(\mathbf{r})$ is certainly a periodic function with period a in x-, y- and z-directions. Let us choose the origin of the potential so as to satisfy the condition

$$\int_{-a/2}^{a/2}\int_{-a/2}^{a/2}\int_{-a/2}^{a/2} V(\mathbf{r})d\mathbf{r} = 0. \tag{5.28}$$

Because of its possession of the periodicity a in real space, the ionic potential can be expanded into a series in terms of the reciprocal lattice vector $\mathbf{g}_n = (2\pi/a)\mathbf{n}$:

$$V(\mathbf{r}) = \sum_{n=-\infty}^{\infty} V_n \exp\left[-i\left(\frac{2\pi}{a}\right)\mathbf{n}\cdot\mathbf{r}\right] = \sum_{n=-\infty}^{\infty} V_n \exp(-i\mathbf{g}_n\cdot\mathbf{r}), \tag{5.29}$$

where the components n_x, n_y and n_z of the vector $\mathbf{n}$ take both positive and negative integers. A set of three integers of the vector $\mathbf{n}$ corresponds to the Miller indices for the relevant lattice planes, as discussed in Section 4.3.[1]

The Bloch wave function $\psi(\mathbf{r})$ given by equation (5.12) is also a periodic

---

[1] Note that the suffix $\mathbf{n}$ in $\mathbf{g}_n$ is also a vector.

function with period a, since it involves the periodic function $u_k(\mathbf{r})$. Thus, $\psi(\mathbf{r})$ is similarly expanded into a series:

$$\psi(\mathbf{r}) = e^{i\mathbf{k}\cdot\mathbf{r}} \sum_{n=-\infty}^{\infty} A_n \exp\left[-i\left(\frac{2\pi}{a}\right)\mathbf{n}\cdot\mathbf{r}\right] = e^{i\mathbf{k}\cdot\mathbf{r}} \sum_{n=-\infty}^{\infty} A_n \exp(-i\mathbf{g}_n\cdot\mathbf{r}), \quad (5.30)$$

where all three components n_x, n_y and n_z cover again both positive and negative integers.

The Fourier coefficients V_n and A_n in equations (5.29) and (5.30) are given by

$$V_n = \left(\frac{1}{a^3}\right) \int_{-a/2}^{a/2} \int_{-a/2}^{a/2} \int_{-a/2}^{a/2} V(\mathbf{r}) \exp(i\mathbf{g}_n\cdot\mathbf{r}) d\mathbf{r} \quad (5.31)$$

and

$$A_n = \left(\frac{1}{a^3}\right) \int_{-a/2}^{a/2} \int_{-a/2}^{a/2} \int_{-a/2}^{a/2} u_k(\mathbf{r}) \exp(i\mathbf{g}_n\cdot\mathbf{r}) d\mathbf{r}. \quad (5.32)$$

Note that the Fourier components V_n and A_n are determined solely from information in the unit cell. The Fourier component V_n can be calculated from equation (5.31) for a given ionic potential. The component A_n is determined by solving the Schrödinger equation, as will be shown below.

The Schrödinger equation is now explicitly rewritten below by inserting equations (5.29) and (5.30) into equation (5.27):

$$\left(\frac{-\hbar^2}{2m}\right)\nabla^2\psi(\mathbf{r}) + \sum_{n=-\infty}^{\infty} V_n \exp(-i\mathbf{g}_n\cdot\mathbf{r})\cdot\exp(i\mathbf{k}\cdot\mathbf{r}) \sum_{n=-\infty}^{\infty} A_n \exp(-i\mathbf{g}_n\cdot\mathbf{r})$$

$$= E \exp(i\mathbf{k}\cdot\mathbf{r}) \sum_{n=-\infty}^{\infty} A_n \exp(-i\mathbf{g}_n\cdot\mathbf{r}), \quad (5.33)$$

where the first term is simply expressed as $\psi(\mathbf{r})$ to avoid a lengthy expression. We pick up the coefficient of $\exp[i(\mathbf{k}-\mathbf{g}_n)\cdot\mathbf{r}]$ in equation (5.33) and obtain an infinite set of linear homogeneous equations for A_n:

$$\left[E - \left(\frac{\hbar^2}{2m}\right)\mathbf{k}'^2\right] A_n = \sum_{n'=-\infty}^{\infty} A_{n'} V_{n-n'}, \quad (5.34)$$

where $\mathbf{k}'^2 = (\mathbf{k} - \mathbf{g}_n)^2$. These equations are solvable, provided that the determinant of the coefficients vanishes. This yields the E–k relation in the same manner as described in the treatment of the Kronig–Penney model. However, we cannot solve the determinantal equation in a rigorous manner, since the variable **n** extends from $-\infty$ to $+\infty$. But it can be easily solved in the case of the periodic empty-lattice model (see Exercise 5.2).

The concept of the pseudopotential will be introduced in Section 8.7. The effective ionic potential $V(\mathbf{r})$ which the conduction electron experiences in metals like Na and Al is so small that the condition $E \gg V(\mathbf{r})$ holds. In such a case, the contribution $A_\mathbf{n}$ due to the ionic potential must be small and the wave function is dominated by the plane wave. This is called the nearly-free-electron model or shortened as the NFE model. As mentioned above, the coefficient $A_\mathbf{n}$ is determined for each set of lattice planes specified by the Miller indices (n_x, n_y, n_z) or the reciprocal lattice vector $\mathbf{g_n} = (2\pi/a)\mathbf{n}$. It was mentioned in Section 4.3 that, the smaller the Miller indices for a given set of lattice planes, the higher is the atomic density in the lattice plane. Hence, we need to consider sets of lattice planes having relatively small Miller indices.

This situation is further simplified such that the Bloch wave consists of only two waves: one the plane wave corresponding to the free-electron state and the other the wave associated with a single set of lattice planes with the Miller indices $(n_x n_y n_z)$ or the reciprocal lattice vector $\mathbf{g_n}$. This is the two-wave approximation. Now the wave function (5.30) is expressed as

$$\psi(\mathbf{r}) = \exp(i\mathbf{k}\cdot\mathbf{r})[A_0 + A_\mathbf{n} \exp(-i\mathbf{g_n}\cdot\mathbf{r})]. \quad (5.35)$$

An infinite number of equations (5.34) are reduced to two equations corresponding to $\mathbf{n} = (0\,0\,0)$ and $\mathbf{n} = (n_x n_y n_z)$. The first equation for $\mathbf{n} = (0\,0\,0)$ is explicitly written as

$$\left[E - \left(\frac{\hbar^2}{2m}\right)\mathbf{k}^2\right]A_0 = \sum_{\mathbf{n}'=0,\mathbf{n}} A_{\mathbf{n}'} V_{-\mathbf{n}'} = A_0 V_0 + A_\mathbf{n} V_{-\mathbf{n}},$$

where V_0 is zero, as defined by equation (5.28). It is noted that, if the ionic potential $V(\mathbf{r})$ is real, its Fourier component satisfies the relation $V_{-\mathbf{n}} = V_\mathbf{n}^*$.[2] As a result, we obtain

$$\left[E - \left(\frac{\hbar^2}{2m}\right)\mathbf{k}^2\right]A_0 - V_\mathbf{n}^* A_\mathbf{n} = 0. \quad (5.36)$$

The second equation for $\mathbf{n} = (n_x n_y n_z)$ is likewise written as

$$\left[E - \left(\frac{\hbar^2}{2m}\right)\mathbf{k}'^2\right]A_\mathbf{n} = \sum_{\mathbf{n}'=0,\mathbf{n}} A_{\mathbf{n}'} V_{\mathbf{n}-\mathbf{n}'} = A_0 V_\mathbf{n} + A_\mathbf{n} V_0,$$

---

[2] $V_{-\mathbf{n}} = V_\mathbf{n}^*$ is immediately derived from equation (5.31), when $V(\mathbf{r})$ is real. In addition, $V_\mathbf{n}$ becomes real if a symmetric potential $V(\mathbf{r}) = V(-\mathbf{r})$ is assumed. Hence, $V_\mathbf{n} = V_{-\mathbf{n}} = V_\mathbf{n}^*$ is obtained.

which is reduced to

$$\left[E - \left(\frac{\hbar^2}{2m}\right)\mathbf{k}'^2\right]A_\mathbf{n} - V_\mathbf{n}A_0 = 0. \tag{5.37}$$

The determinant of the coefficients A_0 and $A_\mathbf{n}$ in equations (5.36) and (5.37) must vanish to obtain non-trivial solutions. This immediately yields the relation

$$E(\mathbf{k}) = \left(\tfrac{1}{2}\right)\left[(E_0 + E_\mathbf{n}) \pm \sqrt{(E_0 - E_\mathbf{n})^2 + 4V_\mathbf{n}^*V_\mathbf{n}}\right], \tag{5.38}$$

where $E_0 = \hbar^2 k^2/2m$ and $E_\mathbf{n} = \hbar^2(\mathbf{k} - \mathbf{g_n})^2/2m$. Equation (5.38) represents the E–$\mathbf{k}$ relation for the conduction electron in the periodic potential due to the set of lattice planes with Miller indices $(n_x n_y n_z)$ or the reciprocal lattice vector $\mathbf{g_n}$.

Let us take the set of lattice planes with the Miller indices (100) in a simple cubic lattice with the lattice constant a. Note that the (100) lattice planes are parallel to the yz-plane and, thus, affect the motion of electrons propagating only along the x-axis, as can be seen from equation (5.35). An infinite number of unperturbed parabola $E_{n00} = \hbar^2(\mathbf{k} - \mathbf{g}_{n00})^2/2m$ can be drawn, which are centered at reciprocal lattice vectors $\mathbf{g}_{n00} = (2\pi/a)(n00)$ with $n = 0, \pm 1, \pm 2, \ldots$. They are shown in Fig. 5.6 as dashed curves.

We examine the E–$\mathbf{k}$ relation given by equation (5.38) for the conduction electron propagating along the x-direction. The contribution of the periodic potential appears as the square-root of $4V_\mathbf{n}^*V_\mathbf{n}$ in equation (5.38). Let us suppose that this term is small and consider first the region near $k = 0$. A comparison of the first term with the second one in the square-root leads to

$$(E_0 - E_{100})^2 = \left(\frac{\hbar^2}{2m}\right)^2\left[k^2 - \left(k - \frac{2\pi}{a}\right)^2\right]^2 \approx \left(\frac{\hbar^2}{2m}\right)^2\left(\frac{2\pi}{a}\right)^4 \gg 4V_{100}^*V_{100},$$

where the suffix x in k_x is omitted. The term $4V_{100}^*V_{100}$ can be neglected and equation (5.38) in the vicinity of $k = 0$ is reduced to

$$E_\pm(k) = \left(\tfrac{1}{2}\right)[(E_0 + E_{100}) \pm |E_0 - E_{100}|].$$

We have two solutions, depending on the choice of a plus or minus sign. As is clear from Fig. 5.6, $E_{100} > E_0$ holds in this region. Thus, the lowest energy state in the vicinity of $k = 0$ is obtained by taking a minus sign in the above solution:

$$E_-(k) = \left(\tfrac{1}{2}\right)[E_0 + E_{100} - E_{100} + E_0] = E_0 = \frac{\hbar^2 k^2}{2m}$$

in good agreement with the free-electron expression given by equation (2.5).

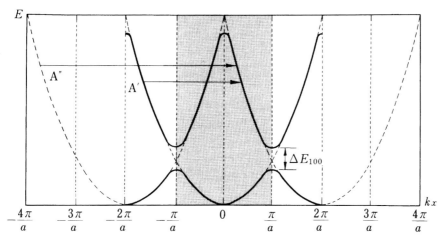

Figure 5.6. E–k relation obtained from equation (5.38) in the nearly-free-electron model. The dashed curve represents the free-electron parabola. ΔE_{100} indicates the energy gap across the {100} zone planes. The shaded area corresponds to the first zone. The electronic states marked by A' and A" can be reduced to the first zone by shifting wave vectors equal to appropriate reciprocal lattice vectors.

We consider next the regions near $k = \pm \pi/a$, where two parabolic bands $E_0 = \hbar^2 k^2/2m$ and $E_{100} = (\hbar^2/2m)[k - (2\pi/a)]^2$ intersect each other. In these regions, the small term $4V^*_{100}V_{100}$ begins to play a dominant role, since $(E_0 - E_{100})^2$ in equation (5.38) becomes negligibly small. At $k = \pm \pi/a$, equation (5.38) is obviously reduced to

$$E_{\pm}(k) = \left(\tfrac{1}{2}\right)[E_0 + E_{100} \pm 2|V_{100}|]. \tag{5.39}$$

Let us denote $2|V_{100}|$ as ΔE_{100}. Then, equation (5.39) is rewritten as

$$E_-(k) = \left(\frac{E_0 + E_{100}}{2}\right) - \left(\frac{\Delta E_{100}}{2}\right)$$

and

$$E_+(k) = \left(\frac{E_0 + E_{100}}{2}\right) - \left(\frac{\Delta E_{100}}{2}\right). \tag{5.40}$$

Here, the first term, representing an average of the two free-electron values, coincides with the free-electron value at $k = \pm \pi/a$. Owing to the second term, the energy state $E_-(k)$ is lowered by the amount $(\Delta E_{100}/2)$ relative to the free-electron value, whereas $E_+(k)$ is raised by the same amount. The E–**k** relation for the conduction electron is, therefore, well approximated by the free-electron parabolic band near $k = 0$ but is strongly perturbed by the presence of the periodic potential in the vicinity of $k = \pm \pi/a$. This is shown by a thick curve in Fig. 5.6.

The deviation from the free-electron parabolic band gradually increases, as k approaches $\pm \pi/a$, and the electron is allowed to take the energy state of either $E_-(k)$ or $E_+(k)$ at $k = \pm \pi/a$. This means that the electron is not allowed to take energies between $E_-(k)$ and $E_+(k)$ and, hence, there appears an energy discontinuity of magnitude ΔE_{100} at $k = \pm \pi/a$. The energy region between $E_-(k)$ and $E_+(k)$ is called the forbidden band and ΔE_{100} is the energy gap. The formation of the energy gap has been already pointed out in the discussion of the Kronig–Penney model in Section 5.4.

We emphasized in Section 5.3 that the wave function in the periodic potential must satisfy the Bloch theorem and that the Bloch state remains unchanged if the wave vector is shifted by an appropriate reciprocal lattice vector. Obviously, the wave function (5.30) employed in the NFE model was chosen to satisfy the Bloch theorem. In Fig. 5.6, we can draw an infinite number of NFE dispersion curves centered at reciprocal lattice vectors $\mathbf{g}_{n00} = (2\pi/a)(n00)$ with $n = 0, \pm 1, \pm 2, \ldots$, all of which are equivalent to one another. For example, the Bloch states of the wave vectors marked as A' and A'' in Fig. 5.6 can be reduced to the states in the first Brillouin zone $-\pi/a < k \leq \pi/a$ by adding the reciprocal lattice vectors $2\pi/a$ and $2(2\pi/a)$, respectively. In this way, the Bloch state having the wave vector outside the first zone can be transferred to an equivalent state in the first zone.

Once the E–$\mathbf{k}$ relation thus obtained is reduced to the first zone (see Fig. 5.21(b)), we obtain a series of E–$\mathbf{k}$ curves stacked in the first Brillouin zone, each being separated by an energy gap. It is seen that an energy gap appears at $k = 0$ in higher energy bands. Indeed, the energy E becomes a multi-valued function of the wave vector in the first zone. Each band is given an index n (note that the band index n takes only a positive integer).

Thus far, we have explained the effect of the periodic potential on the dispersion relation of the conduction electron, by using the set of (100) lattice planes. Our argument can be extended to any set of lattice planes with other Miller indices. As can be seen from equation (5.39), the magnitude of the energy gap $\Delta E_\mathbf{n}$ depends on the Fourier component $V_\mathbf{n}$ of the periodic potential specified by the reciprocal lattice vector $\mathbf{g_n}$. We learned in this way that the effect of the periodic potential is to produce an energy gap in the band structure. In the next section, we consider the formation of this energy gap from the point of view of the diffraction phenomena of the Bloch wave.

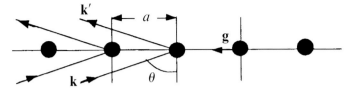

Figure 5.7. Bragg scattering in a one-dimensional lattice. The scattering angle θ is $\pi/2$.

5.6 Energy gap and diffraction phenomena

In Section 4.3, we discussed the Bragg scattering of an x-ray beam incident to lattice planes in a three-dimensional periodic lattice. Here we apply the Laue condition to a Bloch wave incident to the lattice planes. It consists of the two conditions $\mathbf{k'}=\mathbf{k}\pm\mathbf{g}$ and $|\mathbf{k}|=|\mathbf{k'}|$, where $\mathbf{k}$ and $\mathbf{k'}$ are the wave vectors of the Bloch electron before and after the scattering by lattice planes with the reciprocal lattice vector $\mathbf{g}$. The E–$\mathbf{k}$ relation of the Bloch electron is given by an infinite number of equivalent bands separated by the reciprocal lattice vectors, as shown in Fig. 5.6. The two conditions above are combined to a single relation $|\mathbf{k}|=|\mathbf{k}\pm\mathbf{g}|$, which is equivalent to $k^2=(\mathbf{k}\pm\mathbf{g})^2$. We see, therefore, that the Laue condition is fulfilled at the wave vector where two parabolic bands in Fig. 5.6 intersect and an energy gap is produced. As discussed in Section 4.3, this is equivalent to the Bragg condition (see Exercise 5.3).

Let us check the argument above by using a one-dimensional monatomic lattice with the lattice constant a, through which the Bloch wave k propagates. As illustrated in Fig. 5.7, the Bragg condition $2d\sin\theta=n\lambda$ is immediately reduced to $k=n\pi/a$, since $d=a$, $\lambda=2\pi/k$ and $\theta=\pi/2$ hold. This is indeed the wave number at which the energy gap appears (see Fig. 5.6). Thus, we see that the Bragg condition results in the energy gap. It is also noted that $k=n\pi/a$ with $n\geq 2$ or $n\leq -2$ corresponds to the higher-order reflection in the Bragg scattering.

The incident Bloch wave having the wave number $k=\pi/a$ is assumed to propagate to the right by taking a direction to the right as being positive in Fig. 5.7. The wave number k' of the reflected wave is immediately deduced to be $k'=-\pi/a$, since we can choose $g=-2\pi/a$ to satisfy the Laue conditions $\mathbf{k'}=\mathbf{k}+\mathbf{g}$ and $|\mathbf{k}|=|\mathbf{k'}|$. As illustrated schematically in Fig. 5.7, the Bloch wave changes its direction by 180° due to the Bragg reflection and the reflected wave propagates to the left. We may alternatively say that the Bloch electron is elastically backscattered from the periodically arranged ions and that the crystal momentum $\hbar g=-\hbar(2\pi/a)$ is transferred from a whole periodic lattice to the electron to change its direction by 180°

We further discuss why the energy gap appears when the Bragg condition is satisfied at $k = \pm \pi/a$. For simplicity, the Bloch wave is constructed within the two-wave approximation given by equation (5.35). Consider the Bragg condition due to lattice planes having the reciprocal lattice vector $g = 2\pi/a$. The Bloch wave is then expressed by a linear combination of the unperturbed plane wave $A_0 e^{ikx}$ and the wave $A_1 \exp[i\{k - (2\pi/a)\}x]$ perturbed by the lattice planes:

$$\psi(x) = \exp(ikx)\left\{A_0 + A_1\left[\exp-i\left(\frac{2\pi}{a}\right)x\right]\right\}. \tag{5.41}$$

One can easily check from equations (5.36) and (5.37) that the relation $A_0 = \pm A_1$ holds, when $k = \pm \pi/a$. Therefore, we find that the Bloch wave given by equation (5.41) is reduced to the form of either $\sin(\pi x/a)$ or $\cos(\pi x/a)$. This is understood as follows: the running wave $k = \pi/a$ is reflected to the wave $k' = -\pi/a$ by receiving the crystal momentum $g = -2\pi/a$ from the lattice planes and the reflected wave $k = -\pi/a$ is again reflected to the wave $k' = \pi/a$ by receiving the crystal momentum $g = 2\pi/a$ from the lattice planes. This process is infinitely repeated, resulting in a cosine- or sine-type stationary wave. Remember that the wavelength λ of the stationary wave is twice as large as the lattice constant, or $\lambda = 2a$.

The cosine-type stationary wave yields the maximum probability density at $x = 0$ or the center of the ionic potential. The potential at $x = 0$ is the deepest and, hence, the energy of the cosine-type Bloch wave is lowered relative to that corresponding to a uniform density of free electrons. The sine-type stationary wave yields the maximum probability density at the middle of the neighboring ions, where the ionic potential is the highest. Hence, the energy of the sine-type Bloch wave is raised relative to that of the free electron. This is illustrated in Fig. 5.8 [1]. Thus, the difference in energy between these two stationary states must be responsible for the formation of the energy gap ΔE at $k = \pm \pi/a$. It is now clear that $E_-(k)$ and $E_+(k)$ in equation (5.40) possess the cosine- and sine-type wave functions, respectively.

As is clear from the argument above, the Laue or Bragg condition in a three-dimensional periodic lattice is expressed as

$$|\mathbf{k}| = |\mathbf{k} - \mathbf{g_n}|, \tag{5.42}$$

where $\mathbf{g}$ is the reciprocal lattice vector corresponding to the set of lattice planes with Miller indices $(n_x n_y n_z)$. Equation (5.42) is viewed as arising from repeated elastic scattering of the Bloch electron incident on the lattice planes, upon which the momentum $\hbar \mathbf{g_n}$ is transferred back and forth between the electron and the whole lattice. The Bragg condition (5.42) will be employed in the

5.7 Brillouin zone of one- and two-dimensional periodic lattices 105

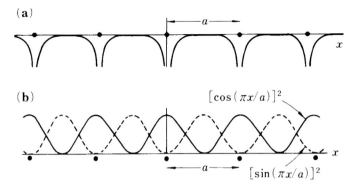

Figure 5.8. (a) The periodic potential with lattice constant *a*. (b) The probability densities of sine- and cosine-type stationary wave functions in the periodic potential. [C. Kittel, *Introduction to Solid State Physics*, (John Wiley & Sons, Inc., Second Edition, 1953)]

following sections to examine the condition for the formation of the energy gap in representative crystal structures.

5.7 Brillouin zone of one- and two-dimensional periodic lattices

As discussed in Sections 5.4 through 5.6, an energy gap appears successively at every $k = n\pi/a$ with $n = \pm 1, \pm 2, \ldots$ in a one-dimensional monatomic periodic lattice. The region centered at $k=0$ is bounded by the first energy gap at $k = \pi/a$ and $k = -\pi/a$. The region $-\pi/a < k \leq \pi/a$ is called the first Brillouin zone. The second, third, fourth, ... Brillouin zones can be successively defined outside the first Brillouin zone, as illustrated in Fig. 5.9. The *n*-th zone with $n \geq 2$ is split into positive and negative regions of equal length. We can easily confirm that its total length is always equal to $2\pi/a$.

The construction of Brillouin zones can be extended to two- and three-dimensional lattices. The Bragg condition (5.42) is most conveniently used for this purpose. As shown in Fig. 5.10, the Bragg condition (5.42) is always satisfied, provided that the wave vector **k** of the electron falls on the plane formed by bisecting perpendicularly the reciprocal lattice vector $\mathbf{g_n}$.

Using this as a guide, we first construct Brillouin zones for a two-dimensional square lattice with the lattice constant *a*. Any lattice vector of this lattice can be expressed as $\mathbf{l} = n_x \mathbf{a}_x + n_y \mathbf{a}_y$. Here $\mathbf{a}_x$ and $\mathbf{a}_y$ are the primitive translation vectors (also called basic vectors), defined as $\mathbf{a}_x = a\mathbf{i}$ and $\mathbf{a}_y = a\mathbf{j}$, where **i** and **j** are unit vectors along the *x*- and *y*-axes. The reciprocal lattice vector can be easily calculated as $\mathbf{g}_n = (2\pi/a)(n_x \mathbf{i} + n_y \mathbf{j})$ by applying equation (4.9) to the two-dimensional lattice, where n_x and n_y are arbitrary integers. The shortest reciprocal vectors among the set of (n_x, n_y) are those with $(1,0)$, $(-1,0)$, $(0,1)$ and

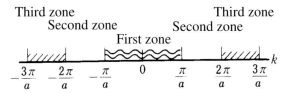

Figure 5.9.   Brillouin zones in a one-dimensional lattice.

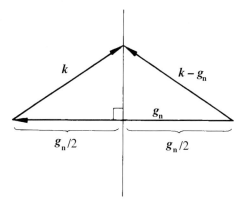

Figure 5.10.   Bragg condition obtained by bisecting perpendicularly the reciprocal lattice vector.

$(0,-1)$. The first Brillouin zone is constructed by bisecting perpendicularly these four reciprocal lattice vectors. We obtain a square with edge length $2\pi/a$ centered at the origin $k=0$. Similarly, the second zone is obtained from the second shortest reciprocal lattice vectors with the set of $(1,1)$, $(1,-1)$, $(-1,1)$ and $(-1,-1)$.

Figure 5.11 illustrates a series of the Brillouin zones for a two-dimensional square lattice, beginning from the first up to the tenth zone. It is interesting to note that each higher zone ($n \geq 2$) is split into small pieces but that they are always summed up to a total area equal to $(2\pi/a)^2$. Hence, all zones have an equal area in a two-dimensional lattice (equal volume in a three-dimensional lattice), regardless of the shape of each piece or the number of pieces from which the Brillouin zone is made up.

5.8 Brillouin zone of bcc and fcc lattices

The bcc and fcc structures, together with the hcp structure, are known to exist most abundantly in real metals and alloys. In this section, we construct the Brillouin zone for the bcc lattice. Figure 5.12 shows the bcc lattice in three different ways. The cubic unit cell with the lattice constant a shown in Fig. 5.12(a) contains two atoms. The volume per atom is, therefore, $a^3/2$. The primitive

5.8 Brillouin zone of bcc and fcc lattices

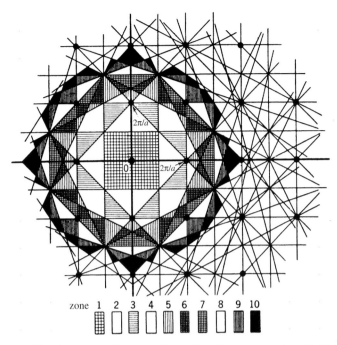

Figure 5.11. Brillouin zones of a two-dimensional square lattice. [L.Brillouin, *Wave Propagation in Periodic Structures*, (Dover Publications, 1953)]

translation vectors $\mathbf{a}_x$, $\mathbf{a}_y$ and $\mathbf{a}_z$ for the bcc lattice are shown in Fig. 5.12 (b) and are given by:

$$\mathbf{a}_x = \left(\frac{a}{2}\right)(-1\mathbf{i} - 1\mathbf{j} + 1\mathbf{k}),$$

$$\mathbf{a}_y = \left(\frac{a}{2}\right)(1\mathbf{i} + 1\mathbf{j} + 1\mathbf{k})$$

$$\mathbf{a}_z = \left(\frac{a}{2}\right)(-1\mathbf{i} + 1\mathbf{j} - 1\mathbf{k}), \tag{5.43}$$

where $\mathbf{i}$, $\mathbf{j}$, $\mathbf{k}$ are unit vectors in cartesian coordinates. The primitive cell formed from the primitive translation vectors is shown in Fig. 5.12 (b). Its volume must be $a^3/2$, since it contains a single atom. This is easily checked by calculating $V = \mathbf{a}_x \cdot (\mathbf{a}_y \times \mathbf{a}_z)$.

There is another representation. A polyhedron is formed by bisecting perpendicularly the shortest lattice vectors drawn from a given atom. It consists of eight {111} planes and six {002} planes in the bcc lattice.[3] A whole space can

---

[3] There are eight equivalent (111), ($1\bar{1}1$), ($11\bar{1}$), ($\bar{1}11$), ($\bar{1}\bar{1}1$), ($\bar{1}1\bar{1}$), ($1\bar{1}\bar{1}$) and ($\bar{1}\bar{1}\bar{1}$) planes. They are altogether expressed as {111} planes. Similarly, (002), (020), (200), ($00\bar{2}$), ($0\bar{2}0$) and ($\bar{2}00$) planes are grouped as {002} planes.

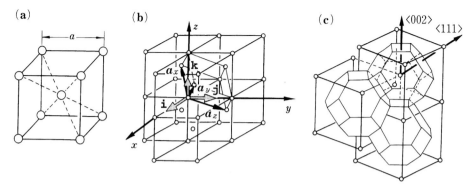

Figure 5.12. Unit cells of the bcc lattice. (a) Cubic unit cell, (b) primitive cell with translation vectors and (c) the Wigner–Seitz cell bounded by eight {111} regular hexagonal planes and six {002} square planes.

be covered by stacking the polyhedra without any overlap or void space. The volume of the polyhedron must be $a^3/2$, since it contains only a single atom at its center. The polyhedron is called the Wigner–Seitz cell. It is noted that the construction of the Wigner–Seitz cell in real space is the same as that of the Brillouin zone in reciprocal space.

We are now ready to construct the Brillouin zone of the bcc lattice. The primitive translation vectors in reciprocal space can be obtained by inserting equation (5.43) into equation (4.9):

$$\mathbf{b}_x = \left(\frac{2\pi}{a}\right)(-1\mathbf{i} + 1\mathbf{k}),$$

$$\mathbf{b}_y = \left(\frac{2\pi}{a}\right)(1\mathbf{j} + 1\mathbf{k})$$

$$\mathbf{b}_z = \left(\frac{2\pi}{a}\right)(-1\mathbf{i} + 1\mathbf{j}). \tag{5.44}$$

Any reciprocal lattice vector for the bcc lattice is now obtained by inserting equation (5.44) into equation (4.8):

$$\mathbf{g}_{hkl} = 2\pi(h\mathbf{b}_x + k\mathbf{b}_y + l\mathbf{b}_z)$$

$$= \left(\frac{2\pi}{a}\right)[(-h-l)\mathbf{i} + (k+l)\mathbf{j} + (h+k)\mathbf{k}]. \tag{5.45}$$

The first Brillouin zone can be constructed from planes normal to the shortest reciprocal lattice vectors in equation (5.45) at the midpoint. They are the following twelve vectors with the choice of ($h = \pm 1$, $k = 0$, $l = 0$), ($h = 0$, $k = \mp 1$, $l = \pm 1$) and so on:

5.8 Brillouin zone of bcc and fcc lattices

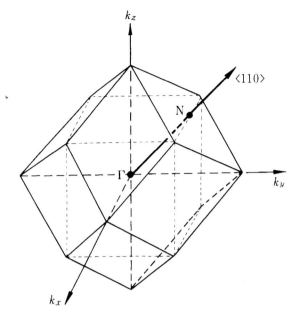

Figure 5.13. The first Brillouin zone of the bcc lattice. The zone is bounded by twelve $\{110\}$ rhombic dodecahedral planes. The origin at $k=0$ and the center of the $\{110\}$ plane are called the Γ and N points, respectively.

$$\mathbf{g}_{nn'0} = \left(\frac{2\pi}{a}\right)(\pm\mathbf{i}\pm\mathbf{j}) \quad (n=1 \text{ or } -1 \text{ and } n'=1 \text{ or } -1)$$

$$\mathbf{g}_{0nn'} = \left(\frac{2\pi}{a}\right)(\pm\mathbf{j}\pm\mathbf{k}) \quad (n=1 \text{ or } -1 \text{ and } n'=1 \text{ or } -1)$$

$$\mathbf{g}_{n0n'} = \left(\frac{2\pi}{a}\right)(\pm\mathbf{i}\pm\mathbf{k}) \quad (n=1 \text{ or } -1 \text{ and } n'=1 \text{ or } -1) \quad (5.46)$$

The twelve equivalent reciprocal lattice vectors are denoted altogether as $\mathbf{g}_{110}$ and the corresponding twelve zone planes as $\{110\}$ planes.[4] The first Brillouin zone thus constructed is the rhombic dodecahedron bounded by twelve $\{110\}$ planes and is shown in Fig. 5.13. This is similar to the Wigner–Seitz cell of the fcc lattice, as will be shown in Fig. 5.14(c). Indeed, it must be noted that the primitive reciprocal translation vectors in equation (5.44) are of the same form as the primitive translation vectors of the fcc lattice in equation (5.48) and that the difference is found only in the coefficients $2\pi/a$ and $a/2$.

Once the volume of the rhombic dodecahedron is obtained, we can easily

---

[4] The suffix (hkl) of the reciprocal lattice vector $\mathbf{g}_{hkl}$ in equation (5.45) refers to the $(\mathbf{b}_x, \mathbf{b}_y, \mathbf{b}_z)$ coordinate system. However, the suffix $(nn'n'')$ in $\mathbf{g}_{nn'n''}$ in equation (5.46) conventionally refers to the $(\mathbf{i}, \mathbf{j}, \mathbf{k})$ coordinate system. The same rule is applied for the fcc lattice (see equation (5.50)).

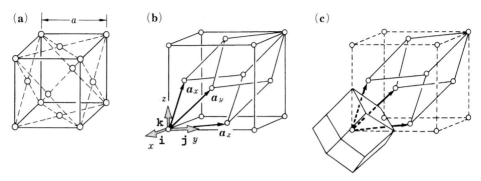

Figure 5.14. Unit cells of the fcc lattice. (a) Cubic unit cell, (b) primitive cell with primitive translation vectors and (c) the Wigner–Seitz cell bounded by twelve {110} rhombic dodecahedral planes.

calculate the number of electrons per atom accommodated in the first Brillouin zone of the bcc lattice. We make use of the fact that the first Brillouin zone of the bcc lattice is similar to the Wigner–Seitz cell of the fcc lattice, the volume of which is obviously equal to $a^3/4$. The translation from real space to reciprocal space can be easily done by substituting $2\pi/a$ for $a/2$. By rewriting the volume $a^3/4$ in the form of $2(a/2)^3$, one can easily deduce the volume of the first Brillouin zone of the bcc lattice as

$$V_B = 2\left(\frac{2\pi}{a}\right)^3. \tag{5.47}$$

Now let us assume a bcc metal cube with edge length L. The reciprocal space is quantized in intervals $2\pi/L$ and, hence, $2(L/2\pi)^3$ electrons can be accommodated in a unit volume of reciprocal space. The factor 2 arises from the degrees of freedom of spin. Since the volume of the first Brillouin zone is given by equation (5.47), we can fit $[2V/(2\pi)]^3 \times 2(2\pi/a)^3 = 4V/a^3$ electrons in the first zone, where $V = L^3$. It is seen that the number of electrons accommodated in the first zone depends on the volume of the metal. Since there exist $2V/a^3$ atoms in a volume V of the bcc metal, we derive the following axiom:

"The first Brillouin zone of the bcc lattice can accommodate 2 electrons per atom."

As a next example, we discuss the Brillouin zone of the fcc lattice. Figure 5.14 shows the fcc lattice in three different ways. The cubic unit cell with the lattice constant a is shown in Fig. 5.14(a). Its volume per atom is $a^3/4$, since the unit cell contains four atoms. Figures 5.14(b) and (c) represent the primitive cell and the Wigner–Seitz cell of the fcc lattice, respectively. The volume of the primitive cell of the fcc lattice is $a^3/4$. It is clear from Fig. 5.14(b) that the primitive translation vectors of the fcc lattice are given by

5.8 Brillouin zone of bcc and fcc lattices

$$\mathbf{a}_x = \left(\frac{a}{2}\right)(-1\mathbf{i} + 1\mathbf{k}),$$

$$\mathbf{a}_y = \left(\frac{a}{2}\right)(1\mathbf{j} + 1\mathbf{k})$$

$$\mathbf{a}_z = \left(\frac{a}{2}\right)(-1\mathbf{i} + 1\mathbf{j}). \tag{5.48}$$

Equation (5.48) is of the same form as equation (5.44) except for the difference in the coefficient. The primitive translation vectors in reciprocal space are calculated by inserting equation (5.48) into equation (4.9):

$$\mathbf{b}_x = \left(\frac{2\pi}{a}\right)(-1\mathbf{i} - 1\mathbf{j} + 1\mathbf{k}),$$

$$\mathbf{b}_y = \left(\frac{2\pi}{a}\right)(1\mathbf{i} - 1\mathbf{j} + 1\mathbf{k})$$

$$\mathbf{b}_z = \left(\frac{2\pi}{a}\right)(1\mathbf{i} + 1\mathbf{j} + 1\mathbf{k}). \tag{5.49}$$

This agrees with the primitive translation vectors of the bcc lattice given by equation (5.43) except for the difference in the coefficient. Any arbitrary reciprocal lattice vector is now expressed as

$$\mathbf{g}_{hkl} = 2\pi(h\mathbf{b}_x + k\mathbf{b}_y + l\mathbf{b}_z)$$

$$= \left(\frac{2\pi}{a}\right)[(-h+k-l)\mathbf{i} + (-h+k+l)\mathbf{j} + (h+k-l)\mathbf{k}]. \tag{5.50}$$

The shortest non-zero reciprocal lattice vectors arise from those like $h=1$, $k=l=0$ and also those like $h=k=l=1$ and $h=k=l=-1$. They are explicitly written as

$$\mathbf{g}_{111} = \left(\frac{2\pi}{a}\right)(1\mathbf{i} + 1\mathbf{j} + 1\mathbf{k}); \quad \mathbf{g}_{11\bar{1}} = \left(\frac{2\pi}{a}\right)(1\mathbf{i} + 1\mathbf{j} - 1\mathbf{k})$$

$$\mathbf{g}_{1\bar{1}1} = \left(\frac{2\pi}{a}\right)(1\mathbf{i} - 1\mathbf{j} + 1\mathbf{k}); \quad \mathbf{g}_{\bar{1}11} = \left(\frac{2\pi}{a}\right)(-1\mathbf{i} + 1\mathbf{j} + 1\mathbf{k})$$

$$\mathbf{g}_{\bar{1}\bar{1}1} = \left(\frac{2\pi}{a}\right)(-1\mathbf{i} - 1\mathbf{j} + 1\mathbf{k}); \quad \mathbf{g}_{1\bar{1}\bar{1}} = \left(\frac{2\pi}{a}\right)(1\mathbf{i} - 1\mathbf{j} - 1\mathbf{k})$$

$$\mathbf{g}_{\bar{1}1\bar{1}} = \left(\frac{2\pi}{a}\right)(-1\mathbf{i} + 1\mathbf{j} - 1\mathbf{k}); \quad \mathbf{g}_{\bar{1}\bar{1}\bar{1}} = \left(\frac{2\pi}{a}\right)(-1\mathbf{i} - 1\mathbf{j} - 1\mathbf{k}). \tag{5.51}$$

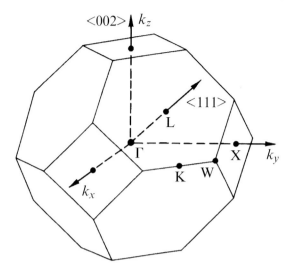

Figure 5.15. The first Brillouin zone of the fcc lattice. The zone is bounded by eight {111} regular hexagonal planes and six {002} square planes. The origin, the center of the {111} plane and the center of the {002} planes are named Γ, L and X points, respectively. Other representative symmetry points K and W are also indicated.

The eight {111} zone planes are formed by bisecting perpendicularly these reciprocal lattice vectors. However, a whole reciprocal space cannot be covered only with these planes. The next shortest reciprocal lattice vectors arise from those like $h=k=1$, $l=0$. They are given by

$$\mathbf{g}_{200} = \left(\frac{2\pi}{a}\right)(2\mathbf{i}); \quad \mathbf{g}_{\bar{2}00} = \left(\frac{2\pi}{a}\right)(-2\mathbf{i}) \text{ and } \mathbf{g}_{020} = \left(\frac{2\pi}{a}\right)(2\mathbf{j})$$

$$\mathbf{g}_{0\bar{2}0} = \left(\frac{2\pi}{a}\right)(-2\mathbf{j}); \quad \mathbf{g}_{002} = \left(\frac{2\pi}{a}\right)(2\mathbf{k}) \text{ and } \mathbf{g}_{00\bar{2}} = \left(\frac{2\pi}{a}\right)(-2\mathbf{k}) \quad (5.52)$$

The six {002} planes can be formed from these reciprocal vectors. A combination of both eight {111} planes and six {002} planes forms the truncated octahedron consisting of fourteen planes in total, as shown in Fig. 5.15. This is the first Brillouin zone of the fcc lattice and is similar to the Wigner–Seitz cell of the bcc lattice shown in Fig. 5.12(c). Therefore, we see that a whole reciprocal space of the fcc lattice can be covered with the repetition of the truncated octahedron.

We can easily calculate the number of electrons per atom accommodated in the first Brillouin zone of the fcc lattice in the same way as in the bcc lattice. The volume of the Wigner–Seitz cell of the bcc lattice is obviously equal to $a^3/2$. This is rewritten as $4(a/2)^3$, into which $2\pi/a$ is inserted in place of $a/2$. The volume of the first Brillouin zone of the fcc lattice is therefore deduced as

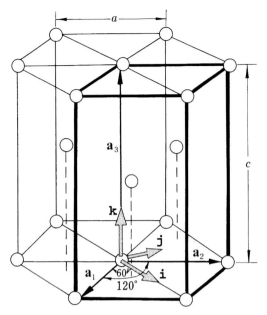

Figure 5.16. Unit cell of the hcp lattice. The primitive cell is shown by thick lines. The angle between the primitive translation vectors $\mathbf{a}_1$ and $\mathbf{a}_2$ makes 120°, while that between $\mathbf{a}_1$ and $\mathbf{i}$ makes 60°. The unit vectors $\mathbf{i}$, $\mathbf{j}$, and $\mathbf{k}$ are perpendicular to each other.

$V_B = 4(2\pi/a)^3$ and the number of electrons accommodated in the first Brillouin zone is $2(L/2\pi)^3 \times 4(2\pi/a)^3 = 8V/a^3$. Since a fcc metal with its volume V contains $4V/a^3$ atoms, we derive another important axiom:

> "The first Brillouin zone of the fcc lattice can accommodate 2 electrons per atom".

5.9 Brillouin zone of hcp lattice

There are many metals and alloys which crystallize into the hexagonal close-packed structure (hereafter abbreviated as hcp). This is the structure obtained by stacking atomic planes consisting of closely packed hard spheres in the sequence ABAB ....[5] The hcp lattice is shown in Fig. 5.16. The basal plane

---

[5] Either hcp or fcc structure can be formed by packing hard spheres as closely as possible. The layer "A" is formed by placing six spheres around the central sphere in contact with each other in a given plane. The second layer "B" is formed by placing a sphere in a hollow formed by three neighboring spheres in the layer A. There are two choices for the position of spheres in the third layer C. One way is to place a sphere in layer C on top of the sphere in layer A, thereby making layer C identical to layer A. This yields the ABAB ... stacking in the hcp lattice. The other way is to place a sphere in layer C on top of the hollow in layer A. This yields the ABCBC ... stacking in the fcc lattice.

with a hexagonal symmetry is called the *ab*-plane and its perpendicular axis the *c*-axis. The atomic distance between the top and bottom planes of the unit cell is often denoted as c and the shortest atomic distance in the *ab*-plane as a. The axial ratio c/a turns out to be $\sqrt{8/3} = 1.633$ for hexagonal closest-packing of hard spheres. There are deviations from the ideal value in real metals: $c/a =$ 1.568 for Be, 1.623 for Mg, 1.856 for Zn and 1.886 for Cd. There exist six atoms in the unit cell shown in Fig. 5.16. Its primitive cell is drawn by thick lines. Four corner atoms in both top and bottom planes are summed up to a single atom belonging to this primitive cell. There is one additional atom inside the primitive cell, but its presence is ignored for the moment. The primitive translation vectors are given by

$$\mathbf{a}_1 = \left(\frac{a}{2}\right)\mathbf{i} - \left(\frac{\sqrt{3}a}{2}\right)\mathbf{j},$$

$$\mathbf{a}_2 = \left(\frac{a}{2}\right)\mathbf{i} + \left(\frac{\sqrt{3}a}{2}\right)\mathbf{j}$$

$$\mathbf{a}_3 = c\mathbf{k}. \tag{5.53}$$

Its volume is easily calculated to be $\sqrt{3}a^2c/2$. The corresponding primitive translation vectors in reciprocal space are obtained as

$$\mathbf{b}_1 = \left(\frac{2\pi}{a}\right)\left[1\mathbf{i} + \left(\frac{-1}{\sqrt{3}}\right)\mathbf{j}\right]$$

$$\mathbf{b}_2 = \left(\frac{2\pi}{a}\right)\left[1\mathbf{i} + \left(\frac{1}{\sqrt{3}}\right)\mathbf{j}\right]$$

$$\mathbf{b}_3 = \left(\frac{2\pi}{c}\right)[1\mathbf{k}]. \tag{5.54}$$

An arbitrary reciprocal lattice vector is then expressed as

$$\mathbf{g}_{n_1 n_2 n_3} = 2\pi(n_1 \mathbf{b}_1 + n_2 \mathbf{b}_2 + n_3 \mathbf{b}_3)$$

$$= 2\pi\left[\left(\frac{1}{a}\right)(n_1 + n_2)\mathbf{i} + \left(\frac{1}{\sqrt{3}a}\right)(-n_1 + n_2)\mathbf{j} + \left(\frac{1}{c}\right)n_3 \mathbf{k}\right]. \tag{5.55}$$

The shortest non-zero reciprocal lattice vectors arise from those like $n_1 = 1$, $n_2 = 0$, $n_3 = 0$ and are explicitly written as[6]

---

[6] In the hcp lattice, the suffix $(n_1 n_2 n_3)$ in the reciprocal lattice vector $\mathbf{g}_{n_1 n_2 n_3}$ in equation (5.56) conventionally refers to the $(\mathbf{b}_1, \mathbf{b}_2, \mathbf{b}_3)$ coordinate system. This is different from the choice of the $(\mathbf{i}, \mathbf{j}, \mathbf{k})$ coordinate system for the bcc and fcc lattices.

5.9 Brillouin zone of hcp lattice

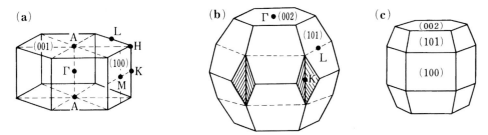

Figure 5.17. Brillouin zones of the hcp lattice. (a) The first zone, and (b) second zone. The dashed area indicates the {100} planes. (c) The Jones zone. The following relations hold. $A\Gamma = \pi/c$, $\Gamma\Gamma = 2\pi/c$, $\Gamma M = 2\pi/\sqrt{3}a$ and $\Gamma L = (2\pi/\sqrt{3}a)\sqrt{1 + (3/4)(a/c)^2}$.

$$\mathbf{g}_{100} = \left(\frac{2\pi}{a}\right)\left[1\mathbf{i} + \left(\frac{-1}{\sqrt{3}a}\right)\mathbf{j}\right]; \quad \mathbf{g}_{\bar{1}00} = \left(\frac{2\pi}{a}\right)\left[-1\mathbf{i} + \left(\frac{1}{\sqrt{3}a}\right)\mathbf{j}\right]$$

$$\mathbf{g}_{010} = \left(\frac{2\pi}{a}\right)\left[1\mathbf{i} + \left(\frac{1}{\sqrt{3}a}\right)\mathbf{j}\right]; \quad \mathbf{g}_{0\bar{1}0} = \left(\frac{2\pi}{a}\right)\left[-1\mathbf{i} - \left(\frac{1}{\sqrt{3}a}\right)\mathbf{j}\right]$$

$$\mathbf{g}_{110} = \left(\frac{2\pi}{a}\right)\left[\left(\frac{2}{\sqrt{3}}\right)\mathbf{i}\right]; \quad \mathbf{g}_{\bar{1}\bar{1}0} = \left(\frac{2\pi}{a}\right)\left[-\left(\frac{2}{\sqrt{3}}\right)\mathbf{i}\right]$$

$$\mathbf{g}_{001} = \left(\frac{2\pi}{c}\right)[1\mathbf{k}]; \quad \mathbf{g}_{00\bar{1}} = \left(\frac{2\pi}{c}\right)[-1\mathbf{k}]. \quad (5.56)$$

The first Brillouin zone of the hcp primitive cell can be constructed from planes bisecting perpendicularly the reciprocal lattice vectors given by equation (5.56). The result is shown in Fig. 5.17(a). It is clear that the first six reciprocal lattice vectors are responsible for creating the side surfaces of the regular hexagonal prism and the last two for its top and bottom surfaces. The six side surfaces are referred to as {100} planes, and the top and bottom surfaces as {001} planes.

The situation in the real hcp lattice is slightly different. As mentioned above, there is an intervening atomic layer through the center of the unit cell. The diffracted wave from the {001} planes is cancelled with that from the intervening planes as a result of the phase difference π between them. Thus, the intensity of the diffracted wave due to the {001} planes is reduced to zero and the energy gap disappears. This is known as the extinction rule mentioned in Section 4.3.

The second Brillouin zone is shown in Fig. 5.17(b). This is constructed from two {002} planes, twelve {101} planes and six {100} planes. In the hcp lattice, the smallest zone bounded by planes having finite energy gaps is not given by the first Brillouin zone because of the absence of the energy gap across the {001} planes but is made up from a combination of the first and second Brillouin zones. As shown in Fig. 5.17(c), it consists of two {002} planes, six {100} planes and twelve {101} planes. This is often called the Jones zone.

One can easily check that the first Brillouin zone for the hcp lattice in Fig. 5.17(a) contains one electron per atom, since the primitive cell contains two atoms. Thus, two electrons per atom can be fitted into the combined first and second zones in Fig. 5.17(b). The number of electrons per atom filled in the Jones zone is calculated to be

$$n = 2 - \left(\frac{3}{4}\right)\left(\frac{a}{c}\right)^2 \left[1 - \left(\frac{1}{4}\right)\left(\frac{a}{c}\right)^2\right]. \quad (5.57)$$

Its derivation is left for the reader as Exercise 5.4.

5.10 Fermi surface–Brillouin zone interaction

We have seen in the preceding section that the Brillouin zone appears in reciprocal space as an assembly of polyhedra bounded by planes normal to the reciprocal lattice vectors at their midpoints and we constructed the Brillouin zone for the bcc, fcc and hcp lattices. A finite energy gap appears across the Brillouin zone plane, unless the extinction rule holds. The presence of an energy gap leads to the deviation of the E–k relation from the free-electron parabolic band in the vicinity of the Brillouin zone. The Fermi surface begins to be distorted from a sphere before making contacts with the Brillouin zone planes. We will study in this section more details about the interaction of the Fermi surface with the Brillouin zone.

According to equation (2.20), the radius of the Fermi sphere is determined by the number of electrons per atom and the volume per atom Ω, since N_0/V in equation (2.20) is equal to $(e/a)/\Omega$. The number of electrons per atom is abbreviated as e/a and is often referred to as the electron concentration. Let us consider a hypothetical simple cubic metal with the lattice constant a and assume the Fermi sphere to expand as a function of only the electron concentration e/a while keeping the lattice constant unchanged.

The first Brillouin zone for the simple cubic lattice is given by a cube with edge length $2\pi/a$. The second and third Brillouin zones are also constructed from planes normal to the second and third shortest reciprocal lattice vectors at their midpoints. They are depicted in Figs 5.18(a), (b) and (c). The third zone looks complex but can be easily understood, if a comparison is made with the Brillouin zone of the two-dimensional square lattice shown in Fig. 5.11. This is because the latter corrresponds to a cross-section of the former cut through

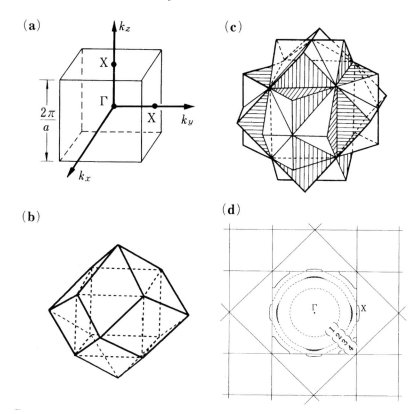

Figure 5.18. Brillouin zones of the simple cubic lattice with lattice constant a. (a) The first zone, (b) the second zone (the first zone is seen as dashed lines), (c) the third zone, and (d) the cross-section of the Fermi surface and the Brillouin zones obtained by cutting through the Γ point parallel to the $k_x k_y$-plane. In (d), the Fermi surface contours with various electron concentrations e/a are shown by dashed curves: (1) spherical surface with $e/a \approx 0.5$, (2) distortion is seen along ΓX directions (distorted area is shown by hatches), (3) the Fermi surface partly touches the first zone, and (4) a part of the Fermi surface appears in the second zone (the Fermi surface with $e/a = 2.0$ is shown by a solid curve).

the origin $k = 0$ parallel to the $k_x k_y$-plane. The origin in reciprocal space is often denoted as Γ in the notation of band calculations.

Figure 5.18(d) shows a series of cross-sections of the Fermi surface having various electron concentrations, together with cross-sections of the first, second and higher Brillouin zones. One can easily check that 2 electrons per atom can be accommodated in each Brillouin zone of the simple cubic lattice. Let us suppose first a monovalent metal in a simple cubic structure. Its Fermi sphere must occupy exactly 50% of the first Brillouin zone in volume, though one can easily calculate the ratio of the Fermi radius over the distance ΓX to be $(3\pi^2)^{1/3}/\pi = 0.98$. Let us assume a hypothetical metal having a much smaller

Fermi sphere, say a metal with $e/a = 0.5$. Then we can draw the free-electron-like spherical Fermi surface for this metal, which is shown by the dashed curve (1) in Fig. 5.18(d). In this case, the effect of the periodic potential on the conduction electron can be fully neglected and the free-electron model holds well.

The metal A with $e/a = 0.5$ is alloyed with a divalent metal B while keeping the crystal structure and its lattice constant unchanged. An average electron concentration for an $A_{1-x}B_x$ alloy is given by $e/a = 0.5(1-x) + 2x = 1.5x + 0.5$. For example, $e/a = 1.25$ is obtained for an equiatomic alloy $A_{0.5}B_{0.5}$. An increase in electron concentration expands the Fermi radius according to equation (2.20) and the Fermi surface approaches six equivalent {100} zone planes. Since the distance ΓX is the shortest from the origin to the zone plane, the Fermi sphere begins to be distorted in the area perpendicular to ΓX as a result of a decrease in slope of the E–k curve relative to the corresponding free-electron value. This gives rise to a swollen Fermi surface with six equivalent swollen areas. A cross-section of this surface is illustrated schematically by the dashed curve (2) in Fig. 5.18(d). (Note that only four of the swollen areas, marked by hatches, are seen in cross-section.)

Further increase in the electron concentration makes the Fermi surface to touch the Brillouin zone boundary, as shown by the dashed curve (3). Note that the Fermi surface (3) becomes discontinuous, being separated into pieces by the zone boundary, since states on the zone are no longer a part of the Fermi surface (note that the Fermi surface refers to a constant energy surface). When the electron concentration increases further and reaches 2.0, the first Brillouin zone is, in principle, fully filled. However, some electrons would jump into the second zone before filling the corner of the first zone, provided that their energies are high enough to overcome the energy gap across the zone plane. This means that a part of the Fermi surface appears in the second zone but the rest remains in the first zone, leaving unoccupied states or "holes" in the first zone. The Fermi surface thus obtained is shown by the solid curve (4) in Fig. 5.18(d).

The E–**k** relations along the $\langle 100 \rangle$ and $\langle 110 \rangle$ directions in Fig. 5.18(d) are depicted schematically in Fig. 5.19. Using the E–**k** relations as a guide, we discuss more about the Fermi surface (4), where some electrons occupy the second zone prior to a complete filling of the first zone. The Fermi level may be viewed as the surface of the water when "water" of electrons is poured into a "pot" formed by the E–k curves. One can easily understand from Fig. 5.19 that, whether or not the "water surface", i.e., the Fermi level, appears in the second zone, certainly depends on the magnitude of the energy gaps across the {100} and {110} planes and their relative positions. A hypothetical simple cubic solid with $e/a = 2.0$ becomes a metal, if the energy gap across the {100}

5.10 Fermi surface–Brillouin zone interaction

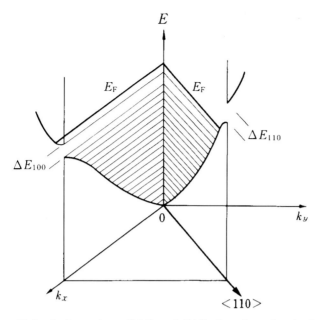

Figure 5.19. E–k relations along ⟨100⟩ and ⟨110⟩ directions for the band structure of the simple cubic lattice shown in Fig. 5.18. It can be seen that electrons are overlapped to the second zone across the {100} planes.

planes is small enough to allow electrons to enter the second zone while leaving some holes in the first zone.

If the energy gap across the {100} planes is very large, electrons can no longer enter the second zone but have to fill the corner of the first zone. The first zone is completely filled with electrons and the Fermi surface disappears when $e/a = 2.0$. Electrons cannot be excited into higher zones, unless the energy supplied from external sources, say by heating or applying an electric field, is high enough for the electrons to overcome a large energy gap. In such a case, our hypothetical solid becomes an insulator.

Whether a given material is a metal or an insulator can be judged by checking the presence or absence of electrons at the Fermi level in both the first and higher zones over all directions in reciprocal space. In other words, a metal is defined as a substance in which the Fermi surface exists and the density of states at the Fermi level is finite at absolute zero. In contrast, an insulator, including a semiconductor, is defined as a substance in which the Fermi surface is absent. The density of states representation is the most convenient means for this purpose. Once the E–**k** relation is calculated along various directions of the reciprocal lattice vectors, the Fermi surface can be constructed in reciprocal space. The density of states curve is then calculated by counting the number of states in the energy interval ΔE.

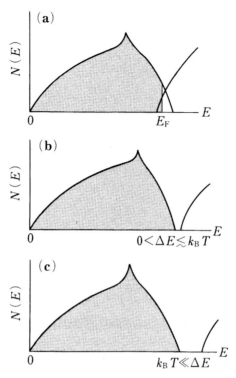

Figure 5.20. Schematic density of states curves. (a) Metals, (b) semiconductors, and (c) insulators.

Figures 5.20(a), (b) and (c) schematically illustrate the density of states curves for a metal, a semiconductor and an insulator, respectively. In the case of (a), the Fermi level sits in the middle of the density of states curve. This is typical of the band structure of a metal. In the cases of (b) and (c), the completely filled density of states, which is often called the valence band, is separated by an energy gap from the completely unoccupied density of states, which is called the conduction band. This is the band structure of a semiconductor and an insulator. As shown in Fig. 5.20(c), the energy gap is fairly large in insulators. In a semiconductor, the energy gap is reasonably small and comparable to the thermal energy $k_B T$. Thus, it is an insulator at low temperatures but becomes conductive at higher temperatures. This is because some electrons begin to be thermally excited into higher zones and excited electrons contribute to the electron conduction. The excitation of electrons into higher zones leaves behind an equal number of holes in the first zone. Such materials are referred to as intrinsic semiconductors (see more details in Section 6.9).

As has been noted, each Brillouin zone generally accommodates two electrons per atom. Hence, a material having an even number of valence electrons per atom tends to become either a semiconductor or an insulator. For example, both Si and Ge possess 4 valence electrons per atom and are indeed known as semiconductors. As another example, the GaAs compound consisting of 50 at.% trivalent Ga and 50 at.% pentavalent As, is also characterized by having 4 valence electrons per atom on average and is typical of a semiconductor. But it must be noted that materials having an even number of valence electrons per atom do not always become a semiconductor or an insulator. For instance, divalent Be, Mg, Zn and Cd, all of which crystallize into the hcp structure, are metals, though the combined first and second zones can just accommodate 2 electrons per atom. As will be shown in Chapter 6, a situation similar to the Fermi surface (4) in Fig. 5.18(d) occurs and the density of states like that in Fig. 5.20(a) is observed in all these divalent metals.

5.11 Extended, reduced and periodic zone schemes

The E–k relation of the conduction electrons in a one-dimensional periodic lattice with the lattice constant a is reproduced in Fig. 5.21(a). The energy gap appears not only at $k = \pm(\pi/a)$ but also at every $k = (\pi/a)n$ with $n \geq 2$ or $n \leq -2$. It is clear that the free-electron parabola is cut into segments at $k = n\pi/a$. This representation is called the extended zone scheme.

There are two alternative representations, which are also equally important in practical use. We have emphasized that the Bloch state k remains unchanged if its wave vector is shifted by an appropriate reciprocal lattice vector. As is clear from Fig. 5.21(a), the E–k curves in the regions $\pi/a < k \leq 2\pi/a$ and $-2\pi/a < k \leq -\pi/a$ belong to the second Brillouin zone. They can be shifted in a negative and a positive direction by the reciprocal lattice vector of $2\pi/a$, respectively, so that they can be reduced into the first zone. Similarly, the n-th zone can be reduced into the first zone. The results after applying this operation to both the second and third Brillouin zones are shown in Fig. 5.21(b). This representation is called the reduced zone scheme. The E–k relation in the reduced zone scheme becomes a multi-valued function of the wave vector. As mentioned in Section 5.5, bands are labelled to distinguish them. The band index $n = 1$ is assigned to the lowest band and $n = 2, 3$ and so on to higher bands.

As the third representation, the n-th zone after reducing to the first zone can be extended periodically with the period of $2\pi/a$ outside the first zone. This representation is called the periodic zone scheme or repeated zone scheme. Figure 5.21(c) shows the E–k relation in the first, second and third zones in the periodic zone scheme.

The choice of zone scheme becomes important when the Fermi surface extends to higher zones. We still assume a simple cubic lattice with the lattice

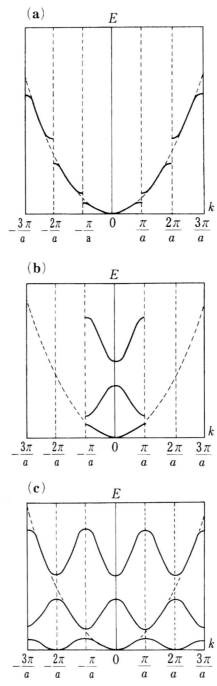

Figure 5.21. *E–k* relations in (a) the extended zone scheme, (b) the reduced zone scheme, and (c) the periodic (or repeated) zone scheme.

5.11 Extended, reduced and periodic zone schemes

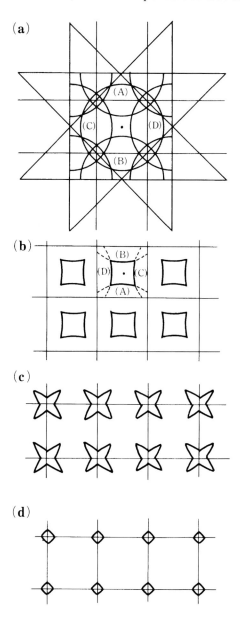

Figure 5.22. The cross-section of the Fermi surface and the Brillouin zones of the simple cubic lattice (see Fig. 5.18(d)). (a) The Brillouin zones and the Fermi surface in the extended zone scheme. The Fermi surface is partly shown in the periodic (or repeated) zone scheme. (b) The Fermi surface of holes in the second zone in the periodic zone scheme. (c) The Fermi surface of electrons in the third zone in the periodic zone scheme. (d) The Fermi surface of electrons in the fourth zone in the periodic zone scheme.

constant a and use the cross-section (see Fig. 5.18(d)) of its Brillouin zones cut through the Γ point parallel to the $k_x k_y$-planes. Furthermore, the periodic empty-lattice model is assumed so that the energy gap across each zone plane is set equal to zero. We consider the situation where the first zone is completely filled with electrons and the Fermi surface exists in the second, third and fourth zones, as shown in Fig. 5.22(a). The cross-section of the Fermi surface is given by a circle because of the absence of the energy gap. We can draw as many Fermi circles centered at reciprocal lattice points as we wish.

Figure 5.22(a) corresponds to the extended zone scheme, in which each Fermi surface is drawn as a continuous circle. But one realizes that the Fermi circle in a higher zone is no longer continuous. For instance, the electron states in the second zone are disconnected by the third and fourth zones and exist as four equivalent segments marked by (A), (B), (C) and (D). Similarly, the electron states in the third and fourth zones are separated into eight and four equivalent segments, respectively, as can be seen from Fig. 5.22(a).

The reduced zone scheme can eliminate this inconvenience. We explain this, using the Fermi surface in the second zone. The portions (A), (B), (C) and (D) are shifted downwards, upwards, to the right and to the left, respectively, by an amount equal to the reciprocal lattice vector $2\pi/a$. This operation brings four separated segments of the electron states into the first zone and, as a result, the Fermi surface becomes connected as a single loop without disruption, as shown in Fig. 5.22(b). This is the Fermi surface of the second zone in the reduced zone scheme. The Fermi surface is no longer a circle but looks like a curved window. Note that the portions (A), (B), (C) and (D) appear along the corner of the first zone and, hence, electrons are present outside the Fermi surface but absent inside. Such a Fermi surface is called the Fermi surface of holes. In Fig. 5.22(b), the reduced Fermi surface is periodically extended to neighboring zones to represent it in the periodic zone scheme.

The Fermi surface in the third zone is more disconnected than that in the second zone. Fig. 5.22(c) in the periodic zone scheme is obtained by reducing the Fermi surface in the third zone. A petal-like Fermi surface emerges under this operation. This is the Fermi surface of electrons, since the states around the Γ point are filled with electrons. Similarly, the Fermi surface in the fourth zone is reduced into the first zone. The results in the periodic zone scheme are shown in Fig. 5.22(d). This is again the Fermi surface of electrons. It is seen that the Fermi surface in the reduced and periodic zone schemes consists of an assembly of segments of the spherical Fermi surface but loses its spherical appearance. Note that the energy gap across the zone is finite in real metals and that the actual Fermi surface in the reduced zone scheme is certainly perturbed by the energy gap and deviates from spherical curvature particularly near the zone planes.

The Fermi surfaces of respresentative metals in the periodic table will be discussed in Chapter 6. We will see that the reduced and periodic zone schemes are frequently used to represent the Fermi surfaces of polyvalent metals.

Exercises

5.1 The wave vector of the Bloch electron is not uniquely determined but involves an arbitrariness associated with any reciprocal lattice vector. Show that this unique property can be explained by expanding the periodic function $u_k(\mathbf{r})$ in equation (5.12) into a series in the same way as equation (5.30).

5.2 Solve equation (5.34) and draw the dispersion relation of electrons in a one-dimensional periodic empty-lattice, where the magnitude of the ionic potential is reduced to be infinitesimally small. Note that, though the periodicity of the lattice remains, the energy dispersion relation is given by $E = k^2$ in atomic units, where $\hbar = 1$ and $m = 1/2$.

5.3 Derive the Bragg condition $2d\sin\theta = m\lambda$ from the Laue condition $\mathbf{k}^2 = (\mathbf{k} - \mathbf{g}_n)^2$, where $\mathbf{g}_n$ is the reciprocal lattice vector for the set of lattice planes with the Miller indices $(n_x n_y n_z)$.

5.4 Calculate the number of electrons per atom filled in (a) the first zone and (b) the Jones zone of the hcp lattice shown in Fig. 5.17.

Chapter Six
Electronic structure of representative elements

6.1 Prologue

The basic ideas and fundamental concepts of the electron theory of metals have been discussed in Chapters 1 to 5. In the present chapter, we present the electronic structure of real metals and semiconductors selected from elements in the periodic table and show how the electronic structure of various elements changes across the periodic table.

6.2 Elements in the periodic table

Table 6.1 lists representative elements in the periodic table, which are classified in terms of the crystal structure and characteristic features, such as valencies. Most elements crystallize into bcc, fcc or hcp structures, though some other elements like Ga, In, Hg and Bi possess more complicated structures. We discuss the electronic structures of representative elements in bcc, fcc and hcp structures whose Brillouin zones have already been discussed in Chapter 5.

6.3 Alkali metals

All the alkali metals Li, Na, K, Rb and Cs are located in the very left-hand column in the periodic table (see Table 1.1) and crystallize in the bcc structure.[1] There is only one outermost electron in the 2s, 3s, 4s, 5s and 6s orbit, respectively, in the free atom. Upon the formation of the bcc metal, they serve to form the valence band containing one electron per atom. This is the reason why each is called a monovalent metal. As has been discussed in Section 5.8, the first Brillouin zone of the bcc lattice can accommodate 2 electrons per atom and,

---

[1] A phase transformation from the bcc to an hcp structure has been reported to occur at 78 and 35 K for Li and Na, respectively.

Table 6.1. *Representative elements in the periodic table*

structure[a]	characteristic features	elements
bcc	monovalent metals	Li, Na, K, Rb, Cs
	transition metals	V, Cr, Fe, Nb, Mo, Ta, W
fcc	monovalent metals	Cu, Ag, Au
	divalent metals	Ca, Sr
	trivalent metals	Al
	tetravalent metals	Pb
	transition metals	Ni, Pd, Pt, Rh, Ir
hcp	divalent metals	Be, Mg, Zn, Cd
	trivalent metals	Tl
	transition metals	Sc, Ti, Y, Zr, La, Hf, Co, Ru, Re, Os
hexagonal	semimetals	C (graphite)
rhombic	semimetals	Bi, Sb
diamond	semiconductors	C (diamond), Si, Ge

Note:
[a] The crystal structure refers to that at room temperature. For example, Fe transforms to the fcc structure known as γ-Fe in the range 1185–1667 K.

hence, one would expect the Fermi surface to remain spherical away from the {110} zone planes of the Brillouin zone (see Fig. 5.13).

The E–$\mathbf{k}$ relations of Na metal have been calculated in the three directions ΓN, ΓP and ΓH in the Brillouin zone of the bcc lattice. The results are shown in Fig. 6.1. Of the three directions, the point N corresponding to the center of the {110} planes is the closest to the origin Γ. Its Fermi level is situated at an energy lower than that marked as N_1', which would be denoted as $E_-(k_N)$ in an expression similar to equation (5.40). This clearly means that the E–$\mathbf{k}$ relations up to the Fermi level are well approximated by the free-electron-like parabola in all directions studied. The energy gap across the {110} planes is calculated to be only 0.018 Ry (see Appendix 1) or 0.25 eV. Indeed, the energy gap denoted as $N_1 - N_1'$ in Fig. 6.1 is hardly visible and the Fermi surface of Na metal is nearly spherical, as illustrated in Fig. 6.2. The de Haas–van Alphen effect measurements, which will be discussed in Section 7.2, proved the presence of a spherical Fermi surface in the bcc Na metal [1,2].[2]

Na metal is known to be almost a free-electron-like metal. This is no longer true in other alkali metals. The E–$\mathbf{k}$ relations calculated for Li metal are shown

---

[2] The de Haas–van Alphen effect, which provides information about the extremal area of cross-section of the Fermi surface (see Section 7.2), requires low temperatures, say, below 4 K. The phase transformation in Na could be suppressed by quenching the sample to low temperatures. The de Haas–van Alphen effect has proved that the distortion of the Fermi surface from a free-electron sphere is less than 0.1% in Na [2]. However, the suppression of the phase transformation has not been successful in Li and, hence, no de Haas–van Alphen effect has been reported for the bcc Li [1].

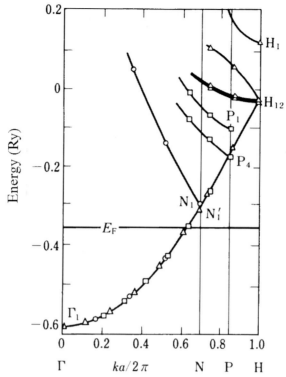

Figure 6.1. E–k relations along ΓN, ΓP and ΓH directions of pure Na metal. Note that the dimensionless wave number $ka/2\pi$ is employed. [F. S. Ham, *Phys. Rev.* **128** (1962) 82]

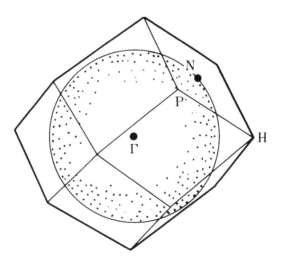

Figure 6.2. The Fermi surface and the first Brillouin zone of bcc Na metal. See also Fig. 5.13. Symmetry points Γ, N, P and H are marked.

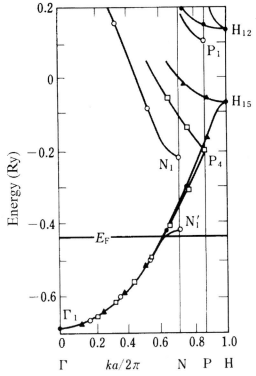

Figure 6.3. *E–k* relations of pure Li metal along ΓN, ΓP and ΓH directions. Note that the dimensionless wave number $ka/2\pi$ is employed. [F. S. Ham, *Phys. Rev.* **128** (1962) 82]

in Fig. 6.3. The Fermi level is positioned at an energy slightly lower than $E_-(k_N)$ marked as N_1' in Fig. 6.3. Hence, its Fermi surface is very close to but has no contact with the {110} zone planes. The energy gap across the {110} zone planes is given by $E_{N_1} - E_{N_1'}$ in Fig. 6.3 and amounts to 0.209 Ry or 2.8 eV. This is fairly large in comparison with that in Na metal. It can be clearly seen that the slope of the *E–k* curve at the Fermi level in the ΓN direction is already well suppressed relative to those in other directions ΓP and ΓH because of the effect of the {110} energy gap. As a consequence, the Fermi surface is no longer spherical but bulges along the ⟨110⟩ direction, as was explained by using the Fermi surface (3) in Fig. 5.18(d).

Figure 6.4 shows the density of states curve of Li metal calculated from the *E–k* relations in Fig. 6.3. A cusp at about 3.7 eV followed by a sharp decline in the density of states is certainly caused by simultaneous contacts of the Fermi surface with the twelve {110} zone planes. The density of states anomaly like the cusp above is often called the van Hove singularity. The Fermi level is located at the position prior to the cusp, where the density of states exhibits a

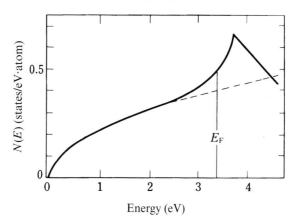

Figure 6.4. Density of states curve of bcc Li metal. A cusp is caused by the interaction with the {110} zone planes of the first Brillouin zone. [F. S. Ham, *Phys. Rev.* **128** (1962) 2524]

sharply increasing slope beyond the free-electron parabola shown by a dashed curve. Thus, we see from Fig. 6.4 that the Fermi surface of Li metal has not touched the Brillouin zone but bulges due to its proximity to the Brillouin zone.

A substantial deviation from the free-electron model in Li metal arises because the Li atom has a very small ion core consisting of only $(1s)^2$ electrons so that the valence electron is much closer to the nucleus than in other alkali metals. Indeed, the lattice constant of Li metal is 0.351 nm at room temperature, which is shorter than that (0.429 nm) of Na metal.

6.4 Noble metals

Cu, Ag and Au are all known as the noble metals having an fcc structure. The outermost 4s, 5s and 6s electrons form the valence band in the respective metals. They constitute, therefore, another group of monovalent metals. Their Fermi surfaces had been believed to be fully contained within the first Brillouin zone of the fcc lattice without contact with the zone planes. However, Pippard discovered in 1957 [3] through the measurement of the anomalous skin effect that the Fermi surface of pure Cu has already touched the {111} zone planes. Subsequently, the detailed structure of the Fermi surface of Cu, Ag and Au has been accurately determined from the measurement of the de Haas–van Alphen effect [1, 4].

Figure 6.5 shows the Fermi surface of pure Cu in the repeated zone scheme. It is seen that the Fermi surface protrudes along the ⟨111⟩ direction and makes simultaneous contacts with eight {111} zone planes. Because of this contact, the Fermi surface is no longer isolated in the first zone but is multiply con-

6.4 Noble metals

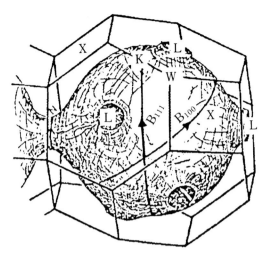

Figure 6.5. The Fermi surface and the first Brillouin zone of pure Cu metal in the periodic zone scheme. See Fig. 5.15 for symbols of symmetry points. [J. M. Ziman, *Adv. Phys.* **10** (1961) 1]

nected to neighboring zones through the center of the hexagonal {111} zone planes in the repeated zone scheme. The narrow region in contact with the {111} planes is called a "neck". In the figure, the orbit marked as "B_{100}" is drawn parallel to the {200} zone planes. This spherical orbit is called the "belly" and corresponds to the maximum circular cross-section of the Fermi surface. The neck diameter is experimentally determined to be $0.19p$, $0.14p$ and $0.18p$ for Cu, Ag and Au, where p refers to the diameter of the belly in the respective metals [1].

The E–$\mathbf{k}$ relations and corresponding density of states curve for pure Cu are shown in Figs. 6.6(a) and (b), respectively. We mentioned above that its Fermi surface has eight necks in the $\langle 111 \rangle$ directions. The location of the Fermi level E_F above the energy $E_-(k_L)$, which is marked as L'_2 in Fig. 6.6(a), confirms the presence of the neck.

As another unique feature of the electronic structure of noble metals, the parabolic E–$\mathbf{k}$ relation is intervened by the less dispersive E–$\mathbf{k}$ curves due to 3d electrons centered at about a few eV below the Fermi level. This means that the 4s and 4p free-electron-like states are hybridized with the 3d states to form a composite band. In spite of strong hybridization, ten 3d electrons per atom are distributed in a rather narrow energy range. This is why the 3d electron density of states is very high, as shown in Fig. 6.6(b).

The Fermi surface of a noble metal should occupy only 50% by volume of the first Brillouin zone of the fcc lattice. Nevertheless, we learned that the Fermi surface of noble metals is in contact with the {111} zone planes. The

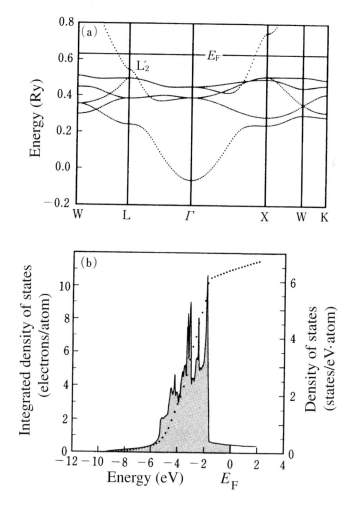

Figure 6.6. (a) $E-k$ relations and (b) the corresponding density of states curve of pure Cu metal. Dotted data points in (b) represent the integrated density of states, i.e., the total number of 3d, 4s and 4p electrons per atom integrated up to the energy E. [V. L. Moruzzi, J. F. Janak and A. R. Williams, *Calculated Electronic Properties of Metals* (Pergamon Press, 1978)]

reason for the contact has been interpreted as arising from the aforementioned hybridization effect of the extended 4s and 4p states with more localized 3d states at energies immediately below the Fermi level.

6.5 Divalent metals

As listed in Table 6.1, Ca and Sr are divalent fcc metals. The electronic configurations of the Ca free atom are composed of $(1s)^2$, $(2s)^2$, $(2p)^6$, $(3s)^2$, $(3p)^6$ and

6.5 Divalent metals

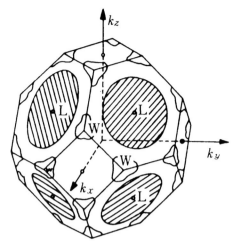

Figure 6.7. The Fermi surface and the first Brillouin zone of pure Ca. The hatched regions represent electrons overlapped into the second zone across the points L. Holes can be seen in the first zone around the points W. [A. P. Cracknell, *The Fermi Surface of Metals* (Taylor & Francis Ltd, 1971)]

$(4s)^2$ electrons. The two outermost 4s electrons per atom contribute to the formation of the valence band of Ca metal. Since the number of electrons per atom, e/a, is equal to 2.0, it is just high enough to fill the first Brillouin zone. However, a Fermi surface does exist, as is depicted in Fig. 6.7 in the extended zone scheme. The Fermi surface marked by hatches indicates electrons overlapped into the second Brillouin zone across the points L. Naturally, the same amount of holes must be left behind in the first zone. This is found at the corners W in the first zone. This is why Ca becomes a metal in spite of $e/a =$ 2.0. A metal is said to be compensated, when there exist an equal number of holes and electrons in the two successive zones.

There is another group of divalent metals in the periodic table. It consists of Be, Mg, Zn and Cd, all of which crystallize into the hcp structure. Here we present the electronic structure of Zn metal. The two outermost 4s electrons form the valence band in Zn metal. It may be noted that its narrow 3d band can be ignored, since it is positioned slightly below the bottom of the valence band and is scarcely hybridized with the valence electrons. The Fermi surface of Zn metal calculated on the basis of the free-electron model is shown in Fig. 6.8 in the extended zone scheme. It is clear that electrons overlap into the higher zone across the points Γ and K.

Figure 6.9 shows the cross-section of constant energy surfaces including the Fermi surface of pure Zn and that of the first and second Brillouin zones cut through the ΓL plane both in (a) the extended and (b) the reduced zone

134   6 Electronic structure of representative elements

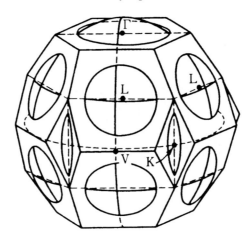

Figure 6.8.   The free-electron Fermi surface and the Brillouin zone of pure Zn in the extended zone scheme. Electrons are overlapped into higher zones across the points Γ, L and K in the free-electron model. [C. H. Barrett and T. B. Massalski, *Structure of Metals and Alloys* (McGraw-Hill, 1966)]

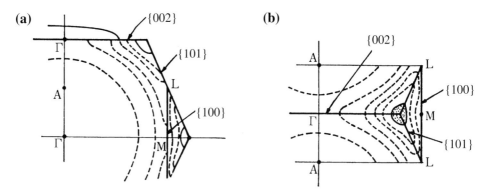

Figure 6.9.   Reduction of the second zone to the first zone in hcp Zn. (a) Cross-section of the constant energy surfaces cut through ΓL in the extended zone scheme. The solid curve corresponds to the Fermi surface of Zn. (b) Reduction of (a) into the first zone. The Fermi surface of holes of Zn is shown by hatches. Electron overlaps across the {101} zone planes are believed to be absent in pure Zn. [T. B. Massalski, U. Mizutani and S. Noguchi, *Proc. Roy. Soc.* (London) **A343** (1975) 363]

schemes. One can easily see how segments of holes in the second zone are folded into the first zone by shifting them by appropriate reciprocal lattice vectors and a continuous Fermi surface of holes is formed. The stereographic Fermi surface of holes in the second zone is shown in Fig. 6.10 in the reduced zone scheme. It is often called the "monster" from its appearance. It is known that the free-electron model holds fairly well in Zn metal. However, it is interesting to note that the Fermi surface in the reduced zone scheme looks significantly different from the free-electron-like sphere.

6.6 Trivalent metals 135

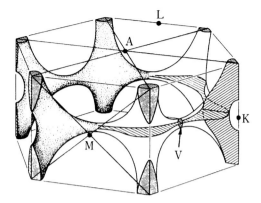

Figure 6.10. Fermi surface of holes in the second zone of pure Zn in the reduced zone scheme. The cross-section shown in Fig. 6.9 corresponds to that at the point V. [C. H. Barrett and T. B. Massalski, *Structure of Metals and Alloys* (McGraw-Hill, New York 1966)]

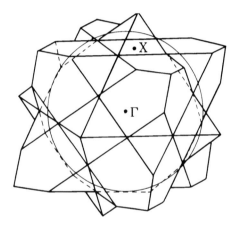

Figure 6.11. The free-electron Fermi surface and the first and second Brillouin zones of fcc Al in the extended zone scheme. Electrons overlap into the third and fourth zones.

6.6 Trivalent metals

Al, Ga, In and Tl are grouped together as IIIB trivalent metals. As their representative, we consider Al metal, in which the free-electron model holds well again. It is a fcc metal and its valence band is formed from two 3s and one 3p electrons per atom. Figure 6.11 shows the free-electron sphere corresponding to $e/a = 3.0$, together with the first and second Brillouin zones. Since its Fermi surface is more expanded than that of the divalent Ca, the first zone is fully occupied by electrons and, thus, the Fermi surface is absent there. However,

136 6 Electronic structure of representative elements

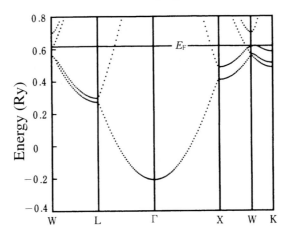

Figure 6.12. *E–k* relations in pure Al. [V. L. Moruzzi, J. F. Janak and A. R. Williams, *Calculated Electronic Properties of Metals* (Pergamon Press, 1978)]

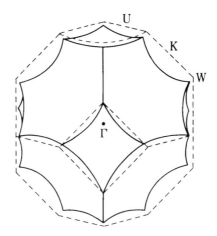

Figure 6.13. The Fermi surface of holes in the second Brillouin zone of pure Al in the reduced zone scheme. [W. A. Harrison, *Pseudopotentials in the Theory of Metals* (W. A. Benjamin, New York, 1966)]

electrons are overlapped into the third and fourth zones, leaving a large number of holes in the second zone.

The *E*–**k** relations obtained from the band calculations for Al metal are shown in Fig. 6.12. It is clear that electrons overlap into higher zones not only across the energy gaps at the points L and X but also across those at the points K and W. Figure 6.13 shows the Fermi surface of holes in the second zone in the reduced zone scheme. Similarly, the Fermi surface of electrons in the third zone, which is also named the "monster", is shown in Fig. 6.14 in the reduced zone scheme.

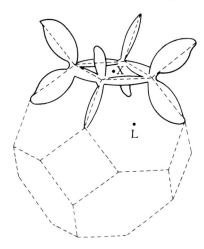

Figure 6.14. The Fermi surface of electrons in the third Brillouin zone of pure Al in the reduced zone scheme. [A. P. Cracknell, *The Fermi Surface of Metals* (Taylor & Francis Ltd, 1971)]

6.7 Tetravalent metals and graphite

Pb is a tetravalent fcc metal, in which two 6s and two 6p electrons per atom form its valence band. The free-electron model is also relatively well applicable to Pb metal. Hence, a Fermi sphere having a slightly larger diameter than that of Al metal shown in Fig. 6.11 can be used as a guide and the Fermi surface is constructed along the same lines as discussed for pure Al. We show in Fig. 6.15 only the Fermi surface of electrons in the third zone in the reduced zone scheme. It can be seen that the Fermi surface is formed along the edge of the Brillouin zone but that its diameter becomes certainly thicker and more uniform than that of Al (see Fig. 6.14) because of an increased electron concentration *e/a* to 4.0. The Fermi surface of electrons of Pb in the third zone is called the "jungle gym" because of the appearance of the multiply-connected tube-like Fermi surface.

We have so far studied how the Fermi surface of representative metals changes with increasing electron concentration *e/a* from 1.0 to 4.0. All of these metals are characterized by the possession of a large Fermi surface. However, there are several elements in the periodic table which exhibit an electronic band structure intermediate between metals and semiconductors. They are called semimetals. Graphite is typical of semimetals and forms its valence band from two 2s and 2p electrons per atom. As shown in Fig. 6.16, the crystal structure of graphite is hexagonal: a regular hexagonal network in the *ab*-plane is dominated by covalent bonding, whereas the layers are bonded by the weak van der Waals interaction.

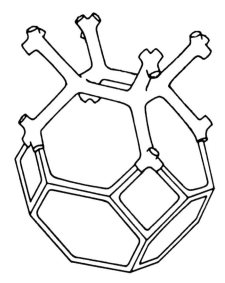

Figure 6.15. The Fermi surface of electrons in the third Brillouin zone of pure Pb. The Fermi surface is called the "jungle gym". [A. P. Cracknell, *The Fermi Surface of Metals* (Taylor & Francis Ltd, 1971)]

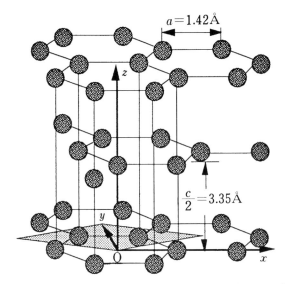

Figure 6.16. Crystal structure of graphite. The unit cell in the two-dimensional lattice model is marked by hatches and contains two atoms.

The essence of the electronic structure of graphite can be well extracted from a two-dimensional lattice model [5]. As shown by hatches in Fig. 6.16, the unit cell in the two-dimensional ab-plane contains two carbon atoms. The rectangular cartesian coordinates are chosen with the origin O at the center of the regular hexagon. The primitive translation vectors are then given by equation

6.7 Tetravalent metals and graphic

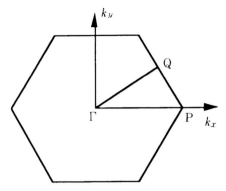

Figure 6.17. The first Brillouin zone in the two-dimensional lattice model for graphite.

(5.53), in which $\mathbf{a}_3 = 0$. The corresponding reciprocal lattice vectors are easily calculated by setting $\mathbf{k} = 0$ in equation (5.54). The first Brillouin zone turns out to be a regular hexagon as expected from Fig. 5.17(a). This is shown in Fig. 6.17.

As noted above, there are two carbon atoms in the two-dimensional unit cell and each carbon atom carries four valence electrons. Hence, we need to consider a total of eight electrons per unit cell. According to the band calculations [5], the 2s, $2p_x$ and $2p_y$ orbitals between the nearest neighbor carbon atoms hybridize with one another to yield the bonding and antibonding bands in the ab-plane (see Section 13.2). They are denoted as either the sp^2 orbital or the σ-band.

The remaining $2p_z$ orbital oriented towards the c-axis can only weakly interact with the neighboring $2p_z$ orbitals in the ab-plane. Here a hybridization with the 2s, $2p_x$ and $2p_y$ orbitals is essentially negligible. This yields relatively narrow bonding and antibonding states, both of which are called the π-band. The resulting E–k relations are shown in Fig. 6.18. The position of the Fermi level is determined by putting six electrons in the bonding σ-band and the remaining two electrons in the bonding π-band, as is shown in Fig. 6.18. The Fermi level is positioned at the energy which separates the bonding and antibonding π-bands. Graphite in the two-dimensional model can be viewed as possessing the band structure characteristic of a semiconductor with a vanishing energy gap.

In real (three-dimensional) graphite, there is a weak interaction between the layers and, hence, the effect along the c-axis must be taken into account. The first Brillouin zone should become a regular hexagonal prism, as indicated in Fig. 5.17(a). The interlayer interaction arises mainly from the $2p_z$ orbitals. The degeneracy in the π-band at the point P in the Brillouin zone is lifted as a result

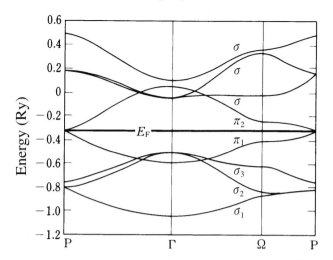

Figure 6.18. E–k relations of graphite obtained from the two-dimensional lattice model. [G. S. Painter and D. E. Ellis, *Phys. Rev.* **B1** (1970) 4747]

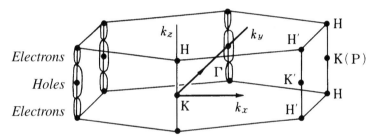

Figure 6.19. The Fermi surface of electrons and holes of pure graphite in the reduced zone scheme. The point K corresponds to the point P in Fig. 6.17. [A. P. Cracknell, *The Fermi Surface of Metals* (Taylor and Francis, Ltd, 1971)]

of the interlayer interaction.[3] Indeed, as shown in Fig. 6.19, small ellipsoidal Fermi surfaces of both electrons and holes have been observed through the measurement of the de Haas–van Alphen effect and independently confirmed by the band calculations [5,6]. A material characterized by the possession of small Fermi surfaces of both electrons and holes is called a semimetal. The calculated density of states curve of graphite is shown in Fig. 6.20. It can be seen that the two bands slightly overlap each other so that electrons and holes coexist at the Fermi level.

---

[3] The states in the π-band are doubly degenerate at the point P in the two-dimensional lattice model. In the three-dimensional analysis, the unit cell becomes a regular hexagonal prism in which four carbon atoms are contained. A new E–**k** relation emerges along the HKH direction as a result of the interlayer interaction. It consists of four bands: non-degenerate E_1 and E_2 bands and a doubly-degenerate E_3 band [5].

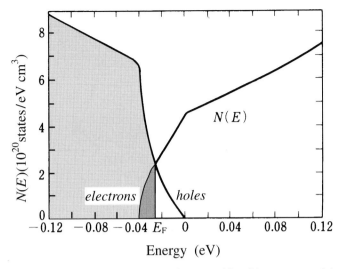

Figure 6.20. Density of states curve near the Fermi level in pure graphite. The Fermi level is located at about -0.02 eV, since the origin of the energy is taken at the point H in the E_3 band. [M. S. Dresselhaus et al., Phys. Rev. **15B** (1977) 3180]

6.8 Pentavalent semimetals

Among elements in group VB in the periodic table, As, Sb and Bi are known as semimetals. Owing to the possession of small Fermi surfaces, de Haas and van Alphen were able to measure in 1930, for the first time, the oscillatory magnetic susceptibility with increasing applied magnetic field for a Bi single crystal, the details of which will be described in Section 7.2. Its electronic structure was analyzed by Mott and Jones in 1936 [7,8]. Thus, Bi was one of the few elements the electronic structure of which was explored both experimentally and theoretically as early as the 1930s.

In spite of the possession of an odd number of electrons per atom, its electrical conductivity is very low. This suggests the presence of a very small Fermi surface in an almost completely filled zone. The crystal structure of Bi belongs to the trigonal space group with two atoms per unit cell but is conveniently constructed from a simple cubic lattice by slight displacement of atoms. We take a rhombohedral unit cell containing two atoms in a simple cubic lattice, as shown in Fig. 6.21, and distort it so as to sharpen the rhombohedral angles from 60° to α. Then, a central atom marked BC is shifted along its diagonal axis. The first Brillouin zone is constructed by choosing the rhombohedral unit cell containing two atoms. Thus, it can accommodate ten electrons per unit cell because of the donation of two 6s and three 6p outermost electrons per atom.

Jones [8] proposed that the zones bounded by six $\{1\bar{1}0\}$ side planes, three

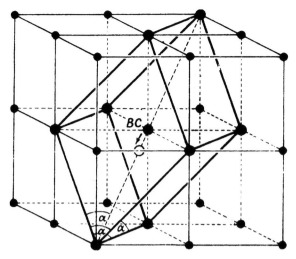

Figure 6.21. Crystal structure of Bi. The rhombohedral lattice can be formed by reducing α from 60° and shifting the body-centered atom (BC) down the diagonal of the rhombohedron. [J. M. Ziman, *Electrons and Phonons* (Clarendon Press, Oxford, 1962)]

{221} top planes and three {$\bar{2}\bar{2}\bar{1}$} bottom planes can accommodate ten electrons per unit cell with considerable energy gaps across these zone boundaries. This is called the Jones zone.[4] Bi would then have become an insulator, provided that ten electrons per unit cell just fill the Jones zone. However, in order to account for a low but finite electrical conductivity of Bi, Jones argued that a small number of electrons overlap outside of the Jones zone across the center of the {$1\bar{1}0$} planes, leaving a compensating number of holes at some corners of the zone, as shown in Fig. 6.22 (see more details in [1]).

Both experimental and theoretical studies of the determination of the Fermi surface topology have been called "Fermiology". As can be understood from the discussion above, the Fermi surface of semimetals like graphite and Bi is much smaller than that of typical metals. The smaller the Fermi surface, the more easily the signal of the de Haas–van Alphen effect measurement can be detected (see Section 7.2). The possession of a small Fermi surface coupled with an ease in the preparation of a single crystal certainly enabled exploration of the electronic structure of Bi in the very early days of investigations on the Fermiology.

---

[4] Certain sets of planes in reciprocal space, which have large energy gaps across them, enclose a symmetrical region containing about the same number of states as the number of valence electrons per atom or unit cell. Such a region is called the Jones zone or large zone, since its volume is usually larger than that of the true Brillouin zone [8]. The Jones zone for the hcp lattice is shown in Fig. 5.17(c). The Jones zone is not necessarily fully filled with an even number of electrons per atom or unit cell.

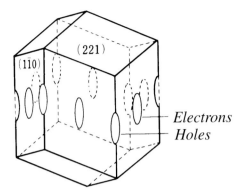

Figure 6.22. The Fermi surface of electrons and holes of pure Bi in the Jones zone. [J. M. Ziman, *Electrons and Phonons* (Clarendon Press, Oxford, 1962)]

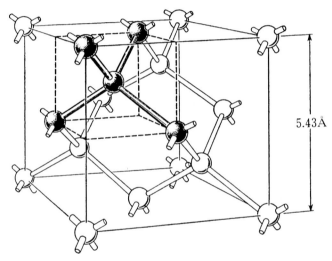

Figure 6.23. Crystal structure of Si in the diamond structure. There are eight atoms in the unit cell. [W. Shockley, *Electrons and Holes in Semiconductors* (van Nostrand Co., 1950)]

6.9 Semiconducting elements without and with dopants

The elements C (diamond), Si, Ge and Sn(α) belong to the group IVB elements in the periodic table and all of them are covalently bonded. These elements possess the diamond structure, as illustrated in Fig. 6.23. It is composed of two interpenetrating fcc lattices and, hence, the structure is fcc but with two atoms in each unit cell. As a result, the first Brillouin zone is the same as that for the ordinary fcc structure. In Si, two 3s and two 3p electrons per atom are hybridized to form directionally-bonded states and the Brillouin zone is completely filled with eight electrons per unit cell. Hence, no Fermi surface exists.

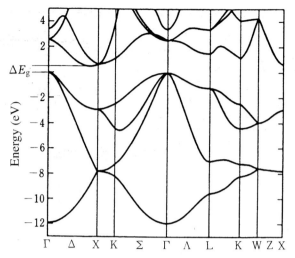

Figure 6.24. *E–k* relations of pure Si. ΔE_g indicates the energy gap. [M. T. Yin and M. L. Cohen, *Phys. Rev.* **B26** (1982) 5668]

The band structure calculations of Si have been performed, using the pseudopotential method described in Section 8.7. The *E–k* relations and the resulting density of states of pure Si are shown in Figs. 6.24 and 6.25, respectively, where the binding energy is scaled relative to the top of the valence band. The valence band is formed from a combination of the bonding orbitals on the two sublattices and is fully filled with eight electrons per unit cell. The conduction band, which is formed from antibonding combinations of the orbital wave functions, is unoccupied and is separated from the valence band by an energy gap ΔE_g. As can be seen from Fig. 6.24, the top of the valence band is found at point Γ but the bottom of the conduction band lies at 0.82X along the ΓX directions. A difference in energy between these two states gives rise to the energy gap $\Delta E_g = 0.48$ eV in the band calculations.[5]

Electrons in the valence band begin to be thermally excited with increasing temperature. Once electrons are excited into the conduction band, an equal amount of holes is left behind in the valence band. The material of this type is called an intrinsic semiconductor. The electronic structure of intrinsic semiconductors is often conventionally illustrated as shown in Fig. 6.26. The Fermi level is located in the middle of the forbidden gap, as long as the mass of the electron in the conduction band is equal to that of the hole in the valence band [9]. The temperature dependence of the conductivity σ is almost linear

---

[5] The calculated energy gap of 0.48 eV is underestimated. Energy gaps of typical semiconductors are as follows: 5.3 eV for C (diamond), 1.1 eV for Si, 0.7 eV for Ge, 2.8 eV for SiC, 4.6 eV for BN and 1.4 eV for GaAs.

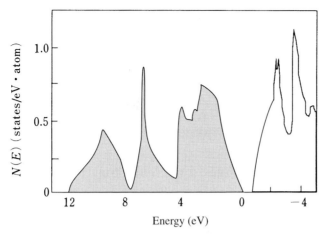

Figure 6.25. Density of states curve of pure Si calculated from the E–k relations in Fig. 6.24. The hatched area represents the valence band filled with electrons. The conduction band is separated from the valence band by the energy gap. [M. T. Yin and M. L. Cohen, *Phys. Rev.* **B26** (1982) 5668]

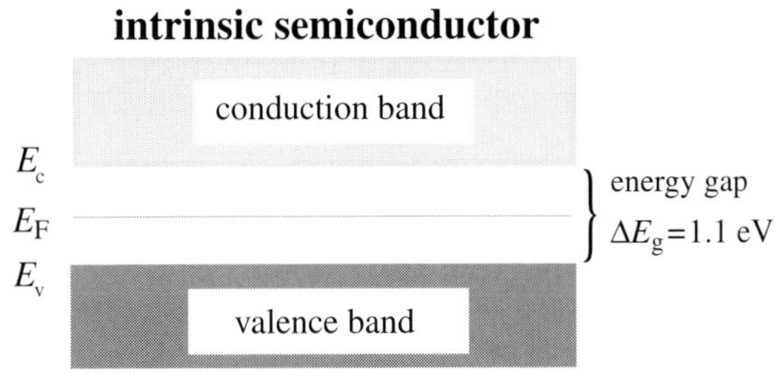

Figure 6.26. Schematic illustration of the band structure of an intrinsic semiconductor. The binding energy is measured in a vertical direction. The valence band and conduction band are separated by the energy gap ΔE_g. It is 1.1 eV for Si. Energies at the top of the valence band and the bottom of the conduction band are denoted as E_v and E_c, respectively.

on the logσ–$1/T$ plot for both Si and Ge over a wide temperature range above room temperature, being typical of intrinsic semiconductors [9].

Semiconducting properties are known to be very susceptible to the addition of very small, controlled amounts of intentionally added "impurities" or alloying elements. Such alloys are called extrinsic semiconductors and such additions dopants. Let us take Si as a host semiconductor. Its semiconducting properties depend significantly on whether pentavalent or trivalent impurities are added.

Pentavalent atoms like P, As and Sb from group VB of the periodic table will have one more valence electron than is required for the covalent bonding. The

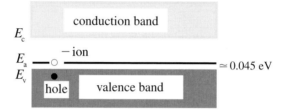

Figure 6.27. Schematic illustration of the band structure of a doped semiconductor. (a) n-type semiconductor. The energy of the donor level is marked E_d. The energy difference $E_c - E_d$ is about 0.045 eV for P in Si. (b) p-type semiconductor. The energy of the acceptor level is marked E_a. The energy difference $E_a - E_v$ is about 0.045 eV for B in Si.

extra electron bound to the impurity atom may be easily excited into the conduction band, provided that the temperature is high enough to promote it there from a level in the gap. This leaves a positive ion behind and four remaining valence electrons which form a covalent bond in the same manner as in the host Si atoms.

As is clear from the argument above, the impurity level is formed just below the bottom of the conduction band and an electron in the impurity level can be easily excited into the conduction band even near room temperature. This is illustrated in Fig. 6.27(a). The impurity level is called the donor level, since it donates electrons to the conduction band. The excited electrons in the conduction band serve as a carrier with a negative charge. Therefore, the semiconductors doped in this way are called n-type semiconductors.

In contrast, trivalent atoms like B, Al, Ga and In from group IIIB in the periodic table possess one electron less than the number required for covalent bonding. Hence, an electron from one of the adjacent Si atoms will enter the orbit about the trivalent impurity atom. This causes the trivalent atom to be negatively charged and the host Si atoms to lose an electron in this process. As illustrated in Fig. 6.27(b), the impurity level is formed just above the top of the valence band and can accept electrons from the valence band, thereby leaving holes in the valence band. Hence, the impurity level is often called the acceptor level. Holes in the valence band now serve as a carrier with a positive charge (see Section 10.5). Thus, the semiconductors containing such dopants are called p-type semiconductors.

It is possible to prepare both n- and p-type regions in a Si single crystal, which are separated from one another by a narrow interface region less than 1 micrometer thick. The interface is called the p–n junction and exhibits very important electrical properties known as rectification and transistor action. Rectification is a phenomenon in which a large current flows when a voltage is applied across the junction in one direction but only a small current will flow when the direction of the voltage is reversed. Its mechanism can be explained on the basis of the electronic structure illustrated in Fig. 6.27 [9].

Chapter Seven

Experimental techniques and principles of electronic structure-related phenomena

7.1 Prologue

Experimental techniques and principles which provide information about the electronic structure of metals are presented in this chapter. They may be classified into three types of experiments: first, Fermiology or the determination of the Fermi surface; second, determination of the energy spectra for occupied and unoccupied electronic states across the Fermi level; and third, measurements associated with the E–$\mathbf{k}$ dispersion relations.

7.2 de Haas–van Alphen effect

There are several experimental techniques for the determination of the Fermi surface topology. Among them, measurement of the de Haas–van Alphen effect (hereafter abbreviated as dHvA effect) has long been recognized as one of the most powerful and accurate methods [1]. The dHvA effect refers to the phenomenon in which the magnetic susceptibility in a single crystal oscillates as a function of the applied magnetic field at low temperatures. The oscillations of the magnetic susceptibility were observed for the first time in 1930 by de Haas and van Alphen for a Bi single crystal cooled to liquid-hydrogen temperatures. As will be described later, an extremal area of the cross-section of the Fermi surface normal to the applied magnetic field can be deduced from the period of the oscillations. In order to understand the physics, we begin by studying the behavior of the conduction electron in magnetic field.

The conduction electron moving with the velocity $\mathbf{v}$ in the presence of electrical and magnetic fields experiences the Lorentz force given by

$$\mathbf{F} = (-e)(\mathbf{E} + \mathbf{v} \times \mathbf{B}) \quad \mathbf{F} = (-e)\left[\mathbf{E} + \left(\frac{\mathbf{v}}{c} \times \mathbf{H}\right)\right] \text{[CGS]}, \tag{7.1}$$

where $(-e)$ is the electronic charge and c is the speed of light. In this section, we ignore the first term and consider only the effect of the magnetic field on the motion of the conduction electron. Equation (7.1) implies that the electron rotates in a closed orbit, if the velocity **v** has no component along the magnetic field **B**. If there is a non-zero component of the velocity in the direction of **B**, the electron will be subjected to a helical motion. In any event, the magnetic field alters the direction of **v** but not its magnitude. Thus, the energy of the electron is kept unchanged.

As will be discussed in Section 10.4, the force **F** exerted on the Bloch electron of wave vector **k** is generally expressed as $\mathbf{F} = \hbar d\mathbf{k}/dt$, where $\hbar$ is the Planck constant divided by 2π. Its insertion into equation (7.1) yields

$$\frac{\hbar d\mathbf{k}}{dt} = (-e)(\mathbf{v} \times \mathbf{B}), \quad \frac{\hbar d\mathbf{k}}{dt} = \left(\frac{(-e)}{c}\right)(\mathbf{v} \times \mathbf{H}) \text{ [CGS]}, \quad (7.2)$$

where $\mathbf{B} \, (= \mu_0 \mathbf{H})$ is the applied field. Thus, the motion of an electron of wave vector **k** can be described in reciprocal space as rotating in a closed orbit on a constant energy surface normal to the magnetic field **B**. By integrating equation (7.2) with respect to time, we obtain

$$\mathbf{k} - \mathbf{k}_0 = \left(\frac{(-e)}{\hbar}\right)(\mathbf{r} \times \mathbf{B}), \quad \mathbf{k} - \mathbf{k}_0 = \left(\frac{(-e)}{\hbar c}\right)(\mathbf{r} \times \mathbf{H}) \text{ [CGS]}, \quad (7.3)$$

where **r** is the position vector of the electron and $\mathbf{k}_0$ is a constant. Equation (7.3) means that the trajectory of the electron in reciprocal space has the same shape as that in real space but is rotated by 90° and scaled by the factor $(-e)B/\hbar$.

The closed orbit of the Bloch electron in the magnetic field is quantized in the same manner as the electron orbit of a free atom. The Bohr–Sommerfeld quantization rule for the motion of the Bloch electron in the magnetic field is explicitly written as

$$\oint (\hbar \mathbf{k} + (-e)\mathbf{A}) d\mathbf{s} = \left(n + \frac{1}{2}\right)\hbar, \quad \oint \left(\hbar \mathbf{k} + \frac{(-e)\mathbf{A}}{c}\right) d\mathbf{s} = \left(n + \frac{1}{2}\right)\hbar \text{ [CGS]}, \quad (7.4)$$

where n is a positive integer including zero and **A** is the vector potential defined as $\mathbf{B} = \mathrm{rot}\mathbf{A}$. The integral in equation (7.4) is carried out over the closed orbit of the electron in real space. Using the Gauss theorem, we can rewrite the second term in equation (7.4) as

$$(-e)\oint \mathbf{A} d\mathbf{s} = (-e) \iint \mathrm{rot}\mathbf{A} d\mathbf{S} = (-e) \iint \mathbf{B} d\mathbf{S} = (-e)BS, \quad (7.5)$$

where S represents the area of the closed orbit in real space. The first term in equation (7.4) is also rewritten by inserting equation (7.3) with $\mathbf{k}_0 = 0$:

$$\oint \hbar \mathbf{k}\, d\mathbf{s} = -(-e)\mathbf{B}\cdot\oint \mathbf{r}\times d\mathbf{s} = -2(-e)BS. \tag{7.6}$$

By combining equations (7.5) and (7.6), we obtain

$$eBS = \left(n+\frac{1}{2}\right)\hbar, \tag{7.7}$$

where the electronic charge e appears as a positive quantity in this equation.

Let us denote the area of the closed orbit in reciprocal space as $A(\varepsilon)$, where ε represents the constant energy of the electron. As is clear from equation (7.3), the area $A(\varepsilon)$ is related to the area S of the closed orbit in real space through the relation

$$A(\varepsilon) = \left(\frac{eB}{\hbar}\right)^2 S. \tag{7.8}$$

Equation (7.7) is rewritten as

$$A_n(\varepsilon) = \left(\frac{2\pi eB}{\hbar}\right)\left(n+\frac{1}{2}\right), \tag{7.9}$$

where the suffix n in $A(\varepsilon)$ emphasizes that the area enclosed by the electron orbit is quantized and can take only discrete sets of values corresponding to the principal quantum number n. We see from equation (7.9) that the enclosed area differs by $2\pi eB/\hbar$, when the electron jumps from a given orbit to an adjacent one.

Suppose that the magnetic field is applied along the z-axis. As described above, the Bloch electron circulates in the $k_x k_y$-plane normal to the magnetic field in reciprocal space. This means that its degree of freedom is reduced to one in the $k_x k_y$-plane and, hence, only a single parameter n is needed to describe its stationary circular motion. In other words, the magnetic field deprives the electron of one degree of freedom in its three-dimensional motion. Indeed, three independent quantum numbers k_x, k_y and k_z of the wave vector $\mathbf{k}$ are reduced to two quantum numbers n and k_z in the presence of a magnetic field along the z-axis.

As a consequence, the spherical Fermi surface of the free electron is replaced by a set of concentric cylinders whose axis is along the z-direction. This is illustrated in Fig. 7.1(a). Note that the quantum number k_z along the z-direction is not affected by the magnetic field at all. The concentric circles specified by the principal quantum number $n = 0, 1, 2, 3, \ldots$ can be obtained by projecting the Fermi surface of the concentric cylinders onto the $k_x k_y$-plane, as shown in Fig. 7.1(b).

We are now ready to study the oscillatory behavior of the magnetic

7.2 de Haas–van Alphen effect

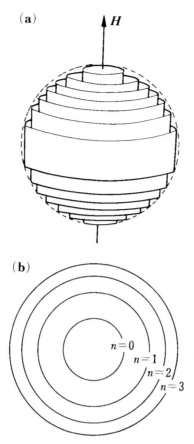

Figure 7.1. (a) The free-electron Fermi surface in the presence of a magnetic field along the z-axis. (b) Its projection onto the $k_x k_y$-plane.

susceptibility, i.e., the dHvA effect. For the sake of simplicity, we assume that the metal we are considering is essentially non-magnetic and obeys the free-electron model. The magnetic field is applied along the z-axis. The energy eigenvalue of the free electron in the magnetic field is obtained as

$$\varepsilon = \left(n + \frac{1}{2}\right)\hbar\omega_c + \frac{\hbar^2 k_z^2}{2m} \quad (n=0, 1, 2, \ldots), \tag{7.10}$$

where n is the principal quantum number and ω_c is defined as $\omega_c = eB/m$ (see Exercise 7.3). The corresponding $\varepsilon - k_z$ relations consist of a discrete set of parabola with different integers n, as shown in Fig. 7.2. They are called the Landau levels. It is clear that each electronic state is specified by a set of two quantum numbers n and k_z.

A change in the energy distribution of the conduction electrons, say at $k_z = 0$, is illustrated schematically in Fig. 7.3, when the applied magnetic field is

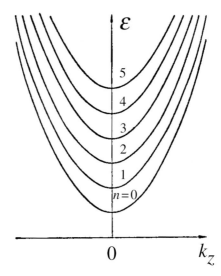

Figure 7.2. ε–k_z relation given by equation (7.10). The Landau level refers to each parabola specified by n.

increased from H_1 to H_3. Since ω_c in equation (7.10) is proportional to the magnetic field, the separation of the adjacent Landau levels is widened with increasing magnetic field. The wider the energy separation, the more electrons should be accommodated in a given level. For instance, the lowest level with $n=0$ is increased with increasing magnetic field from H_1 to H_2. Thus, the energy of electrons with $n=0$ is certainly increased. But the level with $n=0$ at $H=H_2$ can accommodate more electrons, since ω_c is increased. Thus, some of the electrons with $n=1$ at $H=H_1$ fall to the lowest state with $n=0$ at $H=H_2$, contributing to a lowering of the total energy. In this way, the net change in the internal energy with varying magnetic field is expected to be small. Indeed, it has been theoretically proved that the net change in the internal energy vanishes, as long as the Landau levels are below the Fermi level E_F. But the situation at E_F is different, since the cancellation effect mentioned above does not occur.

The internal energy U of the conduction electrons depends critically on the closeness of the highest occupied Landau level to the Fermi level E_F. As shown in Fig. 7.3(b), the internal energy is minimized at $H=H_1$, where the highest occupied Landau level has just passed E_F and is emptied. These electrons must be absorbed in the next highest Landau level below E_F. This is possible, since the area $A_n(\varepsilon)$ in equation (7.9) increases with increasing magnetic field. With increasing magnetic field, the highest Landau level moves up until it coincides with E_F at $H=H_2$. Here the internal energy is a maximum. Further increase in the field raises this Landau level above E_F again so that the level is emptied. As a result, the internal energy is decreased at $H=H_3$. This process is repeated

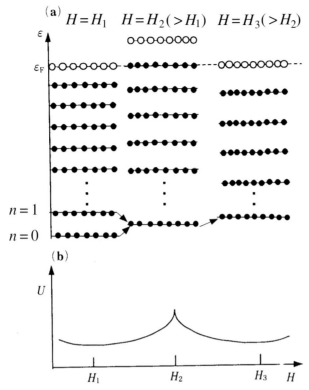

Figure 7.3. (a) Magnetic field dependence of the Landau level at absolute zero. With increasing magnetic field, the separation between the adjacent levels is widened and the population of electrons marked by solid circles in each level increases. The Landau level with $n=0$ is shown to be raised with increasing magnetic field. (b) Corresponding change in the internal energy.

every time the Landau level passes the Fermi level, resulting in oscillations of the internal energy of the conduction electrons as a function of the magnetic field.

The oscillating internal energy of the conduction electron system is reflected in various electronic properties. The magnetic susceptibility is given by the differentiation of the internal energy with respect to the magnetic field. Hence, the magnetic susceptibility oscillates as a function of the magnetic field. This is the dHvA effect. The specific heat is derived from the differentiation of the internal energy with respect to temperature. Hence, the specific heat also oscillates with increasing magnetic field. This is known as magneto-thermal oscillation. The de Haas–Shubnikov effect refers to the oscillations of the electrical resistivity as a function of a magnetic field. Similarly, an oscillatory behavior is observed in the Hall coefficient and the thermoelectric power. The present discussion of the dHvA effect is rather qualitative. A more rigorous

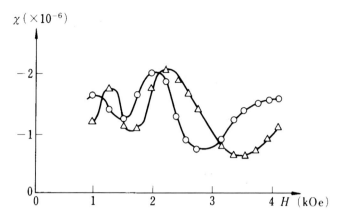

Figure 7.4. Oscillatory behavior of the magnetic susceptibility of a Bi single crystal measured by de Haas and van Alphen in 1930. Open circle and triangle refer to the data when the magnetic field is applied parallel to and perpendicular to the two-fold axis of the sample, respectively. The measuring temperature was 14.2 K. [W. J. de Haas and P. M. van Alphen, *Communications from the Physical Laboratory at the University of Leiden* **212a** (1930) 3]

mathematical treatment of the dHvA effect can been found in the literature [2].

Let us denote magnetic fields as $B_1 (=\mu_0 H_1)$ and $B_2 (=\mu_0 H_2)$, when the two successive Landau levels n and $n-1$ pass the Fermi level. From equation (7.9), we can derive the relation

$$\Delta\left(\frac{1}{B}\right) = \frac{1}{B_1} - \frac{1}{B_2} = \frac{2\pi e}{\hbar A(\varepsilon)}. \tag{7.11}$$

We see that, if the dHvA signal is plotted as a function of an inverse of the magnetic field, an extremal cross-section of the Fermi surface, $A(\varepsilon)$, normal to the applied field can be deduced from the period of oscillations given by equation (7.11) (see Exercise 7.2).

The Fermi surface topology in a given metal or semimetal can be determined by measuring the quantum oscillations in magnetic fields applied along different directions relative to the principal crystal axes of a single crystal. An accurate measurement of the period of the oscillations can be made, even though the signal itself is rather weak. The Fermi radius, say in pure Cu, can be determined with an accuracy of 0.001%. Figure 7.4 shows the first observation of the oscillatory magnetic susceptibility of a Bi single crystal by de Haas and van Alphen in 1930. Note here that the applied field, not its inverse, is taken as the horizontal axis and, hence, the oscillation is not periodic. The data for pure Zn measured in 1962, i.e., 32 years after the pioneering work by de Haas and van Alphen are shown in Fig. 7.5 for comparison.

The measurement of the quantum oscillations discussed above depends

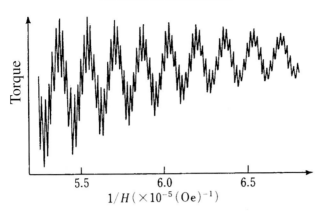

Figure 7.5. dHvA oscillations of pure Zn measured by the torque method. Oscillations with two different periods can be seen. They are attributed to the two extremal areas of the arm in the "monster" shown in Fig. 6.10. [A. S. Joseph and W. L. Gordon, *Phys. Rev.* **126** (1962) 489]

critically on the formation of the Landau level. The level is formed as a result of the revolving motion of the conduction electron caused by the Lorentz force in the magnetic field. The higher the applied magnetic field, the smaller its revolving radius is (see Exercise 7.3). The revolving motion would be disturbed by scattering of the electrons by impurities and phonons. The Landau levels and, hence, the quantum oscillations will disappear if the electron is scattered before the completion of a revolution. Therefore, the electron mean free path must be long enough to guarantee a revolving motion in the magnetic field. It requires a high purity single crystal and the measurement must be made at very low temperatures under high magnetic fields. This certainly limits the materials that can be studied. Indeed, the dHvA effect has been measured exclusively on pure elements and intermetallic compounds with a stoichiometric composition, in which the electron mean free path is very long at low temperatures. Fermiology of pure elements has been well established through extensive studies of the quantum oscillations in the period 1960–70. In contrast, progress in the Fermiology of alloys has been much slower. The Fermi surface of dilute alloys containing less than 1 at.% solute atoms has been studied by the dHvA effect only in the late 1970s and the 1980s.

7.3 Positron annihilation

The positron annihilation experiment is known as an alternative technique to determine the topology of the Fermi surface [3]. For example, the radioactive isotope Na^{22} or Cu^{64} decays by emitting a positron, which can be used as a probe to study the Fermi surface. A positron with high energy, when penetrating into a metal sample, gradually loses its kinetic energy by interacting with

156   7 Principles of measuring electronic structure-related phenomena

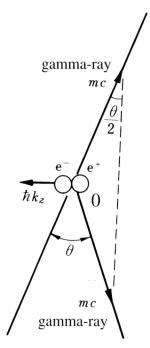

Figure 7.6. A pair of gamma-rays emitted upon the annihilation of the positron and electron. The momenta of the electron, positron and gamma-ray are denoted as $\hbar k_z$, 0 and mc, respectively.

the electron clouds in the metal. Its energy is eventually decreased to the order of the thermal energy $k_B T$, 0.025 eV at 300 K. This low-energy positron, which is called a thermal positron, finally annihilates with a conduction electron in the metal. At this incident, a pair of gamma-rays are emitted in order to conserve the energy and momentum between the annihilating electron and positron. This process is shown schematically in Fig. 7.6. Here the momentum of the thermal positron is ignored relative to $\hbar \mathbf{k}$ of the conduction electron. By denoting the angle between the two gamma-rays as θ, we obtain

$$\hbar k_z = 2mc \sin\left(\frac{\theta}{2}\right) \approx mc\theta \qquad (7.12)$$

from the momentum conservation law. Equation (7.12) indicates that the measurement of the angle θ between the pair of gamma-rays enables us to determine the momentum component k_z of the conduction electron.

A typical positron annihilation setup is illustrated in Fig. 7.7. A sample is mounted between the pole pieces of a magnet, which serves to focus the positrons from a radioactive source like Na^{22} onto the face of the sample. The two gamma-ray detectors equipped with NaI(Tl) scintillators are located behind

7.3 Positron annihilation

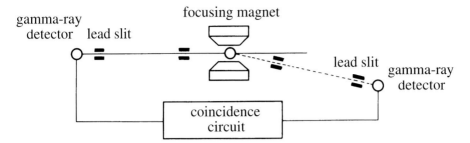

Figure 7.7. Principles of the positron annihilation technique.

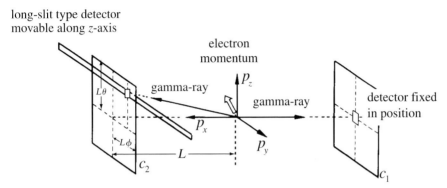

Figure 7.8. A "long-slit" setup for counting a pair of gamma-rays upon annihilation of the positron and electron.

lead slits and are connected through a coincidence circuit in order to detect a pair of gamma-rays emitted simultaneously upon the annihilation of the positron with an electron. The angle between the two photons (or a pair of gamma-rays) is nearly 180° within a few milliradians. Because of such a small deflected angle, a "long-slit" geometry has been devised.

Its principle is illustrated in Fig. 7.8. Suppose that one of the two photons is detected by a counter c_1 and that its partner photon will hit the plane of the counter c_2 at coordinates $L\theta$ and $L\phi$, where L is the distance from the sample. The value of L must be very large. For instance, $L\theta$ amounts to the order of a millimeter when L is 10 m, since the angles θ and ϕ are only a few milliradians. As is clear from equation (7.12), the momentum components p_z and p_y of the annihilating electron can be deduced from the relations $\theta = p_z/mc$ and $\phi = p_y/mc$, respectively.

The usual "long-slit" setup uses two long counters with horizontal slits, both of which subtend a large horizontal angle so that the momentum distribution over the y-component can be integrated. As is clear from the setup, the

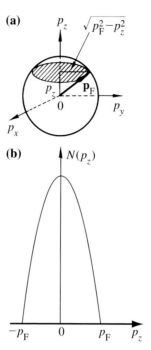

Figure 7.9. (a) The cross-section of the free-electron Fermi sphere cut at a given p_z parallel to the $k_x k_y$-plane. (b) The corresponding angular correlation curve.

x-component is also integrated. We can measure the coincidence rate $N(p_z)$ as a function of p_z by shifting one of the horizontal slits relative to the other along the z-axis. The $N(p_z)$ measured by the long-slit geometry is given by

$$N(p_z) = \iint \rho(\mathbf{p}) dp_x dp_y, \qquad (7.13)$$

where p_x, p_y and p_z are the three components of the momentum of the annihilating electron and $\rho(\mathbf{p})$ is the density of electrons with the momentum $\mathbf{p}$.

For the sake of simplicity, a spherical Fermi surface is assumed. The number of conduction electrons at a given p_z must be proportional to the area $A(p_z)$ of its cross-section normal to the p_z-axis. As is shown in Fig. 7.9(a), we have

$$N(p_z) \propto A(p_z) = \pi(p_F^2 - p_z^2). \qquad (7.14)$$

It is now clear that the resulting $N(p_z)-p_z$ curve in the free-electron model should be parabolic. This is shown in Fig. 7.9(b). The value of p_z at which $N(p_z)=0$ yields the Fermi momentum or the Fermi cut-off. The $N(p_z)-p_z$ spectrum is called the angular correlation curve in positron annihilation measurements.

The angular correlation curve in real metals always deviates from a parabola owing to a departure from the free-electron model. Fujiwara and Sueoka

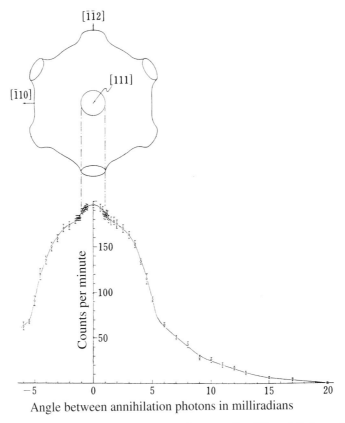

Figure 7.10. (*Lower*) Angular correlation curve for pure Cu. (*Upper*) Schematic view of its Fermi surface. [K. Fujiwara and O. Sueoka, *J. Phys. Soc. Jpn* **21** (1966) 1947]

were the first to confirm the presence of the neck along the ⟨111⟩ direction of a Cu single crystal by positron annihilation measurements. Their data are reproduced in Fig. 7.10. The presence of the neck can be clearly observed in the angular correlation curve.

The positron annihilation measurement is not as accurate as the dHvA effect. However, as is clear from its principle, it does not depend on the mean free path of the conduction electron. Hence, the positron annihilation technique is applicable not only to concentrated crystalline alloys but also to non-periodic substances like liquid metals and amorphous alloys. It is also true that the measurement does not necessarily require low temperatures. There is, however, a serious drawback in positron annihilation experiments because of the possession of a positive charge: the positron tends to be repelled from the vicinity of positive ions and, instead, is attracted to lattice defects like vacancies. Therefore, caution is needed to check if the observed electronic structure is free from disturbances due to such defects.

160   7 Principles of measuring electronic structure-related phenomena

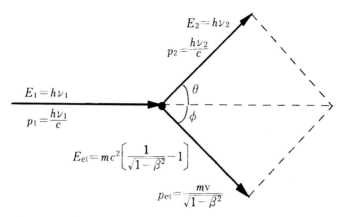

Figure 7.11.   Energy and momentum conservation upon scattering of a photon with an electron at rest.

7.4 Compton scattering effect

The x-ray Compton scattering experiment is known to serve as a tool to extract information similar to the angular correlation curve in the positron annihilation measurement [4]. Compton discovered in 1923 that x-rays scattered from elements like graphite have a wavelength longer than that of the incident beam. This is the Compton effect, as already mentioned in Section 1.2 in connection with the historical survey on the electron theory of metals.

Figure 7.11 illustrates the scattering event of the incident x-ray photon with an electron at rest. The energy conservation law can be stated as

$$E_1 = E_2 + E_{el}, \qquad (7.15)$$

where E_1 is the energy of the incident photon and E_2 and E_{el} are the energies of the photon and electron, respectively, after scattering. This is explicitly written as

$$h\nu_1 = h\nu_2 + mc^2\left(\frac{1}{\sqrt{1-\beta^2}} - 1\right), \qquad (7.16)$$

where β is the ratio of the velocity v of the electron after scattering over the speed of light c. We use the momentum conservation law in two directions: one parallel to the incident x-ray and the other perpendicular to it. They are expressed as

$$\frac{h\nu_1}{c} = \frac{h\nu_2}{c}\cos\theta + \frac{mc\beta}{\sqrt{1-\beta^2}}\cos\phi \qquad (7.17)$$

and

$$0 = \frac{h\nu_2}{c}\sin\theta - \frac{mc\beta}{\sqrt{1-\beta^2}}\sin\phi, \tag{7.18}$$

where θ and ϕ represent the scattering angles of photon and electron relative to the incident beam, respectively, as indicated in Fig. 7.11.

We can rewrite the equations above in terms of the wavelengths $\lambda_1 = c/\nu_1$ and $\lambda_2 = c/\nu_2$ for the incident and scattered photon, respectively. The variable ϕ is eliminated from equations (7.17) and (7.18) by making use of the trigonometric relation $\sin^2\phi + \cos^2\phi = 1$. Then we obtain the relation

$$\frac{h^2}{\lambda_1^2} + \frac{h^2}{\lambda_2^2} - \frac{2h^2\cos\theta}{\lambda_1\lambda_2} = \frac{m^2\beta^2 c^2}{1-\beta^2} = \frac{m^2 c^2}{1-\beta^2} - m^2 c^2. \tag{7.19}$$

Equation (7.16) is also rewritten in terms of the wavelength and is squared:

$$\frac{h^2}{\lambda_1^2} + \frac{h^2}{\lambda_2^2} - \frac{2h^2}{\lambda_1\lambda_2} + 2mch\left(\frac{1}{\lambda_1} - \frac{1}{\lambda_2}\right) + m^2 c^2 = \frac{m^2 c^2}{1-\beta^2}. \tag{7.20}$$

By subtracting equation (7.19) from equation (7.20), we obtain the well-known relation:

$$\Delta\lambda = \lambda_2 - \lambda_1 = \frac{2h}{mc}\sin^2\left(\frac{\theta}{2}\right). \tag{7.21}$$

It is now clear that the wavelength of the x-ray photon, when scattered by an angle θ after collision with an electron at rest, should become longer than that of the incident x-ray by the amount given by equation (7.21). It is also true that the shift of the wavelength depends only on the scattering angle and is independent of the material exposed to the x-ray photons. For instance, the shift is 0.002 43 nm when $\theta = \pi/2$.

As opposed to the prediction from equation (7.21), the "Compton scattered" x-ray line is always broader when measured in a metal. In the treatment described above, the electron is assumed to be at rest. This is certainly not true in a metal. The conduction electron has its own momentum distribution even at absolute zero as a result of the Pauli exclusion principle. A collision of photons with a moving electron can be properly taken into account as a Doppler effect. A shift of the wavelength of the scattered x-ray photons must be given as a sum of equation (7.21) and the term arising from the Doppler effect:

$$\Delta\lambda = \left(\frac{2h}{mc}\right)\sin^2\left(\frac{\theta}{2}\right) + 2(\lambda_1\lambda_2)^{1/2}\left(\frac{p_z}{mc}\right)\sin\left(\frac{\theta}{2}\right), \tag{7.22}$$

where p_z represents the component of the electron momentum along the direction of the scattered photon. The broadening of the Compton scattered line is due to the fact that the scattered x-ray profile reflects the electron momentum distribution through the Doppler term in equation (7.22).

The Compton profile $J(p_z)$ is determined by the probability that the scattered electron has a component of momentum p_z:

$$J(p_z) = \iint \rho(\mathbf{p}) dp_x dp_y, \quad (7.23)$$

where p_x, p_y and p_z are the three components of the electron momentum $\mathbf{p}$ and $\rho(\mathbf{p})$ represents the density of electrons with momentum $\mathbf{p}$. In the case of the free-electron model, equation (7.23) is easily reduced to

$$J(p_z) = p_F^2 - p_z^2 \quad (p \leq p_F)$$

and

$$= 0 \quad (p > p_F), \quad (7.24)$$

where p_F is the Fermi momentum (see Exercise 7.4).

As is clear from the argument above, we see that information similar to the angular correlation curve in the positron annihilation experiment can be derived from the Compton experiment. The determination of the electron momentum distribution from the Compton effect does not rely on the mean free path of the conduction electron. Hence, it can be equally applied to non-periodic substances like amorphous alloys. Recently, a substantial improvement in the resolution has been achieved by making use of very intense x-rays emitted from a synchrotron radiation source. It is also worth noting that, in contrast to the positron annihilation, the Compton measurement utilizes an electromagnetic wave as a probe and, hence, is free from the difficulties associated with the preferential annihilation of positrons with lattice defects in a sample.

7.5 Photoemission spectroscopy

The electronic structure of a solid can be explored by analyzing the energy distribution of photoelectrons emitted upon irradiation of a sample by electromagnetic waves [5–7]. Ultraviolet rays in the energy range 10–50 eV and characteristic x-rays have both been used as irradiating sources in the laboratory setup. Ultraviolet photoemission spectroscopy uses ultraviolet rays as an irradiating source (UPS). A helium gas discharge lamp has often been employed to produce ultraviolet rays with energies of 21.22 and 40.82 eV. Its natural line width is very sharp and is in the neighborhood of 1 meV. The characteristic Kα emission lines of an x-ray tube, especially Mg-Kα (1253.6 eV) and

7.5 Photoemission spectroscopy

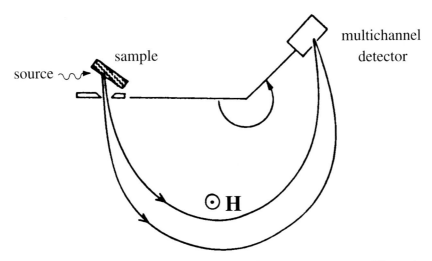

Figure 7.12. Schematic illustration of a photoelectron spectrometer. Photoelectrons emitted from a sample are deflected in roughly circular orbits and focused on a multi-channel detector by which their energies are analyzed.

Al-Kα (1486.6 eV) lines, have also been frequently employed. The technique is referred to as x-ray photoemission spectroscopy and is abbreviated as XPS. A typical natural line width is 1.0–1.4 eV and is much wider than that of the UPS irradiating source. However, a curved quartz crystal monochromater can be fitted to the Al-Kα source to suppress unwanted radiation and reduce its line width to as small as 0.16 eV. The principle of a typical spectrometer is illustrated in Fig. 7.12.

More recently, a very intense radiation with an arbitrary wavelength in the energy range 10 to several 100 eV is available from a synchrotron radiation accelerator. The availability of synchrotron radiation has eliminated the need for the separate disciplines of the UPS and XPS techniques, but both techniques are still frequently employed in many laboratories.

The basic idea of why the photoemission spectrum provides information on the electronic structure of a given substance is illustrated in Fig. 7.13. In both the UPS and XPS techniques, a monochromatized radiation impinges on the surface of a sample kept in a high vacuum of 10^{-9}–10^{-10} Torr. The electron at the Fermi level can escape from a metal, provided that the radiation is strong enough to excite it beyond the work function ϕ. Further higher-excitation energy is needed to "pull out" electrons below the Fermi level. Since the energy $h\nu$ of the exciting photons is kept fixed, the binding energy E_B of the electronic state relative to the Fermi level can be determined by measuring the kinetic energy E_{kin} of the photoelectron:

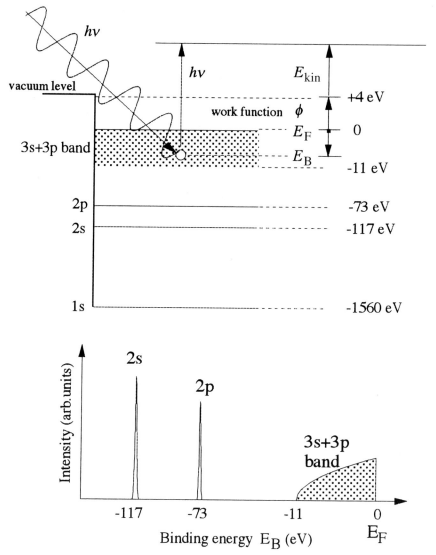

Figure 7.13. Relationship between the electronic structure and the photoemission spectrum. The electronic structure of pure Al is taken as an example. The XPS spectrum is schematically shown in the lower diagram.

$$E_{kin} = h\nu - \phi - E_B. \qquad (7.25)$$

The photoemission process comprises three processes: (1) electrons are first optically excited to states at higher energy, (2) the photoexcited electrons move through the lattice to the surface of a sample and (3) they escape from the surface into vacuum. The number of photoelectrons emitted from the surface of a sample is then expressed as

7.5 Photoemission spectroscopy

$$n(E_{kin}, h\nu) \propto N_i(E) \cdot N_f(E+h\nu) \cdot f(E,T) \cdot \sigma_{opt}(E, h\nu) \cdot P_t(E,h\nu) \cdot P_e(E,h\nu), \quad (7.26)$$

where $N_i(E)$ is the initial density of states at the energy E, $N_f(E+h\nu)$ is the final density of states at the energy $E+h\nu$, $f(E,T)$ is the Fermi–Dirac distribution function at temperature T, $\sigma_{opt}(E, h\nu)$ is the optical transition probability or an average cross-section for all states at the energy E, $P_t(E,h\nu)$ is the electron transport function and $P_e(E,h\nu)$ is the escape function. The measured intensity $I(E_{kin}, h\nu)$ of emitted photoelectrons is proportional to the number of photoelectrons $n(E_{kin}, h\nu)$ but is perturbed by the natural line width of the photon source and the spectrometer resolution. It is therefore expressed as

$$I(E_{kin}, h\nu) \propto [n(E_{kin}, h\nu) \cdot S_A(E_{kin})] \cdot R_A(E_{kin}, h\nu), \quad (7.27)$$

where $S_A(E_{kin})$ denotes the analyzer sensitivity and $R_A(E_{kin}, h\nu)$ the total resolution function determined by the photon source and the spectrometer, which is accounted for by convolution.

We wish to extract information about the initial density of states $N_i(E_B)$ at the binding energy E_B from the measured photoelectron spectra. However, this does not always represent well the $N_i(E_B)$ profile because of the presence of several correction terms in equation (7.26). In particular, the final density of states $N_f(E_B + h\nu)$ significantly affects the measured spectrum when the excitation energy is very low, say below 10 eV. We will discuss this point later.

There are other correction terms in equation (7.26). The measurement is carried out at a finite temperature T. Thus, the Fermi–Dirac distribution function $f(E,T)$ thermally smears the density of states at the Fermi level, as discussed in Section 3.3. The optical transition probability or the photoionization cross-section $\sigma_{opt}(E, h\nu)$ also plays an important role. This term represents the strength of the interaction between the incident photon and the electron in a solid. The subshell photoionization cross-section of various elements in the periodic table has been calculated in the excitation energy range 10.2 to 8047.8 eV by Yeh and Lindau [8]. The observed spectrum should be analyzed, using this tabulated data as a guide.

Because of its low excitation energy, the UPS measurement is well suited to study the valance band structure in the binding energy range 0–20 eV. Its resolution is inherently high and can be improved to 2–10 meV, particularly if special precautions are taken.[1] The effect of the photoionization cross-section and the final density of states on the measured spectrum is of particular importance in UPS measurements.

Figure 7.14(a) shows the atomic subshell photoionization cross-section of Sn. It can be seen that the cross-section for the Sn 5p state exceeds that of the

---

[1] The resolution of the energy analyzer has reached a level of just a few meV. In addition, for a sample cooled down to 10 K to suppress the thermal broadening at the Fermi edge, the overall resolution has been claimed to be reduced to a level less than ± 2 meV.

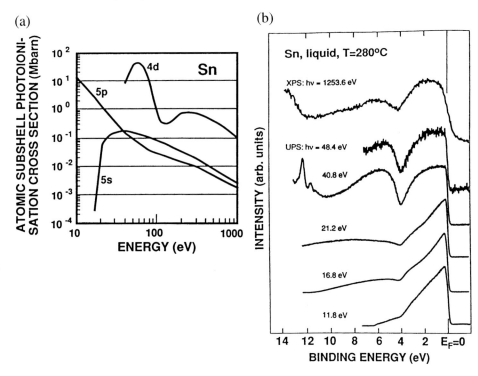

Figure 7.14. (a) Atomic subshell photoionization cross-sections calculated for Sn. [J. J. Yeh and I. Lindau, At.Data and Nucl.Data Tables *32*, (1985) 1]. (b) XPS and UPS valence band spectra of liquid Sn. [G. Indlekofer, PhD thesis, Universität Basel (1987)]

5s state by three orders of magnitude at low excitation energies around 10–15 eV but the situation reverses above about 40 eV. By utilizing this unique excitation energy dependence of the cross-section, one can study the detailed electronic states in the valence band. A typical example is shown in Fig. 7.14(b). The UPS spectra of liquid Sn were taken at different excitation energies. It can be seen that the peak at about 6 eV gradually grows with increasing excitation energy and that a double-peaked structure clearly emerges when the excitation energy reaches 40.8 eV. Based on the energy dependence of the photoionization cross-section, Fig. 7.14(a), we can easily identify the states immediately below the Fermi level and those at higher binding energies as the 5p and 5s states, respectively, in the valence band of liquid Sn. As is clear from the argument above, the energy dependence of the photoionization cross-section is significant and a correct interpretation of the measured spectrum is made possible only if its effect is properly considered. It is also worthwhile noting that the excitation energy dependence of the cross-section is quite common to all s and p valence electrons of polyvalent metals from Al to Bi.

The effect of the final density of states is also important in the interpretation

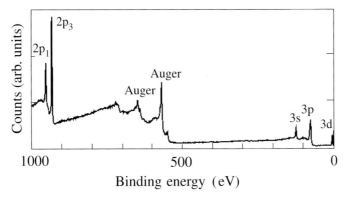

Figure 7.15. XPS spectrum of pure Cu. The Fermi level is taken as the origin. Peaks arising from Auger electrons are also observed.

of UPS spectra, particularly when the excitation energy is lower than 10 eV. This is because the final density of states is not structureless but is perturbed by the periodic potential of the lattice. Hence, the selection rule associated with the optical transition must be taken into account on the basis of given E–$\mathbf{k}$ relations. Indeed, the photoemission spectrum in a crystal is known to depend on whether the optical transition is direct or indirect [9] (see Section 11.9). In the case of disordered materials like liquid metals and amorphous alloys, however, the wave vector $\mathbf{k}$ is no longer conserved upon optical transition. Hence, the UPS spectrum in a disordered system better reflects the initial density of states in excitation energies, at least above about 10 eV.

In contrast to the UPS measurements, the excitation energy in the XPS experiment generally exceeds 1000 eV. Thus, the final density of states is structureless and can be ignored. Core electrons having large binding energies can also be excited. A typical XPS spectrum for pure Cu taken with the monochromatized Al-Kα radiation is shown in Fig. 7.15. It is clear that peaks associated with the core electrons down to the 2p level are observed. By analyzing the position and intensity of the core electron spectrum, one can identify the atomic species and their compositions in an alloy sample. Therefore, the XPS technique has proved to be powerful not only to study the electronic state in a solid but also to carry out the chemical analysis of a sample. In this respect, the XPS technique is sometimes called Electron Spectroscopy for Chemical Analysis (ESCA). However, the resolution in XPS even with monochromatized radiation is, at best, 0.3–0.5 eV and would fail to detect fine structures in the valence band.

In spite of a limited resolution relative to UPS, XPS valence band spectra have been frequently measured by using a monochromatized Al-Kα radiation. The valence band spectrum can be obtained by accumulating signals of photoelectrons emitted from states at low binding energies, say, down to about 15 eV.

168   *7 Principles of measuring electronic structure-related phenomena*

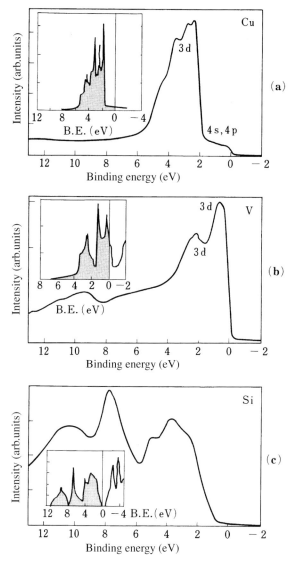

Figure 7.16. XPS valence band spectra for (a) pure Cu, (b) pure V and (c) pure Si. The Fermi level is located at the origin. The XPS apparatus with monochromated Al-$K\alpha$ line is claimed to give a resolution better than 0.25 eV (ESCA-300, Seiko Electronics Industry). Insets show the valence band spectra obtained from band calculations. [V. L. Moruzzi *et al. Calculated Electronic Properties of Metals* (Pergamon Press, 1978)]

Figure 7.16 shows the XPS valence band spectra of pure Cu, V and Si, along with (inset) the density of states curve derived from band calculations. We see that the measured XPS spectra reflect well the calculated valence band structures. The Fermi level is in the middle of the 3d band in pure V, whereas in pure Cu the 3d band is fully filled and the Fermi level is situated in the well-extended valence band consisting of 4s and 4p electrons. As discussed in Section 6.9, Si is a semiconductor. The XPS spectrum clearly shows the absence of a sharp Fermi cut-off in Si in contrast to its presence in pure Cu and V metals.

We see from Fig. 7.16 that the XPS valence band spectra reproduce reasonably well the calculated valence band structures for three representative elements in the periodic table. However, the situation is not so simple in the case of alloys. The photoionization cross-section strongly depends on the atomic species of the constituent elements in an alloy. As a result, the contribution of each element to the valence band spectrum is, in many cases, not proportional to its composition. Hence, the observed XPS valence band spectrum in an alloy does not always resemble the calculated valence band structures (see, for example, Fig. 9.18). Many-body effects also cannot be neglected when interpreting the XPS data in terms of the existing band calculations.[2]

The incident radiation penetrates into a solid to a depth of the order of several tens of atomic layers and interacts with electrons in that region, inducing optical transitions in its immediate vicinity. Once the electron is photoexcited, it must travel to the surface of a sample. The mean escape depth of the photoexcited electron, given by the term $P_t(E, h\nu)$ in equation (7.26), has been evaluated as a few nm over the excitation energy range 10 eV up to above 1000 eV [5, 6]. Hence, the measured spectrum is very sensitive to the surface condition of a sample. Although the measurement is carried out under an ultra-high vacuum, contamination such as oxides on the sample surface must be carefully removed prior to the photoemission measurement. The surface contamination can be removed either by Ar gas sputtering or by cleaving or even polishing mechanically the surface of the sample in a high vacuum. The band structure investigation by means of the XPS and UPS measurements is also free from the mean free path of the electron. Hence, the photoemission measurement can be applied to any material, regardless of the presence or absence of the lattice periodicity.

7.6 Inverse photoemission spectroscopy

Photoemission experiments can be made in a reversed mode by impinging electrons of varying energies onto a sample and detecting the photons that are excited by them. The principle is illustrated schematically in Fig. 7.17. The

---

[2] The removal of an electron from a core level produces a strong perturbation of the remaining core electrons. They will be redistributed to screen the hole. The energy gained in this relaxation process is carried away by the emitted electron. Hence, its energy is influenced. A similar effect has been observed even in the valence electrons, if the band width is narrow. For example, the 3d band in pure Zn is centered at about 8 eV below E_F with a width of only 1 eV. The XPS measurement reveals the Zn 3d band to be centered at 10 eV below E_F. This difference is believed to originate from a change in the total energy of the system associated with the relaxation process. [J. Hafner et al., J. Phys. F: Metal Phys. **18** (1988) 2583]

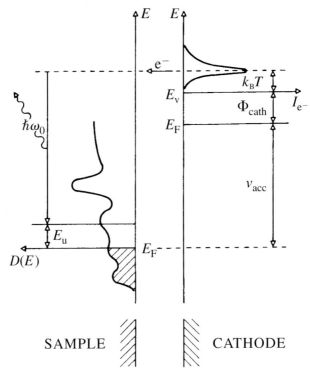

Figure 7.17. Energy diagram of the inverse photoemission process. Note that, by applying the acceleration voltage V_{acc} to the cathode, we can raise its Fermi level by that amount relative to that in the sample. (E_v: vacuum level; I_{e^-}: intensity distribution of emitted electrons; $D(E)$: density of states; E_u: unoccupied state energy; $\hbar\omega_0$: x-ray photon energy; Φ_{cath}: work function of cathode; T: temperature.) [J. K. Lang and Y. Baer, *Rev. Sci. Instrum.* **50** (1979) 221]

incident electrons excite both valence and core electrons in the sample. However, a less-frequent process also occurs, which involves the direct deceleration of an incident electron with simultaneous emission of electromagnetic radiation called Bremsstrahlung. The impinging electron occupies a previously empty state above the Fermi level, as shown in Fig. 7.17. This process may be considered as the inverse of the photoemission process, since the role of initial and final states is exchanged and the initially unoccupied state is filled by the decelerated electron. The unoccupied state above the Fermi level can be chosen by varying the acceleration voltage V_{acc} while keeping $\hbar\omega_0$ constant (isochromat) in Fig. 7.17. This is the reason why this technique is called either Inverse Photoemission Spectroscopy (IPES) or alternatively Bremsstrahlung Isochromat Spectroscopy (BIS).

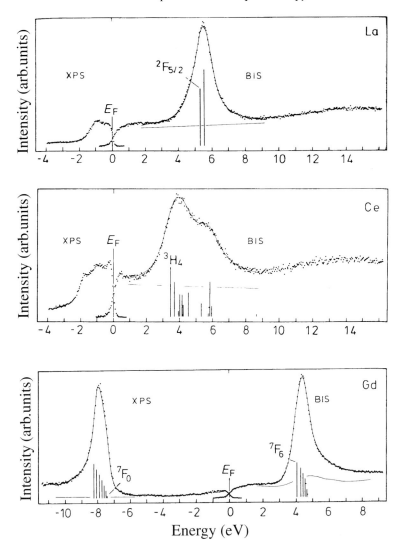

Figure 7.18. Combined XPS and IPES spectra for pure La, Ce and Gd metals. Vertical bars indicate the positions and calculated intensities of the 4f final states. The spin configuration of an atom is often denoted as $^{(2S+1)}L_J$ in the Russell–Saunders nomenclature. Capital letters S, P, D, F are used for $L=0$, 1, 2, 3, respectively. (See Section 13.3.) [J. K. Lang, Y. Baer and P. A. Cox, *J. Phys. F: Metal Phys.* **11** (1981) 121]

It is possible to measure both XPS and IPES spectra in the same apparatus on the same sample by using the combined XPS–IPES spectrometer equipped with a monochromatized Al-Kα x-ray source. In the IPES mode, only photons of 1486.6 eV can be filtered by the bent quartz crystal monochromater designed for the Al-Kα radiation. Its intensity is recorded as a function of the acceleration potential V_{acc} applied to the cathode.

172   *7 Principles of measuring electronic structure-related phenomena*

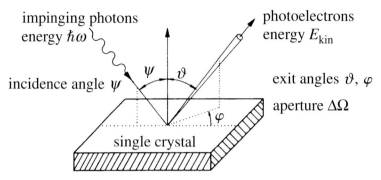

Figure 7.19. Geometry of the ARPES measurements. [S. Hüfner, *Photoelectron Spectroscopy*, Springer Series in Solid State Physics, vol. 82, edited by M. Cardona, (Springer-Verlag, Berlin 1995)]

Figure 7.18 shows the combined XPS–IPES spectra for pure La, Ce and Gd metals [10]. Since the partially filled 4f orbitals maintain their localized character in solids, a multiplet structure is clearly observed in both XPS and IPES spectra. This is different from the situation in the transition metals, where the Fermi level falls in the partially filled 3d, 4d and 5d bands [11]. Consider the Hund rule ground state of the configurations $(4f)^n$. Then one observes the final states of $(4f)^{n-1}$ in XPS and $(4f)^{n+1}$ in IPES (see Sections 13.3 and 14.6), indicating the apparent failure of the one-electron approximation (see Sections 14.1–14.5). For example, the $^7F_0$ and $^7F_6$ states observed in XPS and IPES spectra of Gd correspond to the Hund ground states of Eu and Tb, which are positioned next to Gd in the periodic table and possess one less and one more 4f electron, respectively, than does Gd.

7.7 Angular-resolved photoemission spectroscopy (ARPES)

In addition to the angular-integrated photoemission spectroscopy discussed in Section 7.5, the angular-resolved photoemission spectroscopy technique has been developed as a powerful tool to determine the energy dispersion curves with high accuracy. ARPES uses a single crystal in the geometry shown in Fig. 7.19. The solid angle of detection $\Delta \Omega$ is made small so that photoelectrons in a narrow k-interval only are detected. There are four major parameters in the ARPES experiments: two exit angles ϑ and φ specifying the direction of the photoelectrons, their kinetic energy E_{kin} and the energy of the impinging radiation $\hbar \omega$.

We assume a direct transition between an initial state E_i and a final state E_f, both of which are measured with respect to the Fermi energy, as indicated in Fig. 7.20. The measured quantity is the kinetic energy E_{kin} of a photoelectron detected at an angle ϑ relative to the normal of the sample surface. Knowing

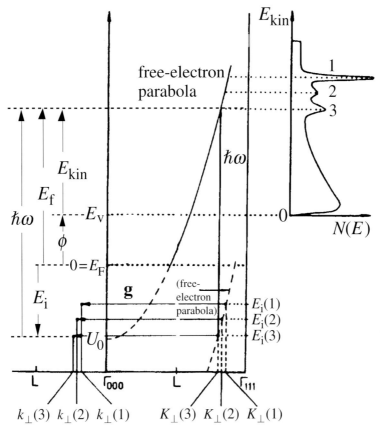

Figure 7.20. Analysis of the ARPES data by assuming the free-electron model for the final states [7]. See text for detailed description. [S. Hüfner, *Photoelectron Spectroscopy*, Springer Series in Solid State Physics, vol. 82, edited by M. Cardona, (Springer-Verlag, Berlin 1995)]

the work function ϕ and radiation energy $\hbar\omega$, one obtains the final and initial state energies from the relations:

$$E_f = E_{kin} + \phi \qquad (7.28)$$

and

$$E_i = E_f - \hbar\omega, \qquad (7.29)$$

where the initial state energy E_i is a negative quantity and is equal to the binding energy E_B if its sign is reversed, i.e., $E_B = -E_i$.

The intensity of a photoelectron emerging in a direction ϑ from a sample is measured as a function of its kinetic energy E_{kin}. In the photoemission process, only the parallel component of the wave vector to the surface of the sample is conserved. Since the photoelectron emerging outside the sample may well be approximated as a free electron with mass m, the parallel component of the wave vector **K** of a photoexcited electron in the crystal is directly determined from the measured kinetic energy E_{kin} through the relation:

$$K_{\parallel}^{pe} = \sqrt{(2m/\hbar^2)E_{kin}}\sin\vartheta = K_{\parallel}, \qquad (7.30)$$

where $K_{\parallel}^{pe}$ and $K_{\parallel}$ are the parallel components of the wave vector of both photoelectron and photoexcited electron in the crystal, respectively, and ϑ is the exit angle shown in Fig. 7.19.

The perpendicular component of the wave vector of the photoelectron is obviously given by

$$K_{\perp}^{pe} = \sqrt{(2m/\hbar^2)E_{kin}\cos^2\vartheta}. \qquad (7.31)$$

However, this is not equal to $K_{\perp}$ of the photoexcited electron in a crystal because of the presence of the work function ϕ. In addition, the value of $K_{\perp}$ cannot be uniquely determined, unless the final state energy E_f is given. As is shown in Fig. 7.20, the final state energy may be approximated by $E_f = (\hbar^2/2m)(K_{\parallel}^2 + K_{\perp}^2) + U_0$ in the free-electron model, where U_0 is the inner potential relative to the Fermi level. An insertion of equations (7.28) and (7.30) leads to

$$K_{\perp} = \sqrt{(2m/\hbar^2)(E_{kin}\cos^2\vartheta + \phi - U_0)}. \qquad (7.32)$$

We see that the wave vector **K** of the photoexcited electron in a crystal can now be determined from equations (7.30) and (7.32) for a given exit angle ϑ, provided that two parameters ϕ and U_0 are reliably fixed.

For the sake of simplicity, we take the normal emission condition $\vartheta = 0$ and try to explain how the E–k relations in Cu are deduced from the measured photoemission spectrum. Let us assume that the photoemission spectrum consisting of three peaks 1, 2 and 3 was obtained by exciting valence electrons to the free-electron states with the radiation $\hbar\omega$. In Fig. 7.20, (upper right), the spectrum is shown as a function of the kinetic energy E_{kin} of the photoelectron, which is measured relative to the vacuum level E_v. The wave vectors of the photoelectron corresponding to the three peaks in the spectrum are denoted as $K_{\perp}^{pe}(i)$ with $i = 1, 2$ and 3. The inner potential U_0 is chosen to be -8.6 eV corresponding to the bottom of the sp band in pure Cu (see the point Γ in Fig. 6.3), whereas the work function ϕ is experimentally determined from equations (7.28) and (7.29).

The determination of the E–$k_{\perp}$ relations is straightforward. First, the energy of the initial state $E_i(i)$ corresponding to the peak i is easily calculated as $-E_i(i) = \hbar\omega - (E_{kin}(i) + \phi)$ from equations (7.28) and (7.29). They are located in Fig. 7.20 as three horizontal dotted lines $E_i(1)$ to $E_i(3)$. Then, the corresponding $K_{\perp}(i)$ values are calculated from equation (7.32) and plotted as three vertical dotted lines in Fig. 7.20. As will be discussed in Section 11.9, the wave vector

7.7 Angular-resolved photoemission spectroscopy (ARPES)

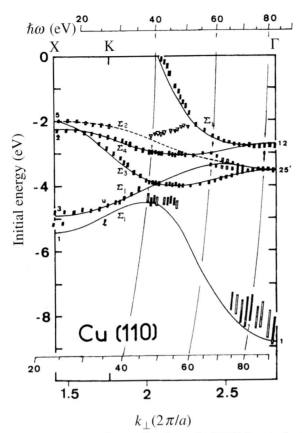

Figure 7.21. Band structure of pure Cu along the ΓKX line derived from the ARPES measurements on a (110) crystal surface. The full curve is due to Burdick's band calculations. The height of the rectangle data points indicates experimental uncertainty. Three parabolic curves represent K-dependence of the initial energy $E_i = (\hbar^2 K^2/2m) + U_0 - \hbar\omega$. [P. Thiry et al., Phys. Rev. Lett. **43** (1979) 82]

is conserved upon optical transition and, hence, the value of $K_\perp(i)$ for the photoexcited electron is equal to $K_\perp(i)$ of the electron in the valence band except for the arbitrariness in the choice of an appropriate reciprocal lattice vector. The resulting three intersections are displaced horizontally to the left by an appropriate reciprocal lattice vector **g**. The resulting three E–$k_\perp$ values are shown in Fig. 7.20.

The full E–$k_\perp$ relations in a given direction can be mapped by repeating the same procedure under different incident photon energies $\hbar\omega$ over the range 10 to 100 eV available in the synchrotron radiation experiment. The E–$k_\perp$ relations along ΓKX of pure Cu thus derived are shown in Fig. 7.21 along with the calculated band structure of Burdick. The agreement between the theory and ARPES experiment is remarkable. More details can be found in the literature [7].

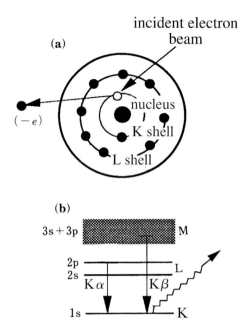

Figure 7.22. Principles of soft x-ray emission spectroscopy. (a) An electron is excited out of the K shell by irradiating the sample with an electron beam, thereby creating a hole in the K shell. (b) The Al-Kβ emission takes place when an electron in the valence band falls into the vacant 1s level. This reflects the energy distribution of the Al-3p electrons because of the selection rule. The Al-Kα transition from 2p to 1s levels is also shown.

7.8 Soft x-ray spectroscopy

The soft x-ray spectroscopy technique has also been frequently employed to explore the electronic structure and can supplement information derived from the photoemission studies discussed in Section 7.5 [12, 13]. Here, soft x-ray refers to an x-ray with a wavelength longer than a few-tenths nm which is emitted or absorbed upon a transition of an electron in a solid. Soft x-ray emission and absorption spectra reflect the occupied and unoccupied density of states below and above the Fermi level, respectively. The soft x-ray emission and absorption spectroscopies are abbreviated as SXES and SXAS, respectively.

We discuss first the SXES experiment, using pure Al as an example. The electronic structure of pure Al is shown in Fig. 7.22(b). An electron beam accelerated to energies of 10–15 keV is imparted onto the surface of an Al sample to excite the 1s core electron in the ground state. The characteristic x-ray is emitted upon the transition of an electron from a higher level to fill the vacant 1s level. Owing to the selection rule, the transition is allowed only when $\Delta \ell = 1$, where $\Delta \ell$ is the difference in the azimuthal quantum number between the initial

7.8 Soft x-ray spectroscopy

and final states. For example, the Al-Kα line refers to the transition from the 2p to the 1s level and is used as an irradiating source in XPS measurements.

Information about the valence band structure can be extracted from the Al-Kβ spectrum in the case of Al metal and Al-based alloys. This corresponds to the transition from the 3p to the 1s level. Remember that the Al-3p electron forms the valence band together with the Al-3s electron. Thus, the Al-Kβ spectrum is no longer a sharp line but is broadened, reflecting the Al-3p electron distribution in the valence band.

Soft x-rays emitted from a sample are diffracted by a crystal spectrometer so that only x-rays with the wavelength λ satisfying the Bragg condition $2d\sin\theta = n\lambda$ are detected. Here d is the grating space of a crystal in the spectrometer. By varying the angle θ of the crystal, we can measure the x-ray intensity as a function of the wavelength λ, which is converted to energy E through the relation $E = h\nu = hc/\lambda$. The Fermi level cannot be determined solely from the SXES measurement. In the case of the Al-Kβ spectrum, we measure first the Al-Kα line, from which the energy difference between Al-2p and Al-1s levels is deduced. Independently, the binding energy of the Al-2p level relative to the Fermi level can be determined from the XPS measurement. A sum of these two enables us to locate the Fermi level in the Al-Kβ spectrum. In contrast to photoemission spectroscopy, the SXES measurement is less sensitive to surface contamination of the sample, since electron irradiation can create holes in the core level of atoms in the region down to a few μm below the surface of a sample and, thus, the emitted soft x-ray reflects well a bulk property.

We need to study more details about the intensity of the x-ray emitted from the sample in order to analyze the SXES spectrum. As mentioned above, a hole is created in the inner core level by the electron irradiation. Now the electron in the higher energy level E_i is allowed to fill the hole in the core level with the energy E_c in accordance with the selection rule. Here the subscripts i and c refer to the initial state and the final core level, respectively. An x-ray photon of energy $\hbar\omega = E_i - E_c$ is emitted. According to radiation theory, the transition probability $I(\omega)$ per unit time for the spontaneous emission can be expressed as

$$I(\omega) = \frac{4e^2\omega^3}{3\hbar c^3} |\langle \psi_c | \mathbf{r} | \psi_i \rangle|^2, \qquad (7.33)$$

where ψ_i and ψ_c represent the electron wave function corresponding to the initial state and the final core level, respectively.[3]

---

[3] When a solid is exposed to radiation, the electron distribution in a given atom is polarized due to the alternating electric field of the electromagnetic wave. The polarized electron distribution is approximated by an electric dipole $(-e)\mathbf{r}$, as long as its wavelength is longer than the orbital radius of a core electron. See, for example, L. I. Schiff, *Quantum Mechanics* (Second Edition, McGraw-Hill, New York, 1955, pp. 251–261).

As noted above, we are interested in the transition of an electron from the valence band into the core level E_c. The wavefunction ψ_i in the initial state must be that of the Bloch electron, which is described in terms of the wave vector **k** and the band index n (see Section 5.5 and also Exercise 5.2). Equation (7.33) is then rewritten as

$$I(\omega) \propto \omega^3 \sum_{n,\mathbf{k}} |\langle \psi_c | \mathbf{r} | \psi_{n,\mathbf{k}} \rangle|^2 \, \delta(E_{n,\mathbf{k}} - E_c - \hbar\omega). \quad (7.34)$$

As will be discussed in Chapter 8, the wave function of the Bloch electron in the valence band can be expanded into the form:

$$\psi_{n,\mathbf{k}}(\mathbf{r}) = N^{-1/2} \sum_{\ell m} b_{n\mathbf{k},\ell m} Y_{\ell m}(\theta, \phi) R_\ell(E_{n\mathbf{k}}, \mathbf{r}), \quad (7.35)$$

where N is the number of unit cells in a crystal, R_ℓ is the radial wave function and $Y_{\ell m}$ is the spherical harmonic function specified by the azimuthal quantum number ℓ and magnetic quantum number m (see equation (8.65)). In this representation, the wavefunction of the Bloch electron is decomposed into the s, p, d, f, ... components having space-symmetries characteristic of electrons in a free atom.

The numbers of electrons having the s, p and d symmetries in the energy interval between E and $E+dE$ are denoted as $n_s(E)dE$, $n_p(E)dE$ and $n_d(E)dE$, respectively. The partial density of states, $n_\ell(E)$, of the quantum number ℓ can be expressed as

$$n_\ell(E) = \sum_{n\mathbf{k}} \sum_{m=-\ell}^{\ell} |b_{n\mathbf{k},\ell m}|^2 \, \delta(E_{n\mathbf{k}} - E), \quad (7.36)$$

where the coefficient $b_{n\mathbf{k},\ell m}$ in equation (7.35) is determined by solving the Schrödinger equation.

The transitions of the electron to the 1s state of the K shell and to the 2p state of the L shell are the most important in SXES measurements. As emphasized above, the selection rule $\Delta \ell = 1$ must be fulfilled to allow these transitions. Therefore, the respective x-ray intensities are given by

$$I_K(\omega) \propto [M_{pK}(E)]^2 \, n_p(E) \delta(E - E_{1s} - \hbar\omega), \quad (7.37)$$

and

$$I_L(\omega) \propto \left\{ [M_{sL}(E)]^2 \, n_s(E) + \frac{2}{5}[M_{dL}]^2 \, n_d(E) \right\} \delta(E - E_{2p} - \hbar\omega), \quad (7.38)$$

where $M_{pK}(E)$, for example, represents the matrix element associated with the transition of the electron in the p states in the valence band to the 1s final state and is proportional to an integral involving the product of these two radial wave functions. Thus, the more these two wave functions overlap, the higher is

7.8 Soft x-ray spectroscopy

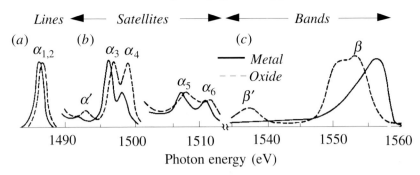

Figure 7.23. SXES spectra for pure Al metal and Al_2O_3. (a) $\alpha_{1,2}$ is the Al-Kα characteristic line and corresponds to the transition from $2p_{1/2}$ and $2p_{3/2}$ to the 1s level. (b) Both $\alpha_3\alpha_4$ and $\alpha_5\alpha_6$ are satellites of the α line. (c) Kβ spectrum corresponds to the transition from the 3p band to 1s level and reflects the Al-3p electron distribution. The Fermi level is located at 1560 eV. See the binding energy of Al 1s level shown in Fig. 7.13.

the x-ray intensity. Equation (7.37) is called the K-transition and the corresponding spectrum provides information about the p partial density of states. Similarly, equation (7.38) refers to the L-transition and provides information about the sum of both s and d partial density of states.

Figure 7.23 shows the SXES emission spectra for pure Al metal and Al_2O_3 oxide. A sharp peak at 1487 eV is identified as the Al-Kα line due to the transition from Al-2p to Al-1s levels. Its wavelength is about 8 Å, being typical of soft x-rays. The broad Al-Kβ spectrum appearing in the energy range 1550–1560 eV is caused by the transition from the 3p states in the valence band to the 1s level and reflects the Al-3p partial density of states. The Fermi level is located at 1560 eV. The intensity of the Al-Kβ spectrum sharply increases below the Fermi level in Al metal but remains low in Al_2O_3. This clearly indicates the difference in the electronic structure between the metal and the insulating oxide.

The Al- and Si-Kβ SXES emission spectra representing the 3p electron distribution are shown in Fig. 7.24. A difference in the valence band structure between the Al metal and semiconducting Si is clearly seen. The SXES spectrum of Al metal is well approximated by the free-electron model, whereas that of Si is characteristic of a semiconducting material, as manifested by the lack of the density of states at the Fermi level and a narrow band width.

At the end of this section, we briefly discuss the SXAS technique,[4] where one

---

[4] EXAFS (extended x-ray absorption fine structure) is known as a tool to determine the atomic structure about the central excited atom, which is deduced by analyzing the oscillatory fine structure in the x-ray absorption spectrum extending over hundreds of electron volts above the edge. Its near-edge structure in the 30–50 eV range, automatically recorded in the EXAFS spectrum, is caused by the electronic process, which is ignored for structural analysis. The SXAS technique discussed here exactly corresponds to this near-edge structure and is alternatively called XANES (x-ray absorption near-edge structure) or NEXAFS (near-edge x-ray absorption fine structure).

180   7 Principles of measuring electronic structure-related phenomena

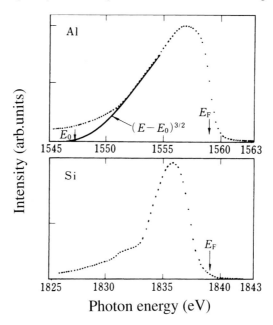

Figure 7.24. Al-Kβ and Si-Kβ SXES spectra of pure Al metal and pure Si crystal. Both reflect the respective 3p electron distribution. The $(E-E_0)^{3/2}$ relation holds in the free-electron model. [K. Tanaka, *Nihon Kinzoku Gakkai Kaiho* **15** (1976) 753 (in Japanese)]

measures the intensity of a monochromatic radiation transmitted through a thickness x of the material being studied.[5] A synchrotron radiation source equipped with a crystal monochrometer is employed for the measurement. The photoabsorption cross-section suddenly increases and the intensity of transmitted radiation sharply drops, when the radiation energy exceeds a threshold corresponding to the excitation of a core electron to unoccupied states above the Fermi level. In the dipole approximation, an increase in the photoabsorption cross-section at the threshold is given by

$$\Delta\sigma(\omega) = \frac{4\pi^2\alpha}{3\varepsilon_0}\hbar\omega \left|\langle\psi_f|\sum_j r_j|\psi_c\rangle\right|^2 n_\ell(E), \qquad (7.39)$$

where the summation is extended over all core electrons in a given atom, $n_\ell(E)$ is the partial density of the final state and α is the fine structure constant [13].[6]

---

[5] The electron-yield detection technique is also employed, particularly when a thin film is not available. It utilizes the non-radiative core hole annihilation process by Auger electron emission, the yield of which is proportional to the x-ray absorption coefficient. The current produced by Auger electrons escaping from the sample upon irradiation is measured.

[6] A summation over all core electrons is needed in a strict sense, since a hole created in a given level perturbs energy states of all remaining core electrons (see footnote 2, p. 169). Equation (7.33) is given in the one-electron approximation and the summation is omitted.

7.9 Electron-energy-loss spectroscopy (EELS)

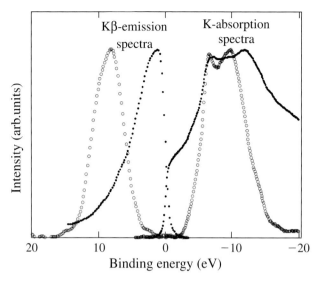

Figure 7.25. Soft x-ray Kβ-emission and K-absorption spectra of pure Al (solid circles) and Al$_2$O$_3$ (open circles) are shown on the same energy scale. The Fermi level is located at $E = 0$. All spectra are normalized to their maximum intensity. [Courtesy Dr Esther Belin-Ferré]

The SXAS spectrum obtained by scanning the radiation energy reflects unoccupied density of states with an appropriate symmetry permitted by the selection rule.

If one measures both SXEX and SXAS data for a given sample, detailed information about the electronic structure across the Fermi level can be extracted. As an example, the Kβ-emission and K-absorption spectra of pure Al and Al$_2$O$_3$ are shown in Fig. 7.25. The difference in the electronic structure between the two solids can be more clearly seen than the data in Fig. 7.23. A sharp Fermi edge is seen in Al metal, whereas an energy gap of 10.8 eV is found in the insulating Al$_2$O$_3$.

The electronic properties, particularly the bonding mechanism among various constituent elements in an alloy or an intermetallic compound, can be more efficiently and thoroughly studied if both the photoemission and soft x-ray spectroscopies are combined and, moreover, if both sets of data are further supplemented with data obtained from electronic specific heat measurements and band calculations [14].

7.9 Electron-energy-loss spectroscopy (EELS)

In the preceding section, we have discussed soft x-ray absorption spectroscopy, from which information about the unoccupied partial density of states above the Fermi level can be deduced. In the transmission electron microscope, electrons generally lose energy in passing through a specimen. As described below, the unoccupied partial density of states can also be studied with a transmission electron microscope combined with a magnetic prism-type spectrometer by measuring the energy loss of incident electrons due to the inelastic scattering by the excitation of core or valence electrons. Basic principles in the electron-energy-loss spectroscopy (EELS) measurements are briefly discussed in this section.

The cross-section for inelastic scattering of an incident electron, which accompanies an excitation of electrons in a solid, is generally given by

$$d\sigma = \frac{2\pi}{\hbar} |\langle F|V(\mathbf{r})|I\rangle|^2 \, \delta(E - E_f + E_i) p^2 dp d\Omega, \tag{7.40}$$

where the initial state $|I\rangle$ and the final state $|F\rangle$ refer to those in a whole system consisting of the incident electron and electrons in a solid, $p^2 dp d\Omega$ is the number of electrons scattered into a solid angle $d\Omega$ in the momentum range p and $p + dp$ and $V(\mathbf{r})$ is the interaction potential of the incident electron with electrons in a solid. The δ-function in equation (7.40) assures energy conservation upon scattering.

The initial state $|I\rangle$ in a whole system is described as the product of the free-electron wave function $e^{i\mathbf{k}_0 \cdot \mathbf{r}}$ of the incident electron and the initial state $|i\rangle$ of the electrons in a solid. The final state $|F\rangle$ is likewise expressed as $e^{i\mathbf{k}\cdot\mathbf{r}}|f\rangle$. By inserting the relation $E = p^2/2m$ or $p^2 dp = m\hbar k \cdot dE$, we obtain

$$\frac{\partial^2 \sigma}{\partial E \partial \Omega} = \frac{m^2}{4\pi^2 \hbar^4} \cdot \frac{k}{k_0} \sum_f \left| \langle f| \int V(\mathbf{r}) \exp(i\mathbf{q}\cdot\mathbf{r}) d\mathbf{r} |i\rangle \right|^2 \cdot \delta(E_f - E_i - E), \tag{7.41}$$

where $\mathbf{q}$ is the momentum transfer vector equal to $\mathbf{q} = \mathbf{k}_0 - \mathbf{k}$. The ratio k/k_0 may be approximated as unity for the fast incident electrons.

The interaction potential $V(\mathbf{r})$ is given by the sum of the Coulomb potentials acting between the incident electron and nucleus and between the incident electron and electron in a solid:

$$V(\mathbf{r}) = -\sum_i \frac{Z_i e^2}{|\mathbf{r} - \mathbf{R}_i|} + \sum_j \frac{(-e)^2}{|\mathbf{r} - \mathbf{r}_j|}, \tag{7.42}$$

where Z_i is the atomic number and $\mathbf{R}_i$ and $\mathbf{r}_j$ are the coordinates of nucleus i and electron j in the solid, respectively. It can be shown that the contribution of the first term to equation (7.41) vanishes because of the orthogonality of the wave functions $|i\rangle$ and $|f\rangle$. Instead, a finite contribution arises from the second term of equation (7.42), when it is inserted into equation (7.41). Since the Fourier component of the second term is easily calculated as

7.9 Electron-energy-loss spectroscopy (EELS)

$$\int \frac{e^2}{|\mathbf{r} - \mathbf{r}_j|} \exp(i\mathbf{q}\cdot\mathbf{r}) d\mathbf{r} = \frac{4\pi e^2}{q^2} \exp(i\mathbf{q}\cdot\mathbf{r}_j), \qquad (7.43)$$

equation (7.41) is reduced to the form:

$$\frac{\partial^2 \sigma}{\partial E \partial \Omega} = \frac{4m^2 e^4}{\hbar^2} \frac{1}{q^4} \sum_f \left| \langle f | \sum_j \exp(i\mathbf{q}\cdot\mathbf{r}_j) | i \rangle \right|^2 \cdot \delta(E_f - E_i - E), \qquad (7.44)$$

where $S(\mathbf{q}, E) = \sum_f \left| \langle f | \sum_j \exp(i\mathbf{q}\cdot\mathbf{r}_j) | i \rangle \right|^2 \cdot \delta(E_f - E_i - E)$ is called the dynamical structure factor (see Section 10.11).

By choosing an appropriate energy range in the spectrometer, one can measure the EELS spectrum associated with the excitation of either valence electrons or core electrons in a solid. If the core electron is excited, the magnitude of the vector $\mathbf{r}_j$ in equation (7.44) is of the order of the radius of the core orbital r_c. As long as the scattering angle is small, the momentum transfer vector $\mathbf{q}$ must be small and the relation $q \ll r_c^{-1}$ holds. Since the dipole approximation is validated in this limit, we see that there is a one-to-one correspondence between the energy loss process of the incident electron and the x-ray absorption process given by equation (7.39). The transition involved is certainly dominated by the dipole selection rule ($\Delta \ell = \pm 1$). The measured intensity of the incident electron in the energy loss spectrum is expressed as

$$I(E) \propto P(E) n(E), \qquad (7.45)$$

where $P(E)$ is the energy-dependent matrix element and $n(E)$ is the partial density of states with an appropriate symmetry. The fine structures in the energy region near the core-edge onset are expected to reflect the unoccupied partial density of states $n(E)$, since $P(E)$ varies slowly as a function of energy.

Figure 7.26 shows the L_2 and L_3 edge EELS spectra of several 3d-transition metals, the data of which were obtained by using thin specimens 100–300 Å in thickness.[7] The L_2 and L_3 edges correspond to a transition from the $2p_{3/2}$ and $2p_{1/2}$ core levels to unoccupied 3d partial density of states above the Fermi level, respectively. It can be seen that, because of an increase in the spin–orbit interaction, the separation of the L_2 and L_3 edges increases across the periodic table from 6 eV in Ti to 20 eV in Cu.

Information about the electronic structure of the 3d transition metals can be extracted from the fine structure in the L_2 and L_3 spectra (Energy Loss

---

[7] An EELS sample must be very thin not only to allow the transmission of the electron beam but also to suppress multiple inelastic scattering.

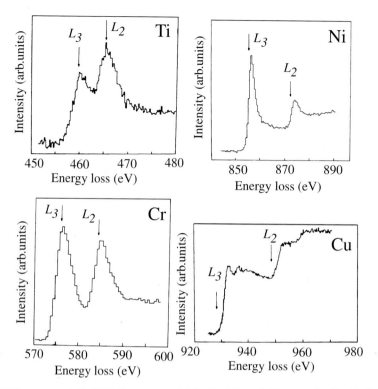

Figure 7.26. L_2 and L_3 EELS spectra of Ti, Cr, Ni and Cu metals. [R. D. Leapman, L. A. Grunes and P. L. Fejes, *Phys. Rev.* **B26** (1982) 614]

Near-Edge Structure or ELNES). Putting detailed discussions aside, we simply focus on the widths of the L_2 and L_3 peaks, both of which should reflect the width of the unoccupied 3d band. This is depicted in Fig. 7.27(a) for four metals Ti, Cr, Ni and Cu. The observed L_3 width decreases from 4.6 eV in Ti to 3.0 eV in Ni in accordance with a decrease in the unoccupied band width, as shown in Fig. 7.27(b). It is noted that Cu has a filled 3d band and, hence, does not exhibit a sharp peak but only steps are seen at the two edges (see Fig. 7.26). The fine structure near the L_2 and L_3 edges in Cu has been discussed with reference to the unoccupied density of states predicted from band calculations [15]. It is also noted that the EELS technique is powerful to single out the unoccupied partial density states of alloys, since the electronic transition involved is specific to each constituent element.

7.10 Optical reflection and absorption spectra

The color of a metal, or more precisely the optical reflection or absorption spectrum, is closely related to its valence band structure. Let us consider why

7.10 Optical reflection and absorption spectra

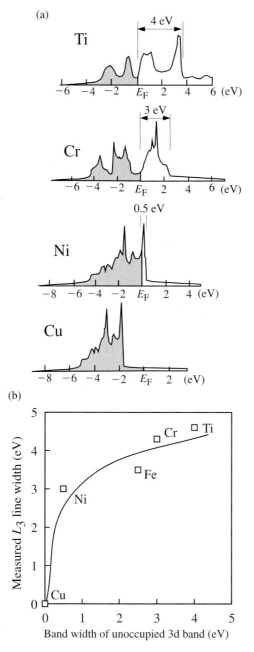

Figure 7.27. (a) Calculated 3d density of states in Ti, Cr, Ni and Cu metals. [D. A. Papaconstantopoulos, *Handbook of the Band Structure of Elemental Solids* (Plenum Press, New York, 1986)] The width of the unoccupied 3d band is marked. (b) Measured L_3 widths read from Fig. 7.26 as a function of the unoccupied 3d band width shown in (a).

186   7 Principles of measuring electronic structure-related phenomena

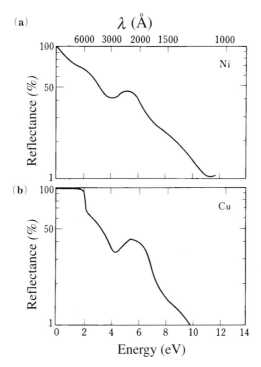

Figure 7.28. (a) Reflectance spectra for (a) pure Ni and (b) pure Cu metals. Both sets of data are shown on the same energy scale. Upper ordinate shows the wavelength converted from energy E through the relation $\lambda(\text{Å}) = 12\,396/E$ (eV). [F. Wooten, *Optical Properties of Solids* (Academic Press, 1972)]

pure Cu metal is red in color. Figure 7.28 indicates the reflection spectra for pure Cu and Ni. It is clear that radiation in the long-wavelength region down to 6000 Å is almost 100% reflected in Cu whereas a decrease in the reflectance is already quite noticeable in pure Ni in this low energy range. We see from Fig. 7.28 that the red color of pure Cu originates from the fact that radiation covering the visible red to infrared region is almost completely reflected from its surface.

As will be described in Section 11.9, conduction electrons in a metal can absorb an electromagnetic wave by resonating with the alternating electric field of the electromagnetic wave. This is absorption of light due to plasma oscillation. The absorption of the electromagnetic wave also occurs upon the interband transition of electrons in a solid. A combination of the plasma oscillation and the interband transition determines the reflection spectrum of a given metal.

The valence band structure of pure Cu was already shown in Fig. 6.6 and also in Fig. 7.16(a). The top of the 3d band lies at about 2 eV below the Fermi

7.10 Optical reflection and absorption spectra

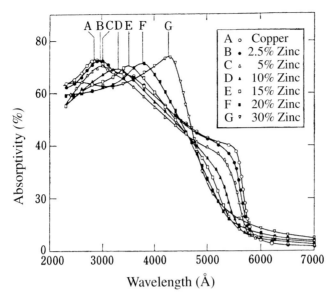

Figure 7.29. Optical absorption spectra of α-phase Cu–Zn alloys. The absorption edge displaces toward shorter wavelengths with increasing Zn content. [M. A. Biondi and J. A. Rayne, *Phys. Rev.* **115** (1959) 1522]

level. A rapid decrease in the reflectance above 2 eV has been attributed to the absorption due to the interband transition from the top of the 3d band to the Fermi level. In other words, electromagnetic waves having energies higher than 2 eV can be absorbed, since 3d electrons are allowed to jump to unoccupied states above the Fermi level. According to the band calculations, a critical wavelength is indeed found to be about 6000 Å. Radiation with wavelengths longer than this has too low an energy to excite the electrons and, hence, is reflected without being absorbed. In contrast, the Fermi level is located in the middle of the 3d band in pure Ni. Thus, there is no threshold in the interband transition and absorption takes place even when the photon energy is small. This is the reason why pure Ni exhibits a gradual decrease in reflectance with increasing photon energy without any threshold value.

Figure 7.29 shows the absorption spectra for a series of α-phase fcc Cu–Zn alloys. Here the absorptivity is defined as $(100 - R)$ in %, where R is the reflectance. In addition, the wavelength is taken as the horizontal axis. Hence, the one-to-one correspondence with Fig. 7.28 can be seen if both ordinate and abscissa are reversed. The absorption edge of pure Cu is located at about 5800 Å. The absorption edge is seen to displace gradually toward a shorter wavelength with increasing Zn content. This implies that visible light up to almost yellow in color begins to be reflected with the addition of Zn to Cu.

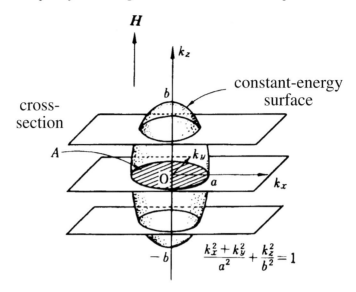

Figure 7Q.1. The Fermi surface of the ellipsoid of revolution about the k_z-axis. [Courtesy Dr Keiichi Ogawa]

Indeed, we know that a brass is yellowish in color, like gold. The electron concentration e/a increases with increasing Zn content. Hence, this results in an increase in the Fermi level and thereby increases the threshold energy beyond that of pure Cu. We see from the discussion above that measurements of optical absorption or reflection spectrum allow us to provide valuable information about the electronic structure of metals and alloys.

Exercises

7.1 The Schrödinger equation for a free electron in a magnetic field is expressed as

$$\frac{1}{2m}[\hbar\nabla + (-e)\mathbf{A}]^2 \psi = E\psi, \tag{7Q.1}$$

where $\mathbf{A}$ is the vector potential. If the magnetic field is applied along the z-axis so that $\mathbf{B} = (0, 0, B)$, then we can choose the vector potential as $\mathbf{A} = (0, xB, 0)$. Show that the Schrödinger equation (7Q.1) is reduced to that for a one-dimensional harmonic oscillator and its eigenvalue is expressed as equation (7.10).

7.2 Consider a metal whose Fermi surface is composed of an ellipsoid of revolution having the z-axis as its long axis. Apply the magnetic field along its z-axis, as shown in Fig. 7Q.1. Explain why the extremal cross-section of the

Fermi surface at $k_z=0$ normal to the magnetic field is obtained from the period of the dHvA periodic oscillations.

7.3 The free electron moving with the velocity **v** in the magnetic field **B** experiences the Lorentz force and rotates in closed orbit in the plane normal to the field. As is clear from Exercise 7.1, the angular frequency ω_c is expressed as $\omega_c = eB/m$ and the radius r of the cyclotron motion is given as

$$r = \frac{v}{\omega_c} = \frac{mv}{eB}. \tag{7Q.2}$$

We also learned from equation (7.10) that the energy separation in the adjacent Landau levels is equal to

$$\Delta E = \hbar\omega_c = \frac{e\hbar B}{m}. \tag{7Q.3}$$

Suppose that we measure the dHvA signals for pure Cu by applying a maximum magnetic field of 50 kOe or 5 tesla and that the Fermi energy of pure Cu is 7.0 eV, as given by the free-electron model (see Table 2.1).

(a) Calculate the radius r in equation (7Q.2) for the electron at the Fermi level and compare it with the mean free path of a conduction electron in high-purity Cu at 4 K. The latter is estimated from its residual resistivity of 7×10^{-9} Ω-cm (see footnote 10 in Chapter 15).
(b) Calculate the separation of the adjacent Landau levels in equation (7Q.3).
(c) Calculate the quantum number n of the Landau level corresponding to the Fermi energy.

7.4 The intensity of the Compton spectrum is given by equation (7.23). Show that it is reduced to equation (7.24) in the free-electron model.

Chapter Eight

Electronic structure calculations

8.1 Prologue

There exist as many conduction electrons as the Avogadro number in a molar quantity of an ordinary metal. Nevertheless, we have employed in previous chapters the one-electron Schrödinger equation as if only a single conduction electron were to exist in a metal, regardless of whether it is a free-electron or the Bloch electron. A conduction electron in a solid should interact not only with an array of positive ions but also with all the other electrons. Can we ignore the electron–electron interaction, which exerts a repulsive Coulomb force between conduction electrons? In this chapter, we begin with the Hamiltonian describing a whole assembly of electrons in the presence of the nuclear potential and discuss the validity of the so-called one-electron approximation. Following the justification of the one-electron approximation, we introduce several band calculation techniques. Our main aim in this chapter is to grasp characteristic features of each band calculation rather than the detailed derivations of equations.

8.2 One-electron approximation

The non-relativistic Schrödinger equation for a system of N electrons in the presence of nuclei at fixed positions is expressed as

$$-\frac{\hbar^2}{2m}\sum_{i=1}^{N}\left(\frac{\partial^2\Psi}{\partial x_i^2}+\frac{\partial^2\Psi}{\partial y_i^2}+\frac{\partial^2\Psi}{\partial z_i^2}\right)+V_{ee}\Psi+V_{en}\Psi=E\Psi, \qquad (8.1)$$

where V_{ee} is the Coulomb potential energy due to electron–electron interaction, V_{en} is that due to electron–nucleus interaction and E is the energy eigenvalue in this system. The Coulomb potential energy summed over N electrons is given by

8.2 One-electron approximation

$$V_{ee} = \frac{(-e)^2}{2} \sum_{i=1}^{N} \sum_{j \neq i}^{N} \frac{1}{|\mathbf{r}_i - \mathbf{r}_j|}, \qquad (8.2)$$

where $(-e)$ is the electronic charge and $\mathbf{r}_i$ is the position vector pointing to the i-th electron. A numerical coefficient $\frac{1}{2}$ appears, since the identical pairs (i,j) and (j,i) are independently counted. Similarly, the Coulomb potential energies over N electrons in the field due to the N_n nuclei with the atomic number Z_α is given by

$$V_{en} = -e^2 \sum_{i=1}^{N} \sum_{\alpha=1}^{N_n} \frac{Z_\alpha}{|\mathbf{r}_i - \mathbf{R}_\alpha|}, \qquad (8.3)$$

where $\mathbf{R}_\alpha$ is the position vector of the α-th nucleus.

Because of the presence of the electron–electron interaction energy V_{ee}, attempts to solve equation (8.1) rigorously are indeed difficult and known as many-body problems. In previous chapters, we adopted the one-electron approximation, in which each electron is treated as an independent particle in a periodic potential, by presuming that other conduction electrons do not affect its motion at all. This does not look to be easily justified. Nevertheless, all band calculations described below rely essentially on the one-electron approximation. Prior to the introduction of various band calculations, we consider in this section how the electron–electron interaction can be properly treated within the context of the one-electron approximation.

The wave function Ψ satisfying equation (8.1) depends on coordinates of all electrons and nuclei:

$$\Psi = \Psi(\mathbf{r}_1, \zeta_1, \mathbf{r}_2, \zeta_2, \ldots, \mathbf{r}_N, \zeta_N, \mathbf{R}_1, \mathbf{R}_2, \ldots, \mathbf{R}_{N_n}), \qquad (8.4)$$

where $\mathbf{r}_i$ and ζ_i are the space and spin coordinates of the i-th electron and $\mathbf{R}_i$ is the fixed coordinate of the i-th nucleus at absolute zero. The spin variable ζ_i takes on the two values $+1$ or -1 for spin-up and spin-down, respectively. The Pauli exclusion principle requires that the wave function (8.4) should be antisymmetric with respect to an interchange in the space and spin coordinates of any pair of electrons:

$$\Psi(\ldots, \mathbf{r}_i, \zeta_i, \ldots, \mathbf{r}_j, \zeta_j, \ldots) = -\Psi(\ldots, \mathbf{r}_j, \zeta_j, \ldots \mathbf{r}_i, \zeta_i, \ldots). \qquad (8.5)$$

It is clear from equation (8.5) that $\Psi = 0$ when $\mathbf{r}_i = \mathbf{r}_j$ and $\zeta_i = \zeta_j$. Thus, we see that the requirement of the antisymmetric wave function is equivalent to the Pauli exclusion principle, which states that not more than two electrons can share the same quantum state.

Let us assume that the minimum energy eigenvalue E_{min} is obtained as a solution of equation (8.1). The ground state energy E_0 of a system is given as

$E_{\min} + V_{nn}$, where V_{nn} is the nuclear potential energy. The cohesive energy of a solid discussed in Section 2.2 is then defined as

$$U = E_0 - \sum_{i=1}^{N_n} E_{0,i}, \qquad (8.6)$$

where $E_{0,i}$ is the ground-state energy of the i-th isolated atom [1].

The wave function in equation (8.4) involves the coordinates of all electrons. We cannot solve the Schrödinger equation in a many-electron system, unless some approximation is made in equation (8.4). The total wave function is assumed to be given by the product of the wave functions, each of which depends only on the coordinates of a given electron:

$$\Psi_H = \psi_1(\mathbf{r}_1)\chi_1(\zeta_1)\psi_2(\mathbf{r}_2)\chi_2(\zeta_2)\psi_3(\mathbf{r}_3)\chi_3(\zeta_3) \cdots \psi_N(\mathbf{r}_N)\chi_N(\zeta_N), \qquad (8.7)$$

where $\psi_i(\mathbf{r}_i)$ and $\chi_i(\zeta_i)$ represent the orbital wave function and spin function of the i-th electron, respectively and the suffix "H" represents Hartree, as will be described below.

Unfortunately, equation (8.7) does not satisfy the antisymmetry requirement of equation (8.5). For the moment, we will ignore this and try to solve equation (8.1) by using equation (8.7) as a trial function. According to the variational principle, the wave functions $\psi_i(\mathbf{r}_i)$ and $\chi_i(\zeta_i)$ are determined so as to minimize the expectation value of a total energy $\int \Psi_H^* H \Psi_H d\mathbf{r}_1 d\mathbf{r}_2 \cdots d\mathbf{r}_N d\zeta_1 d\zeta_2 \cdots d\zeta_N$. Since the Hamiltonian is independent of spin variables, its integration, or more precisely summation, is reduced to unity. Hence, only the integral over the position variables remains. It is shown [2] that the total energy is minimized when the wave function $\psi_i(\mathbf{r}_i)$ satisfies the following Schrödinger equation:

$$-\frac{\hbar^2}{2m}\nabla^2\psi_i + V_{H,i}(\mathbf{r}_i)\psi_i = \varepsilon_i\psi_i, \qquad (8.8)$$

where the potential function $V_{H,i}(\mathbf{r}_i)$ is expressed as

$$V_{H,i}(\mathbf{r}_i) = -e^2 \sum_{\alpha=1}^{N_n} \frac{Z_\alpha}{|\mathbf{r}_i - \mathbf{R}_\alpha|} + (-e)^2 \sum_{j \neq i}^{N} \int \frac{|\psi_j(\mathbf{r}_j')|^2}{|\mathbf{r}_i - \mathbf{r}_j'|} d\mathbf{r}_j'. \qquad (8.9)$$

Here the one-electron energy ε_i is independent of spin states.

Equation (8.8) with a choice of the potential (8.9) was first derived by Hartree in 1928 and is called the Hartree approximation. The first term on the right-hand side of equation (8.9) represents the Coulomb potential energy of the electron at position $\mathbf{r}_i$ arising from interaction with all the nuclei, while the second term represents the Coulomb energy of that electron arising from interaction with all the other electrons, whose probability density at the position $\mathbf{r}_j'$ is given by the absolute square of the wave function $\psi_j(\mathbf{r}_j')$. The integration is carried out over the coordinates of all electrons except the one at $\mathbf{r}_i$. Thus, the

8.2 One-electron approximation

Hartree potential means that the electron under consideration experiences the electrostatic field of the nuclei and a time-averaged electrostatic field created by all remaining electrons. We see that the Hartree field serves as the effective one-electron potential and that the Hartree approximation reduces a many-electron problem to the one-electron Schrödinger equation. The Hartree approximation is known as the best one-electron wave function in the framework of a single product form of wave functions for the ground state.

It is noted that the solution $\psi_j(\mathbf{r}'_j)$ of the Schrödinger equation (8.8) appears in the numerator of the second term in equation (8.9). We are therefore faced with the problem of solving an integro-differential equation self-consistently. First, a set of approximate wave functions $\psi_j(\mathbf{r}_j)$ is chosen as a trial function and the N potential functions are calculated by inserting them into equation (8.9). Then equation (8.8) becomes the ordinary one-electron Schrödinger equation, from which the wave function $\psi_j(\mathbf{r}_j)$ can be calculated. Once the eigenfunctions $\psi_j(\mathbf{r}'_j)$ are derived, we recalculate the potential functions and solve the Schrödinger equation again. A whole cycle of calculations is repeated until a satisfactory self-consistency is achieved.

Let us consider a metal in which N ions each carrying a positive charge ($+e$), are periodically arranged in a volume V. The free-electron model discussed in Chapter 3 is equivalent to replacing the first term in equation (8.9) by a uniform charge distribution with the density $(+e)N/V$. As a natural consequence, the distribution of conduction electrons also becomes uniform. This is called the jellium model. The absolute square of the wave function appearing in the integrand of the second term in equation (8.9) now becomes $|\psi_k(\mathbf{r})|^2 = 1/V$, as is clear from equation (2.13). Thus, the second term is reduced to $(-e)^2 N/V$ and cancels the first term. Therefore, we see that the free-electron model is a special case of the Hartree equation, in which the ionic potential is replaced by a uniform charge distribution.

Once the one-electron wave function $\psi_i(\mathbf{r}_i)$ is determined, the total wave function is given by the product in the form of equation (8.7). As mentioned above, the total wave function must be antisymmetric with respect to any interchange in the space and spin coordinates in accordance with the Pauli exclusion principle. The sign of the total wave function must change upon an odd number of interchanges in the space and spin coordinates. Such an antisymmetric wave function can be expressed as

$$\Psi_{HF} = \begin{vmatrix} \psi_1(\mathbf{r}_1)\chi_1(\zeta_1) & \psi_1(\mathbf{r}_2)\chi_1(\zeta_2) & \cdots & \psi_1(\mathbf{r}_N)\chi_1(\zeta_N) \\ \psi_2(\mathbf{r}_1)\chi_2(\zeta_1) & \psi_2(\mathbf{r}_2)\chi_2(\zeta_2) & \cdots & \psi_2(\mathbf{r}_N)\chi_2(\zeta_N) \\ \vdots & \vdots & \vdots & \vdots \\ \psi_N(\mathbf{r}_1)\chi_N(\zeta_1) & \psi_N(\mathbf{r}_2)\chi_N(\zeta_2) & \cdots & \psi_N(\mathbf{r}_N)\chi_N(\zeta_N) \end{vmatrix} \quad (8.10)$$

Equation (8.10) is called the Slater determinant. The new Schrödinger equation can be obtained if equation (8.10) is employed as a trial function in the variational principle. Now the following term newly appears as an extra potential energy [2]:

$$V_X(\mathbf{r}_i) = -(-e)^2 \sum_{j \neq i}^{N} \left[\int \frac{\psi_j^*(\mathbf{r}_j')\psi_i(\mathbf{r}_i')}{|\mathbf{r}_i - \mathbf{r}_j'|} d\mathbf{r}_j' \right] \psi_j(\mathbf{r}_j) \delta_{\chi_i \chi_j}, \qquad (8.11)$$

where the Kronecker-delta $\delta_{\chi_i \chi_j}$ takes either 1 or zero, depending on whether the spin states χ_i and χ_j are parallel or antiparallel to one another, respectively. Thus, this term remains finite only for electrons with parallel spins. Equation (8.11) is called the exchange integral. The one-electron Schrödinger equation (8.8) with the potential terms given by a sum of equations (8.9) and (8.11) is called the Hartree–Fock equation.

The Coulomb repulsive force acts between any pair of conduction electrons in a real metal and, hence, they cannot behave independently but tend to keep apart from one another. The repulsive electron–electron interaction is called the Coulomb correlation. In the Hartree equation, the electron–electron interaction is treated in such a way that each electron moves in an average field created by all remaining electrons. Thus, the Coulomb correlation is completely ignored. In the Hartree–Fock equation, an average field is employed in the same spirit as in the Hartree equation. However, the Pauli exclusion principle is included so that two electrons with parallel spins are not allowed to occupy the same position. In other words, a kind of repulsive interaction, which is different from the Coulomb correlation, is taken into account between electrons with parallel spins. This is called the exchange energy. Here it is important to remember that the Hartree-Fock equation exaggerates the role of the exchange interaction between electrons with parallel spins, since the Coulomb correlation is completely ignored. The difference between the Coulomb correlation and the exchange energy is referred to as the correlation energy. In other words, we should evaluate properly the Coulomb correlation either by incorporating the correlation energy to the Hartree–Fock solution or by adding both exchange and correlation energies to the Hartree solution.

The evaluation of the correlation energy is a formidable task because the Coulomb interaction decays only as $1/r$ and, hence, remains significant over many atomic distances. Bohm and Pines [3] showed that the Coulomb interaction in a homogeneous electron gas is reduced to the screened Coulomb interaction in the form of $\exp(-\lambda r)/r$, if the long-wavelength contribution is subtracted as the collective motion of electrons, which is known as the plasmon. It is noted that the screened Coulomb interaction is effective only over a short distance. The ground-state energy per electron for a homogeneous electron gas has been calculated by Nozieres and Pines [4] and the result is expressed as

$$\varepsilon_0 = \varepsilon_{\text{kin}} + \varepsilon_X + \varepsilon_C = \frac{2.21}{R_s^2} - \frac{0.916}{R_s} + (0.031\ln R_s - 0.115) \text{ [Ry]} \quad (8.12)$$

where the first two terms ε_{kin} and ε_X represent the kinetic and exchange energies arising from the Hartree–Fock solution, respectively, and the third term is newly derived as the correlation energy ε_C. The parameter R_s is the radius of a sphere containing one electron on average and related to the electron density ρ through the relation $\rho = (4\pi R_s^3/3)^{-1}$. The value of R_s is often expressed in units of the Bohr radius ($a_0 = \hbar^2/me^2 = 0.05292$ [nm] and is calculated as $R_s = 3.21$, 3.96 and 4.8 nm for Li, Na and K, respectively.[1] Equation (8.12) is known to hold well in the range $2 < R_s < 6$, in which the electron density of typical metals falls.

8.3 Local density functional method

Before discussing the local density functional method, we summarize the difficulties in the Hartree and Hartree–Fock approximations. Let us place one spin-up electron at $\mathbf{r} = 0$ in the N-electron system and consider the electron density $\rho(\mathbf{r})$ near $\mathbf{r} = 0$ caused by the remaining $(N-1)$ electrons. In the Hartree approximation, each electron propagates in a time-averaged field created by all the other electrons and, hence, $\rho(\mathbf{r})$ is constant, as illustrated in Fig. 8.1(a). In the Hartree–Fock approximation, the probability of finding a spin-up electron in the vicinity of $\mathbf{r} = 0$ becomes very small because of the Pauli exclusion principle. However, no such restriction exists for the spin-down electron. As a result, the electron distribution in the Hartree–Fock approximation depends on the direction of spin, as shown in Fig. 8.1(b). In a real metal, the Coulomb correlation between electrons should exist regardless of the direction of spin and a true distribution should be like that shown in Fig. 8.1(c).

Equation (8.12) takes into account both the exchange and correlation energies but is valid only for a homogeneous electron gas. The electron distribution in a real metal is no longer homogeneous because of the presence of ionic potentials. The local density functional or LDF method allows us to treat properly the Coulomb correlation within the one-electron approximation even in systems where the electron distribution is no longer homogeneous.

---

[1] In atomic units, we take $\hbar = 1$, $e^2 = 2$ and $m = \frac{1}{2}$. The length is then measured in units of $\hbar^2/me^2$ or 0.05292 nm corresponding to the radius of the 1s orbit of the hydrogen atom in the Bohr model. The energy is in the units of Rydbergs equal to $me^4/2\hbar^2$ or 13.6 eV corresponding to the ionization energy of the hydrogen atom.

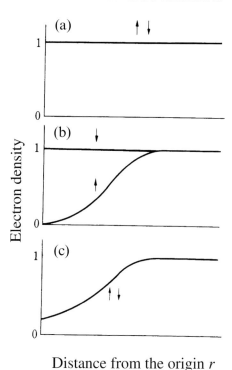

Figure 8.1. Electron density distribution formed by $(N-1)$ electrons when a spin-up electron is placed at $r=0$. (a) Hartree approximation, (b) Hartree–Fock approximation and (c) LDF approximation.

The LDF method was developed by Kohn and Sham in 1965 [5], and Schlüter and Sham [6]. Let us consider a metal in which the ionic potential and electron density at the position $\mathbf{r}$ are $V_{ion}(\mathbf{r})$ and $\rho(\mathbf{r})$, respectively.[2] The potential energy at the position $\mathbf{r}$ is obviously given by $\rho(\mathbf{r})V_{ion}(\mathbf{r})$. The total energy in the ground state of this system is expressed as

$$E=\int V_{ion}(\mathbf{r})\rho(\mathbf{r})d\mathbf{r}+F[\rho], \tag{8.13}$$

where $F[\rho]$ plays the role of the kinetic energy. Here it is assumed that $F[\rho]$ does not explicitly depend on the coordinate of each electron but only on the electron density $\rho(\mathbf{r})$ at the position $\mathbf{r}$. $F[\rho]$ is called the functional of the density $\rho(\mathbf{r})$. Hohenberg and Kohn [7] applied the variational principle and established a very important theorem, which states that the total energy of a system under a given external field $V(\mathbf{r})$ minimizes when a true electron density $\rho(\mathbf{r})$ is found.

---

[2] The ionic potential $V_{ion}(\mathbf{r})$ is obtained by replacing the atomic number Z_α by the valency Z in the first term of the Hartree potential given by equation (8.9). The value of Z is the difference between the nuclear charge and that of the core electrons.

8.3 Local density functional method

This theorem assures that, no matter what electron–electron interaction exists, a true electron density $\rho(\mathbf{r})$ would be determined by the variational principle even in the presence of the ionic potential $V_{\text{ion}}(\mathbf{r})$.

The functional $F[\rho]$ is reduced to the kinetic energy $T_s[\rho]$ of electrons in a non-interacting system or a system in the absence of the electron–electron interaction.[3] In contrast, the functional $F[\rho]$ in the interacting system is assumed to be given by a sum of $T_s[\rho]$ and the exchange and correlation energy $E_{\text{XC}}[\rho]$ [5]:

$$F[\rho] = T_s[\rho] + E_{\text{XC}}[\rho]. \tag{8.14}$$

Our next step is to derive the function $E_{\text{XC}}[\rho]$. Kohn and Sham considered the case, where $\rho(\mathbf{r})$ changes only slowly in space, and approximated $E_{\text{XC}}[\rho]$ as

$$E_{\text{XC}}[\rho] = \int \rho(\mathbf{r}) \varepsilon_{\text{XC}}(\rho) d\mathbf{r}, \tag{8.15}$$

were $\varepsilon_{\text{XC}}(\rho)$ is the sum of the exchange and correlation energies in a homogeneous electron gas with the density ρ. Note that the electron density dependence of $\varepsilon_{\text{XC}}(\rho)$ can be theoretically evaluated for a homogeneous electron gas as in equation (8.12). Hence, $E_{\text{XC}}(\rho)$ in the presence of $V(\mathbf{r})$ can be calculated, once we know $\rho(\mathbf{r})$. The usefulness of the local density functional method stems from the fact that the kinetic energy in the non-interacting system is still employed as a main term in the interacting system and the remaining contributions are all collected together into $E_{\text{XC}}[\rho]$.

A solution for minimizing the total energy can be found by using the local density $\rho(\mathbf{r})$ as a variable in the variational principle [5,6]. It has the form:

$$\int \delta \rho(\mathbf{r}) \left\{ \frac{\delta T_s[\rho]}{\delta \rho(\mathbf{r})} + V_{\text{ion}}(\mathbf{r}) + (-e)^2 \int \frac{\rho(\mathbf{r}')}{|\mathbf{r} - \mathbf{r}'|} d\mathbf{r}' + \mu_{\text{XC}}(\rho(\mathbf{r})) \right\} d\mathbf{r} = 0, \tag{8.16}$$

where

$$\mu_{\text{XC}}(\rho(\mathbf{r})) = \frac{d(\rho \varepsilon_{\text{XC}}(\rho))}{d\rho} \tag{8.17}$$

is called the chemical potential. The second and third terms in equation (8.16) represent the Hartree potential and the fourth term or equation (8.17) corresponds to the contributions from the exchange and correlation energies. In addition to equation (8.16), the following condition is imposed:

---

[3] The kinetic energy in the non-interacting system is given by

$$T_s[\rho] = \int \left(\int_0^{k_F} (\hbar^2 k^2 / 2m) 4\pi k^2 dk \bigg/ \int_0^{k_F} 4\pi k^2 dk \right) \rho(\mathbf{r}) d\mathbf{r} = (\hbar^2/m) \int \frac{3}{10} \left[3\pi^2 \rho(\mathbf{r}) \right]^{2/3} \rho(\mathbf{r}) d\mathbf{r}.$$

$$\int \delta\rho(\mathbf{r})d\mathbf{r} = 0, \qquad (8.18)$$

where the integration is carried out over the whole volume of a specimen. The Schrödinger equation derived from both equations (8.16) and (8.18) can be explicitly written as

$$\left\{-\frac{\hbar^2}{2m}\nabla^2 + V_{\text{ion}}(\mathbf{r}) + (-e)^2\int\frac{\rho(\mathbf{r})}{|\mathbf{r}-\mathbf{r}'|}d\mathbf{r}' + \mu_{\text{XC}}(\rho(\mathbf{r}))\right\}\psi_i(\mathbf{r}) = \varepsilon_i\psi_i(\mathbf{r}), \quad (8.19)$$

where the electron density $\rho(\mathbf{r})$ is given by

$$\rho(\mathbf{r}) = \sum_{i=1}^{N}|\psi_i(\mathbf{r})|^2. \qquad (8.20)$$

Equations (8.19) and (8.20) form two basic equations in the LDF method. In the derivation of its solution, the most appropriate $\rho(\mathbf{r})$ is first assumed and the fourth term in equation (8.19) is calculated by using the $\varepsilon_{\text{XC}}(\rho)$ for a homogeneous electron gas. Now the Schrödinger equation (8.19) can be solved. Once the wave function is obtained, a new electron density $\rho(\mathbf{r})$ is calculated from equation (8.20). The same procedure is repeated by using a newly derived $\rho(\mathbf{r})$ until a self-consistent electron density $\rho(\mathbf{r})$ is deduced.

The local density functional method turned out to be successful in dealing with the exchange and correlation energies of an interacting electron system and has been frequently employed to calculate the ground-state energy in various systems. The ground-state energy is expressed in the form [5]:

$$E = \sum\varepsilon_i - \frac{(-e)^2}{2}\int\int\frac{\rho(\mathbf{r})\rho(\mathbf{r}')}{|\mathbf{r}-\mathbf{r}'|}d\mathbf{r}d\mathbf{r}' + \int\rho(\mathbf{r})\{\varepsilon_{\text{XC}}[\rho(\mathbf{r})] - \mu_{\text{XC}}[\rho(\mathbf{r})]\}d\mathbf{r}. \quad (8.21)$$

Moruzzi et al. (1977) have calculated the cohesive energy of 26 elements in the periodic table via the LDF method and found a good agreement with the experimental data [8]. Lang and Kohn applied the LDF method to calculate the work function and surface energy in various metals and achieved a great success by using the one-electron approximation [9].

In the discussion above, the local electron density $\rho(\mathbf{r})$ is optimized without differentiating the direction of spins. Instead, spin-up and spin-down electrons can be independently treated so that magnetically polarized substances can equally be handled. This is called the local spin-density functional method. Its details are described elsewhere [10].

The exchange energy in equation (8.17), when the correlation energy is ignored, can be written as

$$\mu_x(\rho(\mathbf{r})) = \left(\frac{(-e)^2}{\pi}\right)[3\pi^2\rho(\mathbf{r})]^{1/3}, \qquad (8.22)$$

in the case of a homogeneous electron gas. In 1951, more than 10 years earlier than the development of the LDF method, Slater obtained a similar expression, which is $\frac{3}{2}$ times that of equation (8.22), and expressed the exchange energy in the form of

$$\mu_{X\alpha}(\rho(\mathbf{r})) = \frac{3\alpha}{2}\mu_X(\rho(\mathbf{r})) \tag{8.23}$$

by choosing the parameter α so as to be best appropriate to each substance. This technique is often called the $X\alpha$ method and has been employed for the calculations of the electronic structure of a cluster.

8.4 Band theories in a perfect crystal

A number of band calculation techniques for valence electrons in a periodic potential have been developed on the basis of the one-electron approximation discussed in Sections 8.2–8.3. The one-electron Schrödinger equation is written as

$$-\frac{\hbar^2}{2m}\nabla^2\psi(\mathbf{r}) + V(\mathbf{r})\psi(\mathbf{r}) = E\psi(\mathbf{r}), \tag{8.24}$$

where $V(\mathbf{r})$ is an effective one-electron potential consisting of the Hartree potential and the contribution $\mu_{XC}[\rho(\mathbf{r})]$ from the exchange and correlation energies. The potential $V(\mathbf{r})$ should satisfy the periodic conditions given by equations (5.4) and (5.5) in the case of a crystal. As a consequence, the wave function deduced from equation (8.24) must satisfy the Bloch theorem, which introduces the wave vector $\mathbf{k}$ through the relation

$$\psi_\mathbf{k}(\mathbf{r}+\mathbf{l}) = \exp(i\mathbf{k}\cdot\mathbf{l})\psi_\mathbf{k}(\mathbf{r}), \tag{8.25}$$

where $\mathbf{l}$ is the lattice vector given by equation (4.7). As has been stressed in Sections 5.3–5.4, the wave vector $\mathbf{k}$ plays such a role that, once the wave function at the position $\mathbf{r}$ is given, the wave function at the position $\mathbf{r}+\mathbf{l}$ is determined from equation (8.25). Since the lattice vector applies to all lattice sites in a crystal, this is equivalent to saying that the wave function is extended over the whole crystal. Thus, the energy eigenvalue in equation (8.24) is determined as a function of the wave vector $\mathbf{k}$ for a given crystal. Indeed, the band calculations aim at finding the energy eigenvalue of the Schrödinger equation (8.24) at as many states $\mathbf{k}$ as possible in the irreducible wedge within the Brillouin zone and, once this is done, the occupied and unoccupied density of states and the Fermi surface can be calculated.

The Bloch wave function is alternatively expressed in the form of

$$\psi_k(\mathbf{r}) = \exp(i\mathbf{k}\cdot\mathbf{r})u_k(\mathbf{r}), \tag{8.26}$$

where the function $u_k(\mathbf{r})$ is periodic with the period of a lattice. Hence, $u_k(\mathbf{r})$ can be expanded into the Fourier series:

$$u_k(\mathbf{r}) = \sum_{\mathbf{g}_n} a_k(\mathbf{g}_n)\exp(i\mathbf{g}_n\cdot\mathbf{r}), \tag{8.27}$$

where $\mathbf{g}_n$ is the reciprocal lattice vector of a given crystal. By inserting equation (8.27) into equation (8.26), we obtain

$$\psi_k(\mathbf{r}) = \sum_{\mathbf{g}_n} a_k(\mathbf{g}_n)\exp[i(\mathbf{k}+\mathbf{g}_n)\cdot\mathbf{r}]. \tag{8.28}$$

Equation (8.28) means that an arbitrary Bloch wave function can be expanded in terms of the plane waves having all possible reciprocal lattice vectors $\mathbf{g}_n$. Here the plane waves $\exp[i(\mathbf{k}+\mathbf{g}_n)\cdot\mathbf{r}]$ constitute a complete set of functions and the construction of any Bloch wave function by summing basis functions over all reciprocal lattice vectors as in equation (8.28) is called taking the Bloch sum.

The absolute square of the coefficient $|a_k(\mathbf{g}_n)|^2$ represents the probability density of finding the state of the wave vector $\mathbf{k}+\mathbf{g}_n$ at the position $\mathbf{r}$. As shown in Fig. 5.2, the wave function of the Bloch wave oscillates rapidly in the core region in order to be orthogonal to the wave functions of core electrons. Thus, an extremely large number of plane waves are needed to express such rapidly varying wave functions and the coefficient $a_k(\mathbf{g}_n)$ remains significant up to large reciprocal lattice vectors $\mathbf{g}_n$. Indeed, the nearly-free-electron (NFE) model in Section 5.5 employed plane waves as basis functions in the construction of the Bloch wave function. Obviously, it is not efficient for practical band calculations. We take different basis functions in the realistic band calculations, as will be described below.

8.5 Tight-binding method

The NFE method was formulated by Bethe in 1928. In the same year, Bloch developed a different approach and employed atomic orbitals rather than plane waves as basis functions. The revolving motion of the electron in a free atom is expressed by the Schrödinger equation:

$$-\frac{\hbar^2}{2m}\nabla^2\phi(\mathbf{r}) + [U_a(\mathbf{r}) - E]\phi(\mathbf{r}) = 0, \tag{8.29}$$

where $\phi(\mathbf{r})$ is the atomic orbital wave function and $U_a(\mathbf{r})$ is the potential of a free atom.

The motion of the conduction electron in a crystal is described by the Schrödinger equation:

$$-\frac{\hbar^2}{2m}\nabla^2\psi(\mathbf{r}) + [V(\mathbf{r}) - E]\psi(\mathbf{r}) = 0, \tag{8.30}$$

where $\psi(\mathbf{r})$ is the wave function of the conduction electron moving in a periodic potential $V(\mathbf{r})$. The wave function $\psi(\mathbf{r})$ in a crystal is constructed from the superposition of the atomic orbital wave functions in the form:

$$\psi_\mathbf{k}(\mathbf{r}) = \sum_\mathbf{l} e^{i\mathbf{k}\cdot\mathbf{l}}\phi(\mathbf{r} - \mathbf{l}), \tag{8.31}$$

where the Bloch wave vector $\mathbf{k}$ is introduced to allow the wave function to extend over a whole crystal. Indeed, one can easily confirm that equation (8.31) satisfies the Bloch condition $\psi(\mathbf{r}+\mathbf{l}) = \exp(i\mathbf{k}\cdot\mathbf{l})\psi(\mathbf{r})$. The summation in equation (8.31) runs over all lattice vectors in real space. The band calculation method by which the Bloch function is constructed from atomic orbital wave functions is called the tight-binding method or, alternatively, the linear combinations of atomic orbitals (LCAO) method.

The energy eigenvalue is obtained as an expectation value of the Hamiltonian in equation (8.30) and is calculated by inserting the wave function (8.31) into equation (8.30):

$$E = \int \psi_\mathbf{k}^* H \psi_\mathbf{k} d\mathbf{r} \Big/ \int \psi_\mathbf{k}^* \psi_\mathbf{k} d\mathbf{r}$$

$$= E_0 + \sum_n e^{i\mathbf{k}\cdot\mathbf{l}_n} \int \phi^*(\mathbf{r}+\mathbf{l}_n)[V(\mathbf{r}) - U_a(\mathbf{r})]\phi(\mathbf{r})d\mathbf{r}, \tag{8.32}$$

where E_0 satisfies the Schrödinger equation of a free atom:

$$\left[-\frac{\hbar^2}{2m}\nabla^2 + U_a(\mathbf{r}-\mathbf{l})\right]\phi(\mathbf{r}-\mathbf{l}) = E_0 \phi(\mathbf{r}-\mathbf{l}). \tag{8.33}$$

The integration in equation (8.32) should be carried out over the whole crystal. However, we assume that the overlap of the wave function is so small that it extends only up to the nearest neighbor atoms. Now equation (8.33) is easily reduced to

$$E = E_0 - \alpha - \gamma \sum_n e^{i\mathbf{k}\cdot\mathbf{R}_n}. \tag{8.34}$$

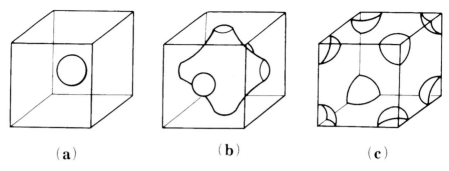

Figure 8.2. The equi-energy surface in the first Brillouin zone of the simple cubic lattice calculated in the tight-binding method: (a) spherical surface near the bottom of the band, (b) the band is half-filled and the Fermi surface makes partial contact with the face of the Brillouin zone, and (c) the Fermi surface when the first zone is almost filled. [From ref. 2.]

Here the summation counts only the nearest neighbor atoms around a given atom and the vector $\mathbf{R}_n$ points to the corresponding lattice sites. The coefficients α and γ are explicitly given by

$$\alpha = -\int \phi^*(\mathbf{r})[V(\mathbf{r}) - U_a(\mathbf{r})]\phi(\mathbf{r})d\mathbf{r} \tag{8.35}$$

and

$$\gamma = -\int \phi^*(\mathbf{r} + \mathbf{R}_n)[V(\mathbf{r}) - U_a(\mathbf{r})]\phi(\mathbf{r})d\mathbf{r}, \tag{8.36}$$

where α and γ are positive, since $[V(\mathbf{r}) - U_a(\mathbf{r})]$ is negative.

For simplicity, we take a simple cubic lattice with the lattice constant a. Since there are six nearest neighbor atoms at positions $\mathbf{R}_n = (\pm a\mathbf{i}, \pm a\mathbf{j}, \pm a\mathbf{k})$, equation (8.34) is reduced to the form:

$$E = E_0 - \alpha - 2\gamma(\cos k_x a + \cos k_y a + \cos k_z a). \tag{8.37}$$

It is clear that the energy is expressed as a function of the wave numbers k_x, k_y and k_z with the period $-\pi/a < k_x \leq \pi/a$, $-\pi/a < k_y \leq \pi/a$ and $-\pi/a < k_z \leq \pi/a$ and that the energy band is extended over the range 12γ. The cube in reciprocal space enclosed by $-\pi/a < k_x \leq \pi/a$, $-\pi/a < k_y \leq \pi/a$ and $-\pi/a < k_z \leq \pi/a$ corresponds to the first Brillouin zone. The energy given by equation (8.37) is found to be proportional to k^2, when the wave vector $\mathbf{k}$ is small.

The equi-energy surfaces of equation (8.37) in the first Brillouin zone are drawn in Fig. 8.2. The Fermi surface is seen to be spherical when $\mathbf{k}$ is small. But as $\mathbf{k}$ increases, the Fermi surface begins to bulge along three orthogonal directions and eventually touches the six equivalent zone planes (see Fig. 8.2(b)). A constant-energy surface always intersects the zone planes perpendicularly. This is easily understood, since the derivative dE/dk_x along the k_x-direction or the normal to the zone plane becomes zero at $k_x = \pm \pi/a$. It is noted that the k^2 dependence holds well up to the region close to $k_x = \pm \pi/a$ in the NFE model, but it holds only in the small-**k** region in the tight-binding model.

8.6 Orthogonalized plane wave method

In 1940, Herring proposed the band calculation technique called the orthogonalized plane wave or OPW method. He considered that a conduction electron would propagate nearly as a free-electron in the intermediate region between neighboring ions in a crystal and that the wave function in this region could be approximated by a single plane wave. On the other hand, the Bloch wave function should rapidly oscillate upon entering the core region, because it corresponds to the core state at higher energies and, hence, must be orthogonal to the wave functions of the existing core electrons. Thus, it is time-consuming to expand such a spatially rapidly changing wave function in terms of plane waves, since too many plane waves are needed. Herring proposed to employ as basis functions the "orthogonalized plane wave", which is defined as

$$X_{\mathbf{k}}(\mathbf{r}) = \frac{1}{\sqrt{N\Omega}} \exp(i\mathbf{k}\cdot\mathbf{r}) - \sum_j \mu_{\mathbf{k},j} \psi_{\mathbf{k},j}(\mathbf{r}) \tag{8.38}$$

where Ω is the volume per atom and $\psi_{\mathbf{k},j}(\mathbf{r})$ appearing in the second term refers to the Bloch wave function in equation (8.32) constructed from the atomic orbitals. The superscript j is used to differentiate the orbital of the core electron and, hence, corresponds to its quantum state, such as 1s, 2s, 2p, ... wave functions. The coefficient $\mu_{\mathbf{k},j}$ is determined so that the OPW wave function is orthogonal to that of each core electron.

We now construct a trial eigenfunction of the Schrödinger equation (8.24) by using as basis functions the OPW functions rather than the plane waves in the NFE model or the atomic orbitals in the LCAO model. As mentioned in Section 8.4, any Bloch wave function of the wave vector **k** can be constructed by a superposition of the basis functions. The OPW functions are chosen as more efficient basis functions and the Bloch sum is taken over all possible reciprocal lattice vectors:

$$\psi_{\mathbf{k}}(\mathbf{r}) = \sum_{\mathbf{g}_n} C(\mathbf{k}+\mathbf{g}_n) X_{\mathbf{k}+\mathbf{g}_n}(\mathbf{r}). \tag{8.39}$$

Obviously, the expansion in terms of the OPW functions would reduce the number of "waves" in comparison with the plane wave expansion in the NFE model or the atomic orbital expansion in the tight-binding method, in order to achieve the same accuracy for the eigenfunction. By inserting equation (8.39) into equation (8.24), we obtain a set of linear homogeneous equations for the coefficient $C(\mathbf{k}+\mathbf{g_n})$:

$$\sum_m \left\{ \left(|\mathbf{k}+\mathbf{g_n}|^2 - E\right)\delta_{mn} + U_{mn} \right\} C(\mathbf{k}+\mathbf{g_m}) = 0, \qquad (8.40)$$

where U_{mn} is given by

$$U_{mn}(E) = V_{mn} + \sum_i (\mu_{\mathbf{k}+\mathbf{g_m},i})^* \mu_{\mathbf{k}+\mathbf{g_n},i} (E - E_i). \qquad (8.41)$$

Note here that $(E - E_i)$ is always positive, since the energy E of the valence electron is always higher than the energy E_i of the core electron. The derivation of equations (8.40) and (8.41) can be done in the same manner as that of equation (5.34) in the NFE model.[4] The set of equations (8.40) has physically meaningful solutions only if the determinant of the matrix of the coefficients is zero. The energy eigenvalue E is determined for a given $\mathbf{k}$ as the roots of the determinantal equation. Since all the matrix elements are linear in energy, this equation can be solved by its diagonalization.

A superposition of only a few OPW functions has been proved sufficient to get reliable energy eigenvalues for certain types of solids. Indeed, attempts to improve the calculation by adding more OPWs have been often unsuccessful. The wave function of the core electron in a crystal is not identical to that of the atomic orbital in a free atom and, hence, the orthogonalization is not rigorous in a solid. The OPW method is more advantageous when the potential functions $V(\mathbf{r})$ overlap between neighboring atoms as in covalent solids like Si and Ge [1]. But it is less effective for transition metals, where the valence band is composed of both sp and d electrons. The motion of the d electrons cannot be well described by the OPW functions, since they are not so tightly bound to the nucleus but yet they cannot be treated as free-electrons. As will be discussed later, the APW and KKR methods are more appropriate in band calculations for transition metals like Cu and Fe.

8.7 Pseudopotential method

The pseudopotential method [11] is an extension of the OPW method. As has been discussed in the preceding section, the wave function of the conduction

---

[4] If we write $e^{i\mathbf{k}\cdot\mathbf{r}} A_{\mathbf{g_n}} \equiv C(\mathbf{k}+\mathbf{g_n})$ in equation (5.30), then equation (5.34) is rewritten as

$$\sum_m \left\{ \left(|\mathbf{k}+\mathbf{g_n}|^2 - E\right)\delta_{mn} + V_{mn} \right\} C(\mathbf{k}+\mathbf{g_m}) = 0$$

in the same form as equation (8.40). Here, δ_{mn} represents the "Kronecker-delta", defined as unity if $\mathbf{m}=\mathbf{n}$ and zero otherwise.

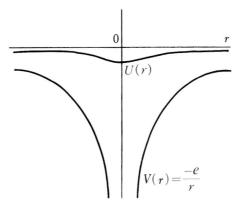

Figure 8.3. Bare Coulomb potential $V(r)$ and pseudopotential $U(r)$.

electron rapidly oscillates upon entering the core region and enhances its kinetic energy as a result of the orthogonalization to the wave functions of core electrons. The ionic potential in equation (8.24) is attractive to the conduction electron and, hence, its Fourier component V_{mn} in the first term of equation (8.41) is always negative. In contrast, the second term is positive and acts as a repulsive potential, as mentioned in Section 8.6. The $U_{mn}(E)$ in equation (8.41), being regarded as the Fourier component of an effective potential the conduction electron experiences in a crystal, becomes very small owing to the cancellation of the two competing terms. The Fourier transform of the $U_{mn}(E)$ into a real space is called the pseudopotential. The pseudopotential thus obtained is illustrated schematically in Fig. 8.3.

Band calculations based on the pseudopotential method have been performed by approximating the pseudopotential in an analytical form. For example, Ashcroft employed the so called empty-core model [11]. As illustrated in Fig. 8.4, the empty-core model assumes the Coulomb potential outside the radius R_M but a constant $-A_0$ inside R_M. The value of R_M is often chosen to be equal to the ionic radius and the depth of the well A_0 to be zero. In the NFE model, the ionic potential was expanded in terms of the plane waves (see equation (5.30)). In the case of the bare Coulomb potential, it is expanded as

$$-\frac{(-e)^2}{|\mathbf{r}_i - \mathbf{r}_j|} = \sum_{\mathbf{q}} V_{\mathbf{q}} e^{i\mathbf{q}\cdot|\mathbf{r}_i - \mathbf{r}_j|} \tag{8.42}$$

with its Fourier component

$$V_{\mathbf{q}} = -\frac{4\pi e^2}{q^2}. \tag{8.43}$$

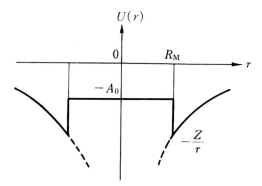

Figure 8.4. Pseudopotential in the empty-core model. [V. Heine and D. Weaire, *Solid State Physics*, **24** (1970) 249].

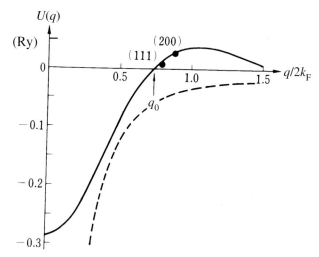

Figure 8.5. Pseudopotential $U(q)$ of pure Al. The abscissa represents the wave number normalized with respect to the Fermi diameter $2k_F$. The two solid circles indicate the values of $U(q)$ deduced from the experimental data at the (111) and (200) reciprocal lattice vectors. The dashed curve shows the bare Coulomb potential given by equation (8.43). The data for pure Al is reproduced from M. L. Cohen and V. Heine, *Solid State Physics*, **24** (1970) 37.

The Fourier transformed pseudopotential for pure Al is calculated using the empty-core model and reproduced in Fig. 8.5, along with the Fourier transform of the bare Coulomb potential given by equation (8.43). The bare Coulomb potential (dashed curve) becomes infinite at $q=0$ and gradually approaches zero while keeping negative values with increasing q. In contrast, the Fourier transform of the pseudopotential is finite at $q=0$, crosses zero at a value q_0 and approaches zero while oscillating about the horizontal axis with

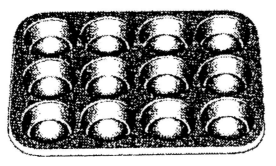

Figure 8.6. Muffin-tin pan. It illustrates well the muffin-tin potential in the APW method.

increasing q. These oscillations are not shown because of the limited range of q.

We noted in Section 5.5 that the energy gap across the Brillouin zone planes is proportional to the Fourier component of the ionic potential (see equation (5.39)). The first Brillouin zone of fcc Al is composed of the (111) and (200) reciprocal lattice vectors. Since the pseudopotential crosses zero near the reciprocal lattice vector (111), as shown in Fig. 8.5, the values of both U_{111} and U_{200} become very small. For example, the value of U_{111} for pure Al is determined as 0.018 Ry or 0.24 eV [11]. Such small values cannot be achieved if the bare Coulomb potential is used.

By using the pseudopotential approach, we can clearly demonstrate that the (111) and (200) reciprocal lattice vectors in fcc metals like Al and Pb fall close to the value of q_0 in the pseudopotential. This leads to fairly small energy gaps across these zone planes. Similar situations occur in other non-transition metals in different crystal structures [11]. This explains why the free-electron model works well for simple polyvalent metals in spite of the presence of ionic potentials.

8.8 Augmented plane wave method

The augmented plane wave or APW method was developed by Slater in 1937. The essence of this technique exists in the approximation of the ionic potential by the so-called muffin-tin potential, which is spherically symmetric within some radius about each lattice site and is constant outside. Its three-dimensional structure may be envisaged by a muffin-tin pan, shown in Fig. 8.6, which is a kitchen utensil used to bake cakes in an oven and has a group of connected cups which, in the figure, are about 5 cm in diameter and 2 cm in depth. A periodic array of cups simulates well the image of the muffin-tin potentials.

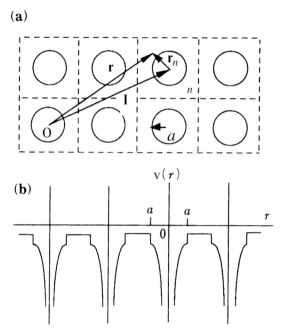

Figure 8.7. (a) Muffin-tin potential in the simple cubic lattice. A dashed square represents the Wigner–Seitz cell. The origin O is taken at an arbitrary lattice site in the lattice. Both lattice vector **l** and position vector **r** are measured relative to the origin O. (b) The muffin-tin potential is spherically symmetric in the range $r \leq a$ and constant outside.

Figure 8.7 illustrates schematically the array of the muffin-tin potentials for a crystal consisting of only one atom species. As mentioned above, the muffin-tin potential in a given cell is spherically symmetric within the region $r \leq a$ and, hence, is expressed as

$$v(\mathbf{r}) = v(|\mathbf{r}|), \tag{8.44}$$

whereas the potential is constant in the region $r > a$. Here the radius a is chosen in such a way that the spherically symmetric potential at a given lattice site does not overlap with the neighboring one. In other words, the diameter $2a$ must be smaller than the edge length of the Wigner–Seitz cell shown by a dashed square in Fig. 8.7(a). The constant potential in the range $r > a$ is called the muffin-tin zero and often is set equal to zero.

Let us consider the muffin-tin potential $v(\mathbf{r} - \mathbf{l})$ at the lattice vector **l**, where the position vector **r** of the electron is measured from the origin O. Now the total potential each electron experiences at the position **r** must be summed over all lattice vectors in a crystal and is expressed as

8.8 Augmented plane wave method

$$V(\mathbf{r}) = \sum_{\mathbf{l}} v(\mathbf{r} - \mathbf{l}). \tag{8.45}$$

For the moment, we consider only a single muffin-tin potential at $\mathbf{l} = 0$. The potential in the region $r \leq a$ maintains spherical symmetry and, hence, the wave function in the Schrödinger equation (8.24) is exactly solved as

$$\psi_{lm}(\mathbf{r}) = Y_{\ell m}(\theta, \phi) R_{\ell}(r) \tag{8.46}$$

in spherical coordinates. Here $Y_{\ell m}(\theta, \phi)$ is a spherical harmonic with the azimuthal quantum number ℓ and magnetic quantum number m. The (θ, ϕ) represents the direction of the vector $\mathbf{r}$. The radial function $R_\ell(r)$ is a solution of the differential equation

$$-\frac{1}{r^2}\frac{d}{dr}\left(r^2 \frac{dR_\ell(r)}{dr}\right) + \left[\frac{\ell(\ell+1)}{r^2} + \left(\frac{2m}{\hbar^2}\right)(V(r) - E)\right]R_\ell(r) = 0, \tag{8.47}$$

where $R_\ell(r)$ must be regular at the origin so that the relation $\lim_{r \to 0} r R_\ell(r) = 0$ should hold.

An augmented plane wave or APW function is now constructed for a system where muffin-tin potentials are periodically arranged. Since the spherical symmetry of the potential is maintained in the region $r \leq a$ at every lattice site, the wave function in the region $r \leq a$ may well be expressed as

$$\chi(E, \mathbf{r}) = \sum_{\ell=0}^{\infty} \sum_{m=-\ell}^{\ell} a_{\ell m} R_\ell(E, r) Y_{\ell m}(\theta, \phi), \tag{8.48}$$

where the azimuthal quantum number ℓ is, in practice, taken up to about 10 or 12 in the APW band calculations. In the region $r \geq a$, the electron sees a constant potential and, thus, the solution of the Schrödinger equation should be written as

$$\chi(\mathbf{k}, \mathbf{r}) = \exp(i\mathbf{k} \cdot \mathbf{r}), \tag{8.49}$$

where $\mathbf{k}$ is the wave vector of the plane wave.

Equation (8.49) is expanded in the spherical harmonics as

$$e^{i\mathbf{k} \cdot \mathbf{r}} = 4\pi \sum_{\ell=0}^{\infty} \sum_{m=-\ell}^{\ell} i^\ell j_\ell(kr) Y^*_{\ell m}(\theta_\mathbf{k}, \phi_\mathbf{k}) Y_{\ell m}(\theta, \phi), \tag{8.50}$$

where $j_\ell(kr)$ is the spherical Bessel function of order ℓ and both (θ, ϕ) and $(\theta_\mathbf{k}, \phi_\mathbf{k})$ represent the polar angles of the vectors $\mathbf{r}$ and $\mathbf{k}$, respectively. Equations (8.48) and (8.49) must be continuous across the boundary $r = a$. By equating equations (8.48) and (8.50), we obtain

$$a_{\ell m} = 4\pi i^\ell Y^*_{\ell m}(\theta_\mathbf{k}, \phi_\mathbf{k}) j_\ell(ka) R_\ell(E, a). \tag{8.51}$$

Note that the wave function becomes continuous across the boundary $r = a$ but there is a possible jump in the first derivative of the wave function there.

The wave function described by equations (8.48) and (8.49) with the coefficients given by equation (8.51) is called the APW function. However, the periodicity of the lattice has not yet been considered. The wave vector **k** in equation (8.49) must be replaced by the Bloch wave vector or $\mathbf{k} + \mathbf{g}_n$, where $\mathbf{g}_n$ is the reciprocal lattice vector. As has been emphasized, any Bloch wave function of a given wave vector **k** can be expanded in terms of basis functions with all possible wave vectors $\mathbf{k} + \mathbf{g}_n$. Here the APW functions are chosen as basis functions. In other words, we need to take the Bloch sum of the APW functions in the same spirit as the OPW method discussed in Section 8.6. A trial Bloch wave function $\psi_k(\mathbf{r})$ is, therefore, written as

$$\psi_k(\mathbf{r}) = \sum_{\mathbf{g}_n} C(\mathbf{k}+\mathbf{g}_n) \chi_{\mathbf{k}+\mathbf{g}_n}(E,\mathbf{r}), \tag{8.52}$$

where the Bloch condition $\chi_k(E,\mathbf{r}) = \sum_\mathbf{l} e^{i\mathbf{k}\cdot\mathbf{l}} \chi(E, \mathbf{r} - \mathbf{l})$ holds.

An insertion of equation (8.52) into equation (8.24) leads to a set of linear homogeneous equations concerning the coefficient $C(\mathbf{k}+\mathbf{g}_n)$:

$$\sum_m \left\{ \left(|\mathbf{k}+\mathbf{g}_n|^2 - E\right)\delta_{mn} + F_{mn} \right\} C(\mathbf{k}+\mathbf{g}_m) = 0. \tag{8.53}$$

This equation is found to be of the same form as equation (8.40) in the OPW method and also equation (5.34) in the NFE method. The Fourier component V_{mn} is now replaced by F_{mn} in the APW method. The function F_{mn} is given by

$$F_{mn}(E) = \frac{4\pi a^2}{\Omega} \{-[(\mathbf{k}+\mathbf{g}_m)\cdot(\mathbf{k}+\mathbf{g}_n) - E]\} \frac{j_\ell(|\mathbf{g}_m - \mathbf{g}_n|a)}{|\mathbf{g}_m - \mathbf{g}_n|}$$

$$+ 4\pi \sum_{\ell m} Y_{\ell m}(\mathbf{k}+\mathbf{g}_m)^* j_\ell(|\mathbf{k}+\mathbf{g}_m|a) \frac{R'_\ell(E,a)}{R_\ell(E,a)} j_\ell(|\mathbf{k}+\mathbf{g}_n|a) Y_{\ell m}(\mathbf{k}+\mathbf{g}_n). \tag{8.54}$$

The physical picture of equation (8.54) is less clear than that of equation (8.41) in the OPW method. It looks as if it does not depend on the muffin-tin potential v(**r**). But it does, since the logarithmic derivative of the radial wave function $R'_\ell(E,a)/R_\ell(E,a)$ in the second term of equation (8.54) is decided by the muffin-tin potential v(**r**).

The determinant of the coefficient $C(\mathbf{k}+\mathbf{g}_n)$ must be zero in order to yield a physically meaningful non-zero solution. As opposed to the OPW or NFE methods, all the matrix elements depend on the energy E either explicitly as in equation (8.53) or implicitly through equation (8.54). Thus, the secular determinant can be computed for a given value of E, which can then be varied until a root is found. Though the non-linear problem requires a great deal of computing time, the APW method has been proved to be very powerful in allowing accurate band calculations for various metals including transition metals like Fe and Cu.

8.9 Korringa–Kohn–Rostoker method

In 1947, Korringa proposed a band calculation technique based on the multiple scattering theory but no practical applications were made at the time because of the lack of large computers and less acquaintance with the multiple scattering theory. In 1954, Kohn and Rostoker [12] proved that the derivation of the energy eigenvalue from the multiple scattering theory is equivalent to that derived from the variational principle. Since then, it has been realized that the application of the multiple scattering theory provides as rigorous a foundation as other band calculations based on the variational principle and that it is substantially efficient from the computational point of view. This technique is called the Korringa–Kohn–Rostoker or KKR method. The APW and KKR methods are now recognized as more accurate techniques in band calculations.

We have two basic equations in the multiple scattering theory:

$$\psi(\mathbf{r}) = \chi(\mathbf{r}) + \sum_n \psi_n^o(\mathbf{r}) \tag{8.55}$$

and

$$\psi_n^i(\mathbf{r}) = \chi(\mathbf{r}) + \sum_{p \neq n} \psi_p^o(\mathbf{r}). \tag{8.56}$$

The first equation (8.55) implies that the total wave function $\psi(\mathbf{r})$ can be given by the unperturbed wave function $\chi(\mathbf{r})$ in the absence of the scatterers plus the sum of the waves $\psi_n^o(\mathbf{r})$ propagating outward from all the scatterers. Similarly, the wave function $\psi_n^i(\mathbf{r})$ incident to the n-th scatterer is given by the unperturbed wave function $\chi(\mathbf{r})$ plus the sum of the waves $\psi_p^o(\mathbf{r})$ propagating outward from all scatterers except the n-th scatterer. One can easily find from these two equations that the total wave function is also expressed as $\psi(\mathbf{r}) = \psi_n^i(\mathbf{r}) + \psi_n^o(\mathbf{r})$.

Let us suppose the wave function $\psi(\mathbf{r}')$ to be scattered at the position $\mathbf{r}'$ by the ionic potential $v(\mathbf{r}' - \mathbf{l})$ with the lattice vector $\mathbf{l}$. For the moment, we assume that a single ionic potential exists at the position $\mathbf{l}$. Now the strength of the scattering of the wave function $\psi(\mathbf{r}')$ by the potential $v(\mathbf{r}' - \mathbf{l})$ would be proportional to $v(\mathbf{r}' - \mathbf{l})\psi(\mathbf{r}')$. In the context of the multiple scattering theory, the wave function $\psi(\mathbf{r})$ at the position $\mathbf{r}$ is considered to be formed by combining the wavelets of strength $v(\mathbf{r}'-\mathbf{l})\psi(\mathbf{r}')$ generated at all different points $\mathbf{r}'$. To formulate this, we introduce the Green function $G(\kappa,\mathbf{r}-\mathbf{r}')$, which is defined as a quantity transmitting the wave functions $\psi(\mathbf{r}')$ at all different positions $\mathbf{r}'$ to the wave function $\psi(\mathbf{r})$ at the position $\mathbf{r}$ through the scattering due to the potential $v(\mathbf{r}'-\mathbf{l})$ and write down in the following form:

$$\psi(\mathbf{r}) = \int G(\kappa,\mathbf{r}-\mathbf{r}')v(\mathbf{r}'-\mathbf{l})\psi(\mathbf{r}')d\mathbf{r}', \tag{8.57}$$

where κ^2 is the kinetic energy E in atomic units.

Equation (8.57) describes the motion of the electron when scattered by the potential $v(\mathbf{r}'-\mathbf{l})$. Hence, it must be related to the corresponding Schrödinger equation. As will be shown below, the Green function in equation (8.57) has to be explicitly written as

$$G(\kappa,\mathbf{r}-\mathbf{r}') = -\frac{1}{4\pi}\frac{\exp(i\kappa|\mathbf{r}-\mathbf{r}'|)}{|\mathbf{r}-\mathbf{r}'|}, \tag{8.58}$$

in order to reconcile equation (8.57) with the Schrödinger equation (8.24).

Let us consider in general the relation $\mathscr{L}\psi(\mathbf{r}) = F(\mathbf{r})$, where $\mathscr{L}$ is an arbitrary linear operator. This relation states that a function $F(\mathbf{r})$ is generated, when $\mathscr{L}$ is operated to a function $\psi(\mathbf{r})$. The Green function $G(\mathbf{r},\mathbf{r}')$ is defined in such a way that the delta function is generated under the operation $\mathscr{L}$ or $\mathscr{L}G(\mathbf{r},\mathbf{r}') = \delta(\mathbf{r}-\mathbf{r}')$. The Schrödinger equation (8.24) is rewritten in atomic units as

$$\{\nabla^2 + \kappa^2\}\psi(\mathbf{r}) = V(\mathbf{r})\psi(\mathbf{r}), \tag{8.59}$$

where the energy E is replaced by $E = \kappa^2$. One can easily show that $\mathscr{L}G(\mathbf{r},\mathbf{r}') = \delta(\mathbf{r}-\mathbf{r}')$ holds, if $\mathscr{L} = \nabla^2 + \kappa^2$ and the Green function is given by equation (8.58) (see Exercise 8.2).

Let us multiply by the Green function $G(\kappa,\mathbf{r}-\mathbf{r}')$ on both sides of equation (8.59) and integrate both sides with respect to $\mathbf{r}'$ over a whole space;

$$\int \{\nabla^2 + \kappa^2\} G(\kappa,\mathbf{r}-\mathbf{r}')\psi(\mathbf{r}')d\mathbf{r}' = \int G(\kappa,\mathbf{r}-\mathbf{r}')V(\mathbf{r}')\psi(\mathbf{r}')d\mathbf{r}'. \tag{8.60}$$

This immediately results in equation (8.57), since the left-hand side is obviously reduced to the wave function $\psi(\mathbf{r})$. We have confirmed in this way that equation (8.57) is indeed the Schrödinger equation in the integral form.

The argument above has been limited to the scattering due to a single ionic potential. It is easily extended to a periodic lattice, where identical muffin-tin potentials given by equation (8.44) are periodically arranged:

$$V(\mathbf{r}) = \sum_{\mathbf{l}} v(\mathbf{r}-\mathbf{l}).$$

8.9 Korringa–Kohn–Rostoker method

The wave function in the periodic lattice is characterized by the Bloch wave vector **k**. Equation (8.57) is therefore extended to the case in a crystal:

$$\psi_{\mathbf{k}}(\mathbf{r}) = \sum_{\mathbf{l}} \int G(\kappa, \mathbf{r} - \mathbf{r}') v(\mathbf{r}' - \mathbf{l}) \psi_{\mathbf{k}}(\mathbf{r}') d\mathbf{r}'$$

$$= \sum_{\mathbf{l}} \int G(\kappa, \mathbf{r} - \mathbf{r}') v(\mathbf{r}' - \mathbf{l}) \exp(i\mathbf{k} \cdot \mathbf{l}) \psi_{\mathbf{k}}(\mathbf{r}' - \mathbf{l}) d\mathbf{r}', \quad (8.61)$$

where the second line is obtained by inserting the Bloch condition (5.14). Equation (8.61) means that the wave function at the position **r** is determined by the sum of the contributions from wavelets scattered from all equivalent muffin-tin potentials with the lattice vectors **l**. Equation (8.61) is often rewritten as

$$\psi_{\mathbf{k}}(\mathbf{r}) = \int G(\kappa, \mathbf{k}; \mathbf{r} - \mathbf{r}'') v(\mathbf{r}'') \psi_{\mathbf{k}}(\mathbf{r}'') d\mathbf{r}'', \quad (8.62)$$

where $G(\mathbf{k}, \mathbf{k}; \mathbf{r} - \mathbf{r}'')$ is called the structure Green function [13] and is defined as

$$G(\kappa, \mathbf{k}; \mathbf{r} - \mathbf{r}'') \equiv \sum_{\mathbf{l}} G(\kappa, \mathbf{r} - (\mathbf{r}'' + \mathbf{l})) \exp(i\mathbf{k} \cdot \mathbf{l})$$

$$= -\frac{1}{4\pi} \sum_{\mathbf{l}} \frac{\exp(i\kappa |\mathbf{r} - (\mathbf{r}'' + \mathbf{l})|)}{|\mathbf{r} - (\mathbf{r}'' + \mathbf{l})|} \exp(i\mathbf{k} \cdot \mathbf{l}). \quad (8.63)$$

The structure Green function is convenient to describe the motion of electrons in a periodic crystal and serves as transmitting all the wavelets at the position **r''** scattered from periodic potentials with lattice vectors **l** to the position **r**.

The Bloch wave function $\psi_{\mathbf{k}}(\mathbf{r})$ can be obtained as a solution of the Schrödinger equation (8.62) in a periodic potential $V(\mathbf{r})$. Equation (8.62) may be solved by using the variational principle. According to Ziman [13], both sides of equation (8.62) are multiplied by $\psi_{\mathbf{k}}^*(\mathbf{r}) v(\mathbf{r})$ with subsequent integraton over the variable **r**. A functional Λ is defined as

$$\Lambda = \int \psi_{\mathbf{k}}^*(\mathbf{r}) v(\mathbf{r}) \psi_{\mathbf{k}}(\mathbf{r}) d\mathbf{r}$$

$$- \sum_{\mathbf{l}} \int \int \psi_{\mathbf{k}}^*(\mathbf{r}) v(\mathbf{r}) G(\kappa, \mathbf{k}; \mathbf{r} - \mathbf{r}'') v(\mathbf{r}'') \psi_{\mathbf{k}}(\mathbf{r}'') d\mathbf{r} d\mathbf{r}''. \quad (8.64)$$

The eigenfunction is found by minimizing Λ. Here we need to integrate only over a single Wigner–Seitz cell, because the muffin-tin potential is zero outside each atomic sphere.

Since the muffin-tin potential is spherically symmetric inside each cell, a trial Bloch function $\psi_{\mathbf{k}}(\mathbf{r})$ is conveniently expressed as

$$\psi_{\mathbf{k}}(\mathbf{r}) = \sum_{\ell m} C_{\mathbf{k},\ell m} R_\ell(E,r) Y_{\ell m}(\theta,\phi), \qquad (8.65)$$

in the same form as equation (8.48). By inserting equation (8.65) into equation (8.64), we obtain a quadratic function in the coefficient $C_{\ell m}$:

$$\Lambda = \sum_{\ell m; \ell' m'} \Lambda_{\ell m; \ell' m'} C_{\ell m} C^*_{\ell' m'} \qquad (8.66)$$

and

$$\Lambda_{\ell m; \ell' m'} = S_{\ell m; \ell' m'} + \kappa \delta_{\ell \ell'} \delta_{mm'} \frac{n'_\ell - n_\ell L_\ell}{j'_\ell - j_\ell L_\ell}, \qquad (8.67)$$

where j_ℓ and n_ℓ are ℓ-th order spherical Bessel functions and spherical Neumann functions, respectively, and $\delta_{\ell \ell'}$ and $\delta_{mm'}$ are Kronecker-deltas. L_ℓ indicates the logarithmic derivative of the radial wave function $R_\ell(r)$ at the muffin-tin radius. Both real and imaginary parts of the complex variable $C_{\ell m}$ are determined so as to minimize the functional Λ. This leads to the determinantal equation $|\Lambda_{\ell m; \ell' m'}| = 0$. By solving this equation, we obtain the band structure, i.e., the wave vector dependence of the energy E.

Finally, we will briefly discuss the first and second terms in equation (8.67). The first term $S_{\ell m; \ell' m'}$ emerges from the Green function in equation (8.64). As is clear from equation (8.63), it does not depend on the wave function but contains information only about the crystal structure, since it takes the sum over the lattice vector **l**. The parameter $S_{\ell m; \ell' m'}$ is called the structure factor and is a function of κ (or energy E) and the wave vector **k**. Once it is tabulated for a given crystal structure, it does not need be calculated again. The second term involves the logarithmic derivative L_ℓ. Since the product $v(\mathbf{r})\psi_\mathbf{k}(\mathbf{r})$ in the first term of equation (8.64) can be replaced by $\{\nabla^2 + \kappa^2\}\psi(\mathbf{r})$, the volume integral is transformed to a surface integral by using the Green theorem. As a result, both the derivative of the wave function and the Green function appear. This explains why L_ℓ appears in the second term of equation (8.67). The characteristic features of a given material are brought in through the logarithmic derivative L_ℓ of the radial wave function. This is because L_ℓ depends on the muffin-tin potential, as is seen from equation (8.54).

The KKR method is alternatively called the Green function method. As is clear from the argument above, the muffin-tin potential is used in the same way as in the APW method. The APW method relies on solving a secular equation for a trial wave function obtained by superimposing the APW functions over reciprocal lattice vectors (see equation (8.52)), whereas the KKR method takes the lattice sum in real space first (see equation (8.63)) and then solves the determinantal equation. The advantage of the KKR method lies in a complete separation of the structure-dependent term $S_{\ell m;\ell'm'}$ from the muffin-tin potential dependent term, thereby contributing to the enhanced efficiency in computation. The KKR method is known to be very accurate for metals but less so for covalent crystals.

8.10 LMTO method

The band calculation techniques so far discussed are divided into two main approaches; one uses a trial wave function, which is formed as linear combinations of basis functions like plane waves in the NFE method and orthogonalized plane waves in the OPW method; and the other expands the wave function into a set of energy-dependent partial waves and applies a matching condition for partial waves at the muffin-tin sphere like the APW and KKR methods. Both approaches have their advantages and disadvantages.

In the former, the application of the variational principle to the Schrödinger equation leads to the eigenvalue equations:

$$(\underline{H} - E\underline{O}) \cdot \mathbf{a} = \mathbf{0} \text{ or } \sum_{\ell'm'} (H^k_{l'm'lm} - E^k_j O^k_{l'm'lm}) a^k_{l'm'lm} = 0, \qquad (8.68)$$

where $[\underline{H}]$ and $[\underline{O}]$ are the Hamiltonian and overlap matrices and $\mathbf{a}$ is a vector consisting of the expansion coefficients. An advantage of using energy-*independent* basis functions as in the NFE and LCAO methods is that all the matrix elements in the secular determinant are *linear* in energy, allowing a fast computation in solving the secular equation. However, a proper choice of the basis functions is important to have a sufficiently small basis set.

In the latter, it is possible to solve the Schrödinger equation *exactly* in terms of the energy-dependent partial-wave expansions like equation (8.48) in the APW method. A matching condition of equation (8.51) eventually leads to equation (8.54) in the APW method. The way in which the matching condition is formulated depends on the partial-wave methods chosen, but the result is reduced to a set of linear homogeneous equations:

$$\underline{M} \cdot \mathbf{a} = \mathbf{0} \text{ or } \sum_{\ell'm'} \left[P_{\ell'm',\ell m}(E) - S^k_{\ell'm',\ell m}(E) \right] a^k_{\ell'm',\ell m} = 0, \qquad (8.69)$$

where the secular matrix $\underline{M}$ has a complicated, *non-linear* energy dependence in contrast to equation (8.68). As a consequence, the partial-wave methods, even for moderately sized matrices, require more computer time than the eigenvalue problem in the fixed-basis approach. However, the partial-wave methods are advantageous in providing solutions of an arbitrary accuracy for a given muffin-tin potential and, as has been emphasized earlier, information about the potential enters the secular equation only through the logarithmic derivative of the wave function at the muffin-tin sphere.

The APW and KKR methods are capable of calculating accurately the band structures for a wide class of materials but have required substantial computational effort to obtain truly self-consistent calculations. The Linear-Muffin-Tin Orbital or LMTO method was developed by Andersen and his group [14, 15] for solving the band structure problem in more efficient and physically more transparent ways. The LMTO method is constructed so as to combine the desirable features of the fixed-basis and partial-wave methods and has been recognized as one of the most efficient techniques to determine the one-electron band structure in crystalline solids with a fast computation time without sacrificing the accuracy.

The outline of the LMTO method will be described by following the scheme developed by Skriver [16]. To begin with, the KKR method combined with the Atomic Sphere Approximation or the KKR–ASA method is discussed to facilitate the understanding of the LMTO method. The Wigner–Seitz polyhedral cell is replaced by an atomic sphere having the same volume. The resulting sphere with the radius r_0 is called the Wigner–Seitz sphere or atomic sphere. This is illustrated in Fig. 8.8. One of the centers is taken as its origin $r=0$. As noted above, the radial wave function $R_\ell(E,r)$ is the exact solution of the Schrödinger equation (8.47), since the spherically symmetric muffin-tin potential is assumed. Now the muffin-tin orbital in the atomic sphere is defined as

$$\chi_{\ell m}(E,\mathbf{r}) = i^\ell Y_{\ell m}(\theta,\phi) \left[R_\ell(E,r) + p_\ell(E) \left(\frac{r}{r_0} \right)^\ell \right] \quad (r<r_0), \quad (8.70)$$

where $Y_{\ell m}(\theta,\phi)$ is the spherical harmonic already appearing in equation (8.46). The second term is ingeniously added, the reason of which will be discussed below.

The potential outside the sphere $r>r_0$ is assumed to be zero and, hence, the tail of the muffin-orbital must be expressed as $(r_0/r)^{\ell+1}$ there. This is a solution of the Laplace equation $\nabla^2 \chi = 0$, which is equivalent to the Schrödinger equation having zero kinetic and potential energies.[5] Thus, the muffin-tin orbital in the region $r>r_0$ can be written as

$$\chi_{\ell m}(E,\mathbf{r}) = i^\ell Y_{\ell m}(\theta,\phi) \left(\frac{r_0}{r} \right)^{\ell+1} \quad (r>r_0). \quad (8.71)$$

The function $p_\ell(E)$ in the second term of equation (8.70) is determined so as to make both (8.70) and (8.71) continuous across $r=r_0$. The result is given by

---

[5] Put $\kappa^2=0$ and $V(\mathbf{r})=0$ in equation (8.59). The radial wave function can be expressed by a linear combination of the spherical Bessel and spherical Neumann functions in spherical corrdinates (see equation (8.87) and also Section 13.8). In the region where r is small, it is approximated as $A(r/r_0)^\ell + B(r_0/r)^{\ell+1}$.

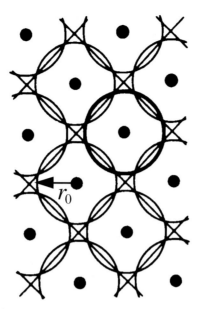

Figure 8.8. Wigner–Seitz cell and its replacement by a Wigner–Seitz sphere with a radius r_0. [From ref. 16].

$$p_\ell(E) = \frac{D_\ell(E) + \ell + 1}{D_\ell(E) - \ell}, \qquad (8.72)$$

where

$$D_\ell(E) = \frac{r_0}{R_\ell(E, r_0)} \cdot \left.\frac{\partial R_\ell(E, r)}{\partial r}\right|_{r=r_0}. \qquad (8.73)$$

This is the matching condition for the partial waves and $D_\ell(E)$ is indeed a quantity similar to L_ℓ in equation (8.67) in the KKR method.

Now we consider the contribution from all the other atomic spheres in a crystal. All the tails arising from the atomic spheres located at the lattice vector **l** are summed up over the whole crystal and the value at the position **r** is reduced to

$$\sum_{\mathbf{l} \neq 0} e^{i\mathbf{k}\cdot\mathbf{l}} \left(\frac{r_0}{|\mathbf{r}-\mathbf{l}|}\right)^{\ell+1} i^\ell Y_{\ell m}(\theta, \phi), \qquad (8.74)$$

where **k** is introduced as the Bloch wave vector. Equation (8.74) can be expanded in a power series about $r = 0$:

$$\sum_{\ell' m'} \frac{-1}{2(2\ell' + 1)} \left(\frac{r}{r_0}\right)^{\ell'} i^{\ell'} Y_{\ell' m'}(\theta, \phi) S^{\mathbf{k}}_{\ell' m', \ell m}, \qquad (8.75)$$

where $S^{\mathbf{k}}_{\ell' m', \ell m}$ is the structure factor which appeared in equation (8.67).

A final wave function extending over the crystal may be constructed by taking the lattice sum of equation (8.70):

$$\sum_{\ell m} a^{j\mathbf{k}}_{\ell m} \sum_{\mathbf{l}} e^{i\mathbf{k}\cdot\mathbf{l}} \chi_{\ell m}(E,\mathbf{r}-\mathbf{l}), \tag{8.76}$$

where the superscript j in the coefficient $a^{j\mathbf{k}}_{\ell m}$ represents the band index. In the region $r < r_0$ of each atomic sphere, the first term $i^\ell Y_{lm}(\theta, \phi) R_\ell(E,r)$ in equation (8.70) is a correct solution to the Schrödinger equation. Thus, the wave function inside the sphere in a crystal should be generalized as

$$\sum_{\ell m} a^{j\mathbf{k}}_{\ell m} i^\ell Y_{\ell m}(\theta,\phi) R_\ell(E,r). \tag{8.77}$$

However, there exists the contribution arising from the second term in equation (8.70), which remains finite inside the sphere at $r=0$. This is explicitly written as

$$\sum_{\ell m} a^{j\mathbf{k}}_{\ell m} i^\ell Y_{\ell m}(\theta,\phi) p_\ell(E) \left(\frac{r}{r_0}\right)^\ell. \tag{8.78}$$

Now the reason for the addition of the second term in equation (8.70) is clear. Equation (8.78) is needed in order to cancel the contribution of equation (8.75) arising from the tails of all the other muffin-tin orbitals. This is the principal idea in the multiple scattering theory employed in the KKR method. The condition for the tail cancellation leads to a set of linear, homogeneous equations:

$$\sum_{\ell m} \left[P_\ell(E) \delta_{\ell'\ell} \delta_{m'm} - S^{\mathbf{k}}_{\ell'm',\ell m} \right] a^{j\mathbf{k}}_{\ell m} = 0, \tag{8.79}$$

where $P_\ell(E)$ is obviously given by

$$P_\ell(E) = 2(2\ell+1) p_\ell(E). \tag{8.80}$$

The determinant of the coefficients $a^{j\mathbf{k}}_{\ell m}$ in equation (8.79) must be zero to have physically meaningful solutions:

$$\det[P_\ell(E) \delta_{\ell'\ell} \delta_{m'm} - S^{\mathbf{k}}_{\ell'm',\ell m}] = 0. \tag{8.81}$$

The first term in equation (8.81) depends only on the potential function $P_\ell(E)$, which is uniquely determined from the muffin-tin potential in an atomic sphere. The second term involves only the structure factor $S^{\mathbf{k}}_{\ell'm',\ell m}$ and is uniquely determined, once a crystal structure is given. Hence, the characteristic feature of the KKR method is preserved in this treatment.

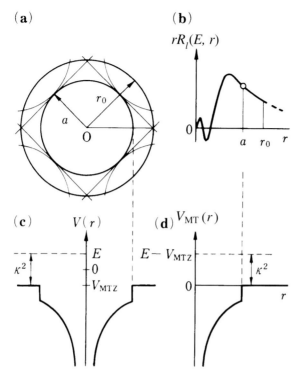

Figure 8.9. (a) Wigner–Seitz atomic sphere with radius r_0 and muffin-tin sphere with radius a. Neighboring Wigner–Seitz spheres are partially drawn so as to make contact with the muffin-tin sphere at the center. (b) The radial wave function. (c) Muffin-tin approximation for the ionic potential. (d) Muffin-tin potential defined by equations (8.82) and (8.83). [From ref. 16].

The KKR–ASA method has employed the energy-dependent orbital (8.70) and results in the non-linear energy dependence in $P_\ell(E)$. To proceed further with our discussion, we introduce the muffin-tin potential shown in Fig. 8.9. A spherically symmetric potential exists in a muffin-tin sphere with a radius a and a constant potential V_{MTZ} called the muffin-tin zero in the intermediate region between the neighboring muffin-tin potentials.

We assume electrons to propagate in the intermediate region with a constant wave number $\kappa = \sqrt{E - V_{MTZ}}$. The muffin-tin potential in the LMTO is defined as

$$V_{MT}(r) = V(r) - V_{MTZ} \qquad (r \leq a) \qquad (8.82)$$

and

$$V_{\text{MT}}(r) = 0 \quad (r > a), \tag{8.83}$$

in which $V(r)$ is spherically symmetric. Note that V_{MTZ} is negative in sign. As can be seen in Fig. 8.9(d), $V_{\text{MT}}(r)$ becomes zero in the region $r > a$ but the electron now possesses the kinetic energy $\kappa^2 = E - V_{\text{MTZ}}$.

The Schrödinger equation in the presence of a single muffin-tin potential is expressed as

$$[-\nabla^2 + V_{\text{MT}}(r) - \kappa^2]\psi_{\ell,m}(E,\mathbf{r}) = 0. \tag{8.84}$$

Its solution $\psi_{\ell,m}(E,\mathbf{r})$ is written as

$$\psi_{\ell,m}(E,\mathbf{r}) = i^\ell Y_{\ell m}(\theta,\phi) R_\ell(E,r), \tag{8.85}$$

where $R_\ell(E,r)$ in the region $r \leq a$ satisfies the radial Schrödinger equation

$$\left[-\frac{d^2}{dr^2} + \frac{\ell(\ell+1)}{r^2} + V_{\text{MT}}(r) - \kappa^2\right] r R_\ell(E,r) = 0. \tag{8.86}$$

$R_\ell(E,r)$ in the region $r > a$, where $V_{\text{MT}}(r) = 0$, is the solution of

$$\left[-\frac{d^2}{dr^2} + \frac{\ell(\ell+1)}{r^2} - \kappa^2\right] r R_\ell(E,r) = 0. \tag{8.87}$$

Equation (8.87) is called the Helmholtz wave equation and its solution is given by a linear combination of the spherical Bessel function $j_\ell(\kappa r)$ and the spherical Neumann function $n_\ell(\kappa r)$. The wave function in the region $r > a$ is expanded in terms of the phase-shifted spherical waves in contrast to the plane wave expansion (8.49) in the APW method.

Based on the arguments above, we newly redefine the muffin-tin orbitals as

$$\chi_{\ell m}(E,\kappa,\mathbf{r}) = i^\ell Y_{\ell m}(\theta,\phi) \kappa n_\ell(\kappa r). \quad (r > a) \tag{8.88}$$

and

$$\chi_{\ell m}(E,\kappa,\mathbf{r}) = i^\ell Y_{\ell m}(\theta,\phi)[R_\ell(E,r) + \kappa \cot(\eta_\ell) j_\ell(\kappa r)] \quad (r \leq a), \tag{8.89}$$

where η_ℓ is the phase shift of the ℓ-th partial wave. The coefficient $\cot(\eta_\ell)$ is chosen in such a way that the partial wave is everywhere continuous and differentiable. The muffin-tin orbitals above are determined so as to be regular both at the origin and infinity.[6] They are a generalization of equations (8.70) and (8.71), which were derived under the condition $\kappa^2 = 0$. As noted in footnote 11 (p. 406) in Chapter 13, the spherical Bessel function $j_\ell(\kappa r)$ and the spherical

---

[6] The radial part of equation (8.88) must be of the form $\kappa[n_\ell(\kappa r) - \cot(\eta_\ell) j(\kappa r)]$. But in order to avoid its divergence at $r \to \infty$, the term $+\kappa \cot(\eta_\ell) j(\kappa r)$ is added. This makes the muffin-tin orbital to be regular at infinity and introduces the term $\kappa \cot(\eta_\ell) j(\kappa r)$ in equation (8.89).

8.10 LMTO method

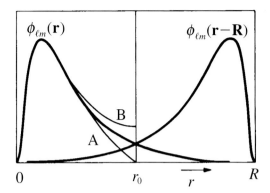

Figure 8.10. $\phi_{lm}(r)$ represents the atomic orbital. The bonding and antibonding molecular orbitals in a homonuclear diatomic molecule are marked as B and A, respectively. r_0 denotes the cell boundary. Molecular orbitals are shown only in the left cell. [From ref. 16.]

Neumann function $n_\ell(\kappa r)$ are approximated as $(r/r_0)^\ell$ and $(r/r_0)^{-\ell-1}$, respectively, when κr is small. Indeed, equations (8.88) and (8.89) are reduced to equations (8.70) and (8.71) when $\kappa r \ll 1$. We realize that $p_\ell(E)$ in equation (8.70) is replaced by $\cot(\eta_\ell)$ in equation (8.89).

The muffin-tin orbitals (8.88) and (8.89) are still energy dependent and, hence, are computationally inefficient. It is necessary to employ a fixed-basis set which leads to a computationally more efficient eigenvalue problem like equation (8.68). Andersen replaced the energy variable E in $R_\ell(E,r)$ by a *fixed but arbitrary energy* E_v and constructed energy-*independent* muffin-tin orbitals composed of the partial radial wave function $R_\ell(E_v,r)$ and their first energy derivative $\dot{R}_\ell(E_v,r)$. We need to know why the energy-independent muffin-tin orbital can be constructed from the sum of $R_\ell(E_v,r)$ and its energy derivative $\dot{R}_\ell(E_v,r)$ [15].

To facilitate the discussion, a homonuclear diatomic molecule is considered. As shown in Fig. 8.10, two atomic orbitals $\phi_{\ell m}(\mathbf{r})$ are positioned at $\mathbf{r}=0$ and $\mathbf{r}=\mathbf{R}$. The bonding and antibonding molecular orbitals may be approximated by using a linear combination of the atomic orbitals (LCAO):

$$\psi_{\text{bonding}}(E_B, \mathbf{r}) \approx \phi_{\ell m}(\mathbf{r}) + (-1)^\ell \phi_{\ell m}(\mathbf{r} - \mathbf{R}) \tag{8.90}$$

and

$$\psi_{\text{antibonding}}(E_A, \mathbf{r}) \approx \phi_{\ell m}(\mathbf{r}) - (-1)^\ell \phi_{\ell m}(\mathbf{r} - \mathbf{R}), \tag{8.91}$$

where E_B and E_A are energies of the bonding and antibonding states.

The muffin-tin potential is spherically symmetric in the region well inside the core in a diatomic molecule. Hence the bonding and antibonding molecular wave functions would be exactly given by

$$\psi_{bonding}(E_B, \mathbf{r}) = \sum_{\ell m} a_{\ell m} R_{\ell m}(E_B, r) Y_{\ell m}(\theta, \phi) \tag{8.92}$$

and

$$\psi_{antibonding}(E_A, \mathbf{r}) = \sum_{\ell m} a'_{\ell m} R_{\ell m}(E_A, r) Y_{\ell m}(\theta, \phi) \tag{8.93}$$

Since the atomic orbital $\phi_{\ell m}(\mathbf{r})$ is reduced to $[\psi_{bonding}(\mathbf{r}) + \psi_{antibonding}(\mathbf{r})]/2$ from equations (8.90) and (8.91), the molecular orbital in a given cell should be augmented as

$$\left(\frac{1}{2}\right)\sum_{\ell m} a_{\ell m}[R_{\ell m}(E_B, r) + R_{\ell m}(E_A, r)] Y_{\ell m}(\theta, \phi) \approx \sum_{\ell m} a_{\ell m} R_{\ell m}(E_v, r) Y_{\ell m}(\theta, \phi), \tag{8.94}$$

where E_v is some energy between E_B and E_A. Similarly, the atomic orbital $\phi(\mathbf{r} - \mathbf{R})$ in the cell is given by $[(-1)^\ell/2][\psi_{bonding}(\mathbf{r}) - \psi_{antibonding}(\mathbf{r})]$ and contributes as a tail at $\mathbf{r} = 0$. Hence, the molecular orbital given by equation (8.94) should be further augmented by the amount:

$$\left[\frac{(-1)^\ell}{2}\right] \sum_{\ell m} a_{\ell m} [R_{\ell m}(E_B, r) - R_{\ell m}(E_A, r)] Y_{\ell m}(\theta, \phi)$$

$$\approx \sum_{\ell m} a_{\ell m} \left.\frac{\partial R_{\ell m}(E, r)}{\partial E}\right|_{E_v} (E - E_v) Y_{\ell m}(\theta, \phi), \tag{8.95}$$

In a solid, a continuous band is formed in the energy range over E_B and E_A and the corresponding wave function is constructed by all possible combinations of bonding and antibonding states between nearest neighbors throughout a crystal. We consider the energy-dependent radial wave function $R_{\ell m}(E, r)$ at any lattice site to be constructed from the sum of the head of the energy-independent orbital and the tails of the energy-independent orbitals from all the other sites in a crystal. In other words, the augmented muffin-tin orbital $R_{\ell m}(E, r)$ is effectively replaced by the sum of equation (8.94) and (8.95) and is expressed as

$$R_{\ell m}(E, r) \approx R_{\ell m}(E_v, r) + (E - E_v) \dot{R}_{\ell m}(E_v, r). \tag{8.96}$$

Here we see that equation (8.96) is the first two terms of the Taylor expansion of $R_{\ell m}(E, r)$. The muffin-tin orbitals thus obtained can be made continuous and differentiable everywhere and orthogonal to the wave functions of the core electrons.

Because of the replacement of the energy E by a fixed energy E_v in the radial wave function, the LMTO method is no longer exact for the muffin-tin potential but the error in the energy is shown to be only of fourth order in the difference $E_j^k - E_v$. The Bloch wave is constructed by taking the Bloch sum of the energy-independent muffin-tin orbitals. A trial function is given by the linear combination of muffin-tin orbitals thus obtained. A use of the variational principle in conjunction with the energy-independent muffin-tin orbitals results in a secular equation of the form (8.68), linear in energy. The energy eigenvalue can be computed by diagonalizing the secular equation in the form of equation (8.68). Readers who need more detailed information about the LMTO method should consult the literature [14–16].

Exercises

8.1 Equation (8.37) is derived for the simple cubic lattice in the tight-binding method. We consider the eight nearest neighbor atoms in the lattice with the lattice constant a. Their atom positions are denoted as $\mathbf{R}_n = (\pm 1/2a, \pm 1/2a, \pm 1/2a)$. Show that equation (8.37) is reduced to

$$E = E_0 - \alpha - 8\gamma \cos\tfrac{1}{2} k_x a \cdot \cos\tfrac{1}{2} k_y a \cdot \cos\tfrac{1}{2} k_z a \tag{8Q.1}$$

8.2 Show that the Green function $G(\kappa, \mathbf{r} - \mathbf{r}') = -(1/4\pi)\exp(i\kappa|\mathbf{r} - \mathbf{r}'|)/|\mathbf{r} - \mathbf{r}'|$ satisfies the Schrödinger equation $\{\nabla^2 + \kappa^2\} G(\kappa, \mathbf{r} - \mathbf{r}') = \delta(\mathbf{r} - \mathbf{r}')$. Here, $\delta(\mathbf{r} - \mathbf{r}')$ represents the delta function, which goes to infinity when $\mathbf{r} = \mathbf{r}'$ and zero otherwise.

Chapter Nine
Electronic structure of alloys

9.1 Prologue

We have so far discussed the electronic structure of metals and semiconductors existing as elements in the periodic table and have assumed a crystal to be ideally perfect without containing any defects like impurities, vacancies, dislocations and grain boundaries. In reality, no metal is perfectly free from such defects, which certainly disturb the periodicity of the lattice and cause scattering of the Bloch electron. Foreign elements can be intentionally added to a given metal, resulting in the formation of an alloy. When the amount of the added element is dilute, the added atoms may be treated as impurities. But when its concentration exceeds several atomic %, the interaction among the added atoms is no longer neglected. In this chapter, we discuss first the effect of an impurity atom on the electronic structure of a host metal and then move on to discuss the electronic structure of concentrated alloys.

9.2 Impurity effect in a metal

Let us consider first a perfect metal crystal consisting of the atom A with the valency Z_1. All atoms become positive ions with the valency $+Z_1$ by releasing the outermost Z_1 electrons per atom to form the valence band. As a result, conduction electrons carrying negative charges are uniformly distributed over any atomic site with equal probability densities and maintain charge neutrality with the array of ions with positive charges. Now we replace the atom A at a given lattice site by the atom B with valency Z_2 ($Z_2 > Z_1$). Effectively, a point charge equal to $\Delta Z = Z_2 - Z_1$ is formed at the atom B and the uniform charge distribution is disrupted.

The excess potential at a distance r away from a point charge e is given by $e^2 \Delta Z/r$ in vacuum. But this is no longer true in a metal, where freely moving

9.2 Impurity effect in a metal

conduction electrons are attracted by the positive point charge at the atom B and thereby it is largely screened. The excess potential is known to be reduced to

$$U(r) = \frac{e^2 \Delta Z \exp(-\lambda r)}{r}, \tag{9.1}$$

where $1/\lambda$ is called the screening radius. We will describe below the derivation of equation (9.1) in the context of the so-called Thomas–Fermi approximation.

The density of conduction electrons $\rho(r)$ at distance r from the atom B certainly deviates from the average density $\rho_0(r)$ existing prior to its introduction. The following Poisson equation holds in the vicinity of atom B:

$$\nabla^2 U(r) = -4\pi e^2 [\rho(r) - \rho_0(r)], \tag{9.2}$$

where $U(r)$ represents the impurity potential caused by atom B with its excess charge. The Fermi energy E_F is related to $\rho_0(r)$ through equation (2.21) in a pure metal. Because of the presence of the potential $U(r)$, equation (2.21) is modified such that

$$E_F + U(r) = \left(\frac{\hbar^2}{2m}\right)[3\pi^2 \rho(r)]^{2/3} \tag{9.3}$$

or

$$\rho(r) = \left(\frac{1}{3\pi^2}\right)\left(\frac{2m}{\hbar^2}\right)^{3/2} E_F^{3/2} \left[1 + \frac{U(r)}{E_F}\right]^{3/2}. \tag{9.4}$$

The term $[1 + (U(r)/E_F)]^{3/2}$ can be expanded in a series, since $U(r) \ll E_F$. Accordingly, equation (9.4) is approximated as

$$\rho(r) = \rho_0 \left[1 + \frac{3}{2}\left(\frac{U(r)}{E_F}\right)\right]. \tag{9.5}$$

An insertion of equation (9.5) into equation (9.2) yields the relation

$$\nabla^2 U(r) = \lambda^2 U(r), \tag{9.6}$$

where λ is given by

$$\lambda = \left(\frac{6\pi e^2 \rho_0}{E_F}\right)^{1/2}. \tag{9.7}$$

Since $U(r)$ in equation (9.6) is spherically symmetric and involves only the radial distance r as a variable, it is reduced to the form:

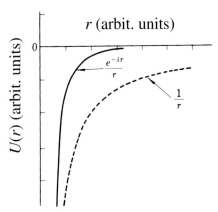

Figure 9.1. Bare Coulomb potential (dashed curve) and screened Coulomb potential (solid curve).

$$\frac{1}{r^2}\frac{d}{dr}\left(r^2\frac{dU(r)}{dr}\right) = \lambda^2 U(r). \tag{9.8}$$

One can easily check that equation (9.1) is the solution of equation (9.8). The screened potential is shown as a solid curve in Fig. 9.1 in comparison with a bare Coulomb potential (dashed curve). It can be seen that the range of the potential the impurity atom exerts is substantially reduced due to the screening by the conduction electrons.[1] The parameter λ is often called the Thomas–Fermi screening parameter.

Consider the case where a single Zn atom is introduced into Cu metal. Obviously, each Cu atom becomes a Cu^{+1} ion by releasing one 4s electron per atom to the valence band. The Zn impurity atom possessing two outermost 4s electrons would become a Zn^{+2} ion by releasing these two electrons. Hence, a positive point charge with the excess valency $\Delta Z = 1$ remains on the Zn atom. The magnitude of the screening radius $1/\lambda$ turns out to be about 0.055 nm, if the values of $E_F = 7$ eV and $\rho_0 \approx 8.5 \times 10^{28}/m^3$ appropriate to pure Cu are inserted into equation (9.7). One can realize how effective the screening effect is, since it is much shorter than the interatomic distance of 0.255 nm in pure Cu.

9.3 Electron scattering by impurity atoms and the Linde law

The screening potential, equation (9.1), is formed in a metal when an impurity atom is introduced. The periodic potential is thereby disturbed and conduction

---

[1] As is clear from equation (9.7), a large number of freely moving conduction electrons must be present in order to screen the excess charge of an impurity ion. This is true in a metal. In the case of a semiconductor, such a screening effect does not occur because of the scarcely populated conduction band and, instead, the impurity level or donor level is formed in the energy gap (see Section 6.9).

electrons are scattered. As will be shown in Section 10.7, only electrons at the Fermi level contribute to the resistivity. Let us assume elastic scattering so that the conduction electron with wave vector **k** on the Fermi surface is scattered into the state of **k**′ on the Fermi surface by the impurity atom. Thus, the relation $|\mathbf{k}|=|\mathbf{k}'|=k_F$ holds, where k_F is the Fermi radius. The differential cross-section upon scattering is expressed as

$$\sigma(\theta) = \left(\frac{2m\Delta Ze^2}{\hbar^2}\right)^2 \frac{1}{(K^2+\lambda^2)^2}, \qquad (9.9)$$

where K represents the magnitude of the scattering vector $\mathbf{K}=\mathbf{k}'-\mathbf{k}$.[2]

The number of impurity atoms per unit volume, N_{imp}, is assumed to be so low that the interaction between the impurities can be neglected. By using the resistivity formula (10.7), we can write the excess resistivity due to the impurity scattering as

$$\Delta\rho = \frac{m}{ne^2\tau} = \frac{mv_F}{ne^2\Lambda}, \qquad (9.10)$$

where Λ is the mean free path of the conduction electron given by the product of the relaxation time τ and the Fermi velocity v_F. Since the mean free path of the conduction electron is given by $1/\Lambda = 2\pi \int_0^\pi (1-\cos\theta)\sigma(\theta)\sin\theta d\theta$ (see Section 10.10), equation (9.10) is rewritten as

$$\Delta\rho = \frac{2\pi N_{imp} m v_F}{ne^2} \int_0^\pi (1-\cos\theta)\left(\frac{2m\Delta Ze^2}{\hbar^2}\right)^2 \frac{\sin\theta d\theta}{(K^2+\lambda^2)^2}$$

$$= \frac{2\pi N_{imp} m v_F}{ne^2}\left(\frac{2m\Delta Ze^2}{\hbar^2\lambda^2}\right)^2 \int_0^1 \frac{8z^3 dz}{[1+(2k_F/\lambda)^2z^2]^2}, \qquad (9.11)$$

where n is the density of conduction electrons in the host metal. The variable θ represents an angle between the wave vectors **k** and **k**′ and is called the scattering angle. The last line in equation (9.11) can be easily obtained by using the relation $|\mathbf{K}|=2k_F\sin(\theta/2)$, which holds in the case of elastic scattering, as shown in Fig. 10.6.

---

[2] The differential cross-section of the conduction electron due to the impurity potential $U(r)$ is expressed in the Born approximation as $\sigma(\theta)=(m/2\pi\hbar^2)^2|U(K)|^2$, where $U(K)$ is the Fourier transform of the excess potential given by $U(K)=\iiint U(\mathbf{r})e^{-i\mathbf{K}\cdot\mathbf{r}}d\mathbf{r}$. An insertion of equation (9.1) leads to $U(K)=-4\pi\Delta Ze^2/(K^2+\lambda^2)$. See more details in L. I. Schiff, "Quantum Mechanics", (McGraw-Hill, New York, 1955), Chapter 8, Section 30.

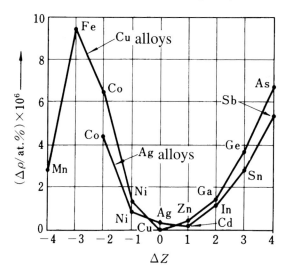

Figure 9.2. The Linde law. The ΔZ dependence of the increment of resistivity upon adding 1 at.% of various elements (as indicated) to Cu and Ag. ΔZ is the difference in valency between the solute and the noble metal matrix. The value of ΔZ is taken to be 0, −1, −2 and −3, for Ni, Co, Fe and Mn, respectively. [F. Seitz, *The Modern Theory of Solids* (McGraw-Hill, New York, 1940)]

Equation (9.11) indicates that the resistivity increase upon impurity scattering is proportional to $(\Delta Z)^2$ or the square of the valency difference between the impurity atom and the host metal. The $(\Delta Z)^2$ dependence is clearly seen in Fig. 9.2, where the resistivity increment upon the addition of one atomic % of various elements to pure Cu or pure Ag, is plotted against ΔZ. This is known as the Linde law. When monovalent Ag or Au is added to Cu or vice versa, the valency difference is obviously zero and, hence, no resistivity increase is expected from equation (9.11). Indeed, the resistivity increment in this case is much smaller than that when the polyvalent element is added. But it is still finite. Even when an element having the same valency as the host metal is added, the difference in ionic potential and atomic size would disturb the periodic structure, thereby resulting in an impurity scattering.

It is worthwhile mentioning that the Linde law no longer holds when the solute concentration exceeds about 5 at.%. Now the impurity–impurity interaction cannot be ignored. We will consider concentrated alloys in the next section.

9.4 Phase diagram in the Au–Cu alloy system and the Nordheim law

We select in this section the Au–Cu alloy system as one of the simplest but representative alloy systems and try to explain how the disruption of the lattice

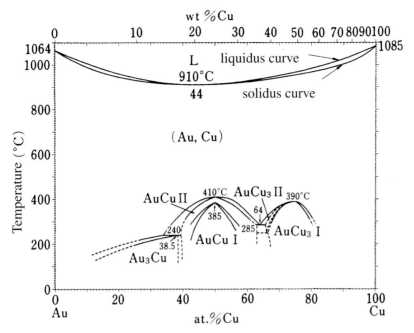

Figure 9.3. Phase diagram of Au–Cu alloy system. [*Binary Phase Diagrams*, edited by T. B. Massalski, H. Okamoto, P. R. Subramanian and L. Kacprzak, (ASM, 1990)]

periodicity gives rise to scattering of the Bloch electrons. First of all, we introduce the phase diagram which shows the phase appearing at a given concentration and at a given temperature in a given alloy system. In the phase diagram for an A–B binary alloy system, the horizontal and vertical axes represent the concentration of the atom A or B and the temperature, respectively. Each equilibrium phase is bounded by phase boundaries. The concentration is expressed either by the atomic % or weight %. Atomic % is more frequently employed in physics. Atomic %A, which is often abbreviated as at.%A, indicates the number of A atoms in 100 atoms and, hence, x at.%A is equal to $(100 - x)$ at.%B in the A–B binary alloy system. Each phase is stabilized by minimizing its free energy at a given temperature and concentration. An equilibrium phase diagram can be constructed by determining the phase boundary by experiments or thermodynamic calculations.

Figure 9.3 shows the equilibrium phase diagram of the Au–Cu system. It can be easily seen that Au and Cu melt at 1064 and 1085 °C, respectively. The region marked as "L" represents the liquid phase. The curve connecting the temperatures at which the liquid begins to solidify, is called the liquidus curve. The liquidus curve in this system is convex downward, indicating that the melting point in the alloys is lowered relative to those of the pure metals. There is

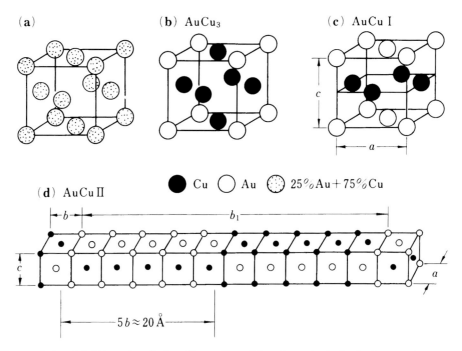

Figure 9.4. (a) Disordered $Au_{25}Cu_{75}$ alloy. The dotted circle indicates that each atomic site contains a Au (or Cu) atom in proportion to the probability of 25 (or 75) %. (b)–(d) Unit cells of $AuCu_3$, AuCu I and AuCu II ordered structures, respectively. The AuCu II phase has a long-period superlattice structure along the b-axis. [C. S. Barrett and T. B. Massalski, *Structure of Metals* (McGraw Hill, New York, 1966)]

another curve underneath, below which the alloy is fully solidified. This is called the solidus curve. The liquidus and solidus curves meet in the pure metals at both ends, indicating that the temperature remains unchanged upon cooling the melt until all of the liquid completes its solidification. In the alloys, with the exception of that containing 44 at.%Cu, both liquid and solid phases coexist in the region bounded by the liquidus and solidus curves.

Solid phases extend below the solidus curve. Here each solid phase occupies its own region in a given phase diagram and its crystal structure is uniquely assigned. Both Au and Cu are fcc. The fcc phase in the Au–Cu system extends from both Au and Cu sides and covers the whole concentration range at high temperatures. Au and Cu atoms are randomly distributed over the fcc lattice with probabilities proportional to their concentration. Though the Au and Cu atoms occupy the lattice sites in an fcc periodic lattice, the conduction electrons are scattered because the periodic potential is disturbed by their random occupation. Such a phase is called disordered. The unit cell of the disordered fcc Au–Cu alloy is shown in Fig. 9.4(a). As mentioned above, at high temperatures

9.4 Phase diagram in Au–Cu alloy system and the Nordheim law

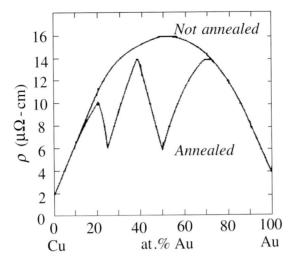

Figure 9.5. Concentration dependence of the resistivity at room temperature in Au–Cu alloy system. "Annealed" means that the sample was heat-treated to form an ordered phase. [F. Seitz, *The Modern Theory of Solids* (McGraw-Hill, New York, 1940)]

the disordered fcc phase extends over the entire concentration range. This is the formation of a complete solid solution.

There exist several intermetallic compounds $AuCu_3$, AuCu I, AuCu II and Au_3Cu in the Au–Cu alloy system below 410 °C. Their unit cells are shown in Fig. 9.4(b)–(d). All these compounds form a completely periodic structure, since the lattice sites which Au and Cu atoms occupy are uniquely assigned. They are each called an ordered alloy or an intermetallic compound. From the point of view of x-ray or neutron diffraction studies, reflections which are independent of the degree of order are called fundamental reflections, as is the case in Fig. 9.4(a), whereas reflections which vanish if the order vanishes are called superlattice reflections, as is the case in (b)–(d). The structure giving rise to the latter is called a superlattice or superstructure.

The disordered Au–Cu alloy can be retained at room temperature by rapidly solidifying the molten alloy. Its resistivity at room temperature exhibits a parabolic concentration dependence with a maximum at 50 at.%Au, as shown in Fig. 9.5. Its behavior is, therefore, well approximated as

$$\rho \propto x(1-x), \tag{9.12}$$

where x is the concentration of the Au. This is known as the Nordheim law, indicating that the degree of the disruption of the periodic potential increases with increasing solute concentration and reaches its maximum at 50 at.%Au. However, it can be seen from Fig. 9.5 that the resistivity drops sharply when an intermetallic compound is formed. This is due to the restoration of the periodic potential, resulting in a substantial reduction in the scattering of conduction electrons.

9.5 Hume-Rothery rule

Figure 9.6 shows the equilibrium phase diagrams for the Cu–Zn, Cu–Ga and Cu–Ge alloy systems. No complete solid solution is formed in these phase diagrams, since Cu is fcc, while its partner elements Zn, Ga and Ge crystallize into hcp, orthorhombic and diamond structures, respectively. As is clearly seen from Fig. 9.6, different phases appear successively with increasing concentration of the partner element. The fcc phase extending from pure Cu is called the α-phase. It is also called a primary solid solution of Cu. Its maximum solubility limit is found to be 38.3 at.%Zn, 19.9 at.%Ga and 11.8 at.%Ge in the Cu–Zn, Cu–Ga and Cu–Ge alloy systems, respectively. Thus, we see that the maximum solubility limit decreases with increasing valency of the partner element.

The β- and β'-phases appear next to the α-phase in the neighborhood of 50 at.%Zn in the Cu–Zn system. The β-phase, Fig. 9.7(a), is disordered bcc and stable at high temperatures but transforms into the β'-phase at low temperatures. The β'-phase has the CsCl-type ordered structure, as shown in Fig. 9.7(b). The β-phase exists at around 25 at.%Ga at high temperatures in the Cu–Ga system. But the hcp ζ-phase appears instead of the β'-phase at low temperatures. The ζ-phase exists in the vicinity of 15 at.%Ge in the Cu–Ge system. All this evidence indicates that the concentration range over which the β- and ζ-phases are stable, moves to lower solute concentrations with increasing valency of the solute elements Zn, Ga and Ge.

Further increase in the solute concentration leads to the formation of the cubic γ-phase in both Cu–Zn and Cu–Ga systems. As shown in Fig. 9.8, the γ-phase is constructed by stacking three bcc cells in x-, y- and z-directions and subsequently removing the center and corner atoms with slight displacements of the remaining atoms. Its unit cell contains a total of 52 atoms. Further increase in solute concentration leads to the formation of the hcp ε-phase. Finally, the phase diagram in the Cu–Zn system is terminated by the hcp η-phase, which is a primary solid solution of Cu in Zn.

We found above that the different phases appear successively and systematically with increasing amount of the element added to the noble metal Cu and that their stable concentration range, including the solubility limit of the primary Cu solid solution, is shifted to lower concentrations with an increase in the valency of the solute element. These features are quite regularly observed

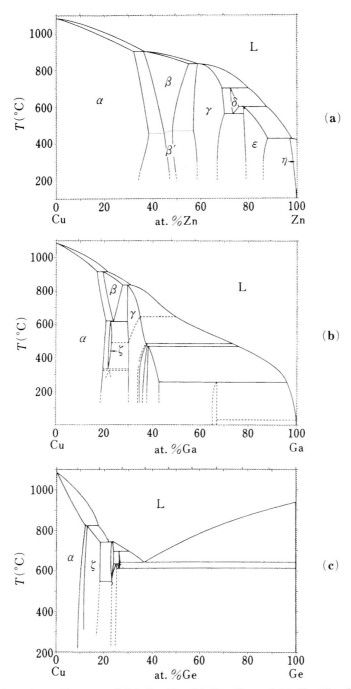

Figure 9.6. Phase diagram of (a) Cu–Zn, (b) Cu–Ga and (c) Cu–Ge alloy systems. [*Binary Phase Diagrams*, edited by T. B. Massalski, H. Okamoto, P. R. Subramanian and L. Kacprzak, (ASM, 1990)]

234　　　　　　　　　　9 *Electronic structure of alloys*

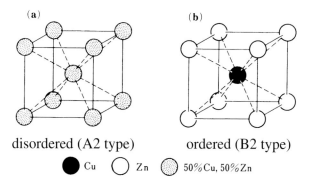

Figure 9.7. Unit cells of (a) disordered bcc (A2 type) and (b) ordered (B2 or CsCl-type) $Cu_{50}Zn_{50}$ alloy. The ordered and disordered alloys are also called the β′- and β-brasses, respectively. Each lattice site in the disordered phase contains a Cu or Zn atom with equal probability. [C. S. Barrett and T. B. Massalski, *Structure of Metals*, (McGraw-Hill, New York, 1966)]

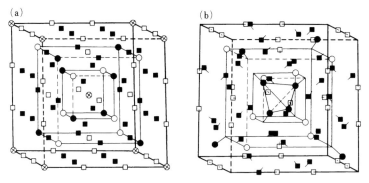

Figure 9.8. Unit cell of the γ-phase Cu–Zn alloy. (a) The structure obtained by stacking together $3 \times 3 \times 3$ bcc unit cells along the x-, y- and z-directions, contains 54 atoms. (b) The unit cell of the γ-phase is obtained from (a) by removing the center and corner atoms, with remaining atoms displaced along the directions marked by arrows. [T. B. Massalski and U. Mizutani, *Prog. Mat. Sci.* **22** (1978) 151]

not only in Cu-based alloys but also in Ag- and Au-based alloys, as long as the partner element is chosen from the polyvalent elements like Mg, Zn, Al, Ga, Sn, Pb and so on in the periodic table. Hume-Rothery revealed empirically that these phases are stabilized at a unique electron concentration or electrons per atom, e/a, regardless of the atom species of the solute element added to the noble metal. This is known as the Hume-Rothery rule and is illustrated in Fig. 9.9.[3]

The fcc α-phase exists at electron concentrations e/a below 1.4 and is

---

[3] Hume-Rothery pointed out several factors responsible for the stabilization of an alloy. (1) Atom size-factor effect. The favorable size-factor is bounded by limits of about 20% or radius ratios 0.8–1.2. (2) The electrochemical factor. An increasing difference between the electronegativities of the two metals enhances the tendency for the formation of an alloy (see footnote 4, p. 443 in Section 14.5 for the definition of electronegativity). (3) A tendency for definite crystal structures to occur at characteristic e/a.

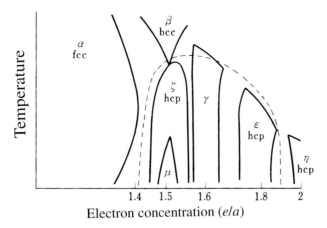

Figure 9.9. Electron concentration dependence of the typical electron phases in Hume-Rothery alloys based on noble metals. [T. B. Massalski and U. Mizutani, *Prog. Mat. Sci.* **22** (1978) 151]

followed by the bcc β-phase in the neighborhood of $e/a = 1.5$ at high temperatures, which is replaced either by its ordered CsCl-type β′-phase or by the hcp ζ-phase at low temperatures. In the vicinity of $e/a = 1.5$, the μ-phase containing 20 atoms in its β-Mn-type cubic unit cell appears in certain alloy systems like Ag–Al and Cu–Si systems. The complex cubic γ-phase is stabilized at about $e/a = 1.6$ and the hcp ε-phase in the range $1.7 < e/a < 1.9$. The hcp η-phase appears as a primary solid solution of Zn and Cd and is centered at $e/a = 2.0$. Because of their locations at particular electron concentrations, these alloys are called the electron compounds or the Hume-Rothery electron phases. Judging from its strong e/a dependence, it has been naturally thought that the interaction of the Fermi surface with the Brillouin zone must play a critical role in stabilizing these electron phases.

9.6 Electronic structure in Hume-Rothery alloys

We learned in the preceding section that alloying destroys the periodicity of the lattice potential and thus results in scattering of the Bloch electron. The Fermi surface is constructed under the assumption that the wave vector **k** of the Bloch electron is a good quantum number. The wave vector **k** changes upon scattering of the Bloch electron. In other words, the lifetime of the Bloch electron becomes finite, when it is scattered by the non-periodic potential. The wave vector **k** can be no longer taken as a good quantum number, if its lifetime becomes too short. In spite of such fundamental difficulties, experimental and theoretical works have provided ample evidence that the concept of the Fermi surface and Brillouin zone is still valid even in concentrated crystalline alloys.

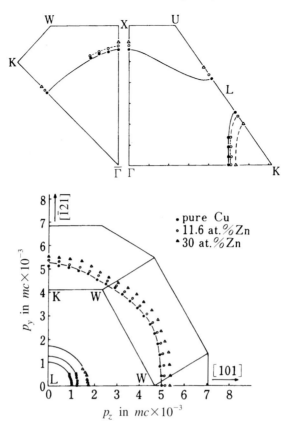

Figure 9.10. Fermi surface of α-phase Cu–Zn alloys revealed by positron annihilation experiments. An expansion of the Fermi surface with increasing Zn concentration can be clearly seen. Here, p_y and p_z represent y- and z-components of the electron momentum (see equation (7.12)). [M. Haghgooie, S. Berko and U. Mizutani, *Proc. of 5th Int. Conf. on Positron Annihilation* (The Japan Institute of Metals, Japan, 1979)]

We will discuss below several important works on the Hume-Rothery electron phases.

Figure 9.10 shows the Zn concentration dependence of the Fermi surface measured by the positron annihilation experiment for α-phase Cu–Zn single crystal alloys. The average electron concentration e/a increases with increasing Zn concentration, since the valencies of Cu and Zn are 1 and 2, respectively. Within the framework of the free-electron model, the radius of the Fermi surface should increase with increasing Zn concentration in accordance with equation (2.20). But, we have to recall that the Fermi surface of pure Cu has already touched the {111} zone planes of its Brillouin zone, as shown in Fig. 6.5. As explained in Section 7.3, positron annihilation measurements would be best suited to examine experimentally how the Fermi surface of pure Cu

9.6 Electronic structure in Hume-Rothery alloys 237

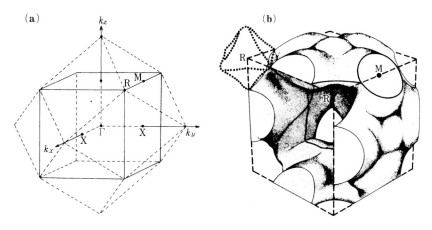

Figure 9.11. (a) Brillouin zone and (b) Fermi surface of the β′-phase Cu–Zn alloy. [T. B. Massalski and U. Mizutani, *Prog. Mat. Sci.* **22** (1978) 151]

changes with increasing Zn concentration in the α-phase. As demonstrated in Fig. 9.10, the expansion of the Fermi surface, including the radius of the neck around the {111} zone planes with increasing e/a, can be clearly observed. Obviously, Fig. 9.10 proves the validity of the concept of the Fermi surface and the Brillouin zone in such concentrated alloys.

Now we discuss the electronic structure of more concentrated alloys or intermediate compounds. Among them, the disruption of the periodic potential in the β′-phase is small because of the ordered structure. As a result, the de Haas–van Alphen measurement is feasible for this particular alloy and its Fermi surface has been accurately determined. The Brillouin zone and the Fermi surface of the β′-phase Cu–Zn alloy are shown in Figs. 9.11(a) and (b), respectively. Since it forms the CsCl-type ordered structure, the cubic zone enclosed by the six {100} planes is newly formed inside the first Brillouin zone of the bcc lattice shown in Fig. 7.10. The Fermi surface makes contacts with the {110} planes and the Fermi surface of holes is left at the corners R of the {100} zone planes. The Fermi surface of holes in the reduced zone scheme is shown in Fig. 9.11(b) as dotted curves.

The construction of the Brillouin zone for the complex cubic γ-phase is not simple, since its unit cell, shown in Fig. 9.8, contains 52 atoms [1]. As discussed in Section 5.6, the Bragg reflection of the Bloch electrons is responsible for the formation of an energy gap across the Brillouin zone planes. There is one-to-one correspondence between the Bragg scattering of the incident x-ray or electron beam and that of the conduction electrons in a periodic lattice. Thus, the family of the crystal planes which yield strong x-ray diffraction lines or strong electron diffraction spots would give rise to the Brillouin zone planes having a

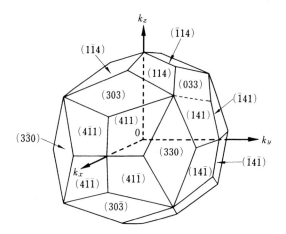

Figure 9.12. The Jones zone of the γ-phase Cu–Zn alloy. [T. B. Massalski and U. Mizutani, *Prog. Mat. Sci.* **22** (1978) 151]

sizable energy gap. The strongest x-ray diffraction lines for the γ-phase arise from the family of the {330} and {411} crystal planes. There are 12 equivalent planes which include (303), (033), etc., in the family of the {330} planes, and 24 equivalent planes in the family of the {411} planes. Hence, the Jones zone (see footnote 4, p. 142 in Chapter 6) consisting of a total of 36 zone planes can be constructed for the γ-phase, as shown in Fig. 9.12.

The volume of the Jones zone is calculated as $45(2\pi/a)^3$, where a is the lattice constant of the γ-phase [2]. Since the unit cell contains 52 atoms, the volume per atom is equal to $a^3/52$. One can easily find that the zone accommodates 1.73 electrons per atom. Since the ratio of the volume of an inscribed sphere to that of the polyhedron is $2\pi/5\sqrt{2} \approx 0.89$, a spherical Fermi surface would touch the Brillouin zone at about 1.54 electrons per atom. The presence of the energy gaps will tend to distort the Fermi surface, resulting in a contact with the zone planes at an e/a value lower than 1.54. The fact that the γ-phase is stable above $e/a = 1.6$ suggests that the Fermi surface is already in contact with the zone planes over the stable concentration range.

It is also important to note that the contact of the Fermi surface with the Jones zone planes occurs simultaneously at many equivalent points. After the contact, electrons would fill the corners of the Jones zone, thereby resulting in a rapid decline in the density of states with increasing electron concentration. Indeed, the electronic specific heat in the γ-phase has been reported to decrease very sharply with increasing electron concentration e/a. As shown in Fig. 9.13, this unique behavior can be understood by assuming that the Fermi surface has touched the {330} and {411} zone planes and that the Fermi level in the γ-phase falls on the subsequent declining slope of the density of states curve.

Finally, we discuss the electronic structure of the ζ- and ε-phase Hume-

9.6 Electronic structure in Hume-Rothery alloys

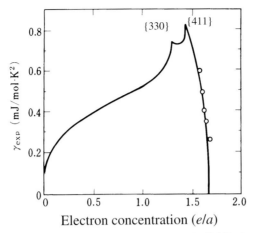

Figure 9.13.   The calculated density of states curve (solid line) and the e/a-dependence of the measured electronic specific heat coefficient (open circles) in the γ-phase Cu–Zn alloys. [T. B. Massalski and U. Mizutani, *Prog. Mat. Sci.* **22** (1978) 151]

Rothery phases, both of which crystallize into the hcp structure. The alloys in this family are existent over a wide electron concentration range, $e/a = 1.4$ to 1.9, as illustrated in Fig. 9.9. Figure 9.14 shows the e/a dependence of the electronic specific heat coefficient for many hcp Hume-Rothery alloys. For comparison, the density of states curves calculated for hcp Zn and Be metals are incorporated. The electronic specific heat coefficient exhibits a large peak centered at $e/a = 1.5$. The formation of the peak can be interpreted as the overlap of electrons across the {100} zone planes coupled with contact with the {101} zone planes in the hcp Brillouin zone shown in Fig. 6.9. Based on the data shown in Fig. 9.14, Massalski and Mizutani proposed in 1978 a possible Fermi surface for the ζ-phase Hume-Rothery alloys. This is illustrated in Fig. 9.15.

Koike *et al.* measured in 1982 the positron annihilation angular correlation curves for the ζ-phase Cu–Ge and Ag–Al alloys. Their results are reproduced in Fig. 9.16. The angular correlation curve clearly shows a bulge in the emitting γ-ray angle ranging from 4 to 6 mrad for the samples with $e/a = 1.45$ (15 at.%Ge) and 1.57 (28.5 at.%Al). They interpreted this bulge as arising from the overlap of electrons into the second zone across the centers M of the {100} zone planes. This is shown in Fig. 9.16(a) and (b). However, the bulge disappears and the Fermi surface becomes more spherical for the sample with $e/a = 1.72$ (36 at.%Al). Thus, the Fermi surface becomes more free-electron-like, as shown in (c). They also revealed that the Fermi surface of electrons in the second zone bulges towards the points L of the {101} zone planes at $e/a = 1.45$ but makes contact with the {101} planes at $e/a = 1.57$ and that the neck exists around the points Γ of the {002} zone planes. All these features are consistent with the diagram shown in Fig. 9.15.

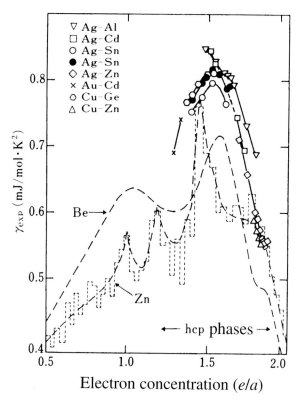

Figure 9.14. Electron concentration dependence of the electronic specific heat coefficient in hcp Hume-Rothery alloys. It can be seen that the data fall on a universal curve, regardless of the alloy system. The peak is formed near $e/a = 1.5$. The calculated density of states curves (dashed lines) for pure Zn and Be are shown for comparison.
[T. B. Massalski and U. Mizutani, *Prog. Mat. Sci.* **22** (1978) 151]

As can be understood from the argument above, the electronic structure of the concentrated Hume-Rothery alloys can be well described in terms of the Fermi surface perturbed by the presence of the Brillouin zone unique to the respective crystal structures. This indicates that the electronic structure of these Hume-Rothery electron compounds can be basically discussed in the framework of the nearly-free-electron model [1].

9.7 Stability of the Hume-Rothery alloys

The discussion of the stability of a given phase at absolute zero has been considered as one of the most important topics in the electron theory of metals. In principle, one has to evaluate not only the energy associated with the conduction electrons but also the energy associated with the ionic lattice. However, as indicated in Fig. 9.9, the concentration range for the existence of the stable

9.7 Stability of Hume-Rothery alloys 241

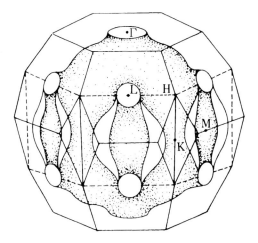

Figure 9.15. Fermi surface of an hcp Hume-Rothery alloy estimated from the data in Fig. 9.14. The {100} overlap and {101} contact are predicted to be present. [T. B. Massalski and U. Mizutani, *Prog. Mat. Sci.* **22** (1978) 151]

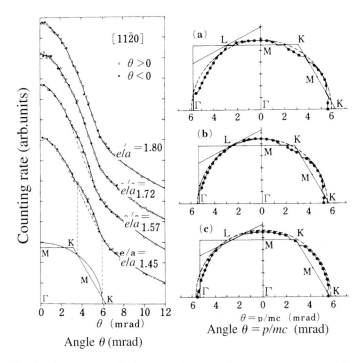

Figure 9.16. Positron annihilation angular correlation curves taken along the ⟨110⟩ direction (ΓK direction in Fig. 5.17(a)) for the ζ-phase Ag–Al (e/a = 1.57, 1.72 and 1.80) and Cu–Ge (e/a = 1.45) alloys. Resulting Fermi surfaces for alloys with e/a = (a) 1.45, (b) 1.57 and (c) 1.72, respectively. [S. Koike, M. Hirabayashi, T. Suzuki and M. Hasegawa, *Phil. Mag.* **B 45** (1982) 261]

Hume-Rothery phases depends strongly on the electron concentration e/a, as if it is independent of the atomic species involved. The Hume-Rothery rule suggests that the energy of the conduction electron system plays a critical role in stabilizing these electron phases. In the present section, we focus on the Hume-Rothery electron phases based on noble metals, in which the nearly-free-electron model is believed to be applicable, and discuss to what extent the stability of the phase can be described in the framework of the Fermi surface–Brillouin zone interaction.

Let us briefly recall the Fermi surface–Brillouin zone interaction discussed in Section 5.10. The E–$\mathbf{k}$ relation becomes flat and the Fermi surface begins to be distorted from a sphere, when it approaches the zone planes (see Fig. 5.19). As a result, the density of states is enhanced relative to the free-electron-like parabolic band, as shown in Fig. 5.20. Further increase in the electron concentration causes the Fermi surface to touch the zone planes. Once the Fermi surface touches the zone planes, the density of states sharply drops, resulting in a cusp or peak called the van Hove singularity in the density of states curve. This feature is already depicted in Fig. 9.13 in relation to the electronic structure of the γ-phase. Therefore, we see that the Fermi surface–Brillouin zone interaction gives rise to a peak and subsequent rapid fall in the density of states curve. The size of this effect depends not only on the magnitude of the energy gap across the zone planes but also on the multiplicity of the equivalent zone planes.

According to equation (2.24), the internal energy of the conduction electron system at absolute zero is expressed as

$$U = \int_0^{E_F} E N(E) dE, \quad (9.13)$$

where $N(E)$ is the electron density of states.

Let us consider the phase competition in the range $1.0 \leq e/a \leq 1.5$ of the Cu–Zn alloy system, in which the α- and β-phases are competing. The electron concentration dependent Fermi surfaces of the α-phase alloys have already been shown in Fig. 9.10. By using this as a guide, one can obtain the density of states curve as a function of e/a or energy for the α-phase alloys. Suppose that we can repeat the same procedure for the competing bcc β-phase. Once the density of states for the phases of interest are obtained, the internal energy can be calculated from equation (9.13) and be plotted against e/a to see if the energy for the fcc α-phase is the lowest in the e/a range below 1.4.

Such calculations were indeed attempted by Jones in 1937 for the competition between the α- and β-phases in the Cu–Zn system. He assumed a spherical Fermi surface for pure Cu and concluded that the Fermi surface in the α-phase touches the {111} zone planes at $e/a = 1.36$ and that the density of states curve begins to drop sharply after passing the peak when e/a increases beyond 1.36. In contrast, he showed that the density of states curve for the competing bcc β-phase causes its first peak at $e/a = 1.48$ due to the contact of the spherical Fermi surface with the {110} zone planes. The density of states curves calculated by Jones for the two competing phases are shown in Fig. 9.17. The calculation of the energy by inserting the density of states in Fig. 9.17 into equation (9.13) led Jones to conclude that the bcc structure is energetically favored relative to the fcc structure, when e/a exceeds 1.4 in the Cu–Zn alloy system. This explained well at that time the Hume-Rothery rule for the competition between the α- and β-phases in the Cu–Zn system. However, as described in Section 6.4, Pippard clearly demonstrated in 1957 that the contact of the Fermi surface with the {111} zone planes has occurred even in pure Cu. This finding invalidated the basic assumption made by Jones.

Heine and Weaire [3] discussed the same problem and evaluated a change in the energy of the conduction electron system when the Fermi surface touches the particular zone plane in the context of the pseudopotential method. According to their calculations, the energy of the conduction electron system, when plotted against e/a, exhibits only a slight change in its slope upon its contact with the zone plane without the formation of a clear minimum. They stressed that the energy gain upon contact with the zone plane is too small to account for the relative stability of the competing phases, particularly when the number of equivalent zone planes responsible for the Fermi surface contact is small, as is the case in fcc and bcc structures. They further noted that the reason why the Cu metal is fcc cannot be explained without taking into account the presence of the 3d band in the middle of its valence band and admitted the difficulty in the theoretical interpretation of the Hume-Rothery rule.

Let us roughly evaluate the energy of the conduction electron system by including the 3d-band contribution in pure Cu. The density of states curve shown in Fig. 6.6 may be employed for this purpose. The total energy calculated from equation (9.13) turns out to be of the order of 10^5 cal/mol. Now consider a hypothetical bcc Cu, in which the position and shape of the 3d band happens to be the same as that in the fcc structure, though this is not physically acceptable. In other words, we assume that the energy difference originates only from the difference in the Fermi surface–Brillouin zone interaction in the same spirit as Jones did. The difference is immediately found to be less than 100 cal/mol. This is a mere 0.1 % of the total energy of the conduction electron system [1]. It is almost impossible to evaluate the total electronic energy for competing phases to an accuracy of 0.1 % using the valence band structure like that shown in Fig. 6.6.

The first-principle band calculations like the LMTO method discussed in

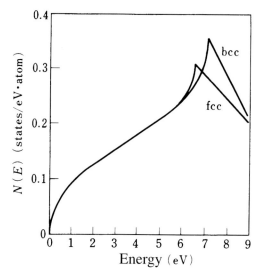

Figure 9.17. The density of states curves for fcc and bcc phases. The peaks in the fcc and bcc structures are caused by Fermi surface contacts with the {111} and {110} zone planes, respectively. [H. Jones, *Proc. Phys. Soc.* **A49** (1937) 250]

Chapter 8 are now believed to be the most reliable and efficient to evaluate the energy of the conduction electron system. Because of the rapid progress in computer science, fairly accurate band calculations have now been made possible at least for pure metals and well-defined intermetallic compounds. As will be described in the next section, great progress has also been made for band calculations of concentrated alloys. But it is still difficult to calculate the electronic energy of concentrated alloys with accuracies of less than 100 cal/mol or 10^{-4} eV/atom. Thus, a theoretical understanding of the Hume-Rothery rule shown in Fig. 9.9 is still far from being achieved.

Though an understanding of the relative stability of phases is extremely difficult, there are a number of works which point to the importance of the contribution of the electronic energy in the stabilization of the phase involved. In particular, when the number of equivalent zone planes are high, the Fermi surface would touch those planes simultaneously and the resulting effect on the electronic energy becomes substantial. The situation shown in Fig. 9.13 for the γ-phase is known as a typical example.

Quasicrystals are characterized by five-fold rotational symmetry incompatible with the translational symmetry in real space and by the possession of many equivalent zone planes in reciprocal space (see Sections 15.11 and 15.12). The question has been addressed as to why a quasicrystal can exist as a stable phase in nature. It has been proposed that the electronic energy can be reduced by the simultaneous contact by its Fermi surface with a large number of equivalent zone planes, thereby contributing to the stabilization of this unique quasiperiodic structure. Band calculations for a quasicrystal are not possible, since its unit cell is infinitely large. Instead, band calculations have been performed for the approximant crystal, which possesses locally the same atomic structure as that of the quasicrystal and is taken as an analog to the quasicrystal. As shown in Fig. 15.22, a deep pseudogap, which is believed to be characteristic of a material possessing icosahedral symmetry, appears near the Fermi level and causes electrons near the Fermi level to push into higher binding energies. This is believed to contribute to a lowering of the electronic energy of a quasicrystal.

9.8 Band theories for binary alloys

All band calculations discussed in Chapter 8 are based on the Bloch theorem and, hence, are applicable only for perfectly periodic metals and semiconductors. In this section, we will discuss band structure calculations for concentrated alloys. As mentioned in Section 9.4, an intermetallic compound having a superlattice structure tends to be formed in a system where the chemical bonding between the unlike atoms A and B is strong. The Au–Cu system in Section 9.4 is indeed typical of such examples.

A disordered alloy is defined as one in which the two different atoms A and B are randomly distributed over the periodic lattice sites at probabilities proportional to their concentration. Since the atoms A and B occupy only lattice sites, the alloy is called substitutional. Even in a substitutional disordered alloy, there exists some tendency for the atomic pair A–B to be formed more preferentially or less preferentially than that expected from the average concentration. This results in short-range order (see Section 15.2), but its incorporation into band calculations is certainly not simple. In contrast to a substitutional alloy, small atoms such as hydrogen, carbon or nitrogen, when dissolved in a metal, are known to occupy tetrahedral or octahedral interstitial sites in the periodic lattice of host atoms. They are often referred to as interstitial alloys and treated separately from substitutional alloys. In the following, we discuss band calculations for a substitutional disordered alloy.

The total potential in a substitutional disordered alloy is expressed as

$$V(\mathbf{r}) = \sum_n v_X(\mathbf{r} - \mathbf{l}_n), \qquad (9.14)$$

where $v_X(\mathbf{r} - \mathbf{l}_n)$ is the muffin-tin potential at the position $\mathbf{r}$ due to the ion X whose position is specified by the lattice vector $\mathbf{l}_n$. The subscript X stands for either ion A or B so that $v_A(\mathbf{r} - \mathbf{l}_n)$ indicates the potential of the ion A.

The rigid-band model is known as the simplest model for such an alloy and assumes $v_A(\mathbf{r}) = v_B(\mathbf{r})$ in equation (9.14) by ignoring the difference in the potentials of ions A and B. Thus, the electronic structure of the pure metal A is assumed to be the same as that of the pure metal B or any compositions in the alloy A–B. The Fermi level is chosen so as to be consistent with the electron concentration of a given alloy. The rigid band model may be appropriate when the two elements A and B possess the same crystal structure and are immediate neighbors in the periodic table. For example, the rigid band model has been applied to the electronic structure of the Cu–Ni and Ag–Pd alloys. However, as will be discussed later in this section, recent photoemission spectroscopy experiments prove that the rigid band model fails even in such binary alloy systems.

The virtual crystal approximation or VCA model is devised as a model superior to the rigid band model. It assumes the periodic potential in an alloy to be given by

$$V_0(\mathbf{r}) = \sum_n v_{av}(\mathbf{r} - \mathbf{l}_n), \qquad (9.15)$$

where $v_{av}(\mathbf{r})$ is expressed as

$$v_{av}(\mathbf{r}) = c_A v_A(\mathbf{r}) + c_B v_B(\mathbf{r}). \qquad (9.16)$$

Here c_A and c_B are concentrations of atoms A and B, respectively, and satisfy the relation $c_A + c_B = 1$. Once the periodic potential (9.15) is employed, all band calculation techniques for pure metals described in Chapter 8 can be applied.

The resulting energy eigenvalue E_0 and wave function $\psi_{0,\mathbf{k}}(\mathbf{r})$ represent the solution of the Schrödinger equation for an average periodic potential but are certainly approximate, since the VCA potential (9.15) is different from the true potential (9.14). Its first-order correction to the approximate energy eigenvalue E_0 is expressed as

$$\Delta E(\mathbf{k}) = \int \psi_{0,\mathbf{k}}(\mathbf{r})^* [V(\mathbf{r}) - V_0(\mathbf{r})] \psi_{0,\mathbf{k}}(\mathbf{r}) d\mathbf{r}, \qquad (9.17)$$

where $[V(\mathbf{r}) - V_0(\mathbf{r})]$ is the difference between the true potential and the VCA potential. However, the application of the Bloch theorem (5.14) to equation (9.15) leads to $\Delta E(\mathbf{k}) = 0$. This indicates that the energy eigenvalue E_0 is a fairly good approximate solution. Though the VCA model is very simple, it has been recognized as a fairly reasonable model for alloys, as long as the difference between $v_A(\mathbf{r})$ and $v_B(\mathbf{r})$ is small or the band structure at energies much higher than the potential is concerned.

A more elaborate model, which is applicable beyond the limit of the VCA model, has been developed. This is the coherent potential approximation or CPA method and has been derived by extending the KKR method based on the multiple scattering theory discussed in Section 8.9. The potentials at all lattice sites except for a given central site are replaced by the effective potential

9.8 Band theories for binary alloys

$w(E, \mathbf{r})$. Now the atom A is located at this central site and its potential is denoted as $v_A(\mathbf{r})$. The conduction electron propagates in a periodic potential

$$V_{CPA}(E, \mathbf{r}) = \sum_n w(E, \mathbf{r} - \mathbf{l}_n) \tag{9.18}$$

as a Bloch electron but is scattered by the potential $v_A(\mathbf{r})$ when it arrives at the central site. The same argument holds for atom B. This is equivalent to impurity scattering in an otherwise perfect crystal.

As mentioned in Section 8.9, the wave $\psi^i(\mathbf{r})$ incident to the ion potential located at the lattice vector $\mathbf{l}_n$ is not independent of the outward wave $\psi^o(\mathbf{r})$ scattered from this atom but is related to it through the relation (8.57):

$$\psi^o(\mathbf{r}) = \sum_n \int G(\kappa, \mathbf{r} - \mathbf{r}') v(\mathbf{r}' - \mathbf{l}_n) \psi^i(\mathbf{r}') d\mathbf{r}', \tag{9.19}$$

where $G(\kappa, \mathbf{r} - \mathbf{r}')$ is the Green function and $v(\mathbf{r}' - \mathbf{l}_n)$ is the potential at the position $\mathbf{l}_n$. The potential $v(\mathbf{r}' - \mathbf{l}_n)$ at each lattice site is replaced by a more general quantity called the t-matrix $T_n(E, \mathbf{r}', \mathbf{r}'')$.[4] Furthermore, the difference potential $[v_\alpha(\mathbf{r} - \mathbf{l}_n) - w(E, \mathbf{r} - \mathbf{l}_n)]$ created by substituting the potential of atom A or B for the effective potential is also replaced by the corresponding t-matrix $\tau_A(E, \mathbf{r}, \mathbf{r}')$ or $\tau_B(E, \mathbf{r}, \mathbf{r}')$. Hence, $\tau_A(E, \mathbf{r}, \mathbf{r}')$ depends not only on v_A but also on $V_{CPA}(E, \mathbf{r})$. In the CPA calculations, the atom A or B is arbitrarily chosen from N atoms and the potential of the remaining $(N-1)$ atoms is replaced by the effective potential. But the effective potential is given by the sum of the concentration-weighted A and B atom potentials. Hence, the sum of the concentration-weighted $\tau_A(E, \mathbf{r}, \mathbf{r}')$ and $\tau_B(E, \mathbf{r}, \mathbf{r}')$ must be zero:

$$c_A \tau_A(E, \mathbf{r}, \mathbf{r}') + c_B \tau_B(E, \mathbf{r}, \mathbf{r}') = 0. \tag{9.20}$$

This relation suppresses the scattering so as to be consistent with the presence of the effective potential $w(E, \mathbf{r})$. The CPA band calculations are known to yield correct results even at low energies comparable to the ion potential, where the VCA method breaks down. It is also consistent with the VCA method at energies higher than the potential. More details of the CPA method can be found in the literature [5].

---

[4] The scattering cross-section is expressed as $\sigma(\theta) = (m/2\pi\hbar^2)^2 |\langle \psi^+(\mathbf{k}')|V(\mathbf{r})|\phi(\mathbf{k})\rangle|^2$, where $\phi(\mathbf{k})$ is the incident wave and $\psi^+(\mathbf{k}')$ is the scattered wave. Since $|\psi^+(\mathbf{k}')\rangle$ and $|\phi(\mathbf{k})\rangle$ cannot be basis vectors in the same representation, it is convenient to introduce a matrix so as to satisfy the relation $\langle \psi^+(\mathbf{k}')|V(\mathbf{r})|\phi(\mathbf{k})\rangle = \langle \phi(\mathbf{k}')|T|\phi(\mathbf{k})\rangle$. The matrix T is called the transition matrix or t-matrix. We may use $\psi^o(\mathbf{r}) = \int G(\kappa, \mathbf{r} - \mathbf{r}')V(\mathbf{r}')\psi^i(\mathbf{r}')d\mathbf{r}'$ in equation (8.57). The $\psi^o(\mathbf{r})$ in the left-hand side is calculated by substituting $e^{i\mathbf{k}\cdot\mathbf{r}}$ for $\psi^i(\mathbf{r}')$. In the next step, the $\psi^o(\mathbf{r})$ thus obtained is substituted for $\psi^i(\mathbf{r}')$ to obtain a new $\psi^o(\mathbf{r})$. This process is iterated until a self-consistent solution is achieved. The resulting $\psi^o(\mathbf{r})$ is employed as $|\psi^+(\mathbf{k}')\rangle$ and inserted into $\langle \psi^+(\mathbf{k}')|V(\mathbf{r})|\phi(\mathbf{k})\rangle = \langle \phi(\mathbf{k}')|T|\phi(\mathbf{k})\rangle$. The t-matrix is now expressed as the power series of $V(\mathbf{r})$. If we take only the first term in this expansion, the scattered wave function $\psi^+(\mathbf{k}')$ is replaced by the incident wave function $\phi(\mathbf{k})$. This is the Born approximation and is valid only if $V(\mathbf{r})$ is small. See more details in ref. [4].

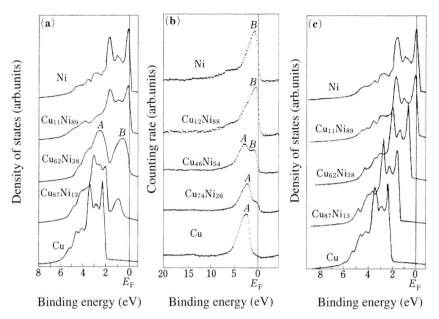

Figure 9.18. The valence band for a series of Cu–Ni alloys (a) calculated by CPA, (b) measured by XPS and (c) calculated by VCA. The peaks A and B in (b) can be ascribed to the Cu-3d and Ni-3d sub-bands. [J. S. Faulkner, *Prog. Mat. Sci.* **27** (1982) 1, for band calculations and S. Hüfner *et al.*, *Phys. Rev. Lett.* **28** (1972) 488, for the XPS data]

CPA band calculations have been applied to many concentrated alloys since its development in 1970 and its usefulness has been proved. As an example, we show in Fig. 9.18 band calculations based on VCA and CPA methods and compare them with the XPS valence band profiles for a series of the Cu–Ni fcc alloys.

Ni and Cu are fcc and located next to each other in the periodic table (see Table 1.1). Furthermore, a complete solid solution is formed in this system. Ni is known to be ferromagnetic with a Curie temperature of 631 K while Cu is known to be non-magnetic. The Curie temperature decreases with increasing Cu concentration and the ferromagnetism disappears at 60 at.%Cu (see Section 13.5). According to the rigid-band model, the addition of Cu would fill the holes of the Ni 3d band and filling is believed to be completed at 60 at.%Cu, since Cu has one more electron than Ni. Hence, the disappearance of the ferromagnetism in the Cu–Ni system has been thought to be taken as validating of the rigid-band model [6]. However, both CPA band calculations and photoemission experiments clearly show that the overall valence band profile drastically changes upon alloying and bands unique to the alloys are formed. One finds that CPA band calculations reproduce the XPS data better than VCA calculations.

Chapter Ten

Electron transport properties in periodic systems (I)

10.1 Prologue

Electron transport properties can be investigated by measuring the response of conduction electrons to a temperature gradient or to external fields such as an electric field, a magnetic field, or a combination of these applied to a specimen. In this chapter, we study basic transport properties with subsequent derivation of the Boltzmann transport equation. The electrical conductivity is then formulated in the light of the Boltzmann transport equation. The temperature-dependent electrical resistivity expression known as the Baym resistivity formula is derived by taking into account the electron–phonon interaction and is applied to obtain the well-known Bloch–Grüneisen law for a crystal metal. The remaining transport properties including the thermal conductivity, the thermoelectric power, the Hall effect and the magnetoresistance will be discussed in Chapter 11.

10.2 The Drude theory for electrical conductivity

In Chapter 2, we learned that electrons on the Fermi surface in a metal like Na carry a Fermi wave number k_F of the order of a few 10 nm^{-1}, which is converted to a Fermi velocity v_F of approximately 10^6 m/s through the relation $v_F = \hbar k_F/m$. In spite of such high velocities of electrons on the Fermi surface, no electrical current can flow, unless an electric field is applied. The reason for this is that the Fermi surface is always symmetric with respect to the origin in reciprocal space and that there always exists an electron with $-v_F$ for the electron with v_F. Obviously, a current flows only when the applied field displaces the Fermi sphere from the origin and its symmetrical geometry breaks.

In the present section, basic properties concerning the electrical conduction are discussed by treating conduction electrons as charged particles obeying the free-electron model. Let us apply the electric field E_x along the x-direction to the assembly of free electrons. The equation of motion for each electron is then expressed as

$$\frac{dp_x}{dt} = \frac{\hbar dk_x}{dt} = (-e)E_x, \qquad (10.1)$$

where the charge of the electron is denoted as $(-e)$ to emphasize the possession of a negative charge. Equation (10.1) is immediately reduced to $k_x = [(-e)E_x/\hbar]t + k_{x_0}$ and the wave number k_x increases indefinitely with increasing time. This means that the Fermi sphere moves as a whole endlessly in a direction opposite to the applied field and the electrical current becomes infinitely large.

This does not happen in a real metal. Instead, a steady current flows, as long as a constant field is applied. A scattering process must be involved so that electrons cannot be endlessly accelerated by the electric field. We introduce the relaxation time τ as a measure of the frequency of the scattering of the conduction electron. The electron is accelerated by the electric field only in the time interval τ before being scattered by lattice vibrations and/or defects like impurity atoms. We consider the steady state to be established after the Fermi sphere is displaced by a certain amount by the electric field. Thus, the flow of a finite current, when the electric field is switched on, is entirely due to the presence of a scattering mechanism.

In order to introduce the scattering term in equation (10.1), we define the drift velocity $\mathbf{v}_D$ as

$$\mathbf{v}_D = \frac{\sum_{i=1}^{n} \mathbf{v}_i}{n}, \qquad (10.2)$$

where the summation is taken over n conduction electrons per unit volume. We see that the drift velocity corresponds to a velocity per electron averaged over a whole assembly of the conduction electrons and that it becomes finite, only when the field is applied and the Fermi sphere is displaced from the origin.

A steady state will be established in a time of the order of τ upon the application of an electric field $\mathbf{E}$. The equation of motion of the conduction electron in the presence of an electric field $\mathbf{E}$ is then expressed as

$$m\left(\frac{d\mathbf{v}_D}{dt} + \frac{\mathbf{v}_D}{\tau}\right) = (-e)\mathbf{E}, \qquad (10.3)$$

where the second term proportional to the drift velocity represents the frictional force and plays a role in resisting the accelerated motion of the electron.

10.2 The Drude theory for electrical conductivity

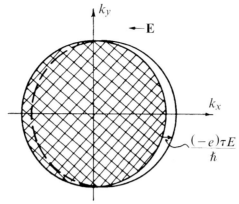

Figure 10.1. The Fermi surface establishes a steady state after a displacement of $(-e)\tau E/\hbar$ in the presence of an electric field **E**.

The reason why τ is called the relaxation time is as follows. Suppose the electric field is turned off. The process for the drift velocity $\mathbf{v}_D$ to decrease to zero will follow the equation of motion:

$$m\left(\frac{d\mathbf{v}_D}{dt} + \frac{\mathbf{v}_D}{\tau}\right) = 0,$$

the solution of which is reduced to $\mathbf{v}_D(t) = \mathbf{v}_D(0)\exp(-t/\tau)$. The parameter τ obviously represents the time for the drift velocity to decay to $1/e$ times $\mathbf{v}_D(0)$.

The relation $d\mathbf{v}_D/dt = 0$ should hold in equation (10.3), when the steady state is reached. Equation (10.3) is immediately solved as

$$\mathbf{v}_D = \frac{(-e)\tau \mathbf{E}}{m}. \tag{10.4}$$

The Fermi sphere is thus displaced by the amount $\Delta k_x = (m/\hbar)\mathbf{v}_D = (-e)\tau E/\hbar$, when the electric field is applied, say, along the x-axis, as illustrated in Fig. 10.1.

Since the electrical current density **J** is defined as $\mathbf{J} = n(-e)\mathbf{v}_D$, equation (10.4) is rewritten as

$$\mathbf{J} = \left(\frac{ne^2\tau}{m}\right)\mathbf{E}, \tag{10.5}$$

where n is the number of conduction electrons per unit volume. Equation (10.5) obviously explains Ohm's law. For simplicity, we assume an isotropic metal, where the electrical conductivity is defined as the ratio of the current density over the electric field, $\sigma = J/E$, in a scalar quantity. The electrical conductivity is explicitly written as

$$\sigma = \frac{ne^2\tau}{m}. \tag{10.6}$$

Table 10.1. *Electron transport properties of metals at 273 K*

element	electrical conductivity, σ ($\times 10^6$ / Ω-m)	resistivity, ρ ($\mu\Omega$-cm)	TCR[a], α ($\times 10^{-3}$/K)
Li	11.8	8.5	4.37
Na	23.4	4.27	5.5
Cu	64.5	1.55	4.33
Ag	66	1.5	4.1
Au	49	2.04	3.98
Mg	25.4	3.94	4.2
Ca	28	3.6	4
Zn	18.3	5.45	4.20
Al	40	2.50	4.67
Pb	5.17	19.3	4.22
Bi	0.93	107	
Ti	2.38	42	5.5
V	0.54	18.2	
Fe	11.5	8.71	6.57
Zr	2.47	40.5	4.0
W	20.4	4.89	4.83

Note:
[a] Temperature coefficient of resistivity (TCR): $\alpha_{273\,K} = (1/\rho)(d\rho/dT)_{T=273\,K}$

The electrical resistivity ρ, being defined as an inverse to the conductivity, is also frequently employed:

$$\rho = \frac{m}{ne^2\tau}. \tag{10.7}$$

The relation (10.6) was first derived by Drude in 1900 before the advent of quantum theory and is often referred to as the Drude conductivity formula. The electrical conductivity and the resistivity at 273 K for representative metals in the periodic table are listed in Table 10.1, along with the temperature coefficient of the resistivity or TCR.

Let us estimate the magnitude of the relaxation time τ of the conduction electron by inserting the measured resistivity into equation (10.7). First, we note that τ is deduced in units of [s], if all relevant quantities in equation (10.7) are inserted in SI units:[1]

---

[1] A multiplication factor 9.0×10^{11} must be used if the resistivity, mass of the electron, its charge and the number density are inserted in units of Ω-cm, g, esu and cm^{-3}, respectively:

$$\frac{m}{\rho n e^2} = \frac{[g]}{[\Omega\text{-cm}][cm]^{-3}[esu]^2} = \frac{[g]}{\left\{\frac{10^7}{(3\times 10^9)^2}\right\}\left(\frac{[erg][s]}{[esu]^2}\right)\frac{[esu]^2}{[cm]^2}} = 9.0\times 10^{11}\,[s].$$

10.2 The Drude theory for electrical conductivity

$$\frac{m}{\rho n e^2} = \frac{[\text{kg}]}{[\Omega\text{-m}][\text{m}]^{-3}[\text{coulomb}]^2} = \frac{[\text{kg}]}{\left[\frac{\text{volt}}{\text{ampere}}\right][\text{m}]^{-2}[\text{coulomb}]^2}$$

$$= \frac{[\text{kg}]}{\left[\left(\frac{\text{joule}}{\text{coulomb}}\right) \Big/ \left(\frac{\text{coulomb}}{\text{s}}\right)\right][\text{m}]^{-2}[\text{coulomb}]^2} = [\text{s}]. \quad (10.8)$$

The value of τ in pure Cu at room temperature is easily deduced to be $\tau = 2.73 \times 10^{-14}$ s by inserting the electron mass $m = 9.1 \times 10^{-31}$ kg, its charge $|e| = 1.6 \times 10^{-19}$ coulomb, the number density $n = 8.9 \times 10^6 \times 6 \times 10^{23}/63.54 = 0.84 \times 10^{29}$ m^{-3} and the observed resistivity $\rho = 1.55 \times 10^{-8}$ Ω-m into equation (10.8). The mean free path given by the product of the relaxation time and the Fermi velocity turns out to be of the order of a few tens nm, since the Fermi velocity is of the order of 10^6 m/s.

It is important to realize how the drift velocity v_D differs from the Fermi velocity v_F in magnitude. The drift velocity v_D is easily calculated to be of the order of 10^{-2} m/s for pure Cu, if a typical value of $E = 10$ [V]/[m] is inserted into equation (10.4), together with the relaxation time obtained above. Thus, we see that the drift velocity v_D is about 10^{-8} times the Fermi velocity v_F.

The drift velocity defined by equation (10.4) depends on the magnitude of the applied electric field and, hence, is not appropriate as a physical quantity specific to a given material. The drift velocity per electric field is defined as the mobility μ and is given by

$$\mu = \frac{v_D}{E} = \frac{(-e)\tau}{m} \quad (10.9)$$

The mobility of the electron becomes negative, since the electronic charge is negative, but an absolute value is conventionally used.

The mobility of pure Cu is immediately calculated to be 4.7×10^{-3} [m]^2/[volt][s] by inserting the relaxation time obtained above into equation (10.9). The mobility is often expressed in practical units of [cm]^2/[volt][s]. Its conversion from CGS units results in a numerical factor 1/300 as shown below:

$$\frac{(-e)\tau}{m} = \left[\frac{[\text{esu}][\text{s}]}{[\text{g}]}\right] = \left[\frac{[\text{esu}][\text{cm}]^2}{[\text{g}][\text{cm}]^2[\text{s}]^{-2}[\text{s}]}\right] = \left[\frac{\left[\frac{\text{esu}}{\text{coulomb}}\right][\text{coulomb}][\text{cm}]^2}{\left[\frac{\text{erg}}{\text{joule}}\right][\text{joule}][\text{s}]}\right].$$

$$= \left[\frac{(3 \times 10^9)^{-1}[\text{coulomb}][\text{cm}]^2}{10^{-7}[\text{joule}][\text{s}]}\right] = \left(\frac{1}{300}\right)\left[\frac{[\text{cm}]^2}{[\text{volt}][\text{s}]}\right] \quad (10.10)$$

The mobility of Cu turns out to be $\mu = 47$ [cm]^2/[volt][s], indicating that the conduction electron drifts 47 cm per second, if 1 volt is applied to a pure Cu rod, 1 cm in length.

Finally, it is worth noting that the mobility is given by the product of the electrical conductivity in equation (10.6) and the Hall coefficient given by equation (11.38):

$$\mu = \sigma R_H, \quad \mu = c\sigma R_H \text{ [CGS]}, \tag{10.11}$$

where c is the speed of light. This relation is often employed to determine the mobility experimentally.

10.3 Motion of electrons in a crystal: (I) – wave packet of electrons

In the preceding section, we discussed very basic properties of electron transport phenomena by treating the electron with a mass m and a charge $(-e)$ in the free-electron model. The equation of motion given by equation (10.3) is based on classical mechanics. This is not satisfactory in a crystal metal, since the conduction electron must be described in terms of the Bloch state of the wave vector **k**. The effect of the band structure on the electron transport properties cannot be taken into account, as long as electrons are treated as classical particles.

In quantum mechanics, a particle of energy ε is equivalent to a wave of angular frequency ω through the well-known relation $\varepsilon = \hbar\omega$.[2] A medium is said to possess a dispersion, if the frequency ω depends on the wave vector **k**. In such a dispersive medium, a wave packet can be constructed by superimposing waves having different wave vectors in the vicinity of a given angular frequency ω. The velocity of the wave packet is called the group velocity and is defined as $\mathbf{v}_k = \nabla_k \omega_k = (1/\hbar)(\partial \varepsilon(\mathbf{k})/\partial \mathbf{k})$. The group velocity in the free-electron model is reduced to $\mathbf{v}_k = \hbar \mathbf{k}/m$, since its energy dispersion is $\varepsilon = \hbar^2 k^2/2m$.

We now construct the wave packet from the free electrons and derive its equation of motion in the presence of a potential $V(x)$ in a one-dimensional system. The time-dependent Schrödinger equation for the electron is given by

$$-\frac{\hbar^2}{2m}\frac{\partial^2 \Psi(x,t)}{\partial x^2} + V(x)\Psi(x,t) = i\hbar\frac{\partial \Psi(x,t)}{\partial t}, \tag{10.12}$$

where $\Psi(x, t)$ is the time-dependent wave function. If the Hamiltonian is independent of time so that the system is conservative, the wave function can be

---

[2] In the remaining chapters, except for Chapter 15, the symbol **E** or E is reserved for the electric field and the energy of the electron is hereafter denoted as $\varepsilon(k)$.

10.3 Motion of electrons in a crystal: (I) – wave packet of electrons

separated into the form $\Psi(x, t) = \psi(x) f(t)$. By inserting this into equation (10.12), we obtain a solution $\Psi(x, t) = \psi_k(x) e^{-i\varepsilon(k)t/\hbar}$. Here $\psi_k(x)$ is a time-independent eigenfunction and $\varepsilon(k)$ is the corresponding eigenvalue of the following Schrödinger equation:

$$-\frac{\hbar^2}{2m}\frac{\partial^2 \psi_k(x)}{dx^2} + V(x)\psi_k(x) = \varepsilon(k)\psi_k(x). \tag{10.13}$$

A general solution of equation (10.12) in a conservative system is written in the form:

$$\Psi(x, t) = \sum_k A(k)\psi_k(x) e^{-i\varepsilon(k)t/\hbar} \tag{10.14a}$$

and, if the energy spectrum is continuous, the sum may be replaced by an integral:

$$\Psi(x, t) = \int A(k)\psi_k(x) e^{-i\varepsilon(k)t/\hbar} dk. \tag{10.14b}$$

Let us assume that the coefficient $A(k)$ is large only in a particular range of the wave number k. For instance, the Gaussian function $A(k) = ce^{-a^2(k-K)^2}$ takes its maximum at $k = K$ and falls to $1/e$ times the maximum at $k - K = \pm 1/a$. Equation (10.14b) can be easily calculated, when $A(k)$, $\psi_k(x)$ and $\varepsilon(k)$ are given by the Gaussian function, the plane wave $e^{i[kx - \omega(k)t]}$ and the dispersion relation $\varepsilon(k) = \hbar^2 k^2 / 2m$ of the free electron, respectively (see Exercise 10.1). One can find that the amplitude of the wave function $\Psi(x, t)$ is virtually zero except at the very center of the wave packet. This is the wave packet of the conduction electrons.

An average of the x-coordinate corresponding to the center of the wave packet is calculated from

$$<x> = \int_{-\infty}^{\infty} \Psi^*(x, t) x \Psi(x, t) dx. \tag{10.15}$$

The group velocity of the wave packet is then obtained by differentiating both sides of equation (10.15) with respect to time:

$$\frac{d<x>}{dt} = \int_{-\infty}^{\infty} \left(x\Psi^*(x, t)\frac{\partial \Psi(x, t)}{\partial t} + x\Psi(x, t)\frac{\partial \Psi^*(x, t)}{\partial t} \right) dx. \tag{10.16}$$

The right-hand side of equation (10.16) can be obtained by multiplying equation (10.12) by $x\Psi^*(x, t)$ and then subtracting its complex conjugate multiplied by $x\Psi(x, t)$ with subsequent integration over x. By repeating integrations by parts twice, we obtain

$$i\hbar\frac{d<x>}{dt} = \frac{\hbar^2}{m}\int_{-\infty}^{\infty} \Psi^*(x, t)\frac{\partial \Psi^*(x, t)}{\partial t} dx, \tag{10.17}$$

where $\lim_{x\to\pm\infty}\Psi(x, t)=0$ is used as a boundary condition. Equation (10.17) is further differentiated with respect to time and then integration by parts leads to the relation:

$$m\frac{d^2\langle x\rangle}{dt^2}=\frac{\hbar}{i}\int_{-\infty}^{\infty}\left(\frac{\partial\Psi^*}{\partial t}\frac{\partial\Psi}{\partial x}-\frac{\partial\Psi}{\partial t}\frac{\partial\Psi^*}{\partial x}\right)dx. \qquad (10.18)$$

After equation (10.12) is inserted into equation (10.18), integration by parts is again carried out to reach the final form:

$$m\frac{d^2\langle x\rangle}{dt^2}=-\frac{\hbar^2}{2m}\int_{-\infty}^{\infty}\left(\frac{\partial^2\Psi^*}{\partial x^2}\frac{\partial\Psi}{\partial x}+\frac{\partial^2\Psi}{\partial x^2}\frac{\partial\Psi^*}{\partial x}\right)dx$$

$$+\int_{-\infty}^{\infty}V(x)\left(\Psi^*\frac{\partial\Psi}{\partial x}+\Psi\frac{\partial\Psi^*}{\partial x}\right)dx$$

$$=-\int_{-\infty}^{\infty}\frac{\partial V}{\partial x}\Psi^*(x, t)\Psi(x, t)dx=-\left\langle\frac{\partial V(x)}{\partial x}\right\rangle. \qquad (10.19)$$

Equation (10.19) indicates that the wave packet follows the Newton equation of motion in the potential field $V(x)$ [1]. Thus, the wave packet can be treated as if it is a classical particle, provided that the wave packet is well localized in reciprocal space.

The wave function of the conduction electron in a crystal should be described by the Bloch wave. The time-dependent Bloch wave may be written as

$$\Psi_{\mathbf{k}}(\mathbf{r}, t)=u_{\mathbf{k}}(\mathbf{r})e^{i\mathbf{k}\cdot\mathbf{r}}e^{-i\varepsilon(\mathbf{k})t/\hbar}. \qquad (10.20)$$

A wave packet can be constructed from Bloch waves in the same way as that from plane waves in equation (10.14) and is expressed as

$$\Psi(\mathbf{r}, t)=\int A(\mathbf{k}')u_{\mathbf{k}'}(\mathbf{r})e^{i[\mathbf{k}'\cdot\mathbf{r}-\varepsilon(\mathbf{k}')t/\hbar]}d\mathbf{k}', \qquad (10.21)$$

where $A(\mathbf{k}')$ has a sharp maximum at $\mathbf{k}'=\mathbf{k}$ and falls rapidly to zero as soon as $\mathbf{k}'$ departs from $\mathbf{k}$.[3]

---

[3] The width Δk of the wave packet should be much smaller than the size of the first Brillouin zone. Hence, $\Delta k \ll 2\pi/a$ should hold, where a is the lattice constant. The uncertainty principle requires $\Delta x\cdot\Delta p\approx\hbar$ or $\Delta x\cdot\Delta k\approx 1$. This leads to $\Delta x\gg a/2\pi$. Hence, the wave packet will extend over several atomic spacings in real space.

The integration can be done only in the vicinity of $\mathbf{k}' = \mathbf{k}$. Hence, the energy eigenvalue $\varepsilon(\mathbf{k}')$ of the Bloch electron can be expanded about $\mathbf{k}' = \mathbf{k}$ as follows:

$$\varepsilon(\mathbf{k}') = \varepsilon(\mathbf{k}) + (\mathbf{k}' - \mathbf{k}) \cdot \nabla_\mathbf{k} \varepsilon(\mathbf{k}) + \cdots . \tag{10.22}$$

Equation (10.21) is now approximated as

$$\Psi(\mathbf{r}, t) = e^{i[\mathbf{k} \cdot \mathbf{r} - \varepsilon(\mathbf{k})t/\hbar]} \int A(\mathbf{k}') u_{\mathbf{k}'}(\mathbf{r}) e^{i[\mathbf{r} - t\nabla_\mathbf{k} \varepsilon(\mathbf{k})/\hbar] \cdot (\mathbf{k}' - \mathbf{k})} d\mathbf{k}'. \tag{10.23}$$

The periodic function $u_{\mathbf{k}'}(\mathbf{r})$ is assumed to vary slowly with $\mathbf{k}'$ in the limited range centered at $\mathbf{k}' = \mathbf{k}$ and is pulled outside the integral. Then we obtain

$$\Psi(\mathbf{r}, t) = \Psi_\mathbf{k}(\mathbf{r}, t) \int A(\mathbf{k}') e^{i[\mathbf{r} - t\nabla_\mathbf{k} \varepsilon(\mathbf{k})/\hbar] \cdot (\mathbf{k}' - \mathbf{k})} d\mathbf{k}'. \tag{10.24}$$

Here $\Psi_\mathbf{k}(\mathbf{r}, t)$ is the Bloch wave given by equation (10.20) and, hence, $|\Psi_\mathbf{k}(\mathbf{r}, t)|^2$ remains unchanged with time. According to equation (10.24), the wave packet centered at $\mathbf{k}' = \mathbf{k}$ displaces its position by $t\nabla_\mathbf{k} \varepsilon(\mathbf{k})/\hbar$ after t seconds. This implies that the group velocity of the wave packet is given by

$$\mathbf{v} = \frac{1}{\hbar} \nabla_\mathbf{k} \varepsilon(\mathbf{k}). \tag{10.25}$$

It is now clear that the effect of the band structure energy $\varepsilon(\mathbf{k})$ of the Bloch electron on electron transport phenomena enters through the group velocity of the wave packet.

10.4 Motion of electrons in a crystal: (II)

We assume that equation (10.25) holds valid when the electric field $\mathbf{E}$ is applied to a metal. Then the work done by the field on a wave packet having group velocity $\mathbf{v}$ is obviously given by

$$(-e)\mathbf{E} \cdot \mathbf{v} = \frac{(-e)}{\hbar} \mathbf{E} \cdot \nabla_\mathbf{k} \varepsilon(\mathbf{k}). \tag{10.26}$$

Since the work must be equal to the change in the electron energy $d\varepsilon/dt$, we obtain

$$(-e)\mathbf{E} \cdot \mathbf{v} = \frac{d\varepsilon}{dt} = \left(\frac{\partial \varepsilon(\mathbf{k})}{\partial \mathbf{k}}\right)\left(\frac{d\mathbf{k}}{dt}\right). \tag{10.27}$$

A comparison of equations (10.26) and (10.27) immediately leads to the well-known relation

$$\frac{d\mathbf{k}}{dt} = \frac{(-e)}{\hbar} \mathbf{E}. \tag{10.28}$$

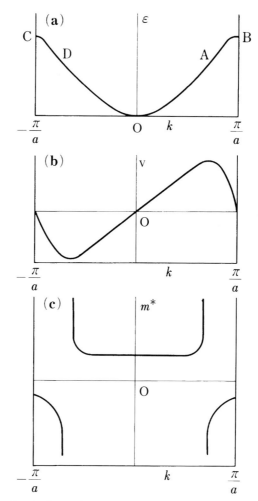

Figure 10.2. (a) ε–k relation of the Bloch electron in the first Brillouin zone, (b) corresponding group velocity and (c) effective mass. Points A and D correspond to inflection points in the ε–k relation.

We see that equation (10.1) is now extended to the Bloch wave in a crystal. In other words, the wave packet centered at the wave vector **k** propagates with a constant velocity in reciprocal space under a given electric field **E**. The discussion above holds true, as long as the concept of the wave packet is justified for the Bloch state. We proceed with our discussion in the framework of this justification and, hence, the conduction electron or the Bloch electron or simply the electron is hereafter meant as the wave packet.

The ε–k relation for a typical metal is depicted in Fig. 10.2(a). The energy gap appears at the zone boundary. Let us consider the motion of the

10.4 Motion of electrons in a crystal: (II)

conduction electron characterized by the band structure energy $\varepsilon(\mathbf{k})$ in the presence of an electric field. First, we take the electron of $\mathbf{k}=0$ and apply the electric field in the negative direction along the x-axis. Then we need to consider only one-dimensional motion along the x-direction. The electron carrying a negative charge experiences the force $(-e)E$ and begins to travel in a positive direction in accordance with equation (10.28). In the case of free electrons without involving any scattering mechanism, the Fermi sphere moves indefinitely under the operation of the electric field (see Section 10.2). But we now have the band structure shown in Fig. 10.2. The electron departs from the origin O and reaches point B after passing the inflection point A in the ε–k relation. However, point B is identical to point C, which is separated by the reciprocal lattice vector from point B. Thus, the electron reaching point B reappears at point C and returns to the origin after passing another inflection point D. This motion will be indefinitely repeated in the first Brillouin zone enclosed by the points B and C.

The Bragg reflection takes place in the periodic potential, whenever the electron reaches point B or C, and the Fermi sphere moves back and forth endlessly within the first Brillouin zone. Of course, the repeated motion of the electron between points B and C has nothing to do with the scattering and, hence, makes no contribution to the resistivity. As mentioned in Section 10.2, the scattering mechanism must be introduced to bring the Fermi sphere to a steady state after a certain displacement upon application of the electric field.

The acceleration of the conduction electron can be calculated by differentiating equation (10.25) with respect to time with a subsequent use of equations (10.27) and (10.28):

$$\frac{d\mathbf{v}}{dt} = \frac{1}{\hbar}\frac{d}{dt}\nabla_\mathbf{k}\varepsilon(\mathbf{k}) = \frac{1}{\hbar}\nabla_\mathbf{k}\left(\frac{d\varepsilon(\mathbf{k})}{dt}\right) = \frac{(-e)}{\hbar^2}\nabla_\mathbf{k}[\mathbf{E}\cdot\nabla_\mathbf{k}\varepsilon(\mathbf{k})] \qquad (10.29)$$

or

$$\frac{dv_i}{dt} = \frac{(-e)}{\hbar^2}\sum_j \frac{\partial^2\varepsilon(\mathbf{k})}{\partial k_i \partial k_j} E_j \qquad (10.30)$$

in the vector component representation.

Now we can find the one-to-one correspondence with the classical particle having its mass m and charge $(-e)$. The Newton equation of motion for the classical particle in an electric field $\mathbf{E}$ is given by

$$\frac{d\mathbf{v}}{dt} = \frac{(-e)}{m}\mathbf{E}. \qquad (10.31)$$

Equation (10.31) may be extended to the form $dv_i/dt = (-e)\sum_j (1/m^*)_{ij} E_j$ by replacing the mass of the free electron by the effective mass tensor m^*_{ij}. A comparison with equation (10.30) leads to the relation:

$$\left(\frac{1}{m^*}\right)_{ij} = \frac{1}{\hbar^2}\frac{\partial^2 \varepsilon(\mathbf{k})}{\partial k_i \partial k_j}, \qquad (10.32)$$

where m^* is called the effective mass of the electron in a crystal and is a tensor. We see that the band structure is also reflected in the effective mass through the second derivative of the ε–k relation.

The group velocity and the effective mass of an electron, whose ε–k relation is shown in Fig. 10.2(a), can be calculated from equations (10.25) and (10.32). The results are shown in Figs. 10.2(b) and (c); the group velocity increases almost linearly in the range OA and reaches its maximum at point A with a subsequent decrease to zero at point B. Then, it reappears at point C and decreases up to the point D, from which its slope changes to a positive sign before returning to the origin O.

Suppose that the conduction electron with wave vector $\mathbf{k}$ is travelling in the periodic potential associated with the family of lattice planes having the reciprocal lattice vector $\mathbf{g}$. The wave function $\psi_\mathbf{k}(\mathbf{r}) = e^{i\mathbf{k}\cdot\mathbf{r}}(1 + \alpha e^{i\mathbf{g}\cdot\mathbf{r}} + \beta e^{-i\mathbf{g}\cdot\mathbf{r}})$ represents the Bloch state.[4] In a one-dimensional periodic potential of lattice constant a, the shortest reciprocal lattice vector $\mathbf{g}$ is equal to $g = 2\pi/a$. Consider first the region $0 < k \leq \pi/a$. The electron wave e^{ikx} describes its motion to the right, whereas the electron wave $e^{i[k-(2\pi/a)]x}$ describes that to the left by receiving the backward crystal momentum $\hbar g$ from the family of lattice planes. In the region OA, the ε–k relation is free-electron-like and, hence, both coefficients α and β are zero and the motion of the electron can be well described by the wave function e^{ikx}. However, once the electron passes point A, it begins to receive the crystal momentum from the periodic potential, resulting in a decrease in the slope of the ε–k relation. This corresponds to an increase in the coefficient β of the electron wave $e^{i[k-(2\pi/a)]x}$ running to the left. At point B, the slope of the ε–k relation is reduced to zero and the energy gap is opened. Indeed, here $\alpha = 0$ and $\beta = 1$ hold and the standing wave $[e^{i(\pi/a)x} \pm e^{-i(\pi/a)x}]$ is formed at point B, where $k = \pi/a$. This corresponds to the Bragg reflection, as discussed in Section 5.6. The group velocity at point B is reduced to zero, as indicated in Fig. 10.2(b).

We have emphasized above that point C is equivalent to point B and, hence,

---

[4] If we write $u_\mathbf{k}(\mathbf{r}) = 1 + \alpha e^{i\mathbf{g}\cdot\mathbf{r}} + \beta e^{-i\mathbf{g}\cdot\mathbf{r}}$, then we can easily confirm the relation $u_\mathbf{k}(\mathbf{r}+\mathbf{l}) = 1 + \alpha e^{i\mathbf{g}\cdot(\mathbf{r}+\mathbf{l})} + \beta e^{-i\mathbf{g}\cdot(\mathbf{r}+\mathbf{l})} = u_\mathbf{k}(\mathbf{r})$ for any lattice vector $\mathbf{l}$.

the electron reaching point B appears at point C. As a matter of fact, the group velocity at point C is also zero. In the region $-\pi/a < k \leq 0$, the electron wave e^{ikx} is running to the left, whereas the wave $e^{i[k+(2\pi/a)]x}$ runs to the right. In the region from point C to point D, the coefficient α in the wave $e^{i[k+(2\pi/a)]x}$ decreases gradually from unity. Once the electron passes point D, both coefficients α and β become zero. Thus, the wave function is reduced to the plane wave e^{ikx} and free-electron behavior is resumed.

As is clear from the argument above, the conduction electron propagating in the periodic potential in the presence of an electric field receives the crystal momentum of $-\hbar g$ or $+\hbar g$ from the lattice planes, when moving through regions AB and CD. The mixing of the electron wave $e^{i(k-g)x}$ with the plane wave e^{ikx} in the region AB is viewed as a braking motion due to the crystal momentum acting against the acceleration of the wave packet by the electric field and eventually forces the electron to form a standing wave at point B. Thus, we see that the electron in regions AB and CD behaves as if it is accelerated by the field in a direction opposite to that for the free electron. This is the reason why the effective mass becomes negative in this region, Fig. 10.2(c).

So far we have discussed only the effect of the electric field on the motion of the conduction electron. Equation (10.28) may be extended to the case where a magnetic field is also present:

$$\frac{d\mathbf{k}}{dt} = \frac{(-e)}{\hbar}(\mathbf{E} + \mathbf{v} \times \mathbf{B}), \quad \frac{d\mathbf{k}}{dt} = \frac{(-e)}{\hbar}\left[\mathbf{E} + \left(\frac{\mathbf{v}}{c} \times \mathbf{H}\right)\right] \text{[CGS]}, \quad (10.33)$$

where the second term has already appeared in equation (7.2). Unfortunately, this term cannot be deduced in the same way as the derivation of equations (10.26) to (10.28), because $d\mathbf{k}/dt$ is no longer parallel to $\mathbf{v}$ in the presence of a magnetic field. We proceed with our discussion by presuming that equation (10.33) can be extended to the Bloch wave in the presence of both electric and magnetic fields. Note that the velocity in equation (10.33) can be calculated from equation (10.25).

10.5 Electrons and holes

In a semiconductor like Si, discussed in Section 6.9, the valence band is separated from the conduction band by an energy gap of the order of 1 eV. At absolute zero, the valence band is completely filled with electrons, whereas the conduction band is completely empty. Its $\varepsilon(k)$–k relation is shown in Fig. 10.3. The electrical current density due to the wave packet centered at the wave vector $\mathbf{k}$ in the valence band is expressed as $\mathbf{j_k} = (-e)\mathbf{v_k}$. Let us consider the

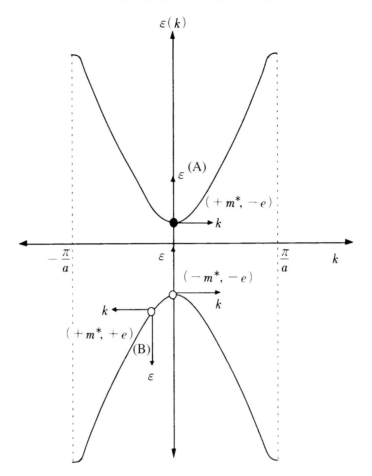

Figure 10.3. ε–k relation of valence and conduction bands in an intrinsic semiconductor. Upward and right directions are taken as positive directions of the energy and wave vector in coordinates (A), whereas downward and left directions as positive directions in coordinates (B), respectively. Coordinates (A) are used to describe the motion of an electron near the bottom of the conduction band. Coordinates (B) are used to describe that of a hole near the top of the valence band.

motion of the remaining electrons in the valence band, when a single electron in the valence band is excited into the conduction band.

If the Brillouin zone corresponding to the valence band is fully filled with electrons, no current can flow and, hence, $\mathbf{J} = \Sigma_{\mathbf{k}} (-e)\mathbf{v}_{\mathbf{k}} = 0$ must hold. Now we assume the situation such that the electron of wave vector $\mathbf{k}$ in the valence band is excited into the conduction band so that the state $\mathbf{k}$ is left vacant. The net current would vanish if the resulting vacancy is filled with an electron carrying the current density $\mathbf{J}_{\mathbf{k}} = (-e)\mathbf{v}_{\mathbf{k}}$, and, hence, the relation

$\left(\Sigma_{\mathbf{k}'\neq\mathbf{k}}(-e)\mathbf{v}_{\mathbf{k}'}\right)+(-e)\mathbf{v}_{\mathbf{k}}=0$ holds. Therefore, the current density due to the vacant state can be expressed as

$$\mathbf{J}_{\mathbf{k}}=\sum_{\mathbf{k}'\neq\mathbf{k}}(-e)\mathbf{v}_{\mathbf{k}}=-(-e)\mathbf{v}_{\mathbf{k}}=(+e)\mathbf{v}_{\mathbf{k}}, \qquad (10.34)$$

indicating that the vacant state in the valence band behaves as if it carries a positive charge.

Let us apply the electric field to a valence band containing one vacant state. All remaining electrons move into a direction opposite to the electric field $\mathbf{E}$ due to the external force $(-e)\mathbf{E}$. The vacant state is also forced to move together with the electrons and follows equation (10.30) in the same way as all the remaining electrons. But we must remember that the $\varepsilon(k)$–k curve of the valence band is convex upward and that the group velocity calculated from equation (10.25) decreases with increasing wave vector $\mathbf{k}$, resulting in a negative effective mass. This refers to the motion in the coordinates marked as $(-m^*, -e)$ in Fig. 10.3. However, the use of a negative effective mass for the vacant state is not convenient. We will employ below the alternative coordinates shown as (B) in Fig. 10.3 to describe the motion of the vacant state.

As discussed above, the motion of the vacant state is expressed in the ordinary coordinates as

$$(-m^*)\frac{d\mathbf{v}}{dt}=(-e)\mathbf{E}, \qquad (10.35)$$

which is rewritten as

$$m^*\frac{d\mathbf{v}}{dt}=-(-e)\mathbf{E}=(+e)\mathbf{E}. \qquad (10.36)$$

Now we see that the vacant state behaves as a particle possessing a positive effective mass with a positive charge in the coordinates (B) marked as $(+m^*, +e)$ in Fig. 10.3. The vacant state is called a positive hole. Note that the positive direction of its energy axis must be taken downward in order to meet the condition $m^*>0$. At the same time, the positive direction of the wave vector $\mathbf{k}$ must be taken toward the left because of the possession of a positive charge $(+e)$ opposite to that of an electron. Note that we have been accustomed to the coordinates (A) to describe the motion of electrons in the conduction band characterized by a parabolic band, as in Fig. 5.6. Both coordinates (A) and (B) are chosen as being symmetric with respect to the origin. A motion of a hole downward along its energy axis raises the energy of the electron system, since it is equivalent to the motion of an electron upward by the same amount.

The valence band of a semiconductor at absolute zero is completely filled with electrons. We can alternatively say that the conduction band is completely filled with positive holes. At finite temperatures, some electrons in the valence band are excited into the conduction band, while holes in the conduction band are excited into the valence band. The $\varepsilon(k)$–k relation in the conduction band corresponds to the excitation of electrons into the Brillouin zone filled with positive holes and constitutes a one-to-one correspondence with the excitation of positive holes in the valence band. The energy axis of the respective excitations is always chosen such that the energy increases with increasing wave number k.

10.6 Boltzmann transport equation

We have learned in the preceding sections that an electron accelerated by external fields establishes a steady state through the scattering process due to disturbances in an otherwise perfectly periodic potential in a crystal. The Boltzmann transport equation is formulated by considering the balance of the distribution function in the steady state brought about by external fields in the presence of the scattering process for an electron or a hole at the position $\mathbf{r}$ represented as the wave packet centered at the wave vector $\mathbf{k}$.

First, we consider a system, in which only a temperature gradient exists and causes the electron to diffuse with the velocity $\mathbf{v}_\mathbf{k}$. Since the electron travels a distance $\mathbf{v}_\mathbf{k}\cdot\Delta t$ after Δt, the electron distribution $f(\mathbf{r}, \mathbf{k}, t)$ at the position $(\mathbf{r}, \mathbf{k})$ in the phase space at a time t would be equal to that at the position $(\mathbf{r}-\mathbf{v}_\mathbf{k}\cdot\Delta t, \mathbf{k})$ at the time $t-\Delta t$.[5] In other words, the relation $f(\mathbf{r}, \mathbf{k}, t)=f(\mathbf{r}-\mathbf{v}_\mathbf{k}\cdot\Delta t, \mathbf{k}, t-\Delta t)$ is assumed to hold. A change in the electron distribution due to diffusion is then approximated as

$$\left(\frac{\partial f(\mathbf{k})}{\partial t}\right)_{\text{diffusion}} = \frac{f(\mathbf{r},\mathbf{k},t)-f(\mathbf{r},\mathbf{k},t-\Delta t)}{\Delta t} = \frac{f(\mathbf{r}-\mathbf{v}_\mathbf{k}\cdot\Delta t,\mathbf{k},t-\Delta t)-f(\mathbf{r},\mathbf{k},t-\Delta t)}{\Delta t}$$

$$= -\mathbf{v}_\mathbf{k}\cdot\frac{\partial f(\mathbf{r},\mathbf{k})}{\partial \mathbf{r}} = -\mathbf{v}_\mathbf{k}\cdot\nabla f(\mathbf{r},\mathbf{k}), \qquad (10.37)$$

where $\mathbf{v}_\mathbf{k}\cdot\nabla f(\mathbf{r}, \mathbf{k})=v_{k_x}\cdot(\partial f/\partial x)+v_{k_y}\cdot(\partial f/\partial y)+v_{k_z}\cdot(\partial f/\partial z)$.

In contrast to a temperature gradient, both electrical and magnetic fields cause the wave vector $\mathbf{k}$ to change in accordance with equation (10.33). In the

---

[5] According to classical mechanics, the motion of an assembly of n particles is completely determined, once their position coordinates $q_1, q_2, q_3, \ldots q_n$ and their conjugate momenta $p_1, p_2, p_3, \ldots p_n$ are given at a time t. Any state is described as a point in the $2n$-dimensional space consisting of mutually perpendicular $q_1, q_2, q_3, \ldots q_n, p_1, p_2, p_3, \ldots p_n$ axes. This is called the phase space. The motion of particles is given by a trajectory in the phase space. In the present section, the wave vector $\mathbf{k}$ is used in place of the momentum $\mathbf{p}$.

10.6 Boltzmann transport equation

same manner as above, the electron distribution $f(\mathbf{r}, \mathbf{k}, t)$ at the position $(\mathbf{r}, \mathbf{k})$ at a time t would be equal to that at the position $(\mathbf{r}, \mathbf{k} - (\partial \mathbf{k}/\partial t)\Delta t, t - \Delta t)$ at the time $t - \Delta t$. By inserting the resulting relation $f(\mathbf{r}, \mathbf{k}, t) = f(\mathbf{r}, \mathbf{k} - (\partial \mathbf{k}/\partial t)\Delta t, t - \Delta t))$ into equation (10.33), we find the electron distribution to change at the rate:

$$\left(\frac{\partial f(\mathbf{k})}{\partial t}\right)_{\text{field}} = \frac{f(\mathbf{r},\mathbf{k},t) - f(\mathbf{r},\mathbf{k},t - \Delta t)}{\Delta t} = \frac{f\left(\mathbf{r},\mathbf{k} - \frac{\partial \mathbf{k}}{\partial t}\Delta t, t - \Delta t\right) - f(\mathbf{r},\mathbf{k},t - \Delta t)}{\Delta t}$$

$$= -\left(\frac{\partial \mathbf{k}}{\partial t}\right)_{\text{field}} \cdot \frac{\partial f(\mathbf{k})}{\partial \mathbf{k}} = -\frac{(-e)}{\hbar}(\mathbf{E} + \mathbf{v}_k \times \mathbf{B}) \cdot \frac{\partial f_\mathbf{k}}{\partial \mathbf{k}}. \tag{10.38}$$

Establishing a steady state means that a change in the electron distribution caused by external fields and/or a temperature gradient is balanced with that of the scattering process $(\partial f/\partial t)$. Since a net change in the electron distribution df/dt is given by the sum of the three contributions:

$$\frac{df}{dt} = \left(\frac{\partial f}{\partial t}\right)_{\text{diffusion}} + \left(\frac{\partial f}{\partial t}\right)_{\text{field}} + \left(\frac{\partial f}{\partial t}\right)_{\text{scatter}}, \tag{10.39}$$

we have the relation $df/dt = 0$ in the steady state. An insertion of equations (10.37) and (10.38) into equation (10.39) gives rise to the Boltzmann transport equation:

$$-\mathbf{v}_k \cdot \nabla f(\mathbf{r}, \mathbf{k}) - \frac{(-e)}{\hbar}(\mathbf{E} + \mathbf{v}_k \times \mathbf{B}) \cdot \frac{\partial f_\mathbf{k}}{\partial \mathbf{k}} = -\left(\frac{\partial f}{\partial t}\right)_{\text{scatter}}. \tag{10.40}$$

The steady state electron distribution function $f(\mathbf{r}, \mathbf{k})$ in the presence of external fields and/or a temperature gradient must deviate from the Fermi–Dirac distribution function $f_0(\varepsilon_\mathbf{k}, T)$ which applies at thermal equilibrium. We write the deviations as

$$\phi(\mathbf{r}, \mathbf{k}) = f(\mathbf{r}, \mathbf{k}) - f_0(\varepsilon_\mathbf{k}, T), \tag{10.41}$$

and assume $\phi(\mathbf{r}, \mathbf{k})$ to be small. Equation (10.40) is now rewritten as

$$-\mathbf{v}_k \cdot \frac{\partial f_0}{\partial T}\nabla T - \frac{(-e)}{\hbar}(\mathbf{E} + \mathbf{v}_k \times \mathbf{B}) \cdot \frac{\partial f_0}{\partial \mathbf{k}}$$

$$= -\left(\frac{\partial f}{\partial T}\right)_{\text{scatter}} + \mathbf{v}_k \cdot \frac{\partial \phi}{\partial \mathbf{r}} + \frac{(-e)}{\hbar}(\mathbf{E} + \mathbf{v}_k \times \mathbf{B}) \cdot \frac{\partial \phi}{\partial \mathbf{k}}$$

by inserting equation (10.41). The term involving the magnetic field in the left-hand side always vanishes, since

$$\mathbf{v_k} \times \mathbf{B} \cdot \frac{\partial f_0}{\partial \mathbf{k}} = \mathbf{v_k} \times \mathbf{B} \cdot \frac{\partial f_0}{\partial \varepsilon_k} \frac{\partial \varepsilon_k}{\partial \mathbf{k}} = (\mathbf{v_k} \times \mathbf{B} \cdot \mathbf{v_k}) \hbar \frac{\partial f_0}{\partial \varepsilon_k} = 0,$$

where $\mathbf{A} \times \mathbf{B} \cdot \mathbf{C}$ is calculated by taking the vector product $\mathbf{A} \times \mathbf{B}$ with subsequent scalar product with the vector $\mathbf{C}$. The term $\mathbf{E} \cdot (\partial \phi / \partial \mathbf{k})$ in the right-hand side is shown to be of the order of E^2, as will be seen from equation (10.46), and is neglected because of the deviation from Ohm's law.

By taking both the temperature and energy derivatives of the Fermi–Dirac distribution function, we easily obtain the following relation:

$$\frac{\partial f_0}{\partial T} = -\left(\frac{\partial f_0}{\partial \varepsilon}\right)\left[\left(\frac{\varepsilon - \zeta}{T}\right) + \frac{\partial \zeta}{\partial T}\right],$$

where $f_0(\varepsilon_k, T) = 1/\{\exp[(\varepsilon_k - \zeta)/k_B T] + 1\}$ and ζ is the chemical potential. Equation (10.40) is now reduced to the form:

$$\left(-\frac{\partial f_0}{\partial \varepsilon}\right) \mathbf{v_k} \cdot \left[-\left(\frac{\varepsilon(\mathbf{k}) - \zeta}{T}\right)\nabla T + (-e)\left(\mathbf{E} - \frac{\nabla \zeta}{(-e)}\right)\right]$$

$$= -\left(\frac{\partial f}{\partial t}\right)_{\text{scatter}} + \mathbf{v_k} \cdot \frac{\partial \phi}{\partial \mathbf{r}} + \frac{(-e)}{\hbar}(\mathbf{v_k} \times \mathbf{B}) \cdot \frac{\partial \phi}{\partial \mathbf{k}}. \quad (10.42)$$

This is the linearized Boltzmann transport equation. The term $\nabla \zeta$ is included as an extra electric field, since it represents an effective field associated with a change in the chemical potential induced by the temperature gradient [2].

We have not yet considered the scattering term in the right-hand side of equation (10.42). The change in the electron distribution due to scattering is generally expressed as

$$\left(\frac{\partial f(\mathbf{k})}{\partial t}\right)_{\text{scatter}} = \sum_{\mathbf{k}'}\{Q(\mathbf{k},\mathbf{k}')f(\mathbf{k}')[1-f(\mathbf{k})] - Q(\mathbf{k}',\mathbf{k})f(\mathbf{k})[1-f(\mathbf{k}')]\}, \quad (10.43)$$

where $Q(\mathbf{k}, \mathbf{k}')$ represents the transition probability in the scattering event. The term $f(\mathbf{k}')[1 - f(\mathbf{k})]$ in the curly bracket indicates that the electron of the state $\mathbf{k}'$ is scattered into the vacant state $\mathbf{k}$ and increases $(\partial f(\mathbf{k})/\partial t)_{\text{scatter}}$, whereas the second term $f(\mathbf{k})[1 - f(\mathbf{k}')]$ decreases it.

The calculation of $(\partial f(\mathbf{k})/\partial t)_{\text{scatter}}$ is a formidable task, since equation (10.43) involves a complicated summation. The relaxation time approximation is frequently employed to avoid this difficulty. The scattering term is then simplified as

$$-\left(\frac{\partial f}{\partial t}\right)_{\text{scatter}} = \frac{f(\mathbf{r},\mathbf{k}) - f_0(\varepsilon_k, T)}{\tau} = \frac{\phi(\mathbf{r},\mathbf{k})}{\tau}, \quad (10.44)$$

10.7 Electrical conductivity formula

We consider in this section the linearized Boltzmann transport equation, where only an electric field is applied to a metal at a constant temperature. Equation (10.42) becomes

$$-\left(\frac{\partial f}{\partial t}\right)_{\text{scatter}} = \left(-\frac{\partial f_0}{\partial \varepsilon}\right) \mathbf{v_k} \cdot (-e)\mathbf{E}. \tag{10.45}$$

Here the term $\mathbf{v_k} \cdot \dfrac{\partial \phi}{\partial \mathbf{r}}$ in the right-hand side of equation (10.42) vanishes, since the system is everywhere at a constant temperature so that $\phi(\mathbf{r})$ is independent of the position vector $\mathbf{r}$. By using the relaxation time approximation, we can rewrite equation (10.45) as

$$\frac{\phi(\mathbf{k})}{\tau} = \left(-\frac{\partial f_0}{\partial \varepsilon}\right) \mathbf{v_k} \cdot (-e)\mathbf{E}. \tag{10.46}$$

In Section 3.7, we obtained the relation $n = (1/4\pi^3) \iiint f_0(\mathbf{k}) d\mathbf{k}$, where n is the number of electrons per unit volume. The current density is then expressed as

$$\mathbf{J} = \frac{(-e)}{4\pi^3} \iiint \mathbf{v_k} f(\mathbf{k}) d\mathbf{k} = \frac{(-e)}{4\pi^3} \iiint \mathbf{v_k} [f(\mathbf{k}) - f_0(\mathbf{k})] d\mathbf{k}$$

$$= \frac{(-e)}{4\pi^3} \iiint \mathbf{v_k} \phi(\mathbf{k}) d\mathbf{k}, \tag{10.47}$$

where $\iiint \mathbf{v_k} f_0(\mathbf{k}) d\mathbf{k}$ is obviously zero.

Equation (10.47) is further rewritten by inserting equation (10.46):

$$\mathbf{J} = \frac{e^2}{4\pi^3} \iiint \tau \mathbf{v_k}(\mathbf{v_k} \cdot \mathbf{E}) \left(-\frac{\partial f_0}{\partial \varepsilon}\right) d\mathbf{k}$$

$$= \frac{e^2}{4\pi^3} \iiint \tau \mathbf{v_k}(\mathbf{v_k} \cdot \mathbf{E}) \left(-\frac{\partial f_0}{\partial \varepsilon}\right) \frac{dSd\varepsilon}{|\nabla_\mathbf{k}\varepsilon|}, \tag{10.48}$$

where the following relation is used:

$$\iiint d\mathbf{k} = \iint dS \int dk_\perp = \iint dS \int \frac{d\varepsilon}{|\partial\varepsilon/\partial k_\perp|} = \iint dS \int \frac{d\varepsilon}{|\nabla_{k_\perp}\varepsilon|}, \tag{10.49}$$

where $\iint dS$ indicates the integral over a constant energy surface and $\int dk_\perp$ is the integral along its normal direction. Now it is important to note that equation (10.48) contains the term $(-\partial f_0/\partial \varepsilon)$, which is finite only in the very vicinity of the Fermi level and behaves like the delta function (see Section 3.3). As a consequence, only electrons at the Fermi level can contribute to the current density, leaving the surface integral over the Fermi surface in equation (10.48).

Equation (10.48) is, therefore, deduced to be

$$\mathbf{J} = \frac{e^2}{4\pi^3 \hbar} \int \frac{\tau \mathbf{v}_k \mathbf{v}_k dS_F}{v_{k_\perp}} \cdot \mathbf{E}, \tag{10.50}$$

where $v_{k_\perp}$ in the denominator represents the component of the velocity of the electron perpendicular to the Fermi surface and is calculated from equation (10.25). It is also noted that there exist two velocity vectors in the numerator and that the one to the left-hand side is parallel to the current density $\mathbf{J}$, whereas the other forms a scalar product with the electric field $\mathbf{E}$. The electrical conductivity tensor is defined in Section 10.2 as $\mathbf{J} = \sigma \mathbf{E}$ or $J_i = \Sigma_j \sigma_{ij} E_j$. A comparison with equation (10.50) leads to

$$\sigma_{ij} = \frac{e^2}{4\pi^3 \hbar} \int \frac{\tau v_i v_j dS_F}{v_{k_\perp}} \tag{10.51a}$$

or

$$\boldsymbol{\sigma} = \frac{e^2}{4\pi^3 \hbar} \int \frac{\tau \mathbf{v}_k \mathbf{v}_k dS_F}{v_{k_\perp}} \tag{10.51b}$$

in the vector representation. The two velocity vectors in the integrand are called the diadic and become a tensor.

For the sake of simplicity, we take an isotropic metal like a bcc or fcc metal. The diagonal and off-diagonal elements of the conductivity tensor satisfy the relations $\sigma_{ii} = \sigma$ and $\sigma_{ij} = 0$ with $i \neq j$. In addition, $v_{k_\perp} = v_F$ holds on the Fermi surface. The electrical conductivity for an isotropic metal is then written as

$$\sigma = \frac{e^2}{4\pi^3} \int \frac{\tau v_i^2 dS_F}{\hbar v_{k_\perp}} = \frac{e^2 \tau}{4\pi^3 \hbar v_F} \cdot \frac{v_F^2}{3} \int dS_F = \frac{e^2 \tau v_F S_F}{12 \pi^3 \hbar}, \tag{10.52}$$

where S_F is the area of the Fermi surface and the relation $v_F^2 = \Sigma_{i=x,y,z} v_i^2 = 3v_i^2$ is used. Following equation (10.49), we can express the electron density of states per unit volume, $N(\varepsilon)$, as

$$\int N(\varepsilon) d\varepsilon = \frac{1}{4\pi^3} \int \int \int dS dk_\perp = \int \left(\frac{1}{4\pi^3} \int \int \frac{dS}{|\partial \varepsilon / \partial k_\perp|} \right) d\varepsilon.$$

This leads to the well-known relation

$$N(\varepsilon)d\varepsilon = \frac{1}{4\pi^3}\int\int \frac{dS}{|\nabla_{k_\perp}\varepsilon|}d\varepsilon. \qquad (10.53)$$

By using equation (10.53), one can alternatively write equation (10.52) as

$$\sigma = \frac{e^2\tau v_F^2}{12\pi^3}\int_{\varepsilon=\varepsilon_F}\frac{dS}{|\nabla_{k_\perp}\varepsilon|} = \frac{e^2}{3}\Lambda_F v_F N(\varepsilon_F), \qquad (10.54)$$

where Λ_F is the mean free path of the conduction electron at the Fermi level and is equal to $\Lambda_F = \tau v_F$. Equation (10.54) indicates that the electrical conductivity is determined only by electrons at the Fermi level and is proportional to the number of electrons at the Fermi level $N(\varepsilon_F)$, their velocity v_F and the mean free path Λ_F.

Equations (10.52), or its alternative, equation (10.54), is frequently used as the conductivity formula for isotropic systems including liquid metals and amorphous metals. We have so far derived two conductivity formulae; one, equation (10.6) in Section 10.2 and the other, equation (10.54). The number of electrons n per unit volume is contained in equation (10.6), so that one may think that all electrons in the Fermi sphere contribute to the electrical conductivity.[6] This is not correct. In the derivation of equation (10.6), we employed the drift velocity defined by equation (10.2). Remember that states having velocities **v** and $-\mathbf{v}$ cancel their contributions and that only states in the very vicinity of the Fermi surface which were previously unoccupied (or occupied) but are newly occupied (or unoccupied) after its displacement due to the electric field, are responsible for the cause of a finite drift velocity. They are indeed electrons at the Fermi level, as is seen in Fig. 10.2.

Equation (10.46) is inserted into equation (10.41). Then we obtain

$$f(\mathbf{r},\mathbf{k}) = f_0(\varepsilon_\mathbf{k}, T) + \left(-\frac{\partial f_0}{\partial \varepsilon}\right)\tau \mathbf{v}_\mathbf{k}\cdot(-e)\mathbf{E}$$

$$= f_0(\varepsilon_\mathbf{k}, T) + \left(-\frac{\partial f_0}{\partial \varepsilon}\right)(-e)\tau \mathbf{v}_\mathbf{k}\cdot\mathbf{E}$$

The last expression can be regarded as the first two terms of the Taylor expansion of the function $f_0(\varepsilon_\mathbf{k} - (-e)\tau \mathbf{v}_\mathbf{k}\cdot\mathbf{E}, T)$. Hence, we have

$$f(\mathbf{r},\mathbf{k}) \cong f_0(\varepsilon_\mathbf{k} - (-e)\tau \mathbf{v}_\mathbf{k}\cdot\mathbf{E}, T). \qquad (10.55)$$

---

[6] We have used the relation $n = \frac{2}{3}E_F N(E_F)$ in the derivation of equation (3.23), which is alternatively expressed as $n = \frac{1}{3}mv_F^2 N(E_F)$ in the free-electron model. An insertion of this relation into equation (10.6) immediately leads to equation (10.54). Therefore, it is important to realize that the relaxation time employed in equations (10.3) and (10.46) is the same but that $v_F N(E_F)$ is not a mere substitute for $v_D n$ in equation (10.6).

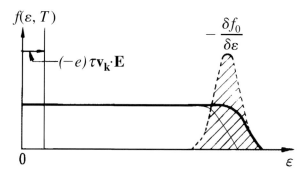

Figure 10.4. The Fermi–Dirac distribution function (thin line) is displaced by an amount $(-e)\tau \mathbf{v}_k \cdot \mathbf{E}$ in the presence of an electric field $\mathbf{E}$. The shaded area represents a derivative of the Fermi–Dirac distribution function and contributes to the electron conduction.

Equation (10.55) implies that the steady state electron distribution can be approximated by displacing the Fermi–Dirac distribution function which applies at thermal equilibrium by an amount equal to $(-e)\tau \mathbf{v}_k \cdot \mathbf{E}$. This is illustrated in Fig. 10.4, where one can clearly see that electrons deep below the Fermi level do not contribute to the electron conduction.

10.8 Impurity scattering and phonon scattering

Let us consider the electron transport phenomenon by taking pure Cu at room temperature as an example. The relaxation time was already calculated in Section 10.2 by using the resistivity value listed in Table 10.1. The mean free path turns out to be about 20 nm from the relation $\Lambda_F = v_F \tau$. Since the lattice constant of pure Cu is 0.36 nm, the mean free path is about 50 times the lattice constant and thus the conduction electron can propagate over several tens of atomic distances without being scattered even at room temperature.

As mentioned in Section 1.2, the mean free path of the conduction electron was calculated to be only a few-tenths nm in the Drude model based on classical mechanics. The Drude theory failed to explain why the conduction electron in a pure metal can travel over many atomic distances without being scattered. This difficulty was resolved by the Bloch theorem discussed in Section 5.3.

We learned in Chapter 5 that the conduction electron propagates in the form of $\psi(\mathbf{r}) = e^{i\mathbf{k}\cdot\mathbf{r}} u_k(\mathbf{r})$ in a periodic potential. The Bloch theorem assures that the wave vector $\mathbf{k}$ remains unchanged in the periodic lattice. This is equivalent to saying that electrons are not scattered, as long as the potential is perfectly periodic. The electrical resistance arises only when its periodicity is disturbed.

There are two sources of disturbance: one, the static source, which includes impurity atoms, vacancies, dislocations and grain boundaries and the other, the dynamical source due to lattice vibrations. The static source of scattering leads to temperature-independent electrical resistivities, as has already been discussed in Sections 9.2 and 9.3, whereas lattice vibrations give rise to temperature-dependent resistivities. The interaction of the conduction electron with lattice vibrations is called electron–phonon interaction and will be a central issue in Sections 10.11 and 10.12.

The total resistivity in metals and alloys can be expressed as the sum of these two contributions:

$$\rho = \rho_{lattice} + \rho_{imp}, \qquad (10.56)$$

where $\rho_{lattice}$ is the resistivity due to lattice vibrations and ρ_{imp} is that due to impurities and defects. The value of $\rho_{lattice}$ in a perfect crystal metal decreases with decreasing temperature and becomes zero at absolute zero when the thermal vibrations cease. In contrast, ρ_{imp} is temperature-independent, as mentioned above.

Equation (10.56) is called the Matthiessen rule. One may measure the temperature dependence of the resistivity at low temperatures to separate $\rho_{lattice}$ from ρ_{imp}. The data for pure Na are shown in Fig. 10.5. It can be seen that the resistivity is almost temperature independent below about 10 K. This is the contribution of ρ_{imp} and is called the residual resistivity. The lower the value of ρ_{imp}, the purer is the metal. The ratio of the resistivity at room temperature over that at 4.2 K, corresponding to the boiling point of liquid helium, $\rho_{300\,K}/\rho_{4.2\,K}$, is referred to as the residual resistivity ratio (RRR or 3R) and is used as a measure to judge the purity of a metal. For instance, a very pure Cu metal whose 3R exceeds 10 000 is commercially available.

10.9 Band structure effect on the electron transport equation

We calculated in Section 9.3 the increment in the resistivity upon adding a single impurity atom or a small number of impurity atoms in an otherwise perfect crystal. In this treatment (see footnote 2, p. 227), the plane wave was used to calculate the scattering probability:

$$U(\mathbf{K}) = \iiint e^{-i\mathbf{k}'\cdot\mathbf{r}} U(\mathbf{r}) e^{i\mathbf{k}\cdot\mathbf{r}} d\mathbf{r} = \iiint e^{i\mathbf{K}\cdot\mathbf{r}} U(\mathbf{r}) d\mathbf{r}, \qquad (10.57)$$

where $U(\mathbf{r})$ is the impurity potential. Remember that the metal of interest is a perfect crystal, unless an impurity atom is added. Then the Bloch wave should be used in place of the plane wave as an unperturbed wave function in the calculation of equation (10.57). Indeed, we have emphasized the band structure

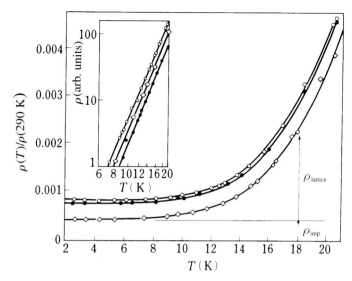

Figure 10.5. Temperature dependence of the electrical resistivity for pure Na at low temperatures. The temperature-dependent part of the resistivity due to electron–phonon interaction is denoted as ρ_{lattice}. Three different data sets yield different residual resistivities owing to a slight difference in purity of the samples. The inset shows a log–log plot for the temperature–dependent ρ_{lattice}. The relation $\rho_{\text{lattice}} = T^5$ holds well for pure Na. [D. K. C. MacDonald and K. Mendelssohn, *Proc. Roy. Soc.* (London) **A202** (1950) 103]

effect of the conduction electron on the electron transport in Sections 10.3–10.4. Admittedly, the calculation of a matrix element like equation (10.57) is laborious, if the Bloch wave has to be used. Instead, it is convenient if the plane wave approximation is justified even for a crystal. The aim of this section is to focus on how the band structure effect is incorporated into the electron transport equation within the framework of the plane wave approximation.

To begin with, the Wannier function is introduced [2]. As shown in Section 8.5, a Bloch wave function can be constructed from a set of atomic orbital wave functions in the tight-binding method. Here, rather than using atomic orbital wave functions, we expand the Bloch wave function in the *n*-th band in the following form:

$$\psi_{\mathbf{k},n} = \frac{1}{\sqrt{N}} \sum_{\mathbf{l}} e^{i\mathbf{k}\cdot\mathbf{l}} a_n(\mathbf{r}-\mathbf{l}), \qquad (10.58)$$

where $a_n(\mathbf{r}-\mathbf{l})$ is the Wannier function at the lattice site $\mathbf{l}$, which is similar to atomic orbital wave functions but is more artificially designed so as to satisfy the orthogonality conditions described below.

Multiplying by $e^{-i\mathbf{k}\cdot\mathbf{l}}$ on both sides of equation (10.58) and subsequently summing over all possible $\mathbf{k}$, we obtain

$$a_n(\mathbf{r}-\mathbf{l}) = \frac{1}{\sqrt{N}} \sum_{\mathbf{k}} e^{-i\mathbf{k}\cdot\mathbf{l}} \psi_{\mathbf{k},n}(\mathbf{r}), \qquad (10.59)$$

where the relation $\sum_{\mathbf{k}} e^{i\mathbf{k}\cdot(\mathbf{l}'-\mathbf{l})} = N\delta_{\mathbf{l}\mathbf{l}'}$ is used. It can be shown that the Wannier function $a_n(\mathbf{r}-\mathbf{l})$ in the n-th band is orthogonal to $a_{n'}(\mathbf{r}-\mathbf{l})$ in the n'-th band, where $n \neq n'$, and that $a_n(\mathbf{r}-\mathbf{l})$ at the lattice site $\mathbf{l}$ is orthogonal to $a_n(\mathbf{r}-\mathbf{l}')$ at the lattice site $\mathbf{l}'$ in the same n-th band, where $\mathbf{l} \neq \mathbf{l}'$. Indeed, one can easily confirm the relation

$$\int a_n^*(\mathbf{r}-\mathbf{l})a_n(\mathbf{r}-\mathbf{l}')d\mathbf{r} = \delta_{\mathbf{l}\mathbf{l}'}, \qquad (10.60)$$

by using the orthogonality condition of the Bloch functions $\psi_{\mathbf{k}',n}(\mathbf{r})$ and $\psi_{\mathbf{k},n}(\mathbf{r})$. As is clear from the argument above, the Wannier function can be expressed as the sum of the Bloch waves in a given single band and is peaked at the individual lattice sites. The Wannier functions constitute the complete orthogonal set of wave functions (see Exercise 10.2). This is the reason why the Wannier function is more convenient than the atomic orbitals, for which the orthogonality condition fails at different lattice sites.

Let us write the Hamiltonian for the conduction electron propagating in the periodic potential $V_0(\mathbf{r})$ as $H_0 = (\hbar^2/2m)\nabla^2 + V_0(\mathbf{r})$ and add to it the perturbing potential $U(\mathbf{r})$ as an impurity potential. The Schrödinger equation is given by

$$[H_0 + U(\mathbf{r})]\psi(\mathbf{r}) = \varepsilon\psi(\mathbf{r}). \qquad (10.61)$$

The wave function in equation (10.61) is expanded in terms of the Wannier functions:

$$\psi(\mathbf{r}) = \sum_{n,\mathbf{l}} f_n(\mathbf{l})a_n(\mathbf{r}-\mathbf{l}), \qquad (10.62)$$

where $f_n(\mathbf{l})$ is reduced to $(1/\sqrt{N})e^{i\mathbf{k}\cdot\mathbf{l}}$, if $\psi(\mathbf{r})$ is the Bloch wave. But we temporarily assume a more general function $f_n(\mathbf{l})$, which is called the envelope function. Equation (10.62) is inserted into equation (10.61) and then the summation is taken over all lattice sites after multiplying by $a_{n'}^*(\mathbf{r}-\mathbf{l}')$ on both sides. We have

$$\sum_{n,\mathbf{l}} \int a_{n'}^*(\mathbf{r}-\mathbf{l}')(H_0 + U)a_n(\mathbf{r}-\mathbf{l})f_n(\mathbf{l})d\mathbf{r} = \varepsilon f_{n'}(\mathbf{l}'). \qquad (10.63)$$

Since H_0 is the Hamiltonian in a completely periodic potential, $H_0\psi_{k,n} = \varepsilon_n(k)\psi_{k,n}$ holds, where $\psi_{k,n}$ is the Bloch wave function and $\varepsilon_n(k)$ is its energy eigenvalue. By using equations (10.59) and (10.58), we obtain

$$H_0 a_n(\mathbf{r}-\mathbf{l}) = \frac{1}{\sqrt{N}} \sum_k e^{-i\mathbf{k}\cdot\mathbf{l}} H_0 \psi_{k,n} = \frac{1}{\sqrt{N}} \sum_k e^{-i\mathbf{k}\cdot\mathbf{l}} \varepsilon_n(\mathbf{k})\psi_{k,n}$$

$$= \frac{1}{N} \sum_k e^{-i\mathbf{k}\cdot\mathbf{l}} \varepsilon_n(\mathbf{k}) \sum_{\mathbf{l}'} e^{i\mathbf{k}\cdot\mathbf{l}'} a_n(\mathbf{r}-\mathbf{l}') = \sum_{\mathbf{l}'} \varepsilon_{n,\mathbf{l}-\mathbf{l}'} a_n(\mathbf{r}-\mathbf{l}'),$$

where $\varepsilon_{n,\mathbf{l}}$ is defined as

$$\varepsilon_{n,\mathbf{l}} = \frac{1}{N} \sum_k e^{-i\mathbf{k}\cdot\mathbf{l}} \varepsilon_n(\mathbf{k}). \tag{10.64}$$

Equation (10.63) is now rewritten as

$$\sum_{n,\mathbf{l}} \{\delta_{nn'} \varepsilon_{n,\mathbf{l}-\mathbf{l}'} + U_{nn'}(\mathbf{l},\mathbf{l}')\} f_n(\mathbf{l}) = \varepsilon f_{n'}(\mathbf{l}'), \tag{10.65}$$

where $U_{nn'}(\mathbf{l},\mathbf{l}')$ is defined as

$$U_{nn'}(\mathbf{l},\mathbf{l}') = \int a_{n'}^*(\mathbf{r}-\mathbf{l}') U(\mathbf{r}) a_n(\mathbf{r}-\mathbf{l}) d\mathbf{r}. \tag{10.66}$$

Obviously, information about the band structure of the conduction electron in the n-th band is contained in the energy eigenvalue $\varepsilon_n(\mathbf{k})$, which appears in the first term of equation (10.65) as $\varepsilon_{n,\mathbf{l}}$. Now a new operator $\varepsilon(-i\nabla)$ is introduced by replacing the wave vector $\mathbf{k}$ in $\varepsilon_n(\mathbf{k})$ by $-i\nabla$ and is operated on the function $f(\mathbf{r})$. We have the following relation:

$$\varepsilon_n(-i\nabla)f(\mathbf{r}) = \sum_\mathbf{l} \varepsilon_{n,\mathbf{l}} e^{\mathbf{l}\cdot(-i\nabla)} f(\mathbf{r})$$

$$= \sum_\mathbf{l} \varepsilon_{n,\mathbf{l}} \left[1 + \mathbf{l}\cdot\nabla + \tfrac{1}{2}(\mathbf{l}\cdot\nabla)^2 + \cdots\right] f(\mathbf{r}) = \sum_\mathbf{l} \varepsilon_{n,\mathbf{l}} f(\mathbf{r}+\mathbf{l}). \tag{10.67}$$

Since we had the relation $\varepsilon_n(-i\nabla)f(\mathbf{l}') = \sum_\mathbf{l} \varepsilon_{n,\mathbf{l}-\mathbf{l}'} f(\mathbf{l})$, equation (10.65) is reduced to

$$\varepsilon_{n'}(-i\nabla)f_{n'}(\mathbf{l}') + \sum_{n,\mathbf{l}} U_{nn'}(\mathbf{l},\mathbf{l}') f_n(\mathbf{l}) = \varepsilon f_{n'}(\mathbf{l}'). \tag{10.68}$$

Equation (10.68) is defined at each lattice vector $\mathbf{l}'$. Let us assume that the conduction electron is in the n-th band and that the potential $U(\mathbf{r})$ varies only slowly over nearest neighbor distances in a given metal. Then the potential $U(\mathbf{r})$ in equation (10.66) may be pulled out from the integral and $U_{nn}(\mathbf{l}, \mathbf{l}') \approx 0$ holds if $\mathbf{l} \neq \mathbf{l}'$ because of the orthogonality condition for the Wannier function. If a discrete variable $\mathbf{l}'$ is replaced by an arbitrary variable $\mathbf{r}$ and the non-vanishing term $U_{nn}(\mathbf{l}', \mathbf{l}')$ by $[U(\mathbf{r})]_{\mathbf{r}=\mathbf{l}'}$, we reach the final expression:

$$\{\varepsilon_n(-i\nabla) + U(\mathbf{r})\} f_n(\mathbf{r}) = \varepsilon f_n(\mathbf{r}). \tag{10.69}$$

Equation (10.69) represents the wave equation for an electron propagating in the perturbing potential field $U(\mathbf{r})$. The band structure effect enters through the operator $\varepsilon_n(-i\nabla)$ and the envelope function serves as the wave function. The envelope function for the Bloch wave is certainly the plane wave, i.e., $f_n(\mathbf{r}) = e^{i\mathbf{k}\cdot\mathbf{r}}$. Therefore, we conclude that the plane wave can be used in place of the Bloch wave and that the band structure effect is incorporated through $\varepsilon_n(-i\nabla)$ in dealing with the scattering phenomenon due to the perturbing potential $U(\mathbf{r})$ in an otherwise perfect crystal.

For example, let us consider the case in which the energy eigenvalue near the bottom of the conduction band is expressed as $\varepsilon(\mathbf{k}) = \hbar^2 k^2 / 2m^*$. The Schrödinger equation (10.69) can be expressed as

$$\left(-\frac{\hbar^2}{2m^*} \nabla^2 + U(\mathbf{r}) \right) \psi(\mathbf{r}) = \varepsilon \psi(\mathbf{r}), \tag{10.70}$$

where the effect of the periodic potential is included in the effective mass m^*. This explains why the plane wave approximation is applicable to the scattering phenomenon in a crystal, as is the case in Section 9.3.

10.10 Ziman theory for the electrical resistivity

Following the discussion in the preceding section, we treat the conduction electron in a crystal as being described by the plane wave of the wave vector $\mathbf{k}$ and consider the situation, where the electron in the state $\mathbf{k}$ is scattered into the unoccupied state $\mathbf{k}'$ due to thermal vibrations of ions. According to equation (10.43), the scattering term is expressed in the integral form:

$$\left(\frac{\partial f(\mathbf{k})}{\partial t} \right)_{\text{scatter}} = \left(\frac{V}{8\pi^3} \right) \int \{ Q(\mathbf{k}' \to \mathbf{k}) f(\mathbf{k}')[1 - f(\mathbf{k})] - Q(\mathbf{k} \to \mathbf{k}') f(\mathbf{k})[1 - f(\mathbf{k}')] \} d\mathbf{k}',$$

where the terms $Q(\mathbf{k}' \to \mathbf{k})$ and $Q(\mathbf{k} \to \mathbf{k}')$ represent the transition probability associated with the scattering of the electron into and out of the state $\mathbf{k}$, respectively. The relation $Q(\mathbf{k}' \to \mathbf{k}) = Q(\mathbf{k} \to \mathbf{k}')$ holds for an isotropic system.

The linearized Boltzmann transport equation (10.45) in the presence of a constant electric field **E** for an isotropic metal is then simplified as

$$\left(-\frac{\partial f_0}{\partial \varepsilon}\right)\mathbf{v_k}\cdot(-e)\mathbf{E} = -\left(\frac{V}{8\pi^3}\right)\int \{f(\mathbf{k'})[1-f(\mathbf{k})] - f(\mathbf{k})[1-f(\mathbf{k'})]\}Q(\mathbf{k},\mathbf{k'})d\mathbf{k'}$$

$$= \left(\frac{V}{8\pi^3}\right)\int \{f(\mathbf{k}) - f(\mathbf{k'})\}Q(\mathbf{k},\mathbf{k'})d\mathbf{k'}. \tag{10.71}$$

In the present section, we assume the scattering potential $U(\mathbf{r})$ to be so weak that the Born approximation is justified. In addition, the scattering involved is assumed to be elastic. Then the transition probability $Q(\mathbf{k},\mathbf{k'})$ is given by

$$Q(\mathbf{k},\mathbf{k'}) = \left(\frac{2\pi}{\hbar}\right)|\langle\mathbf{k'}|U(\mathbf{r})|\mathbf{k}\rangle|^2 \delta(\varepsilon_{\mathbf{k'}} - \varepsilon_{\mathbf{k}}), \tag{10.72}$$

where $\delta(\varepsilon_{\mathbf{k'}} - \varepsilon_{\mathbf{k}})$ assures the energy conservation of the electron in the scattering event. For simplicity, we consider a crystal consisting of a single element at finite temperatures. The scattering potential $U(\mathbf{r})$ in this case is given by the sum of individual pseudopotentials $U_p(\mathbf{r})$ over the whole lattice:

$$U(\mathbf{r}) = \sum_\mathbf{l} U_p(\mathbf{r} - \mathbf{R_l}), \tag{10.73}$$

where $\mathbf{R_l} = \mathbf{l} + \mathbf{u_l}$, $\mathbf{l}$ is the lattice vector defined by equation (4.7) and $\mathbf{u_l}$ is the displacement vector of the ion at $\mathbf{l}$ caused by thermal vibrations.[7] The potential $U(\mathbf{r})$ is no longer periodic because of a finite $\mathbf{u_l}$ at finite temperatures and gives rise to a finite resistivity.[8]

The matrix element involved in equation (10.72) is explicitly written as

$$\langle\mathbf{k'}|U(\mathbf{r})|\mathbf{k}\rangle = \frac{1}{V}\int e^{-i\mathbf{k'}\cdot\mathbf{r}} U(\mathbf{r}) e^{i\mathbf{k}\cdot\mathbf{r}} d\mathbf{r}$$

$$= \frac{1}{V}\sum_\mathbf{l} e^{i(\mathbf{k}-\mathbf{k'})\cdot\mathbf{R_l}} \int e^{i(\mathbf{k}-\mathbf{k'})(\mathbf{r}-\mathbf{R_l})} U_p(\mathbf{r}-\mathbf{R_l}) d(\mathbf{r}-\mathbf{R_l}). \tag{10.74}$$

Since the integral is independent of the position vector $\mathbf{R_l}$, we can rewrite equation (10.74) in the form:

$$\langle\mathbf{k'}|U(\mathbf{r})|\mathbf{k}\rangle = U_{\mathbf{k'k}} = U_p(\mathbf{K})S(\mathbf{K}), \tag{10.75}$$

---

[7] The potential $U(\mathbf{r})$ in equation (10.73) is reduced to $V_0(\mathbf{r})$ in equation (10.61) at absolute zero and, thus, its definition is different from that in equation (10.61). The suffix $\mathbf{l}$ in $\mathbf{u_l}$ is also a vector.

[8] According to equation (10.100), the displacement vector $\mathbf{u_l}$ is expressed in terms of the excitation of phonons. Note that phonons are not created by the zero-point energy at absolute zero. Hence, the zero-point energy does not contribute to the resistivity.

10.10 Ziman theory for the electrical resistivity

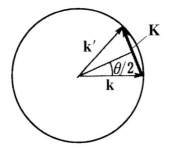

Figure 10.6. Elastic scattering of electron on the Fermi surface.

where $\mathbf{K} = \mathbf{k}' - \mathbf{k}$ is the scattering vector and is related to the scattering angle θ through the relation $\sin(\theta/2) = K/2k_F$ in elastic scattering, as shown in Fig. 10.6.

The functions $U_p(\mathbf{K})$ and $S(\mathbf{K})$ in equation (10.75) are defined as

$$U_p(\mathbf{K}) = \frac{N}{V} \int U_p(\mathbf{r}) e^{-i\mathbf{K}\cdot\mathbf{r}} d\mathbf{r} \tag{10.76}$$

and

$$S(\mathbf{K}) = \left(\frac{1}{N}\right) \sum_l e^{-i\mathbf{K}\cdot\mathbf{R}_l}, \tag{10.77}$$

where N is the number of ions in the volume V. The functions $U_p(\mathbf{K})$ and $S(\mathbf{K})$ are called the atomic form factor and static structure factor, respectively, since $U_p(\mathbf{K})$ involves information only about the single ionic potential $U_p(\mathbf{r})$ whereas the function $S(\mathbf{K})$ only about the spatial distribution of ions. Now equation (10.72) is reduced to the form:

$$Q(\mathbf{k}, \mathbf{k}') = \left(\frac{2\pi}{\hbar}\right) |U_p(\mathbf{K})|^2 \left(\frac{a(\mathbf{K})}{N}\right) \delta(\varepsilon_{\mathbf{k}'} - \varepsilon_{\mathbf{k}}), \tag{10.78}$$

where $a(\mathbf{K})$ is defined as $NS^*(\mathbf{K})S(\mathbf{K})$ and is called the interference function or simply the static structure factor.

We can easily prove the relation

$$f(\mathbf{k}) - f(\mathbf{k}') = \left(-\frac{\partial f_0}{\partial \varepsilon}\right) \tau(-e)\mathbf{E}\cdot(\mathbf{v}_{\mathbf{k}} - \mathbf{v}_{\mathbf{k}'}), \tag{10.79}$$

by using $f(\mathbf{k}) - f_0(\mathbf{k}) = \tau(-\partial f_0/\partial \varepsilon)\mathbf{v}_{\mathbf{k}}\cdot(-e)\mathbf{E}$ derived from equation (10.46) in combination with the relations $f_0(\varepsilon_{\mathbf{k}}) = f_0(\varepsilon_{\mathbf{k}'})$ and $(-\partial f_0/\partial \varepsilon_{\mathbf{k}}) = (-\partial f_0/\partial \varepsilon_{\mathbf{k}'})$, which are valid in elastic scattering. Equation (10.71) is then reduced to

$$\left(-\frac{\partial f_0}{\partial \varepsilon}\right) \mathbf{v}_{\mathbf{k}}\cdot(-e)\mathbf{E} = \left(\frac{V}{8\pi^3}\right) \int \left\{\left(-\frac{\partial f_0}{\partial \varepsilon}\right) \tau(-e)\mathbf{E}\cdot(\mathbf{v}_{\mathbf{k}} - \mathbf{v}_{\mathbf{k}'})\right\} Q(\mathbf{k}, \mathbf{k}') d\mathbf{k}',$$

which is further simplified by eliminating $(-e)(-\partial f_0/\partial \varepsilon)$ from both sides and by assuming that the relaxation time τ is independent of the wave vector $\mathbf{k}'$. We have

$$\frac{1}{\tau} = \left(\frac{V}{8\pi^3}\right)\int\left(\frac{(\mathbf{v_k}-\mathbf{v_{k'}})\cdot\mathbf{E}}{\mathbf{v_k}\cdot\mathbf{E}}\right)Q(\mathbf{k},\mathbf{k}')d\mathbf{k}' = \left(\frac{V}{8\pi^3}\right)\int\left(1-\frac{\mathbf{v_{k'}}\cdot\mathbf{E}}{\mathbf{v_k}\cdot\mathbf{E}}\right)Q(\mathbf{k},\mathbf{k}')d\mathbf{k}', \quad (10.80)$$

where the transition probability of equation (10.78) is given by

$$Q(\mathbf{k},\mathbf{k}')d\mathbf{k}' = Q(\mathbf{k},\mathbf{k}')\left(\frac{\partial k_\perp}{\partial \varepsilon}\right)d\varepsilon dS = \left(\frac{2\pi}{\hbar}\right)|U_p(\mathbf{K})|^2\left(\frac{a(\mathbf{K})}{N}\right)\delta(\varepsilon_{\mathbf{k}'}-\varepsilon_\mathbf{k})\left(\frac{\partial k_\perp}{\partial \varepsilon}\right)d\varepsilon dS. \quad (10.81)$$

If the Fermi surface is spherical, $\mathbf{v_{k'}}//\mathbf{k}'$ and $\mathbf{v_k}//\mathbf{k}$ hold. Then we could confirm the relation $[1-(\mathbf{v_{k'}}\cdot\mathbf{E})/(\mathbf{v_k}\cdot\mathbf{E})]=(1-\cos\theta)$, where θ is the scattering angle between $\mathbf{k}$ and $\mathbf{k}'$ (see Exercise 10.3). By inserting this relation together with equation (10.81) into equation (10.80), we obtain

$$\frac{1}{\tau} = \left(\frac{V}{8\pi^3}\right)\left(\frac{2\pi}{\hbar}\right)\int|U_p(\mathbf{K})|^2\left(\frac{a(\mathbf{K})}{N}\right)(1-\cos\theta)dS\int\delta(\varepsilon_{\mathbf{k}'}-\varepsilon_\mathbf{k})\left(\frac{\partial k_\perp}{\partial \varepsilon}\right)d\varepsilon$$

$$= \left(\frac{V}{8\pi^3 N}\right)\left(\frac{2\pi}{\hbar}\right)\int|U_p(\mathbf{K})|^2 a(\mathbf{K})(1-\cos\theta)dS\int\delta(\varepsilon_{\mathbf{k}'}-\varepsilon_F)\left(\frac{1}{\hbar v_{k_\perp}}\right)d\varepsilon$$

$$= \left(\frac{V}{2\pi N}\right)\left(\frac{1}{\hbar}\right)\left(\frac{3\pi^2(N/V)}{mv_F^2}\right)\int|U_p(\mathbf{K})|^2 a(\mathbf{K})(1-\cos\theta)\sin\theta d\theta,$$

where $dS = 2\pi k_F^2 \sin\theta d\theta$ and $\delta(\varepsilon_{\mathbf{k}'}-\varepsilon_F)$ ensures that the elastic scattering occurs on the Fermi surface. The resistivity formula is finally obtained by inserting the relaxation time thus obtained into equation (10.7):

$$\rho = \left(\frac{3\pi\Omega_0}{2\hbar e^2 v_F^2}\right)\int|U_p(\mathbf{K})|^2 a(\mathbf{K})(1-\cos\theta)\sin\theta d\theta$$

$$= \left(\frac{3\pi\Omega_0}{4e^2\hbar v_F^2 k_F^4}\right)\int_0^{2k_F} a(K)|U_p(K)|^2 K^3 \, dK, \quad (10.82)$$

where Ω_0 is the volume per atom. The relation $\sin(\theta/2) = K/2k_F$ is used to reach the second line. This is the Ziman formula for the electrical resistivity [3].

The weighing factor $(1-\cos\theta)$ in equation (10.82) carries the physical meaning such that the forward scattering with $\theta=0$ makes no contribution to the resistivity, while the back scattering with $\theta=\pi$ makes the largest contribution, equal to 2. In systems where n_{imp} impurities per unit volume are uniformly

10.10 Ziman theory for the electrical resistivity

distributed in an otherwise perfectly periodic lattice, we obtain the following equation:

$$\frac{1}{\Lambda} = 2\pi n_{imp} \int_0^\pi (1 - \cos\theta)\sigma(\theta)\sin\theta \, d\theta, \qquad (10.83)$$

where Λ is the mean free path of the conduction electron and $\sigma(\theta)$ is the differential scattering cross-section for each impurity (see Section 9.3).[9] Note that both equations (10.82) and (10.83) are derived from the Boltzmann transport equation. To validate these transport equations, the mean free path Λ of the conduction electron must be longer than its wavelength λ_F:

$$\Lambda > \lambda_F \text{ or } \Lambda k_F > \frac{1}{2\pi}. \qquad (10.84)$$

This is known as the Ioffe–Regel criterion, since they pointed out, for the first time, that a mean free path of the conduction electron shorter than λ_F is impossible. The condition $\Lambda k_F \approx 1/2\pi$ is equivalent to $\Lambda \approx a$ in ordinary metals [5], where a is an average atomic distance (see Section 15.9).

At this stage, it should be remarked that equation (10.82) cannot be applied to describe the electron transport of a perfect crystal at absolute zero, where $\mathbf{u}_l = 0$ and the potential $U(\mathbf{r})$ given by equation (10.73) resumes a perfectly periodic potential $V_0(\mathbf{r})$. The corresponding structure factor in equation (10.82) is reduced to the delta function $\delta(\mathbf{K} - \mathbf{g})$, where $\mathbf{g}$ is the reciprocal lattice vector. It is obviously zero unless $\mathbf{K} = \mathbf{g}$. Furthermore, the scattering with $\mathbf{K} = \mathbf{g}$ makes no contribution to resistivity, as discussed in Section 10.4. Hence, equation (10.82) seemingly explains a vanishing resistivity at absolute zero. However, this is not a proper argument. A vanishing resistivity for a perfect crystal at absolute zero should be discussed on the basis of the Bloch theorem. Equation (10.82) is valid only when a non-periodic source of scattering is present and can be treated with the second-order perturbation theory. Indeed, Ziman [3] successfully applied equation (10.82) to the resistivity behavior in simple liquid metals like liquid Na, where the periodic lattice vector $\mathbf{l}$ is no longer defined but the assumption of elastic scattering is justified.[10] As will be discussed in Section 15.2, the distribution of ions is by no means periodic in liquid and amorphous metals.

So far we have treated the scattering of the conduction electron as being elastic. Scattering with a static source of disturbances like impurity atoms can

---

[9] The differential scattering cross-section $\sigma(\theta)$ is related to the transition probability $Q(\theta)$ through the relation $\sigma(\theta) = Q(\theta)/v$, where v is the velocity of a particle [4].

[10] At high temperatures, $T > \Theta_D$, elastic scattering dominates. This is the reason for the success of the Ziman theory to simple liquid metals. See footnote 12, p. 284 and Section 11.4.

be treated as being elastic but scattering with lattice vibrations occurs through the exchange of energy with phonons. Thus, consideration of the inelastic electron–phonon interaction is essential in treating electron transport phenomena in both periodic and non-periodic metals at finite temperatures. We will derive the electrical resistivity formula at a finite temperature in Sections 10.11 and 10.12. Readers who are not acquainted with an advanced course of quantum mechanics may wish to skip the next two sections.

10.11 Electrical resistivity due to electron–phonon interaction

In this section, mathematical formulation for the DC electrical conductivity due to inelastic electron–phonon interaction will be given, following the formulation by Itoh [6]. We start again with equation (10.71) based on the Boltzmann transport equation for an isotropic metal consisting of only a single element:

$$\left(-\frac{\partial f_0}{\partial \varepsilon}\right)\mathbf{v_k}\cdot(-e)\mathbf{E} = \left(\frac{V}{8\pi^3}\right)\int \{f(\mathbf{k'})[1-f(\mathbf{k})] - f(\mathbf{k})[1-f(\mathbf{k'})]\} Q(\mathbf{k}, \mathbf{k'}) d\mathbf{k'}. \tag{10.85}$$

The transition probability associated with the scattering of the electron from the state $\mathbf{k}$ to $\mathbf{k'}$ is then expressed as

$$Q(\mathbf{k}\to\mathbf{k'}) = \frac{2\pi}{\hbar}\sum_f \left|\langle \mathbf{k'}, f | \sum_l U_p(\mathbf{r}-\mathbf{R}_l) | \mathbf{k}, i\rangle\right|^2 \delta(\varepsilon_{\mathbf{k'}} - \varepsilon_{\mathbf{k}} + E_f - E_i), \tag{10.86}$$

where the electron in the state $|\mathbf{k}\rangle$ of the energy $\varepsilon_{\mathbf{k}}$ is scattered into the final state $|\mathbf{k'}\rangle$ of the energy $\varepsilon_{\mathbf{k'}}$, which accompanies the transition of a phonon from the state $|i\rangle$ of energy E_i to the final state $|f\rangle$ of energy E_f. The relevant matrix element is calculated by using the plane waves of the wave vectors $\mathbf{k}$ and $\mathbf{k'}$:

$$\langle \mathbf{k'}, f | \sum_l U_p(\mathbf{r}-\mathbf{R}_l) | \mathbf{k}, i\rangle = \langle f | \int e^{-i\mathbf{k'}\cdot\mathbf{r}} \sum_l U_p(\mathbf{r}-\mathbf{R}_l) e^{i\mathbf{k}\cdot\mathbf{r}} d\mathbf{r} | i\rangle$$

$$= \frac{1}{V}\langle f | \sum_l e^{i(\mathbf{k}-\mathbf{k'})\cdot\mathbf{R}_l} | i\rangle \cdot \int e^{i(\mathbf{k}-\mathbf{k'})\cdot(\mathbf{r}-\mathbf{R}_l)} U_p(\mathbf{r}-\mathbf{R}_l) d(\mathbf{r}-\mathbf{R}_l)$$

$$= \frac{1}{N}\langle f | \sum_l e^{-i\mathbf{K}\cdot\mathbf{R}_l} | i\rangle \cdot U_p(\mathbf{K}), \tag{10.87}$$

where $\mathbf{K}$ is the scattering vector defined as $\mathbf{K} = \mathbf{k'} - \mathbf{k}$. As noted in the preceding section, $U_p(\mathbf{K}) = (N/V)\int e^{-i\mathbf{K}\cdot\mathbf{r'}} U_p(\mathbf{r'}) d\mathbf{r'}$ is independent of the position vector $\mathbf{R}_l$, regardless of whether the system is periodic or non-periodic.

10.11 Electrical resistivity due to electron–phonon interaction

We consider the process such that the electron in the state **k** of energy ε_i is scattered into the final state **k**' of energy ε_f by emitting or absorbing a phonon of energy $\varepsilon_f - \varepsilon_i = \hbar\omega$. Note that ω is either positive or negative and that the transition probability $Q(\mathbf{K}, \omega)$ becomes ω-dependent. By inserting equation (10.87) into equation (10.86), we obtain

$$Q(\mathbf{K}, \omega) = \frac{2\pi}{\hbar} \left| U_p(\mathbf{K}) \right|^2 \frac{1}{N^2} \sum_f \langle i | \sum_{l'} e^{i\mathbf{K}\cdot\mathbf{R}_{l'}} | f \rangle \langle f | \sum_{l} e^{-i\mathbf{K}\cdot\mathbf{R}_l} | i \rangle \cdot \delta(\varepsilon_f - \varepsilon_i - \hbar\omega)$$

$$= 2\pi \left| U_p(\mathbf{K}) \right|^2 \left(\frac{a(\mathbf{K},\omega)}{N} \right), \tag{10.88}$$

where $a(\mathbf{K}, \omega) = N S^*(\mathbf{K}, \omega) S(\mathbf{K}, \omega)$ is called the dynamical structure factor [7, 8]. It is reduced to

$$a(\mathbf{K}, \omega) = \frac{1}{N} \sum_{l,l'} e^{i\mathbf{K}\cdot(l-l')} \int_{-\infty}^{\infty} \frac{dt}{2\pi} e^{i\omega t} \left\langle e^{-i\mathbf{K}\cdot\mathbf{u}_{l'}(t)} e^{i\mathbf{K}\cdot\mathbf{u}_l(0)} \right\rangle_T, \tag{10.89}$$

where the bracket $\langle \ldots \rangle_T$ represents a thermal average of the system in equilibrium with the heat bath at temperature T and is explicitly given by

$$\langle A \rangle_T = \frac{\sum e^{-\beta \varepsilon_i} \langle i | A | i \rangle}{\sum e^{-\beta \varepsilon_i}},$$

where $\beta = 1/k_B T$ (see Exercise 10.4).

The scattering rate $(\partial f/\partial t)_{\text{scatter}}$ associated with the absorption and emission of a phonon **q** of energy $\hbar\omega$ is explicitly written as

$$\left(\frac{\partial f(\mathbf{k})}{\partial t} \right)_{\text{scatter}} = \left(\frac{2\pi}{N} \right) \int_{-\infty}^{\infty} d\omega \sum_{\mathbf{k}'} \left| U_p(\mathbf{K}) \right|^2 a(-\mathbf{K}, \omega) f(\mathbf{k}')[1 - f(\mathbf{k})]\delta(\varepsilon_\mathbf{k} - \varepsilon_{\mathbf{k}'} + \hbar\omega)$$

$$- \left(\frac{2\pi}{N} \right) \int_{-\infty}^{\infty} d\omega \sum_{\mathbf{k}'} \left| U_p(\mathbf{K}) \right|^2 a(\mathbf{K}, \omega) f(\mathbf{k})[1 - f(\mathbf{k}')]\delta(\varepsilon_{\mathbf{k}'} - \varepsilon_\mathbf{k} + \hbar\omega), \tag{10.90}$$

where the summation $\sum_{\mathbf{k}'}$ is employed in place of the integration $(V/8\pi^3)\int d\mathbf{k}'$. The scattering processes involved in equation (10.90) are schematically illustrated in Fig. 10.7.

We make use of the relation

$$a(-\mathbf{K}, -\omega) = e^{-\beta\hbar\omega} a(\mathbf{K}, \omega), \tag{10.91}$$

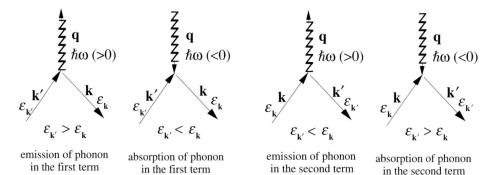

Figure 10.7. Electron–phonon interaction: emission and absorption of phonons due to the first and second terms in equation (10.90).

which is known as the detailed balance condition at temperature T (see Exercise 10.5). The frequency ω is replaced by $-\omega$ in the first term of equation (10.90) and the relation (10.91) is inserted. Then the first term in equation (10.90) is rewritten as

$$\left(\frac{2\pi}{N}\right) \int_{-\infty}^{\infty} d\omega \sum_{\mathbf{k}'} \left|U_p(\mathbf{K})\right|^2 a(-\mathbf{K}, -\omega) f(\mathbf{k}')[1-f(\mathbf{k})]\delta(\varepsilon_\mathbf{k} - \varepsilon_{\mathbf{k}'} - \hbar\omega)$$

$$= \left(\frac{2\pi}{N}\right) \int_{-\infty}^{\infty} d\omega \sum_{\mathbf{k}'} \left|U_p(\mathbf{K})\right|^2 e^{-\beta\hbar\omega} a(\mathbf{K}, \omega) f(\mathbf{k}')[1-f(\mathbf{k})]\delta(\varepsilon_{\mathbf{k}'} - \varepsilon_\mathbf{k} + \hbar\omega),$$

where the relation $\delta(x) = \delta(-x)$ is used. Its insertion into equation (10.89) yields

$$\left(\frac{\partial f(\mathbf{k})}{\partial t}\right)_{\text{scatter}} = \left(\frac{2\pi}{N}\right) \int_{-\infty}^{\infty} d\omega \sum_{\mathbf{k}'} \left|U_p(\mathbf{K})\right|^2 a(\mathbf{K}, \omega)\{e^{-\beta\hbar\omega} f(\mathbf{k}')[1-f(\mathbf{k})]$$
$$- f(\mathbf{k})[1-f(\mathbf{k}')]\} \delta(\varepsilon_{\mathbf{k}'} - \varepsilon_\mathbf{k} + \hbar\omega). \tag{10.92}$$

When both electrons and phonons are in thermal equilibrium, the distribution function $f(\mathbf{k})$ is reduced to the Fermi–Dirac distribution function and one can easily confirm that $(\partial f/\partial t) = 0$.

The Boltzmann transport equation (10.85) in the presence of a constant electric field is now reduced to the form:

$$(-e)(\mathbf{v}_\mathbf{k} \cdot \mathbf{E})\left(-\frac{\partial f_0}{\partial \varepsilon}\right) = \left(\frac{2\pi}{N}\right) \int_{-\infty}^{\infty} d\omega \sum_{\mathbf{k}'} \left|U_p(\mathbf{K})\right|^2 a(\mathbf{K}, \omega)$$
$$\times \{e^{-\beta\hbar\omega} f(\mathbf{k}')[1-f(\mathbf{k})] - f(\mathbf{k})[1-f(\mathbf{k}')]\}$$
$$\times \delta(\varepsilon_{\mathbf{k}'} - \varepsilon_\mathbf{k} + \hbar\omega). \tag{10.93}$$

10.11 Electrical resistivity due to electron–phonon interaction

If the relaxation time approximation is employed again, the Boltzmann transport equation (10.93) is simplified as

$$(-e)(\mathbf{v_k} \cdot \mathbf{E})\left(-\frac{\partial f_0}{\partial \varepsilon_\mathbf{k}}\right) = \frac{2\pi}{\hbar N} \int_{-\infty}^{\infty} d\omega \sum_{\mathbf{k'}} |U_p(\mathbf{K})|^2 a(\mathbf{K}, \omega)$$

$$\times (-e)\{\tau_\mathbf{k}(\mathbf{v_k} \cdot \mathbf{E}) - \tau_{\mathbf{k'}}(\mathbf{v_{k'}} \cdot \mathbf{E})\}$$

$$\times \left(-\frac{\partial f_0}{\partial \varepsilon_\mathbf{k}}\right) \delta(\varepsilon_{\mathbf{k'}} - \varepsilon_\mathbf{k} + \hbar\omega) \beta \omega n(\omega), \quad (10.94)$$

where $n(\omega)$ is the Planck distribution function given by $n(\omega) = 1/(e^{\beta\hbar\omega} - 1)$ (see Exercise 10.6). Since the applied field $\mathbf{E}$ is taken along any direction for an isotropic metal, we choose it to be parallel to the direction of $\mathbf{v_k}$, i.e., $\mathbf{E}//\mathbf{v_k}$. To simplify further, the relaxation time $\tau_\mathbf{k}$ is assumed to be independent of the wave vector $\mathbf{k}$ and is denoted as τ. In addition, $|\mathbf{v_k}| = |\mathbf{v_{k'}}|$ is assumed. Then, the Boltzmann transport equation above is reduced to

$$\frac{1}{\tau} = \left(\frac{2\pi}{N}\right) \int d\omega \sum_{\mathbf{k'}} |U_p(\mathbf{K})|^2 a(\mathbf{K}, \omega)(1 - \cos\theta_{\mathbf{kk'}}) \delta(\varepsilon_{\mathbf{k'}} - \varepsilon_\mathbf{k} + \hbar\omega) \beta \omega n(\omega), \quad (10.95)$$

where $\theta_{\mathbf{kk'}}$ is the angle between the two vectors $\mathbf{v_k}$ and $\mathbf{v_{k'}}$.

An insertion of equation (10.95) into equation (10.7) leads to the resistivity formula due to the inelastic electron–phonon interaction. If a spherical Fermi surface is assumed for an isotropic metal, the summation over the wave vector $\mathbf{k'}$ can be replaced by the integration over the whole of $\mathbf{k'}$-space.[11] Since $\mathbf{v_k}$ is parallel to $\mathbf{k}$, $\theta_{\mathbf{kk'}}$ becomes an angle between $\mathbf{k}$ and $\mathbf{k'}$. The calculations can be performed in the same way as equation (10.82). The resistivity formula for an isotropic metal at a finite temperature is finally deduced to be

$$\rho = \left(\frac{3\pi\Omega_0}{4e^2\hbar v_F^2 k_F^4}\right) \int_0^{2k_F} K^3 |U_p(K)|^2 dK \int_{-\infty}^{\infty} a(K, \omega) \beta \omega n(\omega) d\omega, \quad (10.96)$$

where the upper limit $2k_F$ of the integral corresponds to the maximum momentum transfer and arises when $\theta_{\mathbf{kk'}} = \pi$. Here the electron of wave number k_F is elastically scattered into the state $-k_F$ and, hence, $|K| = |(-k_F) - k_F| = 2k_F$ holds (see Fig. 10.6). Equation (10.96) is regarded as the generalization of equation

---

[11] The following relation holds, when the summation is replaced by an integration: $\Sigma_{\mathbf{k'}} \delta(\varepsilon_{\mathbf{k'}} - \varepsilon_\mathbf{k} + \hbar\omega) \times \cdots = (V/8\pi^3)\int_0^\infty k'^2 dk' \int dS \ldots 1/(\partial \varepsilon_{\mathbf{k'}}/\partial k') \delta(\varepsilon_{\mathbf{k'}} - \varepsilon_F + \hbar\omega)$. An electron in the state $\mathbf{k}$ contributing to the resistivity is on the Fermi surface and, hence, $\varepsilon_\mathbf{k}$ is replaced by the Fermi energy ε_F. Since $\hbar\omega$ is always lower than $k_B\Theta_D$ and is much smaller than ε_F, the last term is well approximated as $\delta(\varepsilon_{\mathbf{k'}} - \varepsilon_\mathbf{k} + \hbar\omega) \cong \delta(\varepsilon_{\mathbf{k'}} - \varepsilon_F)$. Therefore, the same argument as that employed in the derivation of equation (10.82) is applied.

(10.82) and is known as the Baym resistivity formula [9]. This equation will be employed in the next section to discuss the temperature dependence of the electrical resistivity due to electron–phonon interaction for an isotropic metal. As will be discussed below, the Baym formula is applied over a wide temperature range from well below to well above the Debye temperature for both periodic and non-periodic metals.[12]

10.12 Bloch–Grüneisen law

In this section, we apply the Baym resistivity formula (10.96) to a perfect crystal metal at finite temperatures and obtain the theoretical expression for the temperature dependence of the electrical resistivity known as the Bloch–Grüneisen law. For this purpose, we need to evaluate the dynamical structure factor in equation (10.89):

$$a(\mathbf{K}, \omega) = \frac{1}{N} \sum_{\mathbf{l},\mathbf{l}'} e^{i\mathbf{K}\cdot(\mathbf{l}-\mathbf{l}')} \int_{-\infty}^{\infty} \frac{dt}{2\pi} e^{i\omega t} \langle e^{-i\mathbf{K}\cdot\mathbf{u}_{\mathbf{l}'}(t)} e^{i\mathbf{K}\cdot\mathbf{u}_{\mathbf{l}}(0)} \rangle_T,$$

where $\mathbf{l}$ and $\mathbf{l}'$ are the lattice vectors corresponding to equilibrium atom positions at absolute zero and $\mathbf{u}_{\mathbf{l}}$ and $\mathbf{u}_{\mathbf{l}'}$ are the displacement vectors of the relevant atoms. The dynamical structure factor can be quantitatively evaluated in the framework of the harmonic oscillator approximation and can be explicitly written as

$$a(\mathbf{K}, \omega) = \frac{1}{N} \sum_{\mathbf{l},\mathbf{l}'} e^{i\mathbf{K}\cdot(\mathbf{l}-\mathbf{l}')} \int_{-\infty}^{\infty} \frac{dt}{2\pi} e^{i\omega t} e^{-2W_{\mathbf{l}\mathbf{l}'}(\mathbf{K})} \exp\langle(\mathbf{K}\cdot\mathbf{u}_{\mathbf{l}'}(t))(\mathbf{K}\cdot\mathbf{u}_{\mathbf{l}}(0))\rangle_T, \quad (10.97)$$

where $2W_{\mathbf{l}\mathbf{l}'}(\mathbf{K})$ is defined as

$$2W_{\mathbf{l}\mathbf{l}'}(\mathbf{K}) \equiv \tfrac{1}{2}\{\langle(\mathbf{K}\cdot\mathbf{u}_{\mathbf{l}}(0))^2\rangle_T + \langle(\mathbf{K}\cdot\mathbf{u}_{\mathbf{l}'}(t))^2\rangle_T\}, \quad (10.98)$$

and is called the Debye–Waller factor (see Exercise 10.7). Our system is assumed again to be composed of a single element. The Debye–Waller factor $2W_{\mathbf{l}\mathbf{l}'}(\mathbf{K})$ is now independent of the lattice vector $\mathbf{l}$ and, hence, the suffices $\mathbf{l}$ and $\mathbf{l}'$ in $2W_{\mathbf{l}\mathbf{l}'}(\mathbf{K})$ are omitted in the rest of our discussion. Our main objective is, therefore, to calculate a thermal average of $\exp\{\langle[\mathbf{K}\cdot\mathbf{u}_{\mathbf{l}'}(t)][\mathbf{K}\cdot\mathbf{u}_{\mathbf{l}}(0)]\rangle_T\}$ in equation (10.97).

---

[12] At high temperatures $T > \Theta_D$, as in liquid metals, the term $\beta\hbar\omega n(\omega)$ can be approximated as $\beta\hbar\omega n(\omega) = \beta\hbar\omega/(e^{\beta\hbar\omega} - 1) = \beta\hbar\omega/[(1 + \beta\hbar\omega + \cdots) - 1] \approx 1$. If we write $a(K) = \int_{-\infty}^{\infty} a(K,\omega)d\omega$, then we see that the Baym formula is reduced to the Ziman formula of equation (10.82) (see also Section 15.8). Energy transfer by phonons is always limited by $k_B\Theta_D$ and becomes smaller than the spread k_BT of the Fermi–Dirac distribution function at $T > \Theta_D$. This is the reason why electron scattering can be treated quasi-elastically at high temperatures.

10.12 Bloch–Grüneisen law

The exponential term in equation (10.97) may be expanded as

$$\exp\{\langle(\mathbf{K}\cdot\mathbf{u}_{\mathbf{l}'}(t))(\mathbf{K}\cdot\mathbf{u}_{\mathbf{l}}(0))\rangle_T\} = \sum_{m=0}^{\infty} \frac{1}{m!} \{\langle(\mathbf{K}\cdot\mathbf{u}_{\mathbf{l}'}(t))(\mathbf{K}\cdot\mathbf{u}_{\mathbf{l}}(0))\rangle_T\}^m, \quad (10.99)$$

where the m-th term in this expansion is called the m-phonon process. The zero- and one-phonon processes corresponding to $m=0$ and $m=1$ play the dominant role, as discussed below. The displacement of the atom at the lattice site $\mathbf{l}$ is expanded in phonon modes as follows:

$$\mathbf{u}_{\mathbf{l}}(t) = \sum_{\mathbf{q},j} \left(\frac{\hbar}{2NM\omega_\mathbf{q}}\right)^{1/2} \mathbf{e}_\mathbf{q}^j (a_\mathbf{q} e^{i(\mathbf{q}\cdot\mathbf{l}-\omega_\mathbf{q}t)} + a_\mathbf{q}^+ e^{-i(\mathbf{q}\cdot\mathbf{l}-\omega_\mathbf{q}t)}), \quad (10.100)$$

where M is the mass of the constituent atom, $\mathbf{e}_\mathbf{q}^j$ is the j-th polarization vector of the mode $\mathbf{q}$ and $a_\mathbf{q}^+$ and $a_\mathbf{q}$ are the phonon creation and annihilation operators, respectively.

The zero-phonon contribution is obtained by replacing the exponential term in equation (10.97) by unity. This results in

$$a^{(0)}(\mathbf{K},\omega) = \frac{1}{N}\sum_{\mathbf{l},\mathbf{l}'} e^{i\mathbf{K}\cdot(\mathbf{l}-\mathbf{l}')}\delta(\omega)e^{-2W(\mathbf{K})} = \delta(\omega)e^{-2W(\mathbf{K})}a(\mathbf{K}-\mathbf{g}), \quad (10.101)$$

where $a(\mathbf{K}-\mathbf{g}) = (1/N)\sum_{\mathbf{l},\mathbf{l}'} e^{i\mathbf{K}\cdot(\mathbf{l}-\mathbf{l}')} = \delta(\mathbf{K}-\mathbf{g})$ refers to the static structure factor and remains finite only when $\mathbf{K}=\mathbf{g}$ or $\mathbf{k}=\mathbf{k}'+\mathbf{g}$. This means that an electron exchanges its momentum with the lattice by any reciprocal lattice vector $\mathbf{g}$. The delta function $\delta(\omega)$ leads to $\varepsilon_{\mathbf{k}'}=\varepsilon_\mathbf{k}$ and assures the scattering to be elastic. As has been discussed in Section 10.4, an exchange of the momentum equal to any reciprocal lattice vector $\mathbf{g}$ makes no contribution to the resistivity, though the magnitude of $a^{(0)}(\mathbf{K},\omega)$ decreases with increasing temperature due to the temperature dependence of the Debye–Waller factor e^{-2W} (see Exercise 10.8).

The one-phonon contribution with $m=1$ gives rise to a finite electrical resistivity at finite temperatures due to inelastic electron–phonon interaction and can be evaluated by inserting equation (10.100) into the quantity $\langle(\mathbf{K}\cdot\mathbf{u}_{\mathbf{l}'}(t))(\mathbf{K}\cdot\mathbf{u}_{\mathbf{l}}(0))\rangle_T$:

$$\langle(\mathbf{K}\cdot\mathbf{u}_{\mathbf{l}'}(t))(\mathbf{K}\cdot\mathbf{u}_{\mathbf{l}}(0))\rangle_T$$

$$= \sum_{\substack{s,s'\\ \alpha,\alpha'}} \left(\frac{\hbar}{2NM}\right)\sqrt{\left(\frac{1}{\omega_s\omega_{s'}}\right)} \langle(a_s e^{i(\mathbf{q}\cdot\mathbf{l}'-\omega_s t)}+a_s^+ e^{-i(\mathbf{q}\cdot\mathbf{l}'-\omega_s t)})(a_{s'} e^{i\mathbf{q}'\cdot\mathbf{l}}+a_{s'}^+ e^{-i\mathbf{q}'\cdot\mathbf{l}})\rangle_T K_\alpha K_{\alpha'} e_s^\alpha e_{s'}^{\alpha'},$$

$$(10.102)$$

where s is used to abbreviate the pair of indices $(\mathbf{q}, j)$. The following relations hold for a thermal average of the matrix elements involving creation and annihilation operators:

$$\langle a_{s'}^+ a_s \rangle_T = n_s \delta_{s's}, \quad \langle a_s a_{s'}^+ \rangle_T = (1+n_s)\delta_{s's}, \quad \langle a_s^+ a_s^+ \rangle_T = \langle a_{s'} a_s \rangle_T = 0, \quad (10.103)$$

where n_s is the Planck distribution function for phonons of the mode $s=(\mathbf{q}, j)$. The one-phonon contribution to the dynamical structure factor is reduced to

$$a^{(1)}(\mathbf{K}, \omega) = e^{-2W(\mathbf{K})}\left(\frac{\hbar}{2M}\right)$$

$$\times \sum_{\mathbf{l},\mathbf{l}'} \sum_{\alpha,\alpha'} \sum_s K_\alpha K_{\alpha'} e_s^\alpha e_s^{\alpha'} \left(\frac{1}{\omega_s}\right)\left[e^{i(\mathbf{K}-\mathbf{q})\cdot(\mathbf{l}-\mathbf{l}')}\delta(\omega-\omega_s)(n_s+1) + e^{i(\mathbf{K}+\mathbf{q})\cdot(\mathbf{l}-\mathbf{l}')}\delta(\omega+\omega_s)n_s\right]$$

$$= e^{-2W(\mathbf{K})}\left(\frac{\hbar}{2M}\right)\sum_{\mathbf{q}}(\mathbf{K}\cdot\mathbf{e}_\mathbf{q}^j)^2\left(\frac{1}{\omega_\mathbf{q}}\right)[a(\mathbf{K}-\mathbf{q})(\omega-\omega_\mathbf{q})(n_\mathbf{q}+1) + a(\mathbf{K}+\mathbf{q})\delta(\omega+\omega_\mathbf{q})n_\mathbf{q}],$$

(10.104)

where $a(\mathbf{K}\pm\mathbf{q}) = \delta(\mathbf{K}\pm\mathbf{q}+\mathbf{g})$ represents the static structure factor, the details of which are discussed shortly later. The delta function $\delta(\omega\pm\omega_\mathbf{q})$ implies the energy conservation law $\varepsilon_{\mathbf{k}'} - \varepsilon_\mathbf{k} = \pm\hbar\omega_\mathbf{q}$, indicating that the absorption or emission of a phonon of wave vector $\mathbf{q}$ upon the electron scattering from state $\mathbf{k}$ to $\mathbf{k}'$ is accompanied by the energy transfer $\pm\hbar\omega_\mathbf{q}$ between the electron and phonon systems. Hence, the scattering involved is inelastic.

The resistivity of a crystal metal at finite temperatures mainly arises via equation (10.104). The displacement of atoms from their equilibrium positions disrupts the periodicity of the lattice and contributes to the resistivity through inelastic scattering of electrons with the absorption or emission of phonons. Since equation (10.104) involves only a single phonon, the scattering is referred to as the one-phonon process. Higher-order terms in the power series of equation (10.99) give rise to the multiple-phonon processes but their contribution is neglected in the present discussion.

The structure factor $a(\mathbf{K}\pm\mathbf{q})$ in equation (10.104) remains finite when $\mathbf{K}\pm\mathbf{q}=\mathbf{g}$ or $\mathbf{k}-\mathbf{k}'\pm\mathbf{q}=\mathbf{g}$. This represents the momentum conservation law in the one-phonon process. The process with $\mathbf{g}=0$ and, hence, $\mathbf{K}=\pm\mathbf{q}$ is referred to as the normal process. Here an electron of wave vector $\mathbf{k}$ is scattered into the state $\mathbf{k}'$ by emitting or absorbing a phonon of wave vector $\mathbf{q}$, as shown in Fig. 10.7. Obviously, an arbitrariness appears in the momentum conservation law when $\mathbf{g}\neq 0$. The scattering process with $\mathbf{g}\neq 0$ is called the Umklapp process. It is illustrated in Fig. 10.8. The latter becomes important when the temperature

10.12 Bloch–Grüneisen law

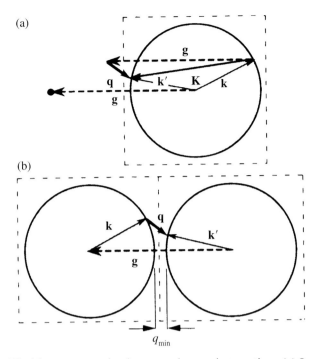

Figure 10.8. Umklapp process in electron–phonon interaction. (a) Large-angle scattering is made possible even at low temperatures with the assistance of the reciprocal lattice vector. Note that the phonon wave vector $\mathbf{q}$ is small at low temperatures. (b) Umklapp process in the repeated zone scheme. A minimum wave number q_{min} is required to allow the Umklapp process to occur. [J. M. Ziman, *Principles of the Theory of Solids* (Cambridge University Press, 1964)]

dependence of the resistivity of a metal needs to be quantitatively evaluated. In the present discussion, we consider only the normal process and discuss qualitatively the temperature dependence of the electrical resistivity of a crystal metal below.

There is another important term $(\mathbf{K} \cdot \mathbf{e}_q^j)^2$ in equation (10.104). Here the polarization vector $\mathbf{e}_q^j$ has three modes with $j = (x, y, z)$. As discussed in Section 4.5, they are two transverse waves and one longitudinal wave for a given wave vector in a three-dimensional system. Since we consider only the normal process, in which the scattering vector $\mathbf{K}$ is equal to the wave vector $\mathbf{q}$, the product $\mathbf{K} \cdot \mathbf{e}_q$ can be replaced by $\mathbf{q} \cdot \mathbf{e}_q$. Thus, this term disappears for the transverse waves, since $\mathbf{q} \perp \mathbf{e}_q$. In other words, we learn that only the longitudinal wave with $\mathbf{q} // \mathbf{e}_q$ can contribute to the resistivity in the one-phonon normal process.

The one-phonon normal term $a^{(1)}(\mathbf{K}, \omega)$ in equation (10.104) is now inserted into the Baym resistivity formula given by equation (10.96). The frequency-dependent part is explicitly written as

$$\int_{-\infty}^{\infty} a^{(1)}(\mathbf{K}, \omega)\beta\omega n(\omega)d\omega$$

$$= \left(\frac{K^2}{2M}\right)\int_{-\infty}^{\infty} d\omega\beta\omega n(\omega)\sum_{\mathbf{q}}\omega_{\mathbf{q}}^{-1}\{(n_{\mathbf{q}}+1)\delta(\mathbf{q}-\mathbf{K})\delta(\omega-\omega_{\mathbf{q}})+n_{\mathbf{q}}\delta(\mathbf{q}+\mathbf{K})\delta(\omega+\omega_{\mathbf{q}})\}$$

$$= \left(\frac{q^2}{2M}\right)\beta\{\omega_{\mathbf{q}}[n(\omega_{\mathbf{q}})+1]n(\omega_{\mathbf{q}})\omega_{\mathbf{q}}^{-1}+(-\omega_{-\mathbf{q}})n(\omega_{-\mathbf{q}})n(-\omega_{-\mathbf{q}})\omega_{-\mathbf{q}}^{-1}\}$$

$$= \left(\frac{q^2}{2M}\right)\beta\{(n_{\mathbf{q}}+1)n_{\mathbf{q}}+n_{\mathbf{q}}(n_{\mathbf{q}}+1)\}$$

$$= \left(\frac{q^2}{M}\right)\beta(n_{\mathbf{q}}+1)n_{\mathbf{q}}, \tag{10.105}$$

where the relations $\omega_{\mathbf{q}} = \omega_{-\mathbf{q}}$ and $n(-\omega_{\mathbf{q}}) = -[n(\omega_{\mathbf{q}})+1]$ are used. Now the resistivity due to the inelastic one-phonon normal process is deduced to be

$$\rho = \left(\frac{3\pi\Omega_0}{4e^2\hbar v_F^2 k_F^4}\right)\left(\frac{1}{M}\right)\int_0^{q_{max}} q^5|U_p(q)|^2\beta n(\omega_{\mathbf{q}})[n(\omega_{\mathbf{q}})+1]dq, \tag{10.106}$$

where the upper limit of the integral is replaced by the maximum phonon wave number q_{max}, since the variable K in the integral is equal to q in the normal process.

Let us take the ratio of equation (10.106) at temperature T over that at the Debye temperature Θ_D. A new variable $x = \beta\hbar\omega$ is rewritten as $x = \hbar\omega/k_B T = (\omega/\omega_D)/(T/\Theta_D) = (q/q_D)(\Theta_D/T)$ in the Debye model. We see that the maximum value of x at any temperature T is replaced by Θ_D/T, since $q_{max} = q_D$. The ratio is then easily calculated as

$$\frac{\rho(T)}{\rho(\Theta_D)} = \left(\frac{T}{\Theta_D}\right)^5 \frac{\int_0^{\Theta_D/T}|U_p(x)|^2 x^5 \frac{e^x}{(e^x-1)^2}dx}{\int_0^1 |U_p(x)|^2 x^5 \frac{e^x}{(e^x-1)^2}dx}. \tag{10.107}$$

This indicates that the temperature-dependent resistivity can be normalized with respect to that at the Debye temperature. This is known as the Bloch–Grüneisen law and reminds us of equation (4.41) for the lattice specific

10.12 Bloch–Grüneisen law

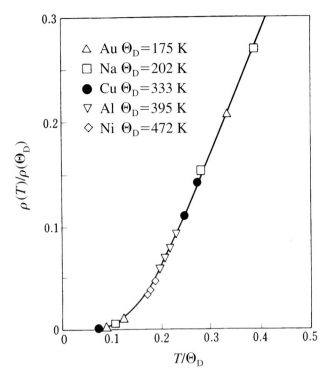

Figure 10.9. Temperature dependence of the electrical resistivity for various metals. The Debye temperature is deduced so as to fit the experimental data to equation (10.107). There is reasonable agreement with the Debye temperature deduced from the low-temperature specific heat (see Table 3.1). [J. Bardeen, *J. Appl. Phys.* **11** (1940) 88]

heat discussed in Section 4.8. Indeed, as shown in Fig. 10.9, the resistivity data for various simple metals fall on a universal Bloch–Grüneisen curve when $\rho(T)/\rho(\Theta_D)$ is plotted as a function of T/Θ_D.

At high temperatures $T \gtrsim \Theta_D$, all phonon modes are excited. Since the relation $\hbar\omega \leq \hbar\omega_D < k_B T$ or $\beta\hbar\omega < 1$ holds, we have

$$\beta n(\omega_q)\{n(\omega_q)+1\} = \frac{\beta}{e^{\beta\hbar\omega}-1} \cdot \frac{e^{\beta\hbar\omega}}{e^{\beta\hbar\omega}-1} = \left(\frac{\beta}{e^{\beta\hbar\omega}-1}\right)^2 \cdot \frac{e^{\beta\hbar\omega}}{\beta} \approx \left(\frac{1}{\hbar\omega}\right)^2 k_B T.$$

Hence, equation (10.106) at $T \gtrsim \Theta_D$ can be approximated as

$$\rho = \frac{3\pi\Omega_0}{4e^2\hbar v_F^2 k_F^4}\int_0^{q_D} q^5 |U_p(q)|^2 \frac{k_B T}{(\hbar s q)^2} dq = \frac{3\pi\Omega_0}{4e^2\hbar v_F^2 k_F^4}\left(\frac{q_D^2 T}{k_B \Theta_D^2}\right)\int_0^{q_D} q^3 |U_p(q)|^2 dq \propto \frac{T}{\Theta_D^2},$$

(10.108)

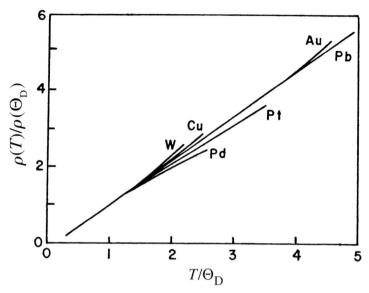

Figure 10.10. Normalized electrical resistivity versus normalized temperature for pure metals in high-temperature range. [K. Schröder, *Handbook of Electrical Resistivities of Binary Metallic Alloys* (CRC Press, Florida, 1983)]

where the upper limit is replaced by the Debye radius corresponding to the available maximum wave number in the Debye model (see Section 4.8).[13] We see that the resistivity of a crystal metal increases linearly with increasing temperature in the range $T \geq \Theta_D$. Indeed, as shown in Fig. 10.10, a linear temperature dependence of resistivity holds well for typical metals like Cu, Au and Pt over a wide temperature range. For example, pure Cu with $\Theta_D = 333$ K exhibits an almost linear temperature dependence in the temperature range from 80 K up to about 800 K.

With decreasing temperature below Θ_D, phonons having smaller wave vectors become dominant and, hence, the magnitude of the scattering vector **K** is also decreased in the one-phonon normal process. Thus, the resistivity at $T \leq \Theta_D$ turns out to be

$$\rho = \frac{3\pi\Omega_0}{4e^2\hbar v_F^2 k_F^4}\left(\frac{1}{M}\right)\left(\frac{T}{\Theta_D}\right)^6\left(\frac{\hbar}{k_B T}\right) q_D^6 |U_p(0)|^2 \int_0^{\Theta_D/T} x^5 \frac{e^x}{(e^x - 1)^2} dx, \quad (10.109)$$

where $|U_p(K)|^2$ is replaced by $|U_p(0)|^2$, since $|U_p(K)|^2$ is almost independent of K at low K values (see Fig. 15.10). If the upper limit Θ_D/T is further replaced

---

[13] Note the difference from footnote 12 on p. 284, where the high-temperature limit is taken in the Baym resistivity formula. Instead, equation (10.108) is derived by incorporating only the one-phonon normal process into the Baym formula.

by infinity at $T \ll \Theta_D$, then the integral becomes a numerical constant equal to

$$\int_0^\infty x^5 \frac{e^x}{(e^x-1)^2} dx = \int_0^\infty x^5 e^{-x}(1 + 2e^{-x} + 3e^{-2x} + 4e^{-3x} + \cdots) dx$$

$$= 5!\left(\frac{1}{1^5} + \frac{1}{2^5} + \frac{1}{3^5} + \frac{1}{4^5} + \cdots\right) = 5!\zeta(5) \cong 120 \times 1.0369 = 124.428, \quad (10.110)$$

where $\zeta(5)$ is the Riemann zeta function (see equation (4.44)). Consequently, we find that the resistivity of a metal increases with increasing temperature in proportion to T^5 at temperatures $T \ll \Theta_D$ or normally below about 20 K. This behavior is roughly seen in Fig. 10.9 in the range $T/\Theta_D \leq 0.15$ for various metals. More straightforward demonstration for the T^5-law was already shown in the inset to Fig. 10.5 for pure Na metal below 20 K.

Only lattice waves with long wavelengths or shorter wave vectors can survive in a perfect crystal at low temperatures $T \ll \Theta_D$. As a result, the scattering angle becomes smaller and smaller with decreasing temperature. Thus, the T^5-law at low temperatures is the manifestation of the survival of only small-angle scattering. The Bloch–Grüneisen law, particularly the T^5-law, is a phenomenon characteristic of a crystalline metal, where the mean free path of the conduction electron is much longer than the atomic distance. The Bloch–Grüneisen law breaks down in non-periodic solids like amorphous alloys and quasicrystals, where the mean free path of the conduction electron is shortened because of the failure of the Bloch theorem and often becomes comparable to an average atomic distance. However, we will show that the Baym resistivity formula can be equally applied to relatively low-resistivity non-periodic metals over a wide temperature range from $T \ll \Theta_D$ to $T \geq \Theta_D$, provided that $\Lambda > \lambda$ in equation (10.84) is still satisfied. More details will be discussed in Section 15.8.

Exercises

10.1 Let us assume that $A(k)$ in equation (10.14b) is given by the Gaussian function $A(k) = ce^{-a^2(k-K)^2}$, where c, a and K are constants. In the free-electron model, the wave function and the energy eigenvalue are given by $\psi_k(x) = e^{i[kx-\omega(k)t]}$ and $\omega(k) = \varepsilon(k)/\hbar = \hbar k^2/2m$, respectively. Calculate the wave packet and its probability density at time t. Show that its group velocity is equal to $v = \hbar K/m$.

10.2 The Bloch wave $\psi_k(\mathbf{r}) = (1/\sqrt{N})u(\mathbf{r})e^{i\mathbf{k}\cdot\mathbf{r}}$ is assumed for an electron in a simple cubic lattice with the lattice constant d.

(a) Show that the Wannier function centered at the origin is expressed as

$$a(\mathbf{r}) = u(\mathbf{r}) \frac{\sin\left(\dfrac{\pi x}{d}\right) \cdot \sin\left(\dfrac{\pi y}{d}\right) \cdot \sin\left(\dfrac{\pi z}{d}\right)}{\left(\dfrac{\pi x}{d}\right) \cdot \left(\dfrac{\pi y}{d}\right) \cdot \left(\dfrac{\pi z}{d}\right)}. \tag{10Q.1}$$

(b) Draw the Wannier function and show that the Wannier functions centered at different lattice sites are mutually orthogonal.

10.3 Prove the relation $[1 - (\mathbf{v}_{\mathbf{k}'} \cdot \mathbf{E})/(\mathbf{v}_{\mathbf{k}} \cdot \mathbf{E})] = 1 - \cos\theta$ in equation (10.80). Since the relaxation time τ is assumed to be independent of the wave vector $\mathbf{k}$, one can choose the vector $\mathbf{k}$ parallel to the electric field $\mathbf{E}$ without losing its generality. Then $(1 - \cos\theta)$ is immediately obtained. Show this relation by using spherical trigonometry, when the direction of the vector $\mathbf{k}$ is arbitrarily chosen relative to $\mathbf{E}$.

10.4 Show that the dynamical structure factor is given by equation (10.89).

10.5 Derive equation (10.91).

10.6 Derive equation (10.94).

10.7 Derive the dynamical structure factor given by equation (10.97) from equation (10.89).

10.8 Calculate the temperature dependence of the Debye–Waller factor e^{-2W} by using the Debye model for lattice vibrations.

Chapter Eleven

Electron transport properties in periodic systems (II)

11.1 Prologue

In this chapter, electron transport properties other than the electrical resistivity and conductivity are presented. Included are the thermal conductivity, the thermoelectric power, the Hall coefficient and the optical conductivity, all of which will be discussed again within the framework of the linearized Boltzmann transport equation in combination with the relaxation time approximation. At the end of this chapter, the basic concept of the Kubo formula is introduced.

11.2 Thermal conductivity

The thermal conductivity κ is defined as the ratio of the thermal current density or the flow of heat $\mathbf{U}$ over the temperature gradient ∇T across a specimen;

$$\mathbf{U} = \kappa(-\nabla T). \tag{11.1}$$

In contrast to the electrical conduction, the thermal current is conveyed by both electrons and phonons. The total thermal conductivity κ in a metal is given by the sum of conductivities, not resistivities, due to both carriers:

$$\kappa = \kappa_{el} + \kappa_{ph}, \tag{11.2}$$

where κ_{el} and κ_{ph} are the electronic and lattice thermal conductivities, respectively.

The inverse, $1/\kappa$, is called the thermal resistivity and denoted as W. In the same manner as the derivation of equation (10.56), known as the Matthiessen rule, the total electronic thermal resistivity W_{el} or $1/\kappa_{el}$ is given by

$$W_{el} = W_{el}^{imp} + W_{el}^{lattice}, \tag{11.3}$$

Table 11.1. *Thermal conductivity data for metals, alloys and insulators*

substance	maximum thermal conductivity κ_{max} (watt/cm·K)	temperature at κ_{max} (K)
Cu	20–40	20–30
$Cu_{99.9}Zn_{0.1}$	8.7	30
$Cu_{99}Zn_{1}$	3	40
$Cu_{90}Zn_{10}$	1	60
$Cu_{74.5}Zn_{25.5}$	0.7	100
graphite	0.1	60
stainless steel	0.1–0.2	100–300
sapphire	60	30–40
nylon	0.001	20

where W_{el}^{imp} and $W_{el}^{lattice}$ represent the thermal resistivities due to impurities and lattice vibrations, respectively. Likewise, the phonon thermal resistivity W_{ph} or $1/\kappa_{ph}$ consists of contributions due to imperfections, phonon–electron and Umklapp phonon–phonon interactions [1–4].[1] The thermal conductivity data for representative metals, alloys and insulators are listed in Table 11.1. It is clear that the thermal conductivity of metals is generally much higher than that of insulators, indicating that conduction electrons carry more heat than phonons.[2]

According to the kinetic theory of gases, the thermal conductivity κ can be expressed as

$$\kappa = \frac{1}{3} C v \Lambda, \qquad (11.4)$$

where C is the specific heat of carriers per unit volume, v is the average velocity and Λ is the mean free path [1–3]. We first discuss the temperature dependence of the electronic thermal conductivity in metals by applying equation (11.4) to the electron gas in a metal.

There are two different scattering sources in electronic thermal conduction, as indicated in equation (11.3). The mean free path Λ of the conduction electron due to impurity scattering is obviously temperature independent. In addition, we are well aware that the electronic specific heat C_{el} is proportional to the absolute temperature T, whereas the Fermi velocity v_F is independent of T. Therefore, κ_{el} is expected to be proportional to the absolute temperature

---

[1] The normal phonon–phonon interaction does not contribute to the thermal resistivity [1].
[2] As is listed in Table 11.1, sapphire is a very good thermal conductor, particularly below 50 K, though it is an insulator.

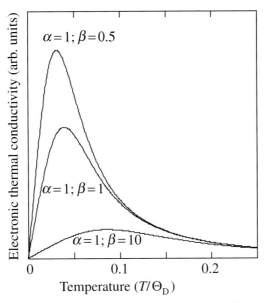

Figure 11.1. Temperature dependence of the electronic thermal conductivity in a metal. A decrease in β indicates an increase in the purity of a specimen.

and, hence, the resistivity W_{el}^{imp} would behave as $W_{el}^{imp} = \beta/T$. The more impure the sample, the shorter the mean free path and, hence, the larger the coefficient β.

The temperature dependence of the electronic thermal resistivity due to lattice vibrations can also be deduced without difficulty. In the Debye model, the number of phonons decreases in proportion to T^3 with decreasing temperature in the range $T \ll \Theta_D$.[3] Thus, the mean free path of the conduction electron would increase as T^{-3} with decreasing temperature. By inserting $C_{el} \propto T$, $\Lambda \propto T^{-3}$ and the temperature-independent Fermi velocity into equation (11.4), we obtain the relation

$$W_{el}^{lattice} = \alpha T^2, \tag{11.5}$$

where α is a numerical coefficient. The temperature dependence of the electronic thermal resistivity is, therefore, described as $W_{el} = \alpha T^2 + (\beta/T)$. The total thermal conductivity κ at low temperatures in a metal is then written as

$$\kappa = \kappa_{el} + \kappa_{ph} = (1/W_{el}) + \kappa_{ph} = \frac{1}{\alpha T^2 + (\beta/T)} + \kappa_{ph}. \tag{11.6}$$

[3] The number of phonons per unit volume can be calculated from $n = \int_0^{\omega_D} D(\omega)n(\omega,T)d\omega$, where $D(\omega)$ is the phonon density of states and $n(\omega,T)$ is the Planck distribution function (see Section 4.8). Since $D(\omega) \propto \omega^2$ holds in the Debye model, $n \propto T^3 \int_0^{\omega_D} x^2 dx/(e^x - 1)$ is immediately deduced.

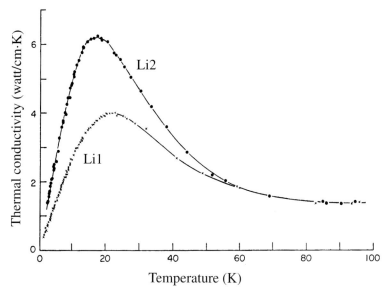

Figure 11.2. Temperature dependence of the thermal conductivity of two lithium metal samples. [Reproduced from K. Mendelssohn and H. M. Rosenberg, *Solid State Physics*, edited by F. Seitz and D. Turnbull, (Academic Press, New York, 1961) vol.12]

Figure 11.1 illustrates the temperature dependence of the electronic thermal conductivity for three sets of the parameters α and β. The purer the sample of a metal, the smaller is the impurity coefficient β so that the maximum in the electronic thermal conductivity will be more enhanced. For instance, the temperature dependence of the thermal conductivity of two lithium metal samples is shown in Fig. 11.2. It can be seen that the data reflect well the temperature dependence of the electronic thermal conductivity and that the Li2 sample must be purer than the Li1 sample.

We learned above that the electronic thermal conductivity rapidly decreases with increasing solute concentration in an alloy. The phonon thermal conductivity also decreases but its rate of decrease is much slower than that of κ_{el} so that the ratio κ_{ph}/κ sharply increases on alloying. This means that the relative contribution of the phonon thermal conductivity to the total increases with increasing solute concentration in an alloy. Here we simply note that the ratio κ_{ph}/κ at about 10 K is only 0.002 for pure Cu but becomes 0.3 for the $Cu_{80}Zn_{20}$ alloy [2].

11.3 Electronic thermal conductivity

In this section, we formulate the electronic thermal conductivity by using the linearized Boltzmann transport equation in combination with the relaxation time approximation. Both the electrical current density **J** and thermal current density **U** are expressed as linear functions of the electric field **E** and temperature gradient ∇T in the following forms:

$$\mathbf{J} = L_{EE}\mathbf{E} + L_{ET}\nabla T \tag{11.7a}$$

and

$$\mathbf{U} = L_{TE}\mathbf{E} + L_{TT}\nabla T, \tag{11.7b}$$

where L_{EE} is the electrical conductivity, L_{ET} and L_{TE} are interconnected through the relation $L_{ET} = -L_{TE}/T$.[4] The thermal conductivity is measured under the condition $\mathbf{J} = 0$. Hence, we can find the thermal conductivity to be equal to $\kappa = -(L_{TT} - L_{TE} \cdot L_{ET}/L_{EE})$ from equations (11.7a) and (11.7b).

At first sight, one may think from the analogy with the expression for the electrical current density that the thermal current density **U** is simply given by $\mathbf{U} = (1/4\pi^3)\int \mathbf{v}_\mathbf{k}\varepsilon(\mathbf{k})f(\mathbf{r},\mathbf{k})d\mathbf{k}$. Indeed, the thermal current is equal to the energy current, as long as $\mathbf{J} = 0$. However, there is a phenomenon like the Peltier effect, in which both heat and electrical charge are carried simultaneously. Here the thermal current is no longer equal to the energy current. The degenerate Fermi gas in the case of $\mathbf{J} > 0$ gives rise to an energy flow equal to $[\mathbf{J}/(-e)]\zeta$, since we have already learned from equation (10.54) that electrons at the Fermi level $\varepsilon_F(\equiv \zeta)$ convey the current density **J**. This must be subtracted from the energy current to derive the flow of heat. The thermal current density is, therefore, expressed as

$$\mathbf{U} = \frac{1}{4\pi^3}\int \mathbf{v}_\mathbf{k}\varepsilon(\mathbf{k})f(\mathbf{r},\mathbf{k})d\mathbf{k} - \left(\frac{\mathbf{J}}{(-e)}\right)\zeta$$

$$= \frac{1}{4\pi^3}\int [\varepsilon(\mathbf{k}) - \zeta]\mathbf{v}_\mathbf{k}[f(\mathbf{r},\mathbf{k}) - f_0(\mathbf{k})]d\mathbf{k}, \tag{11.8}$$

where the current density was already defined as $\mathbf{J} = ((-e)/4\pi^3)\int \mathbf{v}_\mathbf{k} f(\mathbf{r},\mathbf{k})d\mathbf{k}$. The second line in equation (11.8) is easily derived, since $\int \mathbf{v}_\mathbf{k} f_0(\mathbf{k})d\mathbf{k} = 0$ and $\int \varepsilon(\mathbf{k})\mathbf{v}_\mathbf{k} f_0(\mathbf{k})d\mathbf{k} = 0$.

We try to solve the linearized Boltzmann transport equation under the condition that only the temperature gradient ∇T exists in a metal. The relaxation time approximation given by equation (10.44) is again employed and incorporated into equation (10.42). The resulting equation becomes

$$f(\mathbf{r},\mathbf{k}) - f_0(\varepsilon_\mathbf{k}, T) = \left(-\frac{\partial f_0}{\partial \varepsilon}\right)\tau \mathbf{v}_\mathbf{k} \cdot \left[-\left(\frac{\varepsilon(\mathbf{k}) - \zeta}{T}\right)\nabla T + (-e)\left(\mathbf{E} - \frac{\nabla \zeta}{(-e)}\right)\right], \tag{11.9}$$

---

[4] The relation $L_{ET} = -L_{TE}/T$ is known as the Kelvin relation and proved in relation to the Onsager relations [1]. The use of the linearized Boltzmann transport equation with the relaxation time approximation does not violate the Kelvin relation. This is easily confirmed by inserting equation (11.9) into equations (10.47) and (11.8).

where **E** is not zero, since the thermal conductivity is measured for a specimen on an open electrical circuit. As discussed above, the term $(\varepsilon(\mathbf{k}) - \zeta)$ appears to represent the flow of heat. The term $\nabla\zeta/(-e)$ implies that the chemical potential gradient gives rise to an additional field to induce a diffusional current. Thus, we consider $\{\mathbf{E} - [\nabla\zeta/(-e)]\}$ to serve as an effective electric field.

The thermal conductivity can be evaluated by inserting the first term in the right-hand side of equation (11.9) into equation (11.8);

$$\mathbf{U} = -\frac{1}{4\pi^3} \int [\varepsilon(\mathbf{k}) - \zeta]^2 \tau(\mathbf{k}) \mathbf{v}_\mathbf{k} \mathbf{v}_\mathbf{k} \left(-\frac{\partial f_0}{\partial \varepsilon}\right) d\mathbf{k} \cdot \left(\frac{\nabla T}{T}\right),$$

which can be rewritten as

$$\mathbf{U} = -\frac{1}{4\pi^3} \int\int \mathbf{v}_\mathbf{k} \mathbf{v}_\mathbf{k} \tau(\mathbf{k}) [\varepsilon(\mathbf{k}) - \zeta]^2 \frac{dS}{\hbar v_\perp} \left(-\frac{\partial f_0}{\partial \varepsilon}\right) d\varepsilon \cdot \left(\frac{\nabla T}{T}\right). \quad (11.10)$$

Since equation (11.10) involves the derivative of the Fermi–Dirac distribution function, the expansion theorem of equation (3.15) can be applied. Then we obtain

$$\mathbf{U} = -\left[\frac{[\varepsilon(\mathbf{k}) - \zeta]^2}{4\pi^3\hbar} \int \mathbf{v}_\mathbf{k} \mathbf{v}_\mathbf{k} \tau(\mathbf{k}) \frac{dS}{v_\perp} + \right.$$

$$\left. \frac{\pi^2}{6}(k_B T)^2 \frac{\partial^2}{\partial \varepsilon^2} \left\{\frac{[\varepsilon(\mathbf{k}) - \zeta]^2}{4\pi^3\hbar} \int \mathbf{v}_\mathbf{k} \mathbf{v}_\mathbf{k} \tau(\mathbf{k}) \frac{dS}{v_\perp}\right\} + \cdots\right]_{\varepsilon=\zeta} \cdot \left(\frac{\nabla T}{T}\right)$$

or

$$\mathbf{U} = -\left[\frac{[\varepsilon(\mathbf{k}) - \zeta]^2 \sigma(\varepsilon)}{(-e)^2} + \frac{\pi^2(k_B T)^2}{6} \frac{\partial^2}{\partial \varepsilon^2} \left\{[\varepsilon(\mathbf{k}) - \zeta]^2 \frac{\sigma(\varepsilon)}{(-e)^2}\right\} + \cdots\right]_{\varepsilon=\zeta} \cdot \left(\frac{\nabla T}{T}\right), \quad (11.11)$$

where $\sigma(\varepsilon)$ is given by equation (10.51) and represents the electrical conductivity when its Fermi energy is ε. A straightforward calculation of equation (11.11) with subsequent insertion of $\varepsilon = \zeta$ leads to a very simple expression:

$$\mathbf{U} = -\left(\frac{\pi^2 k_B^2 T \sigma}{3(-e)^2}\right) \nabla T. \quad (11.12)$$

As is clear from the derivation above, we have calculated the coefficient L_{TT} in equation (11.7b). The thermal conductivity was defined as $\kappa = -L_{TT}(1 - L_{TE} \cdot L_{ET}/L_{EE} \cdot L_{TT})$. However, one can easily show the correction term $L_{TE} \cdot L_{ET}/L_{EE} \cdot L_{TT}$ to be of the order of $(T/T_F)^2$, if the free electron model is applied to a metal having the Fermi temperature T_F (see Exercise 11.1). Since $(T/T_F)^2$ is negligibly small in ordinary metals, the electronic thermal conductivity is well represented by

$$\kappa_{el} = \frac{\pi^2 k_B^2 T}{3(-e)^2} \sigma(T), \quad (11.13)$$

where $\sigma(T)$ is the electrical conductivity at temperature T. This leads to the Wiedemann–Franz law mentioned in Section 1.2. Here it is important to discuss the universality of the relation (11.13).

An insertion of the expression (10.54) for the electrical conductivity into equation (11.13) introduces the electron density of states at the Fermi level, which is alternatively expressed in terms of the electronic specific heat coefficient given by equation (3.22). Now one can immediately see that equation (11.13) is reduced to the same form as equation (11.4). Equation (11.13) is derived without any special assumption about the band structure but is derived in the framework of the linearized Boltzmann transport equation coupled with the relaxation time approximation. A more rigorous treatment can prove that equation (11.13) is valid when the scattering of electrons is elastic and the relaxation time is independent of energy [6].

We have formulated, in Sections 10.11 and 10.12, the Bloch–Grüneisen law by incorporating the inelastic electron–phonon interaction into the Boltzmann transport equation, and pointed out that the inelastic scattering effect is of fundamental importance in discussing electron transport in the temperature range $T < \Theta_D$ (see Section 10.11). This indicates that the Wiedemann–Franz law would fail at temperatures below Θ_D. Indeed, one cannot evaluate the electronic thermal resistivity due to lattice vibrations simply by inserting into equation (11.13) the temperature dependence of the electrical resistivity given by equations (10.108) and (10.109).

11.4 Wiedemann–Franz law and Lorenz number

As mentioned in Section 1.2, the Wiedemann–Franz law states that the ratio of the electronic thermal conductivity over the electrical conductivity, say, at room temperature becomes constant, regardless of the metal concerned. Indeed, the following relation is immediately derived from equation (11.13):

$$L_0 \equiv \frac{\kappa_{el}}{\sigma T} = \frac{\pi^2}{3} \frac{k_B^2}{(-e)^2}, \qquad (11.14)$$

where L_0 is called the limiting Lorenz number and is a universal constant equal to 2.45×10^{-8} volt^2/K^2. The measured Lorenz number and thermal conductivity for typical metals at 273 K are listed in Table 11.2. One can clearly see that the deviation of the measured Lorenz number L from the limiting value L_0 is rather small at 273 K in most metals.

The Wiedemann–Franz law would fail when the temperature is lowered below 273 K, since the inelastic scattering effect becomes substantial. We show below that inelastic scattering contributes to the electrical and thermal conductions in a different way so that its role can be qualitatively extracted by studying the

Table 11.2. *Thermal conductivity κ and measured Lorenz number L for typical metals at 273 K*

element	κ (W/m·K)	L (10^{-8} V²/K²)
Ag	436	2.34
Al	236	2.10
Au	318	2.39
Ca	186	2.13
Cs	37	2.5
Cu	404	2.27
Fe	80	2.57
K	98	2.24
Li	65	2.05
Mg	151	2.29
Na	142	2.23
Ni	93	2.19
Pb	36	2.50
Pd	72	2.57
Pt	72	2.59
Rb	56	2.30
Ru	131	2.52
Sn	62	2.5
Zn	127	2.60

temperature dependence of the measured Lorenz number over a wide temperature range.

The electron distributions in both electrical and thermal conduction processes are illustrated in Fig. 11.3 [1]. First, the electrical conduction is reviewed. As shown in Figs. 11.3(a), the Fermi surface of a metal specimen is shifted as a whole when an electric field is applied. This implies that more electrons travel to the right than to the left. Scattering processes are needed to establish a steady state. We learned in Section 10.12 that the scattering angle involved becomes smaller and smaller with decreasing temperature due to the fact that phonons of only small wave vectors remain active. At low temperatures, therefore, electrons cannot be transferred from one side of the Fermi surface to the other in a single jump. This leads to the well-known T^5-law for the electrical resistivity at low temperatures.

Electron distributions are different in the thermal conduction process. As shown in Fig. 11.3(b), the temperature gradient across a sample causes more electrons to be distributed above the Fermi level on the right and more electrons below it on the left. This means that we have more "hot" electrons on the

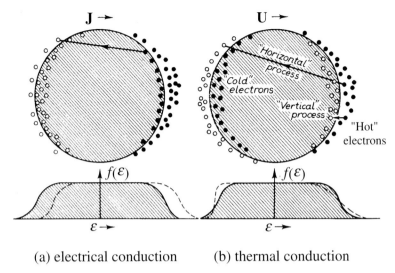

(a) electrical conduction   (b) thermal conduction

Figure 11.3. Electron distributions and scattering process in (a) electrical conduction and (b) thermal conduction. The dashed curve in each represents the electron distribution in the presence of external fields: (a) **E** and (b) ∇T. [Reproduced from J. M. Ziman, *Principles of the Theory of Solids* (Cambridge University Press, 1964)]

right-hand side of the Fermi surface and more "cold" electrons on the left-hand side, resulting in the flow of heat. We can see from Fig. 11.3(b) that there are horizontal and vertical scattering processes. Obviously, the horizontal process favors large-angle scattering, which is abundant at high temperatures but becomes scarce at low temperatures in both electrical and thermal conduction. The vertical process is unique to thermal conduction and requires a single jump with a small scattering angle but the energy must be exchanged with phonons so that the scattering involved must be inelastic. At high temperatures $T > \Theta_D$, however, the vertical process is no longer well defined, since smeared region $k_B T$ across the Fermi sphere exceeds the maximum phonon energy of $k_B \Theta_D$. We call such scattering of electrons *quasi-elastic*. In such a quasi-elastic scattering regime, the vertical process is ineffective and the ratio L/L_0 tends to unity, leading to the validity of the Wiedemann–Franz law.

With decreasing temperature below Θ_D, the vertical process begins to play its unique role in thermal conduction and the thermal current is more substantially reduced than expected from the Wiedemann–Franz law. This results in a lowering of the ratio L/L_0 below unity and the Wiedemann-Franz law gradually breaks down with decreasing temperature, as shown in Fig. 11.4. As another unique feature, we point to the dependence of the ratio L/L_0 on the purity of a specimen in the temperature range well below Θ_D. Consider, first, an ideally pure metal free from impurity scattering at low temperatures

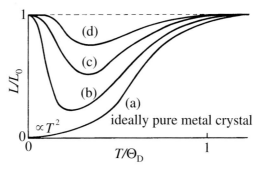

Figure 11.4. Temperature dependence of L/L_o, where L_o is the limiting value given by equation (11.14). Curve (a) refers to an ideally perfect crystal metal with zero residual resistivity. The residual resistivity increases in the sequence (b) to (d). [Reproduced from H. M. Rosenberg, *Low Temperature Solid State Physics*, (Clarendon Press, Oxford 1963)]

$T \ll \Theta_D$. We know that the number of phonons decreases with decreasing temperature and eventually vanishes at absolute zero. As discussed in Section 11.2, the electronic thermal resistivity due to lattice vibrations decreases as T^2, while the electrical resistivity ρ_{lattice} decreases as T^5 with decreasing temperature. Thus, we see that the ratio L/L_0 tends towards zero as T^2 in a pure metal, as illustrated in Fig. 11.4, curve (a). If a specimen is impure, curves (b)–(d), impurity scattering dominates at low temperatures $T \ll \Theta_D$. Since the impurity scattering is elastic, the ratio approaches unity again and the Wiedemann–Franz law revives in the range where the temperature-independent residual resistivity is observed. Therefore, Fig. 11.4 is instructive to see the role of the elastic and inelastic scattering of conduction electrons over a wide temperature range for a given metal specimen.

Before ending this section, we summarize the conditions for the Wiedemann–Franz law to be valid. (1) The phonon thermal conductivity can be ignored relative to the electronic thermal conductivity or is subtracted from the total thermal conductivity. (2) The elastic scattering should dominate. This is realized either at high temperatures $T > \Theta_D$ or at low temperatures where only the residual resistivity is observed. (3) The relaxation time involved in the electronic thermal conductivity is the same as that in the electrical conductivity so that the same scattering mechanism must be responsible for both of them.

11.5 Thermoelectric power

The linearized Boltzmann transport equation (11.9) is inserted into the expression for the electrical current density given by equation (10.47):

11.5 Thermoelectric power

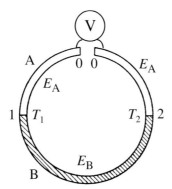

Figure 11.5. Circuit for the measurement of the Seebeck effect.

$$\mathbf{J} = \frac{e^2 \tau}{4\pi^3 \hbar} \int \int \mathbf{v_k v_k} \left(-\frac{\partial f_0}{\partial \varepsilon} \right) \frac{dS}{v_{k_\perp}} d\varepsilon \cdot \mathbf{E}'$$

$$+ \frac{(-e)\tau}{4\pi^3 \hbar} \int \int \mathbf{v_k v_k} \left(\frac{\varepsilon - \zeta}{T} \right) \left(-\frac{\partial f_0}{\partial \varepsilon} \right) \frac{dS}{v_{k_\perp}} d\varepsilon \cdot (-\nabla T), \quad (11.15)$$

where $\mathbf{E}' = \mathbf{E} - [\nabla \zeta/(-e)]$. We see from equation (11.15) that the relation $\sigma = L_{EE}$ holds under the isothermal condition $\nabla T = 0$, since the electrical conductivity σ is defined as $\mathbf{J} = \sigma \mathbf{E}$. The second term in equation (11.15) indicates that the temperature gradient can also induce an electrical current. This is the thermoelectric effect.

Let us assume that a metal is in a temperature gradient ∇T but is electrically open so that $\mathbf{J} = 0$ holds. Equation (11.7a) leads to the relation

$$\mathbf{E} = Q \nabla T, \quad (11.16)$$

where

$$Q = -\left(\frac{L_{ET}}{L_{EE}} \right). \quad (11.17)$$

The coefficient Q is called the absolute thermoelectric power or the Seebeck coefficient.

Equation (11.16) implies that an electric field is generated due to a temperature gradient across a specimen. In order to observe this effect, one has to set up a closed circuit consisting of two metals A and B with junctions at different temperatures T_1 and T_2. This is illustrated in Fig. 11.5. The electromotive force ϕ around the circuit is calculated by integrating the electric field $\mathbf{E}$ along the wire:

$$\phi = \int_0^1 E_A dx + \int_1^2 E_B dx + \int_2^0 E_A dx = \int_2^1 Q_A \left(\frac{\partial T}{\partial x}\right) dx + \int_1^2 Q_B \left(\frac{\partial T}{\partial x}\right) dx$$

$$= \int_{T_2}^{T_1} Q_A dT + \int_{T_1}^{T_2} Q_B dT = \int_{T_1}^{T_2} (Q_B - Q_A) dT. \tag{11.18}$$

We find that the voltage generated in the circuit is obtained by integrating the difference in the thermoelectric power of the two metals between the temperatures T_1 and T_2 at the two junctions. This is known as the Seebeck effect.

The expression for the thermoelectric power Q is derived as follows. The numerator $-L_{ET}$ in equation (11.17) is explicitly calculated from equation (11.15) in the form:

$$-L_{ET} = \frac{(-e)\tau}{4\pi^3 \hbar} \iint \mathbf{v_k v_k} \left(\frac{\varepsilon - \zeta}{T}\right) \left(-\frac{\partial f_0}{\partial \varepsilon}\right) \frac{dS}{v_{k_\perp}} d\varepsilon,$$

which can be expanded around the Fermi energy in the same way as in equation (11.9):

$$= \frac{(-e)\tau}{4\pi^3 \hbar} \cdot \frac{1}{T} \left\{ \left[\int \mathbf{v_k v_k} (\varepsilon - \zeta) \frac{dS}{v_{k_\perp}}\right]_{\varepsilon=\zeta} \right.$$

$$\left. + \frac{\pi^2}{6}(k_B T)^2 \left[\int (\varepsilon - \zeta) \frac{\partial^2}{\partial \varepsilon^2}\left(\mathbf{v_k v_k} \frac{dS}{v_{k_\perp}}\right) + 2\int \frac{\partial}{\partial \varepsilon}\left(\mathbf{v_k v_k} \frac{dS}{v_{k_\perp}}\right)\right]_{\varepsilon=\zeta} + \ldots \right\}.$$

Here the terms involving $(\varepsilon - \zeta)$ disappear at the Fermi energy ζ. Thus, $-L_{ET}$ is reduced to

$$-L_{ET} = \frac{\pi^2}{3}(k_B T)^2 \cdot \frac{1}{T}\left\{\frac{(-e)\tau}{4\pi^3 \hbar}\left[\int \frac{\partial}{\partial \varepsilon}\left(\mathbf{v_k v_k} \frac{dS}{v_{k_\perp}}\right)\right]_{\varepsilon=\zeta}\right\}. \tag{11.19}$$

As mentioned earlier, the denominator L_{EE} represents the electrical conductivity and is explicitly written as

$$L_{EE} = \left[\frac{(-e)^2 \tau}{4\pi^3 \hbar} \int \mathbf{v_k v_k} \frac{dS}{v_{k_\perp}}\right]_{\varepsilon=\zeta} = [\sigma(\varepsilon)]_{\varepsilon=\zeta}. \tag{11.20}$$

Therefore, the thermoelectric power Q is formulated as

$$Q = \frac{\pi^2}{3(-e)} k_B^2 T \left[\frac{\partial \ln \sigma(\varepsilon)}{\partial \varepsilon}\right]_{\varepsilon=\zeta}, \tag{11.21}$$

11.5 Thermoelectric power

where $\sigma(\varepsilon)$ is the electrical conductivity when the Fermi level is ε. Thus, the thermoelectric power can be calculated, once the energy dependence of the conductivity is given.

Equation (11.16) implies that a potential difference $\int_{x_1}^{x_2} E dx$ is generated, when a temperature difference $\Delta T = T_2 - T_1$ exists between the two points x_1 and x_2 in a metal bar. The potential difference is often expressed as $\sigma_T \Delta T$ (in volt), where σ_T is called the Thomson coefficient. By using the Thomson coefficient σ_T, the Seebeck coefficient in equation (11.16) can be expressed as $Q = \int_0^T (\sigma_T/T) dT$.[5] If an electronic charge $(-e)$ flows up through the temperature gradient dT, a heat equal to $(-e)\sigma_T dT$ must be evolved per electron. Hence, we obtain the relation $\sigma_T = C_{el}/n(-e)$, where C_{el} is the electronic specific heat per unit volume and n is the number of electrons involved. Its insertion into the relation above immediately results in

$$Q = \left[\frac{1}{n(-e)}\right] \int_0^T \frac{C_{el}}{T} dT = \frac{s}{n(-e)}, \qquad (11.22)$$

where s is the electronic entropy density. We see, therefore, that the thermoelectric power can also be discussed in terms of the carrier entropy in the system.

Let us apply the free-electron model to the electrical conductivity formula given by equation (10.52). Since the relation $\sigma(\varepsilon) \propto \varepsilon^{3/2}$ holds, its insertion into equation (11.21) results in

$$Q_{\text{free}} = \frac{\pi^2 k_B T}{2(-e) T_F} \approx -4.25 \times 10^2 \frac{T}{T_F} \; [\mu V/K], \qquad (11.23)$$

where T_F is the Fermi temperature. Note that the thermoelectric power for ordinary metals is fairly small in magnitude, since T/T_F is only 0.001–0.005 at room temperature. Equation (11.23) is equally obtained, if the free-electron expression $N(\varepsilon_F) = 3n/2k_B T_F$ is inserted into equation (11.22).

As a typical example, we show in Fig. 11.6 the temperature dependence of the thermoelectric power for a well-annealed strain-free pure Al metal

---

[5] In order to find the interrelations among the Seebeck coefficient Q, the Thomson coefficient σ_T and the Peltier coefficient Π, we consider a closed circuit consisting of two metals A and B with junctions at different temperatures $T_1 = T$ and $T_2 = T + dT$. The total e.m.f. $d\phi$ generated in the circuit due to the temperature difference dT between the junctions is given by $d\phi = \Delta\Pi + \Delta\sigma dT$, where $\Delta\Pi = \Pi_A - \Pi_B$ is heat liberated or absorbed at the junctions and $\Delta\sigma = \sigma_A - \sigma_B$. The thermocouple thus formed is considered as a heat engine, which works between two temperatures T_1 and T_2 and produces the electrical energy $d\phi$. Its efficiency is defined as dT/T, which is alternatively expressed as the ratio of the electrical energy $d\phi$ available for external work over the heat $\Delta\Pi$. Hence, we have $dT/T = d\phi/\Delta\Pi$ or $d\phi = (\Delta\Pi/T)dT$. Its comparison with the relation above leads to $\Delta\sigma = \Delta\Pi/T$ or $\sigma_T = \Pi/T$. By using equation (11.18), we obtain $dQ/dT = -d^2\phi/dT^2 = -d(\Pi/T)/dT = \Pi/T^2 = \sigma_T/T$.

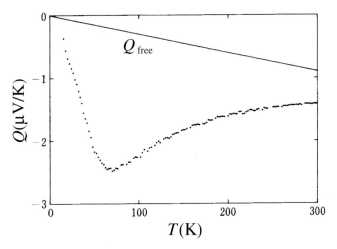

Figure 11.6. Temperature dependence of the thermoelectric power of pure Al. The solid line represents the free-electron behavior obtained by inserting into equation (11.23) the Fermi temperature of 1.35×10^5 K. [Courtesy Dr T. Matsuda]

(99.999%), together with the free-electron behavior obtained by inserting the Fermi temperature $T_F = 1.35 \times 10^5$ K of pure Al into equation (11.23). Obviously, the experimental data deviate substantially from the free-electron model and exhibit a minimum at about 70 K. The formation of the minimum has been attributed to the phonon drag effect unique to a crystal metal, where the phonon mean free path is long. The phonon drag effect will be discussed in the next section.

There is another thermoelectric phenomenon known as the Peltier effect. Two different metals A and B are joined and connected to a battery, as shown in Fig. 11.7. An electrical current density **J** is fed through the circuit while the circuit is maintained at a uniform temperature. Now equations (11.7a) and (11.7b) are reduced to $\mathbf{U} = L_{TE}\mathbf{E}$ and $\mathbf{J} = L_{EE}\mathbf{E}$, from which we obtain

$$\mathbf{U} = \Pi \mathbf{J}, \tag{11.24}$$

where

$$\Pi = \frac{L_{TE}}{L_{EE}}. \tag{11.25}$$

The coefficient Π is called the Peltier coefficient and is related to the thermoelectric power Q through the relation:

$$Q = -\left(\frac{L_{ET}}{L_{EE}}\right) = \left(\frac{L_{TE}/T}{L_{EE}}\right) = \frac{\Pi}{T}. \tag{11.26}$$

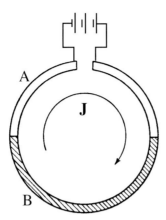

Figure 11.7. Circuit for the measurement of the Peltier effect.

Equation (11.24) implies that an electrical current fed to the circuit generates thermal currents $\mathbf{U}_A = \Pi_A \mathbf{J}$ and $\mathbf{U}_B = \Pi_B \mathbf{J}$ in the metals A and B, respectively. Thus, a heat flux $(\Pi_A - \Pi_B)\mathbf{J}$ will be emitted at one junction and absorbed at the other junction. As a consequence, the one junction becomes hotter, the other junction colder. This is the Peltier effect.

11.6 Phonon drag effect

As has been discussed in the preceding section, a voltage is generated between the two ends of a sample across which a temperature gradient ∇T exists. However, there exists no current flow due to conduction electrons because of an open circuit. Instead, phonons at the high-temperature end are driven to the colder end under a finite temperature gradient. If the mean free path of the phonon is very long, then the collision of one phonon with other phonons is so scarce that its energy cannot be released to the lattice system. Instead, phonons can exchange their energy with electrons, since the relaxation time for the phonon–electron interaction is much shorter than that for the phonon–phonon interaction. This means that the extra local energy carried by a phonon is fed back to the electron system, resulting in a new extra electric field because of $\mathbf{J} = 0$. The generation of the electric field in the electron system due to the flow of the non-equilibrium phonon is called the phonon drag effect, where "phonon drag" refers to the situation where electrons are carried along by the flow of phonons caused by the temperature gradient.

We discuss the phonon drag effect following Rosenberg [3]. As shown in Fig. 11.8, a temperature gradient dT/dx is assumed along the axis of a metal bar with a unit cross-section. The temperature at the center of the region 2Λ,

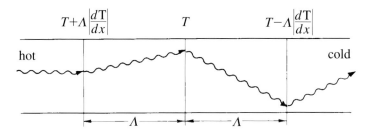

Figure 11.8. Phonon drag effect. [Reproduced from H. M. Rosenberg, *Low Temperature Solid State Physics* (Clarendon Press, Oxford, 1963)]

corresponding to twice the mean free path, is T. Let us suppose that the thermal current density of the phonon is equal to $U(T+\Lambda|dT/dx|)$ at the "hot" end, where the temperature is $T+\Lambda|dT/dx|$. Similarly, the thermal current density is $U(T-\Lambda|dT/dx|)$ at the "cold" end, where the temperature is $T-\Lambda|dT/dx|$. The difference in thermal energy in the region over 2Λ will be

$$U\left(T+\Lambda\left|\frac{dT}{dx}\right|\right) - U\left(T-\Lambda\left|\frac{dT}{dx}\right|\right) = 2\Lambda C_{\text{lattice}} \left|\frac{dT}{dx}\right|, \quad (11.27)$$

where C_{lattice} is defined as $C_{\text{lattice}} = dU/dT$ and represents the lattice specific heat per unit volume. This extra energy has to be absorbed in this region, where the only sink available is that provided by $2\Lambda n$ electrons. Thus, the extra energy must be converted into an electric field ΔE and its magnitude is derived from the relation:

$$2\Lambda n(-e)\Delta E = 2\Lambda C_{\text{lattice}} \left|\frac{dT}{dx}\right| = 2\Lambda C_{\text{lattice}} \nabla T$$

yielding,

$$\Delta E = \frac{C_{\text{lattice}}}{(-e)n} \nabla T. \quad (11.28)$$

The thermoelectric power due to phonon drag, $\Delta E = Q_{\text{ph. drag}} \nabla T$, is given by

$$Q_{\text{ph. drag}} = \frac{C_{\text{lattice}}}{(-e)n}. \quad (11.29)$$

One can easily calculate the thermoelectric power due to phonon drag at $T>\Theta_D$ to be -86 μV/K by inserting $C_{\text{lattice}}=3R$ expected from the Dulong–Petit law into equation (11.29). This is much larger than the measured value. This indicates the failure of equation (11.29) at $T>\Theta_D$. Indeed, the mean free path of the phonon becomes so short that the phonon drag effect is known to become unimportant at such high temperatures. On the other hand, the lattice specific heat decreases as T^3 below about 20 K and, hence, the phonon drag effect becomes ineffective again at low temperatures. This means that it is most significant in the intermediate temperature range around $T/\Theta_D \approx 0.2$ and is responsible for the formation of a deep valley like that shown in Fig. 11.6 for pure Al.

The valley becomes shallower in alloys because of the shortening of the mean free path of phonons due to the disruption of the periodic lattice. The phonon drag effect is essentially absent in amorphous alloys because of the lack of lattice periodicity. Hence, the temperature dependence of the thermoelectric power in amorphous alloys is attributed to other effects like the inelastic electron–phonon interaction and the energy dependence of the relaxation time [5].

11.7 Thermoelectric power in metals and semiconductors

The interpretation of the measured thermoelectric power is not straightforward even in simple metals [6,7]. For example, the sign of the thermoelectric power Q in the alkali metals cannot be correctly predicted from the free-electron model. As listed in Table 11.3, its sign is positive for Li but is negative for Na and K, though all these metals possess a single-electron Fermi surface. The diversity of its trends has been discussed in relation to the energy dependence of the relaxation time and inelastic electron–phonon interaction [7]. A positive thermoelectric power has also been observed in monovalent noble metals. As shown in Fig. 11.9, its temperature dependence is quite complex and a positive thermoelectric power dominates over the whole temperature range. Contact of the Fermi surface with the {111} zone planes has been suggested to play an important role in its behavior.

Setting aside such difficulties in the theoretical interpretation, much attention has been directed to synthesize thermoelectric device materials to convert efficiently heat to electricity or vice versa. As is clear from the argument above, ordinary metals possess the Fermi temperature of 10^4–10^5 K and, thus, the resulting thermoelectric power is, at most, 10–20 μV/K. A large value of Q should be achievable, not in metals, but in heavily doped semiconductors. Figure 11.10 illustrates the carrier concentration dependence of the thermoelectric power or carrier entropy together with that of the resistivity, thermal conductivity and dimensionless figure of merit for an idealized semiconductor [8]. Here the dimensionless figure of merit ZT is defined as $ZT = TQ^2/\kappa\rho$, where T is the absolute temperature, κ is the thermal conductivity and ρ is electrical resistivity, and represents the heat–current conversion efficiency for a thermoelectric material.

Table 11.3. *Absolute thermoelectric power of various substances*

substance	Q (μV/K) at 273 K[a]
Ag	1.38
Al	−1.6
Au	1.74
Ca	10.3
Cd	2.56
Cs	−0.9
Cu	1.5
Fe	16.2
K	−12.8
Li	10.6
Mg	−1.47
Na	−5.8
Pb	−1.25
Pd	−9.7
Rb	−9.47
Bi_2Te_3	+250 (p); −300 (n)
$FeSi_2$	+250 (p); −250 (n)

Note:
[a] The measuring temperatures for Bi_2Te_3 and $FeSi_2$ are 300–500 K and above 900 K, respectively.

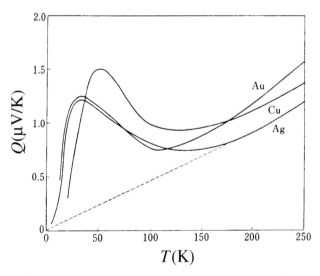

Figure 11.9. Temperature dependence of the thermoelectric power of the noble metals Cu, Ag and Au. [D. K. C. MacDonald, *Principles of Thermoelectricity* (John Wiley & Sons, Inc., New York 1962) p. 71]

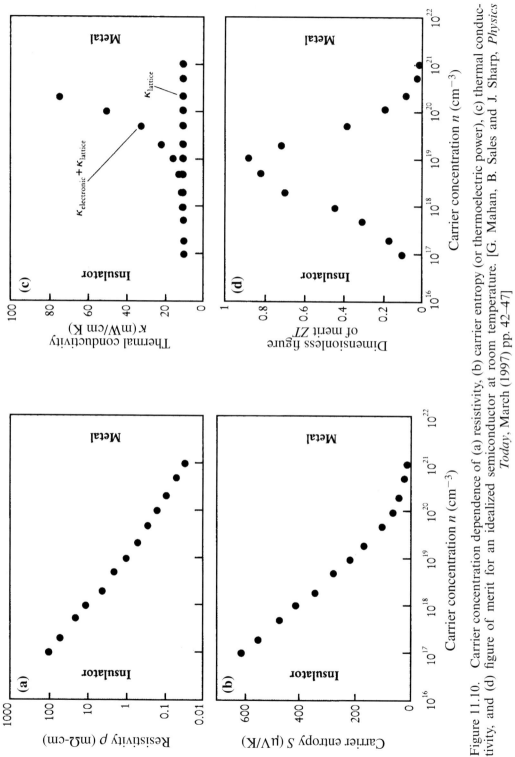

Figure 11.10. Carrier concentration dependence of (a) resistivity, (b) carrier entropy (or thermoelectric power), (c) thermal conductivity, and (d) figure of merit for an idealized semiconductor at room temperature. [G. Mahan, B. Sales and J. Sharp, *Physics Today*, March (1997) pp. 42–47]

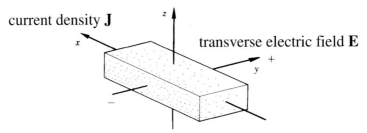

Figure 11.11. The geometry for the Hall effect measurement.

The thermoelectric power Q is expected to reach a value as high as 500–600 μV/K for a carrier concentration of 10^{17}–10^{18} cm. As shown in Table 11.3, compounds like Bi_2Te_3 and $FeSi_2$ exhibit a thermoelectric power of a few hundreds μV/K and are considered as the most efficient thermoelectric device materials available at present. Further increase in Q beyond several hundreds μV/K, while suppressing the electrical resistivity and thermal conductivity to be as low as possible, is of urgent need from the viewpoint of practical applications. It is also interesting from the fundamental point of view to pursue if there is any theoretical limit to the upper value of Q in the metallic regime. It is indeed quite challenging to explore marginal solids characterized by a deep pseudogap at the Fermi level near the metal–insulator transition (see section 15.14), in which the electronic entropy is expected to increase through interaction with excitations like phonons and spin fluctuations.

11.8 Hall effect and magnetoresistance

The Hall effect is a phenomenon observed in the presence of both electric and magnetic fields and its measurement is generally carried out using the configuration shown in Fig. 11.11. An electrical current is fed along the x-direction in a rectangular specimen and a magnetic field is applied along the z-direction. Since no current flows along the y- and z-directions, the condition $J_y = J_z = 0$ holds. We will learn below that the transverse electric field E_y is developed along the y-direction. This is called the Hall effect.

The Hall effect can be treated by incorporating two external forces $(-e)\mathbf{E}$ and $(-e)\mathbf{v_k} \times \mathbf{B}$ into the linearized Boltzmann transport equation. Taking the directions of the external fields as $\mathbf{E} = (E, 0, 0)$ and $\mathbf{B} = (0, 0, B)$, we can write equation (10.42) as

$$(-e)E\mathrm{v}_x \left(-\frac{\partial f_0}{\partial \varepsilon} \right) = \frac{\phi}{\tau} - \frac{(-e)}{\hbar} B \left(\mathrm{v}_x \frac{\partial \phi}{\partial k_y} - \mathrm{v}_y \frac{\partial \phi}{\partial k_x} \right). \qquad (11.30)$$

11.8 Hall effect and magnetoresistance

where we have also employed the relaxation time approximation. Since the magnetic field affects only the x- and y-components of the wave vectors, the function ϕ in equation (10.44) may be assumed to have a form [7]:

$$\phi = ak_x + bk_y. \tag{11.31}$$

Equation (11.31) and the free-electron relations $v_x = \hbar k_x/m$ and $v_y = \hbar k_y/m$ are inserted into equation (11.30). This allows us to determine the coefficients a and b and then the components σ_{xx} and σ_{xy} of the conductivity tensor (see Exercise 11.2). Other components, like σ_{yy} and σ_{zz}, are similarly calculated by applying the external electric field $\mathbf{E}$ along y- and z-directions.

The resulting conductivity tensor defined as $\mathbf{J} = \sigma \mathbf{E}$ for electrons under the condition $\mathbf{B} = (0, 0, B)$ is explicitly written as

$$\sigma_{ij} = \frac{n(-e)^2 \tau}{m} \begin{pmatrix} \frac{1}{1+\alpha^2} & \frac{-\alpha}{1+\alpha^2} & 0 \\ \frac{\alpha}{1+\alpha^2} & \frac{1}{1+\alpha^2} & 0 \\ 0 & 0 & 1 \end{pmatrix}, \tag{11.32}$$

where $\alpha = \omega_c \tau$ and $\omega_c = (+e)B/m$. The resistivity tensor defined as $\mathbf{E} = \rho \mathbf{J}$ for electrons under the condition $\mathbf{B} = (0, 0, B)$ is obtained by inversion of equation (11.32):

$$\rho_{ij} = \frac{m}{n(-e)^2 \tau} \begin{pmatrix} 1 & \alpha & 0 \\ -\alpha & 1 & 0 \\ 0 & 0 & 1 \end{pmatrix}. \tag{11.33}$$

Since the Hall measurement is carried out by feeding a current perpendicular to the magnetic field, we must employ equation (11.33) instead of equation (11.32). It is clear from equation (11.33) that the current density $\mathbf{J} = (J_x, 0, 0)$ gives rise to a y-component of the electric field E_y, which is perpendicular to the directions of both current and magnetic field. This is explicitly written as

$$E_y = -\alpha \left(\frac{m}{n(-e)^2 \tau} \right) J_x = -\left(\frac{(+e)\tau}{m} \right) \cdot \left(\frac{m}{n(-e)^2 \tau} \right) B_z J_x = \frac{1}{n(-e)} B_z J_x,$$

where $\alpha = (+e)B_z \tau/m$. The Hall coefficient R_H is defined as the coefficient of $B_z J_x$ in the transverse electric field E_y:

$$R_H = \frac{1}{n(-e)}. \tag{11.34}$$

It is clear that R_H in the free-electron model is independent of the magnetic field and depends only on the number of electrons per unit volume.

The Hall effect of either electrons or holes may be discussed in a simpler but less rigorous way by replacing the right-hand side of equation (10.3) by the Lorentz force given in equation (7.1):

$$m^*\left(\frac{d\mathbf{v}_D}{dt} + \frac{\mathbf{v}_D}{\tau}\right) = (\mp e)(\mathbf{E} + \mathbf{v}_D \times \mathbf{B}) \quad m^*\left(\frac{d\mathbf{v}_D}{dt} + \frac{\mathbf{v}_D}{\tau}\right) = (\mp e)\left(\mathbf{E} + \frac{1}{c}\mathbf{v}_D \times \mathbf{H}\right)[\text{CGS}], \tag{11.35}$$

where $\mp$ refers to the sign of the charge of the respective carriers. Since the condition $d\mathbf{v}_D/dt = 0$ holds in the steady state, we get

$$\mathbf{v}_D = \left(\frac{(\mp e)\tau}{m^*}\right)(\mathbf{E} + \mathbf{v}_D \times \mathbf{B}). \tag{11.36}$$

As shown in Fig. 11.11, we choose the z-axis as the direction of the magnetic field and use the relations $\omega_c = (+e)B_z/m^*$ and $\mathbf{J} = n(\mp e)\mathbf{v}_D$.[6] Now the following relations, consistent with equation (11.32), are immediately obtained from equation (11.36):

$$J_x = \frac{n(\mp e)^2 \tau}{m^*(1 + \omega_c^2 \tau^2)}(E_x \mp \omega_c \tau E_y)$$

$$J_y = \frac{n(\mp e)^2 \tau}{m^*(1 + \omega_c^2 \tau^2)}(\omega_c \tau E_x \pm E_y) \tag{11.37}$$

$$J_z = \frac{n(\mp e)^2 \tau}{m^*} E_z.$$

Since no current flows along the y- and z-directions in the configuration shown in Fig. 11.11, the condition $J_y = J_z = 0$ holds. This leads to $E_y = \mp \omega_c \tau E_x$ or $R_H = 1/n(\mp e)$ for electrons and holes, respectively.

The Hall coefficient for either the electron or hole is expressed as

$$R_H = \frac{1}{nq} \quad R_H = \frac{1}{nqc}[\text{CGS}], \tag{11.38}$$

where q is the electric charge of the carriers and n is the number of carriers per unit volume. The carrier is an electron if $q = (-e)$, while it is a hole if $q = (+e)$.

The Hall coefficient is negative and temperature independent in the free-electron model. The density of the conduction electron n is related to the Fermi

---

[6] $\omega_c = (+e)H/m^*c$ in CGS units, where c is the speed of light.

Table 11.4. *Hall coefficients in pure elements*

metal	e/a	Hall coefficient at 300 K $R_H^{300\,K}$ ($\times 10^{-11}$ m³/A·s)	Hall coefficient from the free-electron model R_H^{free} ($\times 10^{-11}$ m³/A·s)	$R_H^{300\,K}/R_H^{free}$
Li	1	−15	−13.4	1.12
Na	1	−25.8	−24.6	1.05
K	1	−35	−47.1	0.74
Cu	1	−5.07	−7.35	0.69
Ag	1	−8.8	−10.6	0.83
Au	1	−7.08	−10.6	0.67
Mg	2	−8.3	−7.23	1.14
Ca	2	−17.8	−13.5	1.31
Zn	2	5.5	−4.74	−1.16
Cd[a]	2	3.9 (13.9)	−6.74	−0.57 (−2.0)
Al	3	−3.44	−3.45	0.99
In	3	−0.216	−5.43	0.04
Sn	4	−0.22	−4.21	0.05
Pb	4	0.098	−4.59	−0.02
As		450		
Sb		−198		
Bi		−54 000		

Note:
[a] The values of 3.9 and (13.9) are obtained with the magnetic field parallel to and perpendicular to the *c*-axis, respectively.

radius k_F through equation (2.20) in the free-electron model. Thus, the value of n or k_F can be experimentally derived by measuring the Hall coefficient for metals where the free-electron model holds well. The Fermi diameter $2k_F$ is obtained from

$$2k_F = 1.139 \times 10^{-3} (|R_H|)^{-1/3}, \qquad (11.39)$$

where the Hall coefficient and the Fermi diameter are in the units of m³/A·s and (Å)$^{-1}$, respectively. The Hall coefficient in the free-electron model may be simply calculated from the relation:

$$|R_H^{free}| = 1.036 \times 10^{-11} \left[\frac{A}{d \cdot (e/a)}\right], \qquad (11.40)$$

where A is the atomic weight in g, d is the mass density in (g/cm³) and e/a is the number of carriers per atom. The Hall coefficients at room temperature in low magnetic fields for typical metals are listed in Table 11.4, along with the corresponding free-electron values. A close agreement with the free-electron value is observed only in limited number of metals, for example Na. The sign of the Hall coefficient in the divalent metals Zn and Cd is positive and a significant deviation from the free-electron model is apparent.

As discussed in Chapter 6, both electron and hole Fermi surfaces coexist in polyvalent metals like Zn, Al and Pb. Let us assume that we have two types of carriers, i.e., electrons and holes. In the configuration such that $\mathbf{J} \perp \mathbf{B}$ and $\mathbf{E} \perp \mathbf{B}$, the current density for each of the carriers is reduced to the form [1]:

$$\mathbf{J}_i = \frac{\sigma_i}{1+\beta_i^2 B^2}\mathbf{E} - \frac{\sigma_i \beta_i}{1+\beta_i^2 B^2}\mathbf{B}\times\mathbf{E}, \quad (11.41)$$

where σ_i is the conductivity of the i-th carrier and $\beta_i = q_i \tau_i / m_i$ (see Exercise 11.3).[7] The total current is obviously equal to

$$\mathbf{J} = \mathbf{J}_1 + \mathbf{J}_2 = \left(\frac{\sigma_1}{1+\beta_1^2 B^2} + \frac{\sigma_2}{1+\beta_2^2 B^2}\right)\mathbf{E} - \left(\frac{\sigma_1 \beta_1}{1+\beta_1^2 B^2} + \frac{\sigma_2 \beta_2}{1+\beta_2^2 B^2}\right)\mathbf{B}\times\mathbf{E}. \quad (11.42)$$

The Hall coefficient at low magnetic fields is then deduced to be

$$R = \frac{\sigma_1^2 R_1 + \sigma_2^2 R_2}{(\sigma_1 + \sigma_2)^2}, \quad (11.43)$$

where R_1 and R_2 are the Hall coefficient for the respective carriers (see Exercise 11.4). We see that the sign of the Hall coefficient is determined by the balance between the electrons and holes having $R_1 < 0$ and $R_2 > 0$, respectively. A positive Hall coefficient in divalent metals like Zn and Cd means that the hole contribution dominates over the electron contribution. In such two-carrier metals, the Hall coefficient often exhibits a strong temperature dependence. In contrast to a single-carrier metal, the relaxation time is involved in the Hall coefficient through σ_i and its temperature dependence is believed to be responsible for that of the Hall coefficient. The temperature dependence of the Hall coefficient for typical metals is shown in Fig. 11.12. The presence of the Brillouin zone in crystal metals yields electrons and holes having different effective masses and relaxation times. The Hall coefficient reflects the anisotropy of the electronic structure and exhibits substantial deviation from the free-electron behavior. In contrast, the Hall coefficient in amorphous alloys is essentially temperature independent because of the lack of anisotropy of the Fermi surface (see Section 15.6).

Finally, the magnetoresistance is briefly discussed. It is clear from equation (11.33) that the electric field in the direction of the current density $\mathbf{J} = (J, 0, 0)$ is not affected by the magnetic field. This implies that the magnetoresistance

---

[7] $B_i = q_i \tau_i / m_i c$ in CGS units.

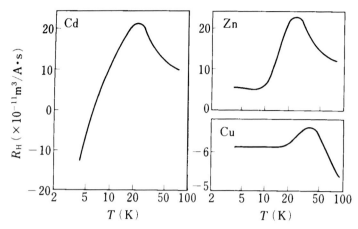

Figure 11.12. Temperature dependence of the Hall coefficient of pure metals Cd, Zn and Cu. [K. E. Saeger and R. Lück, *Phys. Kondens. Materie* **9** (1963) 91]

$\Delta\rho_{xx}$ defined as $[\rho_{xx}(B) - \rho_{xx}(0)]$ is equal to zero. In real metals, however, $\Delta\rho_{xx}$ is finite and increases with increasing magnetic field. This is the magnetoresistance effect and can be explained in the two-band model [1]. The transverse magnetoresistance refers to the resistivity defined as the component of **E** along **J** in the presence of magnetic field perpendicular to both **E** and **J** and is deduced to be

$$\frac{\Delta\rho_{xx}}{\rho_{xx}(0)} \equiv \frac{\rho_{xx}(B) - \rho_{xx}(0)}{\rho_{xx}(0)} = \frac{\sigma_1\sigma_2(\beta_1 - \beta_2)^2 B^2}{(\sigma_1 + \sigma_2)^2 + B^2(\beta_1\sigma_1 + \beta_2\sigma_2)^2}. \quad (11.44)$$

Equation (11.44) indicates that $\Delta\rho_{xx}$ is always positive but disappears when $\beta_1 = \beta_2$ and that it is proportional to B^2, as long as the magnetic field is low.

11.9 Interaction of electromagnetic wave with metals (I)

The optical reflectance or absorption spectrum of a material can be measured by subjecting it to electromagnetic radiation. Information on the electronic structure of a solid can be extracted from the spectrum itself or even more specifically from the optical conductivity deduced from the spectrum [9,10].

Different optical excitations are illustrated in Fig. 11.13. The transitions marked 1 or 2 refer to a direct transition or vertical transition, where an electron is excited vertically from one band to another without altering the wave vector of the electron. A direct transition occurs, since the momentum of the absorbed photon is very small compared with the shortest reciprocal lattice vector in the Brillouin zone and the change in the electron wave vector can be ignored (see Fig. 11.14). The "oblique" transitions marked 3 and 4 are called

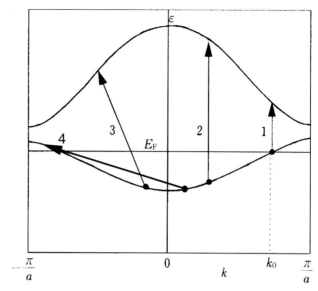

Figure 11.13. Interband (1 to 3) and intraband (4) transitions due to optical excitations.

indirect transitions, since an additional momentum is needed to satisfy the momentum conservation law. An indirect transition is made possible by creating a phonon of wave vector **q** during the transition so that the relation $\mathbf{k}_{electron} + \mathbf{q}_{phonon} = \mathbf{q}_{photon} \approx 0$ holds. There is no threshold energy for indirect transitions. The transitions marked 1, 2 and 3 are induced between different bands and are called interband transitions. The transition marked 4 occurs within a given band and is called an intraband transition.

We have already shown the absorption spectra for Cu–Zn alloys in Section 7.10 and discussed the shift of the absorption edge towards shorter wavelengths due to an increase in the Zn concentration. It was mentioned that the absorption edge corresponds to the transition of electrons at the top of the Cu 3d band to unoccupied states immediately above the Fermi level. This is an interband transition induced by the electromagnetic wave. In addition to the interband transitions, the optical spectra reflect the motion of both bound electrons and conduction electrons driven by the electromagnetic field. We will study in this section basic properties of electrons in a metal interacting with electromagnetic waves.

The propagation of the electromagnetic wave in a metal can be described in terms of the Maxwell equations. We assume that the metal is isotropic and that the dielectric constant ε in the electric displacement $\mathbf{D} = \varepsilon\varepsilon_0 \mathbf{E}$ and the permeability μ in the magnetic flux density $\mathbf{B} = \mu\mu_0 \mathbf{H}$ are independent of the position

11.9 Interaction of electromagnetic wave with metals (I)

vector **r** throughout the metal, where ε_0 and μ_0 are those in vacuum.[8] Since no external charge exists, the Maxwell equations are explicitly written as

$$\text{div}\mathbf{E} = 0 \qquad \qquad \text{div}\mathbf{E} = 0 \text{ [CGS]} \qquad (11.45\text{a})$$

$$\text{rot}\mathbf{E} = -\mu\mu_0 \frac{\partial \mathbf{H}}{\partial t} \qquad \text{rot}\mathbf{E} = -\frac{\mu}{c}\frac{\partial \mathbf{H}}{\partial t} \text{ [CGS]} \qquad (11.45\text{b})$$

$$\text{div}\mathbf{H} = 0 \qquad \qquad \text{div}\mathbf{H} = 0 \text{ [CGS]} \qquad (11.45\text{c})$$

$$\text{rot}\mathbf{H} = \sigma\mathbf{E} + \varepsilon\varepsilon_0 \frac{\partial}{\partial t}\mathbf{E} \quad \text{rot}\mathbf{H} = \frac{4\pi}{c}\sigma\mathbf{E} + \frac{\varepsilon}{c}\frac{\partial \mathbf{E}}{\partial t} \text{ [CGS]}, \qquad (11.45\text{d})$$

where σ is the optical conductivity. By using the vector identity rot(rot**E**) = grad(div**E**) − ∇^2**E**, we eliminate the magnetic field **H** from equations (11.45b) and (11.45d) and obtain

$$\nabla^2 \mathbf{E} = \varepsilon\varepsilon_0 \mu\mu_0 \frac{\partial^2 \mathbf{E}}{\partial t^2} + \sigma\mu\mu_0 \frac{\partial \mathbf{E}}{\partial t} \quad \nabla^2 \mathbf{E} = \frac{\varepsilon\mu}{c^2}\frac{\partial^2 \mathbf{E}}{\partial t^2} + \frac{4\pi\sigma\mu}{c^2}\frac{\partial \mathbf{E}}{\partial t} \text{ [CGS]}. \quad (11.46)$$

In contrast to the polarization **P**, the magnetization **M** can resonate with external rf-fields of a few 10^9 Hz (1 Hz = 1 s^{-1}), which is much lower than the frequency of radiation in the optical region (see Fig. 11.14). Hence, the interaction of **M** with radiation in the optical region can be ignored. For simplicity, we assume hereafter non-magnetic metals, where $\mu = 1$ or **M** = 0.

Let us consider an incident wave running in the x-direction in vacuum. Its y-component is obviously expressed as $E_y(\text{incident}) \propto \exp[i(\hat{q}x - \omega t)]$, where $\hat{q}$ is the x-component of the wave vector of the radiation. The wave is attenuated, when it is transmitted in a medium. This means that the wave vector involves an imaginary part and is expressed in a complex quantity as

$$\hat{q} = \hat{n}q = (n + i\kappa)q, \qquad (11.47)$$

where $\hat{n}$, n and κ are called the complex refractive index, refractive index and extinction coefficient, respectively. The transmitted wave is then written as

$$E_y(\text{trans}) \propto \exp\{i[(n + i\kappa)qx - \omega t]\} = \exp(-\kappa qx)\exp\{i(nqx - \omega t)\}. (11.48)$$

Thus we see that, when the wave propagates in a medium with refractive index n and extinction coefficient κ, its velocity is reduced to $1/n$ and the wave is damped by a fraction $\exp(-2\pi\kappa/n)$ per wavelength.

---

[8] $\mathbf{D} = \varepsilon\varepsilon_0\mathbf{E} = \varepsilon_0\mathbf{E} + \mathbf{P}$ and $\mathbf{B} = \mu\mu_0\mathbf{H} = \mu_0\mathbf{H} + \mathbf{M}$ in SI units. Note also $c^2 = \varepsilon_0\mu_0$ in SI units where $\mu_0 = 4\pi \times 10^{-7}$ henry/m and $\varepsilon_0 = 10^7/4\pi c^2 = 8.85 \times 10^{-12}$ farad/m. In CGS units, $\varepsilon_0 = 1$ and $\mu_0 = 1$.

11 Electron transport properties in periodic systems (II)

The dispersion relation is obtained by inserting $\mathbf{E}=\mathbf{E}_0\exp[i(\hat{\mathbf{q}}\cdot\mathbf{r}-\omega t)]$ into equation (11.46):

$$\hat{q}^2=\mu_0\omega^2\left(\varepsilon(\omega)\varepsilon_0+i\frac{\sigma(\omega)}{\omega}\right) \qquad \hat{q}^2=\frac{\omega^2}{c^2}\left(\varepsilon(\omega)+i\frac{4\pi\sigma(\omega)}{\omega}\right) \text{ [CGS]}. \quad (11.49)$$

A comparison of equation (11.49) with equation (11.47) leads to

$$n^2-\kappa^2=\varepsilon(\omega) \qquad\qquad n^2-\kappa^2=\varepsilon(\omega) \text{ [CGS]} \quad (11.50)$$

and

$$2n\kappa=\frac{\sigma(\omega)}{\varepsilon_0\omega} \qquad\qquad 2n\kappa=\frac{4\pi\sigma(\omega)}{\omega} \text{ [CGS]}, \quad (11.51)$$

where $\omega=q/\sqrt{\varepsilon_0\mu_0}$ in SI units ($\omega=cq$ in CGS units). Though n, κ and $\varepsilon(\omega)$ are dimensionless, $\sigma(\omega)/\varepsilon_0$ and $\sigma(\omega)$ are in units of s^{-1} in SI and CGS units, respectively.[9] We see that both dielectric constant and optical conductivity can be determined from the equations above by measuring the refractive index and extinction coefficient of the electromagnetic wave either reflected from a metal or transmitting through a thin film.

Equation (11.45d) is often rewritten as

$$\text{rot}\mathbf{H}=\frac{\partial\mathbf{D}^{\text{tot}}}{\partial t}=\hat{\varepsilon}(\omega)\varepsilon_0\frac{\partial\mathbf{E}}{\partial t} \qquad \text{rot}\mathbf{H}=\frac{1}{c}\frac{\partial\mathbf{D}^{\text{tot}}}{\partial t}=\frac{\hat{\varepsilon}(\omega)}{c}\frac{\partial\mathbf{E}}{\partial t} \text{ [CGS]}, \quad (11.52)$$

where the complex dielectric constant $\hat{\varepsilon}(\omega)$ is defined as $\mathbf{D}^{\text{tot}}=\hat{\varepsilon}(\omega)\varepsilon_0\mathbf{E}$. The total electric displacement $\mathbf{D}^{\text{tot}}$ includes not only contributions from "free" and "bound" electrons in a solid but also $\varepsilon_0(\partial\mathbf{E}/\partial t)$ in vacuum [10].[10] Since the electric field $\mathbf{E}$ varies as $e^{-i\omega t}$ with time, a comparison with equation (11.45d) results in

$$\hat{\varepsilon}(\omega)=\varepsilon(\omega)+i\left(\frac{\sigma(\omega)}{\varepsilon_0\omega}\right) \qquad \hat{\varepsilon}(\omega)=\varepsilon(\omega)+i\left(\frac{4\pi\sigma(\omega)}{\omega}\right) \text{ [CGS]}. \quad (11.53)$$

Equation (11.45d) is alternatively expressed as

$$\text{rot}\mathbf{H}=\mathbf{J}^{\text{tot}}+\varepsilon_0\frac{\partial\mathbf{E}}{\partial t}=\hat{\sigma}(\omega)\mathbf{E}+\varepsilon_0\frac{\partial\mathbf{E}}{\partial t} \qquad \text{rot}\mathbf{H}=\frac{4\pi}{c}\sigma(\omega)\mathbf{E}+\frac{1}{c}\frac{\partial\mathbf{E}}{\partial t} \text{ [CGS]}, \quad (11.54)$$

---

[9] The conductivity σ in the units of $[\Omega]^{-1}[\text{m}]^{-1}$ is converted to that of $[\text{s}]^{-1}$ by dividing σ by $4\pi\varepsilon_0$ in SI units, whereas it is directly converted to $[\text{s}]^{-1}$ in CGS units:

$$\left[\frac{\sigma}{4\pi\varepsilon_0}\right]=\frac{9\times 10^9[\text{m}]}{[\text{farad}]}\cdot\frac{1}{[\Omega][\text{m}]}=\frac{9\times 10^9[\text{volt}][\text{m}]}{[\text{coulomb}]}\cdot\frac{[\text{coulomb}]^2}{[\text{joule}][\text{s}][\text{m}]}=\frac{9\times 10^9}{[\text{s}]} \text{ [SI]}$$

$$[\sigma]=\frac{1}{[\Omega][\text{m}]}=\frac{[\text{coulomb}]^2}{[\text{joule}][\text{s}][\text{m}]}=\frac{(3\times 10^9)^2[\text{esu}]^2}{10^7[\text{erg}][\text{s}]10^2[\text{cm}]}=\frac{9\times 10^9[\text{dyne}][\text{cm}]^2}{[\text{erg}][\text{s}][\text{cm}]}=\frac{9\times 10^9}{[\text{s}]} \text{ [CGS]}.$$

[10] Free and bound electrons are intentionally put in quotation marks because of the difficulty in separating them in high-frequency AC fields, as is discussed at the end of this section.

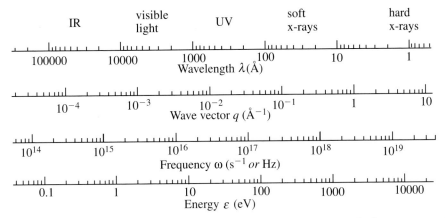

Figure 11.14. Wavelength, wave vector, frequency and energy of electromagnetic wave in a vacuum. The following relations hold among the relevant quantities: $\omega = \varepsilon/\hbar = 1.5193 = 10^{15}\varepsilon$ [s^{-1} or Hz], $q = \omega/c = 5.06 \times 10^{-4}\varepsilon$ [Å]$^{-1}$ and $\lambda = 2\pi/q = 1.24 \times 10^{4}/\varepsilon$ [Å], where ε is the energy in eV.

where the complex optical conductivity $\hat{\sigma}(\omega)$ is defined as $\mathbf{J}^{tot} = \hat{\sigma}(\omega)\mathbf{E}$. Here the second term $\varepsilon_0(\partial \mathbf{E}/\partial t)$ must be added, since the total current density $\mathbf{J}^{tot}$ describes currents only from electrons in a solid. In the same way as above, we obtain

$$\frac{\hat{\sigma}(\omega)}{\varepsilon_0} = \frac{\sigma(\omega)}{\varepsilon_0} + i\{\omega[1-\varepsilon(\omega)]\} \qquad \hat{\sigma}(\omega) = \sigma(\omega) + \frac{i}{4\pi}\{\omega[1-\varepsilon(\omega)]\} \text{ [CGS]}. \quad (11.55)$$

We find that the response of a medium to an oscillatory electric field can be described by the ω-dependent dielectric constant $\varepsilon(\omega)$ or the conductivity $\sigma(\omega)$. In the DC limit, the conductivity strictly describes the motion of conduction electrons and the dielectric constant the displacement of bound electrons. Such a clear distinction is blurred in the case of the AC field at high frequencies where $\omega\tau \gg 1$ is satisfied. However, we conventionally reserve $\sigma(\omega)$ as a quantity to describe the response to electrons in partially filled bands and $\varepsilon(\omega)$ that of bound electrons or those in completely filled bands [6].

11.10 Interaction of electromagnetic wave with metals (II)

Both the dielectric constant and optical conductivity are functions of the wave vector $\mathbf{q}$ and angular frequency ω and denoted as $\varepsilon(\mathbf{q}, \omega)$ and $\sigma(\mathbf{q}, \omega)$, respectively. The $\mathbf{q}$ dependence certainly results from their spatial dispersion. Figure 11.14 depicts the relationship between the energy of the electromagnetic wave in a vacuum and the corresponding wavelength, wave vector and frequency. The measurement of optical properties is generally carried out at wavelengths in the infrared (IR)-to-ultraviolet (UV) region or 0.1 to about 50 eV in energies. Note that the magnitude q of the wave vector in this energy range is less than 0.1 Å^{-1} and is much smaller than the shortest reciprocal lattice vector in the Brillouin zones of an ordinary metal. Hence, the **q** dependence is generally neglected (see also Section 11.9).

Now we study the response of the conduction electron upon exposure to an electromagnetic wave with an oscillating electric field $\mathbf{E}=\mathbf{E}_0\exp(-i\omega t)$ in the high-frequency range $\omega\tau\gg 1$ so that an electron is not allowed to make collisions within the period of the rapidly oscillating field. Since $\partial f/\partial t$ is not zero, the linearized Boltzmann transport equation (10.42) is replaced by

$$\frac{\partial f}{\partial t}-(-e)\mathbf{v}\cdot\mathbf{E}\frac{\partial f_0}{\partial \varepsilon}-\frac{(-e)}{\hbar}\mathbf{v}\times\mathbf{B}\cdot\frac{\partial f}{\partial \mathbf{k}}=-\frac{\phi}{\tau} \tag{11.56}$$

and, as before, we have employed the relaxation time approximation. In response to the oscillating field $\mathbf{E}=\mathbf{E}_0\exp(-i\omega t)$, the distribution function would also oscillate as $\phi=f-f_0=\phi_0\exp(-i\omega t)$. Further, the third term involving the magnetic field is neglected, since a change in the electron wave vector can be ignored.[11] Therefore, an insertion of this relation into equation (11.56) results in

$$\phi=\frac{(-e)\mathbf{v}\cdot\mathbf{E}\tau}{1-i\omega\tau}\cdot\frac{\partial f_0}{\partial \varepsilon}. \tag{11.57}$$

This is an extension of the relation for a static electric field, equation (10.46). The conductivity due to the oscillating electric field can be easily obtained by following the same procedure as described in Section 10.7. The complex optical conductivity is now given by

$$\hat{\sigma}(\omega)=\frac{n(-e)^2}{m}\frac{1}{1-i\omega\tau}. \tag{11.58}$$

Equation (11.58) is valid in the range $\omega\tau\gg 1$ and is known as the Drude expression for the optical conductivity or AC conductivity.

The complex conductivity given by equation (11.58) is decomposed into its real and imaginary parts:

$$\hat{\sigma}(\omega)=\frac{\sigma(0)}{1-i\omega\tau}=\frac{\sigma(0)}{1+\omega^2\tau^2}+i\frac{\sigma(0)\omega\tau}{1+\omega^2\tau^2}, \tag{11.59}$$

where $\sigma(0)=n(-e)^2\tau/m^*_{\text{opt}}$ is the electrical conductivity at zero frequency or the DC electrical conductivity given by equation (10.6). The parameter m^*_{opt}

---

[11] It is also noted that $\mathbf{v}\times\mathbf{B}\cdot(\partial f_0/\partial\mathbf{k})=\mathbf{v}\times\mathbf{B}\cdot(\partial f_0/\partial\varepsilon)(\partial\varepsilon/\partial\mathbf{k})\propto\mathbf{v}\times\mathbf{B}\cdot\mathbf{v}=0$ holds in the case of the equilibrium distribution function.

appearing in the denominator is often called the optical effective mass of the conduction electron.

The complex optical conductivity is alternatively expressed as

$$\hat{\sigma}(\omega) = \sigma_1(\omega) + i\sigma_2(\omega). \tag{11.60}$$

A comparison with equations (11.55), (11.59) and (11.60) leads to the following expression for its real and imaginary parts:

$$\sigma_1(\omega) = \sigma(\omega) = \frac{\sigma(0)}{1+\omega^2\tau^2} = \frac{\varepsilon_0 \omega_p^2 \tau}{1+\omega^2\tau^2} \quad \sigma_1(\omega) = \frac{\sigma(0)}{1+\omega^2\tau^2} = \frac{\omega_p^2}{4\pi}\frac{\tau}{1+\omega^2\tau^2} \, \text{[CGS]} \tag{11.61}$$

and

$$\sigma_2(\omega) = \omega\varepsilon_0[1-\varepsilon(\omega)] = \sigma(0)\left(\frac{\omega\tau}{1+\omega^2\tau^2}\right) = \frac{\varepsilon_0\omega_p^2\omega\tau^2}{1+\omega^2\tau^2} \quad \sigma_2(\omega) = \frac{\omega_p^2}{4\pi}\frac{\omega\tau^2}{1+\omega^2\tau^2}\, \text{[CGS]}, \tag{11.62}$$

where ω_p is called the plasma frequency defined as

$$\omega_p = \sqrt{\frac{n(-e)^2}{\varepsilon_0 m^*_{opt}}} \quad \omega_p = \sqrt{\frac{4\pi n(-e)^2}{m^*_{opt}}}\, \text{[CGS]}. \tag{11.63}$$

The plasma frequency is associated with the cooperative oscillations of the assembly of conduction electrons driven by the alternating electric field. It is clear from equation (11.63) that ω_p depends on the number of conduction electrons n per unit volume and the effective mass m^*_{opt}. For instance, an insertion of an appropriate number density n into equation (11.63) yields $\omega_p = 9.2 \times 10^{15}$ s^{-1} ($= 6$ eV) and 2.4×10^{16} s^{-1} ($= 14.9$ eV) for monovalent Na and trivalent Al metals, respectively.

The complex dielectric constant $\hat{\varepsilon}(\omega)$ is also often expressed as $\hat{\varepsilon}(\omega) = \varepsilon_1(\omega) + i\varepsilon_2(\omega)$. Then, we obtain from equations (11.53) and (11.62)

$$\varepsilon_1(\omega) = \varepsilon(\omega) = 1 - \frac{\sigma_2(\omega)}{\varepsilon_0 \omega} \quad \varepsilon_1(\omega) = \varepsilon(\omega) = 1 - \frac{4\pi\sigma_2(\omega)}{\omega}\, \text{[CGS]} \tag{11.64}$$

and

$$\varepsilon_2(\omega) = \frac{\sigma_1(\omega)}{\varepsilon_0 \omega} \quad \varepsilon_2(\omega) = \frac{4\pi\sigma_1(\omega)}{\omega}\, \text{[CGS]}. \tag{11.65}$$

Equations (11.50) and (11.51) are now rewritten as

$$n^2 - \kappa^2 = \varepsilon(\omega) = 1 - \sigma(0)\left(\frac{\tau}{1+\omega^2\tau^2}\right) = 1 - \frac{\omega_p^2 \tau^2}{1+\omega^2\tau^2} \tag{11.66}$$

and

$$2n\kappa = \varepsilon_2(\omega) = \left(\frac{1}{\omega}\right) \cdot \frac{\sigma(0)}{1 + \omega^2 \tau^2} = \left(\frac{\tau}{\omega}\right) \cdot \left(\frac{\omega_p^2}{1 + \omega^2 \tau^2}\right). \tag{11.67}$$

The electron scattering has been treated above in terms of the relaxation time approximation but, more precisely, the AC conductivity has to be treated by using the linear-response theory known as the Kubo formula, the outline of which will be described in Section 11.12. For example, the following formula known as the sum rule is derived [11]:

$$\int_0^\infty \varepsilon_2(\omega) \omega d\omega = \int_0^\infty \frac{\sigma_1(\omega)}{\varepsilon_0} d\omega = \frac{\pi \omega_p^2}{2} \qquad \int_0^\infty \varepsilon_2(\omega) \omega d\omega = 4\pi \int_0^\infty \sigma_1(\omega) d\omega = \frac{\pi \omega_p^2}{2} \text{ [CGS]}. \tag{11.68}$$

11.11 Reflectance measurement

In this section, we describe the reflectance measurement and the method to extract the optical constants like the dielectric constants and optical conductivity from the measured spectrum. A sample with a well-polished surface is placed in a vacuum and the reflectance is measured as a function of the frequency ω by directing an electromagnetic wave (again in the IR-to-UV region) onto the surface. Let us assume the incident light to be perpendicular to the surface of the sample. We introduce the ratio $\hat{r}(\omega)$ defined as

$$\hat{r}(\omega) = E_{\text{ref}}/E_{\text{in}} = \rho(\omega) e^{i\theta(\omega)}, \tag{11.69}$$

where E_{in} and E_{ref} are the electric field of the incident and reflected light and $\rho(\omega)$ and $\theta(\omega)$ are the amplitude and the phase of $\hat{r}(\omega)$, respectively. They are related to the optical constants n and κ through the relation [9]:

$$\rho(\omega) e^{i\theta(\omega)} = \frac{1 - n - i\kappa}{1 + n + i\kappa}. \tag{11.70}$$

Equation (11.70) indicates that the optical constants n and κ can be determined, once the amplitude and phase of the reflectance are known. Accordingly, the dielectric constant $\varepsilon(\omega)$ and optical conductivity $\sigma(\omega)$ are deduced from equations (11.50) and (11.51).

In the reflectance measurement, one can measure only the ratio of the intensity of the reflected wave over that of the incident one. The measured reflectance R is therefore given as functions of the optical constants as follows:

$$R = \hat{r}^* \hat{r} = \rho^*(\omega) \rho(\omega) = \frac{(n-1)^2 + \kappa^2}{(n+1)^2 + \kappa^2}, \tag{11.71}$$

11.12 Reflectance spectrum and optical conductivity

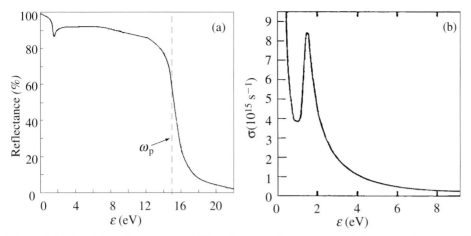

Figure 11.15. (a) Reflectance and (b) optical conductivity spectra of pure Al. A small dip at 1.5 eV in (a) is due to the interband transition, which is enhanced in (b).
[H. Ehrenreich, H. R. Philipp and B. Segall, *Phys. Rev.* **132** (1963) 1918]

where $\hat{r}^*$ and ρ^* represent the complex conjugates of $\hat{r}$ and ρ. As is clear from equation (11.71), the phase cannot be experimentally determined. This hampers the determination of the dielectric constant solely from reflectance data.

The amplitude and phase are, however, not independent of each other but are linked through the Kramers–Kronig relation:

$$\theta(\omega) = -\frac{2\omega}{\pi} \wp \int_0^\infty \frac{\ln \rho(\omega')}{(\omega')^2 - \omega^2} d\omega', \tag{11.72}$$

where $\wp$ indicates the principal integration [9]. We measure, first, the reflectance spectrum over as wide a frequency range as possible, for example, 0.1 eV up to 50 eV. By inserting the resulting data into equation (11.72), the frequency dependence of the phase can be determined. Once the amplitude and phase are determined for a given frequency, then the optical constants n and κ can be calculated from equation (11.70).

11.12 Reflectance spectrum and optical conductivity

We consider the reflectance spectrum and the resulting optical conductivity spectrum for pure Al and graphite as representatives of a typical metal and semimetal. The reflectance spectrum for pure Al is shown in Fig. 11.15(a). The reflectance exceeds 80% in the region $\omega < \omega_p$ ($\omega_p = 14.7$ eV) but drops sharply to almost zero above ω_p. This is a feature characteristic of a metal and is called a

Drude-type spectrum. The abrupt change in reflectance at $\omega \approx \omega_p$ is called the Drude edge. A small minimum at about 1.5 eV is due to the interband transition discussed in section 11.9.

Let us consider first the high-frequency range $\omega\tau \gg 1$, where the electron relaxation time τ is so long for fixed ω that electrons are simply accelerated by the radiation without collisions. Equations (11.66) and (11.67) can be approximated as

$$n^2 - \kappa^2 \approx 1 - \frac{\omega_p^2}{\omega^2} \quad (11.73)$$

and

$$2n\kappa = \left(\frac{1}{\omega\tau}\right) \cdot \left(\frac{\omega_p^2}{\omega^2}\right) \approx 0. \quad (11.74)$$

In the case of pure Al, the relaxation time τ of the conduction electron is of the order of 10^{-14} s, as deduced from its DC conductivity of $\sigma(0) \approx 10^{17}$ s^{-1} (see Section 10.2).[12] Hence, the condition $\omega\tau \gg 1$ is satisfied for electromagnetic waves in the range $\omega > 10^{15}$ s^{-1} corresponding to the visible ultraviolet region (see Fig. 11.14). If $\omega > \omega_p$ ($\omega_p \approx 10^{16}$ s^{-1} for pure Al), κ must be reduced to zero. This means that the radiation can propagate through a metal without attenuation and, hence, the metal becomes transparent in this range. However, when ω decreases well below ω_p, $n \ll \kappa$ should hold from equation (11.73). Under such circumstances, the reflectance approaches unity, as can be seen from equation (11.71). This explains why the reflectance exceeds 80% below ω_p but drops to zero in the range $\omega > \omega_p$.

When the condition $\omega\tau \ll 1$ is satisfied, the conduction electron is scattered by defects like lattice vibrations in a time shorter than a single oscillation induced by the oscillating electric field. We can no longer rely on equations (11.66) and (11.67). Instead, the dielectric constant $\varepsilon(\omega)$ in equation (11.49) may be ignored in the approach to the DC limit and equations (11.50) and (11.51) are then approximated as

$$n \approx \kappa \quad (11.75)$$

and

$$2n\kappa = \frac{\sigma(0)}{\omega}. \quad (11.76)$$

where $\sigma(0)$ is in units of s^{-1}. Let us use again the values of $\tau \approx 10^{-14}$ s and $\sigma(0) \approx 10^{17}$ s^{-1} appropriate for pure Al. Since $\omega \ll 1/\tau \approx 10^{14}$ s^{-1}, the ratio $\sigma(0)/\omega$

---

[12] The value of $\sigma = 4.0 \times 10^7/\Omega$-m for pure Al in Table 10.1 yields $\sigma \approx 10^{17}$ s^{-1} (see footnote 9, p. 320).

in equation (11.76) would be very large and, hence, $n\kappa \gg 1$ holds. This means that both n and κ must be large so that the reflectance R in equation (11.71) is very close to unity and total reflection occurs in the infrared region, where $\omega\tau \ll 1$.

The optical conductivity spectrum of pure Al shown in Fig. 11.15(b) is typical of a metal with a high conductivity. Its extrapolation to zero frequency corrresponds to the DC conductivity. The value of $\sigma(0) = 1.55 \times 10^{17}$ s^{-1} thus obtained is in reasonable agreement with the DC conductivity of 3.59×10^{17} s^{-1} for bulk Al. A rapid decay in the optical conductivity with increasing frequency reflects the $(1+\omega^2)^{-1}$ dependence of the Drude relation given by equation (11.61). A deep minimum at an energy of 1.5 eV is due to the interband transition associated with the $W_2 \rightarrow W_1$ in the energy band of pure Al [12].

The dielectric constants $\varepsilon_1(\omega)$ and $\varepsilon_2(\omega)$ and the optical conductivity $\sigma(\omega)$ for pure graphite were determined by application of the Kramers–Kronig relation to the measured reflectance spectrum shown in Fig. 11.16(a). The π-bands (see Section 6.7) are mainly responsible for the intra- and interband transitions in the range below 9 eV and a broad peak near 15 eV is ascribed to an interband transition involving 3 electrons per atom in the σ-band.

The optical conductivity spectrum is shown in Fig. 11.16(b). Drude-type behavior is no longer observed. The optical conductivity extrapolated to zero energy or frequency is about 9×10^{14} s^{-1}, which is 1/6000 that of pure Al. Note that the scale of the ordinate in Fig. 11.16(b) is 1/10 that in Fig. 11.15(b). This is consistent with the DC electrical resistivity data: the value in the *ab*-plane of pure graphite at room temperature is of the order of 10^3 $\mu\Omega$-cm and is about 1000 times that of pure Al.

Since transitions involving the 1s core level can be neglected in pure graphite, the effective number of electrons n_{eff} per atom can be calculated by inserting the measured $\varepsilon_2(\omega)$ into equation (11.68) in the integral range from 0 to an energy $\hbar\omega$. The result is shown in Fig. 11.17. A plateau appears at $n_{eff} = 1$ electron/atom with subsequent saturation to an n_{eff} value of 4 electrons/atom. This clearly indicates that only electrons in the π-band contribute to the optical excitation up to about 9 eV whereas the transitions involving σ-electrons set in at higher energies. Because of this apparent separated behavior, two plasma frequencies ω_p are deduced: one associated with only π-electrons characterized by $\omega_p = 12.7$ eV and the other with combined π- and σ-electrons by $\omega_p = 25.2$ eV [13]. It is important to note here that σ-electrons well below the Fermi level can contribute to the plasma oscillations as well. It is true that electrons even in an insulator respond as if they were free electrons, provided that the frequency ω of the electromagnetic wave exceeds ω_0 corresponding to the band gap. This is because the photon energy becomes greater than the binding energy of the electrons and the insulator exhibits metallic reflectance in this energy region.

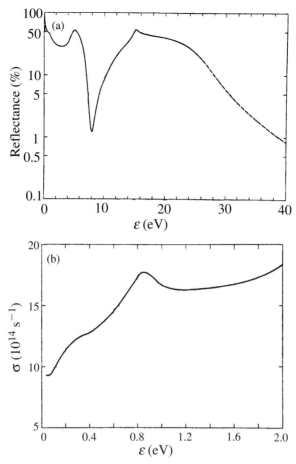

Figure 11.16. (a) Reflectance and (b) optical conductivity spectra of graphite. Note that reflectance (%) is plotted on a logarithmic scale. [From E. A. Taft and H. R. Philipp, *Phys. Rev.* **138** (1965) A197]

11.13 Kubo formula

All the discussions so far developed in Chapters 10 and 11 are based on the Boltzmann transport equation, which is derived from the local balance in the steady state of the electron distribution in the phase space without taking into account the microscopic structure of a solid. Furthermore, a rigorous solution of the Boltzmann transport equation is not generally achieved because of its complicated integro–differential equation. To circumvent this difficulty, we have simplified the situation such that the deviation from an equilibrium state is small enough to linearize the Boltzmann transport equation. Indeed, the linearized Boltzmann transport equation (10.42) was employed to discuss the various transport properties in the preceding sections.

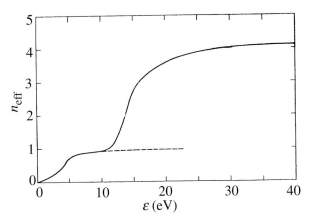

Figure 11.17. Effective number of electrons per atom versus energy, obtained from numerical integration of experimental $\varepsilon_2(\omega)$ for pure graphite. [From E. A. Taft and H. R. Philipp, *Phys. Rev.* **138** (1965) A197]

The Kubo formula [14] is constructed on a more general ground than the Boltzmann transport equation and derived by employing as its basis the theory of Brownian motion put forward by Einstein in 1905 [15]. Prior to the derivation of the Kubo formula, we will study first the Einstein relation. As discussed in Section 10.2, the equation of motion for the free electron in the presence of the electric field **E** was expressed in equation (10.3) as

$$m\left(\frac{d\mathbf{v}_D}{dt} + \frac{\mathbf{v}_D}{\tau}\right) = (-e)\mathbf{E},$$

where $\mathbf{v}_D$ is the drift velocity. The conductivity formula of equation (10.6) was derived by assuming the steady state condition given by $d\mathbf{v}_D/dt = 0$. Strictly speaking, however, the motion of the Brownian particles cannot be rigorously described in terms of equation (10.3), since each particle experiences complicated forces from its surroundings and maintains an average kinetic energy consistent with the equipartition law of $k_B T/2$. If we denote this random force as $\mathbf{F}(t)$, equation (10.3) may be replaced by

$$m\left(\frac{d\mathbf{v}_D}{dt} + \frac{\mathbf{v}_D}{\tau}\right) = (-e)\mathbf{E} + \mathbf{F}(t), \tag{11.77}$$

where $\mathbf{F}(t)$ is fluctuating with time t but its time-average must be zero. In other words, equation (10.3) can be regarded as the equation of motion of a particle after averaging over many particles.

We are now interested in the motion of the assembly of the Brownian particles in the absence of the electric field **E** in equation (11.77):

$$m\left(\frac{d\mathbf{v}_D}{dt} + \frac{\mathbf{v}_D}{\tau}\right) = \mathbf{F}(t), \tag{11.78}$$

from which we can deduce the famous Einstein relation

$$D = \frac{\tau}{m}k_BT = \frac{\mu}{(+e)}k_BT, \qquad (11.79)$$

where D is called the diffusion coefficient of the Brownian particle carrying the electronic charge $(+e)$ and μ is the mobility defined as $\mu = (+e)\tau/m$ in equation (10.9) (see Exercise 11.5).[13]

Let us consider the physical meaning of the diffusion coefficient D of particles subjected to the Brownian motion. Suppose that the distribution of particles is perturbed at a given instant to cause some heterogeneity in an otherwise homogeneous system. Such a perturbation will be smeared out by the diffusion of particles through the random motion of each particle. The flow of the Brownian particles due to diffusion is driven by the concentration gradient:

$$-D\frac{\partial n(x)}{\partial x}, \qquad (11.80)$$

where $n(x)$ is the number density of the Brownian particles. A minus sign indicates that the particles are driven to diffuse in the direction to reduce the concentration gradient. An increasing rate of the number density in a given volume element must be equal to the difference in the number of particles entering and leaving this element:

$$\frac{\partial n}{\partial t} = -\frac{\partial}{\partial x}\left(-D\frac{\partial n}{\partial x}\right) = D\frac{\partial^2 n}{\partial x^2}. \qquad (11.81)$$

Equation (11.81) is known as the Fick equation.

The diffusion coefficient D in the left-hand side of equation (11.79) characterizes the Brownian motion of particles drifting in the system and, hence, is associated with statistical fluctuations. On the other hand, the term τ/m in its right-hand side represents the frictional effect of particles driven under the action of an external force, such as an electric field for charged particles, and, hence, is associated with the dissipation observed after smoothing out the statistical fluctuations. We see, therefore, that the Einstein relation (11.79) connects two quantities stemming from different origins, i.e., fluctuations and dissipation.

Now we are ready to study the Kubo formula. First, let us assume that a particle at position $x(0)$ at a time $t=0$ diffuses to the position $x(t)$ at the time $t=t$. The diffusion coefficient D is formulated as

$$D = \lim_{t\to\infty}\frac{1}{2t}\langle\{x(t)-x(0)\}^2\rangle = \lim_{t\to\infty}\frac{\langle\Delta x^2\rangle}{2t}, \qquad (11.82)$$

---

[13] The diffusion coefficient D involves no activation energy. Since the mobility μ is related to D by the Einstein relation, the product μT or D can be finite at $T=0$.

11.13 Kubo formula

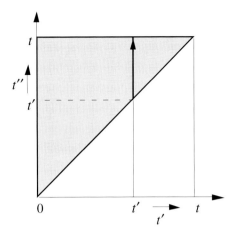

Figure 11.18. Integral region in equation (11.83). Integration can be done in the shaded region $0 \leq t' \leq t$ and $t' \leq t'' \leq t$. Results obtained must be doubled.

where $\langle \rangle$ represents an average over the particle ensemble. The square of the displacement Δx of the particle in a time t is expressed in the integral form as

$$\langle \Delta x^2 \rangle = \left\langle \int_0^t v(t')dt' \cdot \int_0^t v(t'')dt'' \right\rangle = \int_0^t dt' \int_0^t \langle v(t')v(t'') \rangle dt''$$

$$= 2\int_0^t dt' \int_{t'}^t \langle v(t')v(t'') \rangle dt'' = 2\int_0^t dt' \int_0^{t-t'} \langle v(t')v(t'+\tau) \rangle d\tau,$$

where the second line is easily obtained by changing the integration path as indicated in Fig. 11.18. Now D is rewritten as

$$D = \lim_{t \to \infty} \frac{\langle \Delta x^2 \rangle}{2t} = \lim_{t \to \infty} \frac{1}{t} \int_0^t dt' \cdot \int_0^{t-t'} \langle v(t')v(t'+\tau) \rangle d\tau = \int_0^\infty \langle v(t)v(t+\tau) \rangle d\tau. \quad (11.83)$$

Equation (11.83) holds true, provided that the integral $\int_0^{t-t'} \langle v(t')v(t'+\tau) \rangle d\tau$ converges more rapidly than the increasing rate of t. Thus, the diffusion coefficient is expressed in terms of the correlation function of the velocity of particles (see Exercise 11.5). By inserting the Einstein relation (11.79) and the diffusion coefficient in equation (11.83) into equation (10.6) for the Drude expression of the conductivity, we obtain

$$\sigma = \frac{ne^2}{k_B T} \int_0^\infty \langle v(0)v(t) \rangle dt. \quad (11.84)$$

The electrical conductivity can be more generally expressed in the form of a tensor:

$$\sigma_{ij} = \frac{1}{k_B T} \int_0^\infty \langle J_i(0) J_i(t) \rangle dt. \tag{11.85}$$

where the current density J given by $J = (-e) \sum_i v_i$ is used instead of the velocity. As is clear from the argument above, equation (11.85) may be considered as the application of the Einstein relation to the conduction electron system.

Suppose that the conduction electron system in the absence of the external electric field is in thermal equilibrium. The current density J in the system is fluctuating with time as a natural motion of the conduction electrons. We can evaluate the conductivity tensor by integrating the correlation function $\langle J_i(0) J_i(t) \rangle$ over the time-dependent fluctuations. Equation (11.85) can be further extended to the system where an alternating electric field with the frequency ω is applied. The resulting AC conductivity is expressed as

$$\sigma_{ij}(\omega) = \frac{1}{k_B T} \int_0^\infty \langle J_i(0) J_i(t) \rangle e^{-i\omega t} dt. \tag{11.86}$$

Equation (11.86) can be further extended to the situation where the transport phenomenon is treated in quantum mechanics. The Kubo formula for the electrical conductivity is finally obtained in the following form:

$$\sigma_{ij}(\omega) = \int_0^\infty \left(\int_0^{1/k_B T} \mathrm{Tr} \rho J_i(-i\hbar\lambda) J_j(t) d\lambda \right) e^{-i\omega t} dt, \tag{11.87}$$

where ρ is the density matrix of the system in thermal equilibrium, J_i is the current operator and Tr stands for the trace of the matrix [16].

The Boltzmann transport equation introduced in Section 10.6 refers to the transport equation determining the steady state electron distribution in the presence of external fields. In its practical use, the deviation from the equilibrium state is assumed to be so small that the linearized approximation is validated. As has been emphasized, its validity is lost when the scattering becomes strong and the mean free path of the conduction electron becomes comparable to an atomic distance. On the other hand, the Kubo formula is rigorous in the framework of the linear-response theory. The Kubo formula can be applied to systems like transition metals and their alloys, where scattering of conduction electrons is strong.

Exercises

11.1 The thermal conductivity is given by $\kappa = -L_{TT}(1 - L_{TE}L_{ET}/L_{EE}L_{TT})$. Show that the correction term $L_{TE}L_{ET}/L_{EE}L_{TT}$ is of the order of $(T/T_F)^2$ for ordinary metals, where T_F is the Fermi temperature. Use equations (11.12), (11.19) and (11.20) and the relation $\sigma(\varepsilon) \propto \varepsilon^{3/2}$ in the free-electron model.

11.2 Derive the components σ_{xx} and σ_{xy} of the conductivity tensor by inserting equation (11.31) into equation (11.30) under the conditions $\mathbf{E} = (E, 0, 0)$ and $\mathbf{B} = (0, 0, B)$.

11.3 In the presence of both an electric field $\mathbf{E}$ and a magnetic field $\mathbf{B}$, the relation $\mathbf{E} = (\mathbf{J}/\sigma) + \beta \mathbf{B} \times \mathbf{J}/\sigma$ holds, indicating that the electric field to produce the current density $\mathbf{J}$ has two components [1]. In the Hall effect measurement, we choose the configurations $\mathbf{J} \perp \mathbf{B}$ so that the second term gives rise to a transverse electric field. This is the Hall effect. Note that $\mathbf{E}$ and $\mathbf{J}$ are in the same plane but are neither parallel nor perpendicular to each other.

Show that the inversion of the relation above results in

$$\mathbf{J} = [\sigma/(1 + \beta^2 B^2)]\,\mathbf{E} - [\sigma\beta/(1 + \beta^2 B^2)]\mathbf{B} \times \mathbf{E}.$$

Use a right-angled triangle for the relevant vectors $\mathbf{E}$, $\mathbf{J}/\sigma$ and $\beta \mathbf{B} \times \mathbf{J}/\sigma$ [1].

11.4 Prove equation (11.43) by using equation (11.42).

11.5 Derive the Einstein relation from equation (11.78) by using the velocity correlation function defined as $\phi(t) = \langle v(0)v(t) \rangle$.

Chapter Twelve
Superconductivity

12.1 Prologue

We learned in Chapter 10 that lattice vibrations in a metal always give rise to a finite resistivity and that it disappears only when the metal resumes perfectly periodic ion potentials at absolute zero. However, there are many metals the resistivity of which completely vanishes at finite temperatures. Kamerlingh Onnes from the Netherlands, is famous for his success in liquefying helium for the first time in 1908. During his extensive studies on the electrical resistivity of various metals by immersing them in liquid helium, he happened to discover in 1911 [1] that the resistivity of mercury suddenly drops to zero at 4.2 K, the boiling point of liquid helium at 1 atmospheric pressure. This is the superconducting phenomenon discovered three years after his helium liquefaction.

Since then, superconductivity has been discovered in many metals, alloys and compounds. As listed in Table 12.1, the value of the superconducting transition temperature T_c of elements in the periodic table is always less than 10 K. Many researchers have attempted to synthesize new superconductors with as high a T_c value as possible. In 1986, Bednorz and Müller [2] revealed that the electrical resistivity of La–Ba–Cu–O sharply dropped at about 35 K and vanished below about 13 K and pointed out the possibility of synthesizing a new high-T_c superconducting oxide. Their work opened up a new era for the research of high-T_c superconductors and the Nobel Prize in physics was awarded to them in 1987 for their discovery.

The mechanism of superconductivity had remained unsettled for many years as one of the most inexplicable phenomena in physics until Bardeen, Cooper and Schrieffer put forward the historic theory in 1957 [3]. The BCS theory successfully accounted for various observed properties inherent in the superconducting state and has been regarded as a milestone in the development

Table 12.1. *Superconducting properties of pure metals*

element	superconducting transition temperature T_c (K)	critical magnetic field extrapolated to 0 K H_c (Oe)
Al	1.18	99
Ga	1.09	51
Hg	4.15	412
In	3.4	293
La	5.9	1600
Mo	0.92	98
Nb	9.2	1950
Pb	7.2	803
Ru	0.49	66
Sn	3.7	309
Ta	4.39	830
Ti	0.39	100
V	5.3	1020
Zn	0.85	53
Zr	0.55	47

of the electron theory of metals. In this chapter, we discuss the electron theory of superconductivity by introducing first London's phenomenological theory and then describing how experimental and theoretical works have been condensed into the successful construction of the BCS theory. Basic properties of the DC and AC Josephson effects, high-T_c superconducting oxides and application-oriented studies of the type-II superconductors are also discussed.

12.2 Meissner effect

The mechanism of superconductivity cannot be resolved only from the resistanceless phenomenon. The understanding of the behavior of a superconductor upon the application of a magnetic field is crucial. The superconducting state is lost and the normal state resumed if an applied magnetic field exceeds some critical value H_c.[1] The field H_c is called the critical magnetic field. The value of H_c is zero at T_c but increases rapidly with decreasing temperature, as shown in Fig. 12.1. The value of H_c obtained by extrapolation to absolute zero for various superconducting elements in the periodic table is listed in Table 12.1, together with the value of T_c. It can be seen that the lower the T_c, the lower the H_c and that the value of H_c in elements is generally lower than about 2 kOe. It

---

[1] The non-superconducting state is referred to as the normal state in this chapter.

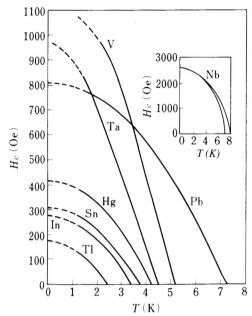

Figure 12.1. Temperature dependence of the critical magnetic field in pure metals. [D. Shoenberg, *Superconductivity* (Cambridge University Press, 1952)]

is known that the value of H_c depends on the purity of a material and the higher the purity of the material, the lower the value of H_c. Some superconducting compounds discussed later exhibit values of H_c exceeding 200 kOe or 20 tesla.

Since the superconducting state is perfectly conductive, the electric field **E** inside a superconductor must be zero. Thus, the Maxwell equation yields

$$-\left(\frac{\partial \mathbf{B}}{\partial t}\right) = \text{rot}\mathbf{E} = 0 \qquad -\frac{1}{c}\left(\frac{\partial \mathbf{B}}{\partial t}\right) = \text{rot}\mathbf{E} = 0 \text{ [CGS]}, \qquad (12.1)$$

where **B** is the applied magnetic flux density.

Let us assume the situation such that the magnetic field is applied to a specimen above T_c and then its temperature is lowered below T_c. This sequence of operations is called field cooling and abbreviated as FC. When a specimen is above T_c, the magnetic field uniformly penetrates through it, as shown in Fig. 12.2(a). When the temperature is lowered below T_c in the presence of a magnetic field ($H < H_c$), the specimen enters the superconducting state. The Maxwell equation (12.1), however, implies that the magnetic flux **B** inside the specimen will remain unchanged with time. If this were true, the magnetic field would have remained penetrating in the specimen, as shown in Fig. 12.2(b). A simple extension of this conjecture would result in a permanent magnet as shown in Fig. 12.2(c), if the applied field is switched off.

12.2 Meissner effect

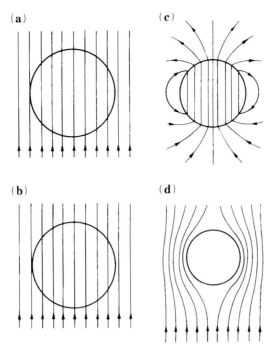

Figure 12.2. (a) Magnetic flux distribution in the presence of a magnetic field at $T > T_c$. The specimen is in the normal state. (b) Subsequent flux distribution, when the temperature in (a) is lowered below T_c. (c) Flux distribution, when the magnetic field is switched off in (b). (d) Flux distribution, when the magnetic field is applied after cooling below T_c. Cases (b) and (c) are not observed but case (d) is observed, regardless of the FC or ZFC mode.

Alternatively, we may cool the specimen below T_c and subsequently apply a magnetic field in what is called the zero field cooling or ZFC mode. Since there was no magnetic flux in the superconductor, equation (12.1) requires the absence of flux in the specimen when the magnetic field is applied, as shown in Fig. 12.2(d). So, the application of the Maxwell equation to a perfect conductor would lead to a strange situation such that the magnetic field distribution in the superconductor depends on the sequence of application of the magnetic field and cooling of the specimen.

To ascertain this, Meissner and Ochsenfeld [4] measured the magnetic field distribution around a superconductor in 1933. They found that the magnetic field inside the superconductor always remains zero, as in Fig. 12.2(d), regardless of whether the FC or ZFC mode is employed. This has ruled out the situation in the FC mode shown in Fig. 12.2(b) and (c). Indeed, the magnetic fluxes are always expelled from the specimen, as long as the applied field is lower than the critical magnetic field H_c. This is called the Meissner effect and has been considered as one of the most fundamental properties of a superconductor, together with the phenomenon of zero resistance.

The Meissner effect looks to be at variance with the Maxwell equation (12.1). This difficulty can be resolved if a superconductor is magnetized in an opposite direction to cancel the external field $\mathbf{H}_{ext}$. Once the superconducting state is attained, surface currents are induced such that the resultant field in the sample is zero, i.e., $\mathbf{B}=0$. Since $\mathbf{B}=\mu_0\mathbf{H}_{ext}+\mathbf{M}$ holds, the magnetic susceptibility is immediately deduced to be $\chi=\mathbf{M}/\mu_0\mathbf{H}_{ext}=-1$.[2] As is clear from this, a superconductor exhibits an extremely large diamagnetism (recall the magnitude of the magnetic susceptibility in non-magnetic metals to be of the order of 10^{-6}/mol, as listed in Table 3.2). Because of this, a superconductor is indeed a perfectly diamagnetic material. The surface current continues to flow, as long as the superconducting state is maintained. There must be some small but finite surface layer in which the current flows. Therefore, the condition $\mathbf{B}=0$ fails in this surface layer. The depth of the layer is often referred to as the penetration depth.

12.3 London theory

In 1935, F. London and H. London [5] considered the superconducting state to be characterized by both zero electrical resistance and the Meissner effect and expressed these two phenomena in terms of the Maxwell equations. As has been noted, the electric field $\mathbf{E}$ inside a superconductor must be zero, but let us assume that $\mathbf{E}$ is instantly generated. Then, the superconducting electron would be subjected to the equation of motion described by equation (10.3). But the second term, corresponding to the friction, would not exist because of zero resistance. Hence we obtain

$$m_s \frac{d\mathbf{v}_s}{dt} = q_s \mathbf{E}, \qquad (12.2)$$

where m_s, $\mathbf{v}_s$ and q_s are the mass, drift velocity and charge of the superconducting electron, respectively. The superconducting current density $\mathbf{J}_s$ is then given by

$$\mathbf{J}_s = n_s q_s \mathbf{v}_s, \qquad (12.3)$$

where n_s is the number of superconducting electrons per unit volume.[3] An insertion of equation (12.3) into equation (12.2) leads to

---

[2] Since $\mathbf{B}=\mathbf{H}+4\pi\mathbf{M}$ in CGS units, $\chi=\mathbf{M}/\mathbf{H}_{ext}=-1/4\pi$ holds.
[3] We will learn in Section 12.11 that the "superconducting electron" is, in fact, a pair of electrons, called the Cooper pair with mass $m_s=2m$ and electron charge $q_s=(-2e)$. Its number density n_s is one-half the number of electrons n per unit volume. Hence, the current density is expressed as $\mathbf{J}_s = n_s q_s \mathbf{v}_s = (n/2)(-2e)\mathbf{v}_s = n(-e)\mathbf{v}_s$, provided that all n electrons form the pair

12.3 London theory

$$\frac{\partial \mathbf{J}_s}{\partial t} = \left(\frac{n_s q_s^2}{m_s}\right)\mathbf{E}. \qquad (12.4)$$

As discussed in the preceding section, the Meissner effect describes the phenomenon manifested by a superconductor in a magnetic field. The magnetic field is related to the electric field and the current through the Maxwell equations:

$$\text{rot}\mathbf{E} = -\frac{\partial \mathbf{B}}{\partial t} \qquad \text{rot}\mathbf{E} = -\frac{1}{c}\frac{\partial \mathbf{B}}{\partial t} \quad [\text{CGS}], \qquad (12.5)$$

$$\text{rot}\mathbf{H} = \mathbf{J} + \frac{\partial \mathbf{D}}{\partial t} \qquad \text{rot}\mathbf{H} = \frac{4\pi}{c}\mathbf{J} + \frac{1}{c}\frac{\partial \mathbf{D}}{\partial t} \quad [\text{CGS}] \qquad (12.6)$$

and

$$\text{div}\mathbf{B} = 0. \qquad (12.7)$$

We ignore the time derivative of the displacement current $\mathbf{D}$ in the steady state and solve the Maxwell equations in the region very close to the surface of a superconductor. The permeability μ is approximated as that in a vacuum so that $\mathbf{B} = \mu_0 \mathbf{H}$ holds.[4]

Equation (12.6) is therefore written as

$$\text{rot}\mathbf{B} = \mu_0 \mathbf{J}_s \qquad \text{rot}\mathbf{B} = \frac{4\pi}{c}\mathbf{J}_s \quad [\text{CGS}]. \qquad (12.8)$$

On the other hand, an insertion of equation (12.4) into equation (12.5) yields

$$\frac{\partial \mathbf{B}}{\partial t} = \left(-\frac{m_s}{n_s q_s^2}\right)\text{rot}\left(\frac{\partial \mathbf{J}_s}{\partial t}\right) \qquad \frac{\partial \mathbf{B}}{\partial t} = \left(-\frac{m_s c}{n_s q_s^2}\right)\text{rot}\left(\frac{\partial \mathbf{J}_s}{\partial t}\right) \quad [\text{CGS}].$$

If we replace $\frac{\partial \mathbf{J}_s}{\partial t}$ by $\frac{1}{\mu_0}\text{rot}\left(\frac{\partial \mathbf{B}}{\partial t}\right)$ from equation (12.8), we obtain

$$\frac{\partial \mathbf{B}}{\partial t} = \left(-\frac{m_s}{n_s q_s^2 \mu_0}\right)\text{rot rot}\left(\frac{\partial \mathbf{B}}{\partial t}\right) \qquad \frac{\partial \mathbf{B}}{\partial t} = \left(-\frac{m_s c^2}{4\pi n_s q_s^2}\right)\text{rot rot}\left(\frac{\partial \mathbf{B}}{\partial t}\right) \quad [\text{CGS}]. \qquad (12.9)$$

By using the vector identity $\text{rot rot}\dot{\mathbf{B}} = \text{grad div}\dot{\mathbf{B}} - \nabla^2 \dot{\mathbf{B}}$, together with equation (12.7), we can rewrite equation (12.9) as

$$\nabla^2\left(\frac{\partial \mathbf{B}}{\partial t}\right) = \left(\frac{n_s q_s^2 \mu_0}{m_s}\right)\left(\frac{\partial \mathbf{B}}{\partial t}\right) \qquad \nabla^2\left(\frac{\partial \mathbf{B}}{\partial t}\right) = \left(\frac{4\pi n_s q_s^2}{m_s c^2}\right)\left(\frac{\partial \mathbf{B}}{\partial t}\right) \quad [\text{CGS}]. \qquad (12.10)$$

---

[4] In the preceding section, we stressed that the magnetic susceptibility χ is equal to -1 inside a superconductor, i.e., the permeability $\mu = 0$. This is true well inside a superconductor where the surface current completely dies away. In the present section, we are solving the Maxwell equations in the surface layer where the shielding effect increases from zero ($\mu = \mu_0$) to that producing a perfect diamagnetism ($\mu = 0$). In the surface layer, μ is approximated as μ_0.

For the sake of simplicity, we consider a semi-infinite superconductor having a plane boundary and apply a magnetic field parallel to this boundary. The direction normal to the boundary is taken as the x-axis. Equation (12.10) is now reduced to

$$\frac{\partial^2 \dot{B}_x}{\partial x^2} = \left(\frac{n_s q_s^2 \mu_0}{m_s}\right) \dot{B}_x \qquad \frac{\partial^2 \dot{B}_x}{\partial x^2} = \left(\frac{4\pi n_s q_s^2}{m_s c^2}\right) \dot{B}_x \text{ [CGS]}.$$

The solution is immediately found to be

$$\dot{B}_x(x) = \dot{B}_{x,0} \exp\left(-\frac{x}{\sqrt{a}}\right), \qquad (12.11)$$

where $a = m_s/n_s q_s^2 \mu_0$. Equation (12.11) indicates that, on account of its exponential decay, the time derivative of B_x cannot penetrate deeply inside the superconductor.

Equation (12.11) is derived by using only the zero resistance phenomenon. The Meissner effect must be included, which imposes stricter restrictions such that the magnetic flux density **B** itself should vanish inside a superconductor. F. London and H. London [5] suggested that the Meissner effect might be correctly described in terms of equation (12.10) with the variable **B** instead of the derivative $\partial \mathbf{B}/\partial t$. Now equation (12.10) is phenomenologically replaced by

$$\nabla^2 \mathbf{B} = \left(\frac{n_s q_s^2 \mu_0}{m_s}\right) \mathbf{B} \qquad \nabla^2 \mathbf{B} = \left(\frac{4\pi n_s q_s^2}{m_s c^2}\right) \mathbf{B} \text{ [CGS]}. \qquad (12.12)$$

Equation (12.12) indicates that the magnetic flux density **B** should fall off exponentially in the superconductor and describes well the Meissner effect (see Exercise 12.1). All $\partial \mathbf{B}/\partial t$ appearing in equations (12.5) to (12.10) are now replaced by **B**. In place of equation (12.9), we obtain

$$\mathbf{B} = \left(-\frac{m_s}{n_s q_s^2}\right) \text{rot} \mathbf{J}_s \qquad \mathbf{B} = \left(-\frac{m_s c}{n_s q_s^2}\right) \text{rot} \mathbf{J}_s \text{ [CGS]}. \qquad (12.13)$$

This is known as the London equation.

As is clear from the argument above, the London equation cannot explain the mechanism of superconductivity but imposes some limitation to the Maxwell equations and successfully describes the most fundamental features of superconductivity, i.e., zero resistance and the Meissner effect. The solution of equation (12.12) indicates that the value of **B** decreases by $1/e$ at $x = \sqrt{a}$ measured from the surface of a superconductor. The characteristic distance λ_L is given by

$$\lambda_L = \sqrt{\frac{m_s}{\mu_0 n_s q_s^2}} \qquad \lambda_L = \sqrt{\frac{m_s c^2}{4\pi n_s q_s^2}} \text{ [CGS]}, \qquad (12.14)$$

which is often referred to as the London penetration depth.

Let us assume that one electron per atom contributes as a superconducting electron in a simple cubic lattice with the lattice constant of 0.3 nm. Since its density is $n_s = 3.7 \times 10^{28} \text{m}^{-3}$, the penetration depth λ_L turns out to be about 10^{-8} m from equation (12.14). We see how small the penetration depth is. Equation (12.14) also indicates that the penetration depth increases with increasing temperature towards T_c because of a decreasing number of superconducting electrons n_s (see Section 12.12). This is consistent with the observed temperature dependence of the penetration depth. But the actual penetration depth is not precisely described by the London theory because of its oversimplification.

12.4 Thermodynamics of a superconductor

Prior to the discussion of the microscopic theory of superconductivity, it is instructive to describe the thermodynamics of a superconductor based on the discussions developed in the preceding section. The superconducting state is uniquely determined if the temperature T and magnetic field **H** are fixed as external parameters. Let us denote the free energies for the superconducting and normal states as $G_s(T,\mathbf{H})$ and $G_n(T,\mathbf{H})$, respectively. Since the superconducting state is more stable in the temperature range $T < T_c$ in zero field, we must have $G_s < G_n$ in this range. When a magnetic field **H** is applied to a superconductor, the shielding surface current flows and a perfect diamagnetism appears. The resulting magnetic energy per unit volume is expressed as

$$-\int_0^H \mathbf{M} \cdot d\mathbf{H} = \int_0^H \mu_0 \mathbf{H} \cdot d\mathbf{H} = \frac{\mu_0 H^2}{2} \qquad -\int_0^H \mathbf{M} \cdot d\mathbf{H} = \left(\frac{1}{4\pi}\right)\int_0^H \mathbf{H} \cdot d\mathbf{H} = \frac{H^2}{8\pi} \quad [\text{CGS}],$$

(12.15)

where $\mathbf{M} = -\mu_0 \mathbf{H}$ ($\mathbf{M} = (-1/4\pi)\mathbf{H}$ in CGS units) is inserted. Hence, we obtain

$$G_s(T,\mathbf{H}) = G_s(T,0) + \frac{\mu_0 H^2}{2} \qquad G_s(T,\mathbf{H}) = G_s(T,0) + \frac{H^2}{8\pi} \quad [\text{CGS}]. \quad (12.16)$$

A superconductor generally exhibits only a weak para- or diamagnetism in its normal state. As discussed in Section 3.6, its magnetic susceptibility is so small in the normal state that we can ignore any contribution made by the presence of a magnetic field to the free energy. Thus we have

$$G_n(T,\mathbf{H}) = G_n(T,0). \qquad (12.17)$$

Since the free energy of the superconducting state should coincide with that of the normal state at the critical field H_c, the relation $G_s(T,H_c) = G_n(T,H_c)$ holds. Therefore, we obtain

$$G_s(T,0) = G_n(T,0) - \frac{\mu_0 H_c^2(T)}{2} \qquad G_s(T,0) = G_n(T,0) - \frac{H_c^2(T)}{8\pi} \text{ [CGS]}. \qquad (12.18)$$

A differentiation of both sides of equation (12.18) with respect to temperature yields the entropy in the respective states:

$$S_s = -\frac{\partial G_s}{\partial T} = S_n + \mu_0 H_c \frac{dH_c}{dT} \qquad S_s = S_n + \left(\frac{H_c}{4\pi}\right)\frac{dH_c}{dT} \text{ [CGS]}. \qquad (12.19)$$

As is clear from Fig. 12.1, dH_c/dT is always negative, resulting in the relation $S_s < S_n$. This means that in the range $T < T_c$ the entropy of the superconducting state is always lower than that of the normal state, i.e., the superconducting state is more ordered than the normal state.

The specific heat is calculated by differentiating equation (12.19) with respect to temperature and then multiplying by T:

$$C_s - C_n = \mu_0 T \left[H_c \frac{d^2 H_c}{dT^2} + \left(\frac{dH_c}{dT}\right)^2 \right]$$

$$C_s - C_n = \left(\frac{T}{4\pi}\right) \left[H_c \frac{d^2 H_c}{dT^2} + \left(\frac{dH_c}{dT}\right)^2 \right] \text{ [CGS]}. \qquad (12.20)$$

Equation (12.20) at $T = T_c$ is reduced to

$$\Delta C(T_c) = \mu_0 T_c \left[\left(\frac{dH_c}{dT}\right)^2 \right]_{T=T_c} \qquad \Delta C(T_c) = \left(\frac{T_c}{4\pi}\right) \left[\left(\frac{dH_c}{dT}\right)^2 \right]_{T=T_c} \text{ [CGS]}, \qquad (12.21)$$

since $H_c = 0$ at $T = T_c$. Equation (12.21) is always positive, indicating that, at $T = T_c$, the specific heat in the superconducting state is higher than that in the normal state. Indeed, this behavior has been confirmed by experiments, as will be shown in Fig. 12.3, Section 12.7.

We see from equation (12.18) that the free energy in the superconducting state is smaller by $H_c^2/8\pi$ than that in the normal state. For example, the value of H_c for pure Al metal is 99 Oe, as listed in Table 12.1. Thus, the free energy difference is about 390 erg/cm^3. We know that pure Al metal possesses three conduction electrons per atom. If all of them serve as superconducting electrons, then the free energy difference per electron is deduced to be 2.7×10^{-9} eV. According to the BCS theory, which will be discussed in Section 12.12, this energy is found to be of the order of $(k_B T_c)^2/\varepsilon_F$. By inserting $T_c = 1.19$ K from Table 12.1 and $\varepsilon_F = 11.6$ eV from Table 2.1 into the BCS expression, we obtain

$$\frac{(k_B T_c)^2}{\varepsilon_F} = \frac{(1.19 \times 1.38 \times 10^{-16})^2 [\text{erg}]^2}{11.6 \times 1.6 \times 10^{-12} [\text{erg}]} = 3.91 \times 10^{-21} [\text{erg}] = 2.4 \times 10^{-9} [\text{eV}],$$

which is in good agreement with that estimated from equation (12.18) based on thermodynamics. In general, the energy per particle associated with the phase transition is known to be of the order of $k_B T_c$, where T_c is the phase-transition temperature. Thus the energy involved in the superconducting transition is smaller by $k_B T_c / \varepsilon_F$ than that in the ordinary phase transition. This is a feature unique to the superconducting phenomenon.

12.5 Ordering of the momentum

We are well aware that, when a particle with electronic charge q_s moves with velocity **v** in the presence of a magnetic flux density **B**, the Lorentz force $q_s vB$ is exerted on the particle in a direction perpendicular to the plane formed by the two vectors **v** and **B**. The Lorentz force does not affect the magnitude of the velocity but changes the direction of the motion of the particle. As a consequence, the charged particle moves on a circle in the plane perpendicular to the direction of the magnetic field. At first sight, one might think that the momentum remains unchanged, since the magnitude of the velocity is unchanged. But this is not true. The effect of a change in the direction of the velocity caused by the magnetic field must be included. The total momentum of the superconducting electron moving in the magnetic field must be given by the sum of the ordinary kinetic momentum $m\mathbf{v}_s$ and the momentum associated with the magnetic field:

$$\mathbf{p}_s = m\mathbf{v}_s + q_s \mathbf{A}, \quad (12.22)$$

where **A** is the vector potential defined as $\mathbf{B} = \mathrm{rot}\mathbf{A}$. By taking the rotation of equation (12.22) and inserting it into the London equation (12.13), then we obtain

$$\mathrm{rot}\mathbf{p}_s = 0. \quad (12.23)$$

There exists some arbitrariness in the choice of the vector potential **A**. We can impose the relation $\mathrm{div}\mathbf{A} = 0$ and the normal component of both the vector potential **A** and $\mathbf{v}_s$ to diminish at the surface of an isolated superconductor as additional constraints.[5] Apart from this, the equation of continuity assures the relation $\mathrm{div}\mathbf{J}_s = 0$. Now the divergence of equation (12.22) immediately results in the relation $\mathrm{div}\mathbf{p}_s = 0$. We find the important relation $\mathbf{p}_s = 0$ as a solution subjected to the conditions above.

---

[5] By using the vector potential **A** and scalar potential ϕ, one can write the electric field **E** and magnetic flux density **B** as $\mathbf{E} = -\mathrm{grad}\phi - \partial\mathbf{A}/\partial t$ and $\mathbf{B} = \mathrm{rot}\mathbf{A}$, respectively. However, the vector potential **A** and scalar potential ϕ are not uniquely decided. They can be written as $\mathbf{A}' = \mathbf{A} + \mathrm{grad}\varphi$ and $\phi' = \phi - (\partial\varphi/\partial t)$ without altering values of **E** and **B**, where φ is an arbitrary scalar function. Hence, an additional restriction can be imposed. We choose the London gauge where the conditions $\mathrm{div}\mathbf{A} = 0$ and $A_\perp = 0$ hold on any external surface of an isolated superconductor.

The London equation now leads to the important conclusion that the momentum of the superconducting electron in the magnetic field remains zero. Since the London equation describes the superconducting state in terms of macroscopic quantities based on the drift velocity in equation (10.2), the above conclusion should imply that an average of the momenta over a large number of superconducting electrons is zero.

F. London proposed a new concept of the long-range order for the average momentum of the superconducting electrons [6]. His proposal is stated such that all the superconducting electrons are in the state $\mathbf{p}_s = 0$, regardless of the presence or absence of a magnetic field. The first term $m\mathbf{v}_s$ in equation (12.22) is postulated to be zero in the absence of a magnetic field. When the magnetic field is applied, the superconducting electrons begin to move coherently with the velocity $\mathbf{v}_s = -(q_s/m_s)\mathbf{A}$ so as to maintain the condition $\mathbf{p}_s = 0$ inside the superconductor. This is nothing but the induced surface current discussed in Section 12.2. His proposal is certainly in conflict with the Pauli exclusion principle. However, soon after it was proved that he had indeed pointed to the essence of the superconducting mechanism.

12.6 Ginzburg–Landau theory

Ginzburg and Landau [7] constructed a phenomenological theory to account for the ordered state of a superconductor on the basis of the Landau theory as it relates to a second-order phase transition. Following Landau, the free energy of the superconducting state just below the transition temperature T_c is assumed to be only slightly lowered relative to that of the normal state and the difference is expanded in powers of an order parameter ψ:

$$G_s = G_n + \alpha(T)|\psi|^2 + \frac{\beta(T)}{2}|\psi|^4 + \cdots, \qquad (12.24)$$

where ψ is assumed to vanish at $T = T_c$. The parameter α is further assumed to be a linear function of temperature, i.e.,

$$\alpha = \alpha_0(T - T_c) \qquad (12.25)$$

in the temperature range not too far below T_c. Since the relation $|\psi|^2 = \alpha_0(T_c - T)/\beta$ is deduced below T_c from the equilibrium condition $\partial G_s/\partial|\psi| = 0$, a comparison of equation (12.24) with equation (12.18) immediately results in the relation $H_c(T) = (2\alpha_0^2/\beta\mu_0)^{1/2}(T_c - T)$.

Equation (12.24) is limited to the case where the order parameter is constant throughout a superconductor. Ginzburg and Landau took into account the

12.6 Ginzburg–Landau theory

spatial variation in the order parameter and added the term $\gamma|\nabla\psi|^2$ to equation (12.24). They could derive two sets of the Ginzburg–Landau equations by minimizing the free energy with respect to the order parameter ψ and the vector potential **A** for the magnetic field. The so-called GL equations have led to various interesting solutions. For example, an obvious solution $\psi \equiv 0$ describes the normal state whereas the solution of $\psi = (-\alpha/\beta)^{1/2}$ and $\mathbf{A} = 0$ reproduces well the Meissner effect. For weak magnetic fields, the solution is reduced to the same form as the London equation.

Let us briefly discuss the physical implication of the order parameter $\psi(\mathbf{r})$. The phase in the wave function of the conduction electron in the normal state is random and, hence, no interference occurs among wave functions of the normal electrons. A different situation exists in the superconducting state. The ordered state with $\mathbf{p}_s = 0$, which is deduced from the London equation, would correspond to the state where all wave functions of the superconducting electrons are coherent and characterized by a single amplitude and phase.

If a large number of superconducting wave functions are coherent, the sum of the amplitude would be increased in proportion to the number of superconducting electrons. Thus, the superconducting state can be described by a macroscopic wave function. Ginzburg and Landau intuitively considered the order parameter $\psi(\mathbf{r})$ as a kind of a "wave function" for a "particle" in the superconducting state. If the wave function is expressed as $\psi(\mathbf{r}) = |\psi|e^{i\eta}$, then $|\psi(\mathbf{r})|^2$ would be proportional to the density of superconducting electrons at the position **r**. Suppose that the phase η follows the relation $\eta = kx - \eta_0$ along the x-direction. Now a spatial change in the phase yields the momentum of the superconducting electron because of the obvious relation $d\eta/dx = k = p_s/\hbar$.

The state with $\mathbf{p}_s = 0$ is, therefore, equivalent to the possession of a spatially independent phase and amplitude of the wave function. As a result, $\psi(\mathbf{r})$ becomes uniform throughout a superconductor. When a magnetic field is applied, the second term in equation (12.22) becomes finite. But, as discussed in the preceding section, the velocity $\mathbf{v}_s = -(q_s/m_s)\mathbf{A}$ is induced so as to maintain $\mathbf{p}_s = 0$. Therefore, we see that the Meissner effect can be interpreted as a phenomenon which makes $\psi(\mathbf{r})$ uniform throughout a superconductor even in the presence of the magnetic field.

As will be discussed in Section 12.15, Abrikosov extended the GL theory and could successfully provide the theoretical basis for type-II superconductors (see Section 12.15). In the following sections, we focus on the experimental studies showing evidence of the energy gap in a superconductor and the isotope effect on the transition temperature, both of which served as very important observations in understanding the mechanism of superconductivity.

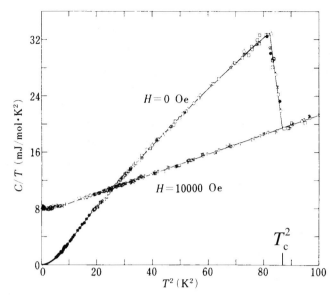

Figure 12.3. Temperature dependence of the low-temperature specific heat in the form of C/T versus T^2 for pure Nb in the presence and absence of a magnetic field H. [H. A. Leupold and H. A. Boorse, *Phys. Rev.* **134** (1964) A1322]

12.7 Specific heat in the superconducting state

Figure 12.3 shows the low-temperature specific heat data in the form of C/T against T^2 for pure Nb metal. The Nb metal undergoes a superconducting transition at 9.2 K. Thus, the data below T_c in the absence of a magnetic field represent the specific heat behavior in the superconducting state. Since the critical magnetic field H_c of pure Nb is 1950 Oe, as is listed in Table 12.1, the magnetic field of 10 000 Oe can completely suppress the superconducting transition. Indeed, the C/T against T^2 data fall on a straight line, indicating restoration of the normal state. The electronic specific heat coefficient can be obtained by extrapolating the linear trend of the data in the normal state to absolute zero.

As is clear from Fig. 12.3, when approaching T_c from higher temperatures the specific heat in the absence of a magnetic field jumps at T_c and subsequently decreases gradually toward zero with decreasing temperature. The electronic specific heat in the superconducting state can be fitted to an $\exp(-2\Delta/k_B T)$-type temperature dependence in contrast to the γT dependence in the normal state. Indeed, the exponential temperature dependence can be deduced from the BCS theory, which will be described in Section 12.12. The fact that the data in Fig. 12.3 no longer fall on a straight line but approach zero exponentially with decreasing temperature can be taken as evidence that the superconducting electrons do not obey the Fermi–Dirac statistics.

The specific heat jump at T_c without involving any latent heat indicates that the phase transition from the normal to superconducting state is of the second-order. A low-temperature phase is generally more ordered relative to a high-temperature phase. The motion of conduction electrons in the normal state is of course random. Instead, the motion of electrons in the superconducting state must be ordered, as has been discussed in previous sections. In addition, the $\exp(-2\Delta/k_B T)$-type temperature dependence indicates the presence of an energy gap in the superconducting phase.

12.8 Energy gap in the superconducting state

The presence of an energy gap in a superconductor has been also confirmed by other experiments. For example, the absorption of an infrared electromagnetic wave by a superconductor occurs only when its frequency exceeds some critical value v_0. The value of v_0 for typical superconducting metals lies in the range of 10^{11} Hz, from which the energy gap is estimated to be of the order of 10^{-1} meV. Note that the energy gap in a superconductor is about 1/1000 that in a semiconductor (for example, 1 eV for Si).

The tunneling experiment has proved to be very powerful in determining the magnitude of the energy gap. As shown in Fig. 12.4(a), a very thin insulating layer of 1–2 nm thickness is sandwiched between a superconductor and a normal metal. The current–voltage characteristics for this device are shown in Fig. 12.4(c). We see that the tunneling current begins to flow only when the bias voltage exceeds some critical value V_0. This is the demonstration for the presence of the energy gap 2Δ in a superconductor. Indeed, the value of the energy gap can be obtained from the relation $\Delta = eV_0$ into which the measured V_0 value is inserted. The energy gaps 2Δ determined by this technique for pure metals like Al, V and Nb are in the range 0.5–1 meV.

12.9 Isotope effect

The superconducting transition temperature T_c is found to depend on the mass M of the isotope for a given element. This is called the isotope effect and has played a crucial role in the construction of the BCS theory. For instance, when the Hg isotope changes its mass from 199.5 to 203.4 g, the value of T_c changes from 4.185 to 4.146 K. This is shown in Fig. 12.5. The experimental data can be fitted to the relation:

$$M^\alpha T_c = constant, \tag{12.26}$$

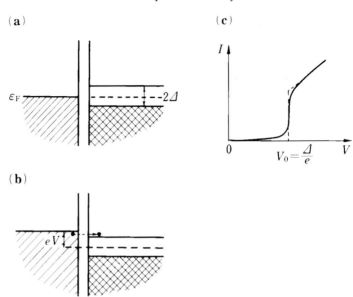

Figure 12.4. Measurement of the energy gap in a superconductor. A thin insulating layer is sandwiched between a metal and superconductor. (a) Energy levels before applying a voltage. (b) A voltage is applied to overcome the energy gap so that electrons in the normal metal can tunnel to the superconductor. (c) The resulting I–V characteristic. The energy gap is obtained from the voltage $V_0 = \Delta/e$.

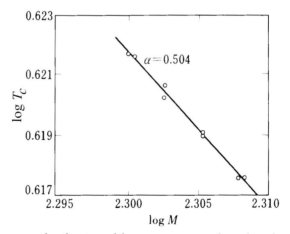

Figure 12.5. Superconducting transition temperature plotted against the mass of the isotope in mercury on a log–log diagram. [C. A. Reynolds, B. Serin and L. B. Nesbitt, *Phys. Rev.* **84** (1951) 691]

where α is deduced to be 0.504. According to equations (4.17) and (4.40), the Debye temperature Θ_D is proportional to $1/\sqrt{M}$. Hence, the ratio T_c/Θ_D becomes constant. This finding clearly indicates the importance of the interaction of the conduction electron with lattice vibrations in the superconducting state.

12.10 Mechanism of superconductivity–Fröhlich theory

We have so far presented the most important experimental evidences which have served as clues to resolve the mechanism of superconductivity. In this section, our attention is directed to the theoretical progress and the Fröhlich theory is first outlined. In Sections 10.8 to 10.10, it was emphasized that a finite electrical resistivity originates from the static and dynamical disruptions of the periodic potentials and that the Bloch–Grüneisen law describing the temperature dependence of the resistivity in normal metals can be formulated by incorporating the electron–phonon interaction into the Boltzmann transport equation. This clearly means that the electron–phonon interaction determines the electrical resistivity at finite temperatures. However, as demonstrated by the isotope effect, the interaction of electrons with lattice vibrations must play a crucial role in the resistanceless phenomenon of superconductivity. Fröhlich [8] pointed out, for the first time, in 1950, that this seemingly paradoxical matter is not impossible.

Fröhlich considered that the region left behind along the passage of a conduction electron would become slightly excessive in positive charges as a result of the displacement of neighboring ions. If this is so, another electron may be naturally attracted by this positively charged region. This process would give rise to a weak but attractive interaction between the two electrons and is taken as a possible mechanism to stabilize the superconducting state.

The attractive electron–electron interaction proposed by Fröhlich is schematically illustrated in Fig. 12.6. We consider two electrons having wave vectors $\mathbf{k}_1$ and $\mathbf{k}_2$ and assume that their states change into $\mathbf{k}'_1$ and $\mathbf{k}'_2$ after the interaction involving a phonon of wave vector $\mathbf{q}$. The whole process may be divided into two halves. In the first half of the process, a phonon of the wave vector $\mathbf{q}$ is emitted when the electron $\mathbf{k}_1$ is scattered into $\mathbf{k}'_1$. But this phonon is immediately absorbed in the second half of the process, in which another electron $\mathbf{k}_2$ is scattered into $\mathbf{k}'_2$. The momentum conservation law assures $\mathbf{k}_1 - \mathbf{q} = \mathbf{k}'_1$ and $\mathbf{k}_2 + \mathbf{q} = \mathbf{k}'_2$ in the first and second halves of the process, respectively. We immediately obtain $\mathbf{k}_1 + \mathbf{k}_2 = \mathbf{k}'_1 + \mathbf{k}'_2$, indicating that the momentum of the two electrons is conserved in this process.

What about the energy conservation law? The emitted phonon can survive

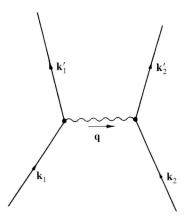

Figure 12.6. Electron–electron interaction mediated by the emission and absorption of a virtual phonon.

only in the intermediate state and its lifetime Δt is very short. Thus, the energy conservation law no longer holds between the initial and the intermediate states or between the intermediate and final states because of the uncertainty principle in energy $\Delta E \cdot \Delta t \approx \hbar$. Such a short-lived phonon is called the virtual phonon and the scattering involved the virtual phonon process. The second-order perturbation theory [8] gives rise to an energy change given by

$$\langle \mathbf{k}_1 - \mathbf{q}, \mathbf{k}_2 + \mathbf{q} | V | \mathbf{k}_1, \mathbf{k}_2 \rangle = \frac{2\hbar\omega_q |M_{\mathbf{k}-\mathbf{q},\mathbf{k}}|^2}{\{[\varepsilon(\mathbf{k}) - \varepsilon(\mathbf{k} - \mathbf{q})]^2 - (\hbar\omega_q)^2\}}, \quad (12.27)$$

where $M_{\mathbf{k}-\mathbf{q},\mathbf{k}}$ satisfying the relation $M_{\mathbf{k}_1-\mathbf{q},\mathbf{k}_1} = M_{\mathbf{k}_2+\mathbf{q},\mathbf{k}_2}$, is the matrix element associated with the absorption or emission of a phonon of wave vector $\mathbf{q}$, $\varepsilon(\mathbf{k})$ and $\varepsilon(\mathbf{k} - \mathbf{q})$ are the energies of the electron before and after the emission of the phonon, respectively, and $\hbar\omega_q$ is the energy of the emitted phonon. As mentioned above, the denominator is not zero. The interaction energy becomes negative for the narrow range of energy where $|\varepsilon(\mathbf{k}) - \varepsilon(\mathbf{k} - \mathbf{q})| < \hbar\omega_q$ holds. This leads to an attractive electron–electron interaction. Once this mechanism works, the total energy of the system is lowered relative to that in the normal state at absolute zero. Fröhlich considered this energy gain to be responsible for the stabilization of the superconducting state.

The total energy gain would be roughly given by $N(\varepsilon_F)(\hbar\omega_D)^2$, since $\Delta\varepsilon$ in equation (12.27) involves an average phonon energy $\hbar\omega_D$ per electron and the number of electrons involved in this process must be $N(\varepsilon_F)\hbar\omega_D$. Since $N(\varepsilon_F) = \frac{3}{2}(n/\varepsilon_F)$ holds in the free-electron model, where n is the number of conduction electrons per unit volume, we obtain the energy gain per electron as

$$\frac{\Delta\varepsilon}{n} \approx \frac{N(\varepsilon_F)(\hbar\omega_D)^2}{n} \approx \frac{n}{\varepsilon_F} \cdot \frac{(\hbar\omega_D)^2}{n} = \frac{(\hbar\omega_D)^2}{\varepsilon_F} = \frac{(k_B\theta_D)^2}{\varepsilon_F}. \quad (12.28)$$

12.11 Formation of the Cooper pair

We immediately obtain $\Delta\varepsilon/n \propto 1/M$, since $\Theta_D \propto 1/\sqrt{M}$. This explains the isotope effect discussed in Section 12.9. However, we noted in Section 12.4 that the energy gain per electron is $(k_B T_c)^2/\varepsilon_F$ rather than $(k_B \theta_D)^2/\varepsilon_F$. Thus, we see that the Fröhlich theory overestimates it by an amount equal to $(\Theta_D/T_c)^2$.

12.11 Formation of the Cooper pair

Since the advent of the Fröhlich theory, people have tended to believe that the stability of the superconducting state is brought about by a lowering of the total energy in the electron system through the attractive interaction between the two electrons mediated by the virtual phonon. However, we have also emphasized in Sections 12.5 and 12.6 that the average momentum of superconducting electrons is zero and that the long-range order of the momentum must be a characteristic feature in the superconducting state.

Conduction electrons are subject to the Pauli exclusion principle. But how can this obvious fact be reconciled with a state of zero momentum? A particle with zero spin obeys the Bose statistics. For example, the He^4 atom, which is typical of the Bose particle, enters a superfluid state below 2.19 K and can pass through a hole, however small its size. London has shown that the superfluidity of He^4 particles arises as a result of a condensation into a state of zero momentum. Condensation into a zero momentum state may be realized if the two electrons form a bound state via the attractive electron–electron interaction and the resulting pair of electrons behaves as a single particle obeying the Bose statistics. Now the question arises as to what electrons can form such a pair. In 1956 Cooper pointed out that, if two electrons interacting with an attractive force are placed immediately above the Fermi sphere at absolute zero, the two electrons form a bound state and their total energy is lowered relative to $2\varepsilon_F$, even though the attractive interaction is very weak.

We assume that the phonon-mediated electron–electron interaction is responsible for the attractive interaction between the two electrons. The energy states below the Fermi energy ε_F are completely occupied by electrons at absolute zero. Two electrons with energies ε_1 and ε_2 are added immediately above ε_F so as not to violate the Pauli exclusion principle. Then the following relation holds:

$$\varepsilon_1 - \varepsilon_2 = \frac{\hbar^2}{2m}[(\mathbf{k}_F + \Delta\mathbf{k}_1)^2 - (\mathbf{k}_F + \Delta\mathbf{k}_2)^2] \approx \frac{\hbar^2 k_F \Delta k}{m} \approx \hbar\omega_q, \quad (12.29)$$

where $\Delta k = |\Delta\mathbf{k}_1 - \Delta\mathbf{k}_2|$ and the relation $\varepsilon_1 - \varepsilon_2 \approx \hbar\omega_q$ is used. Equation (12.29) indicates that the momenta of the two electrons must be confined within the range $\Delta k = m\omega_q/\hbar k_F$ across the Fermi level. This is illustrated schematically in Fig. 12.7.

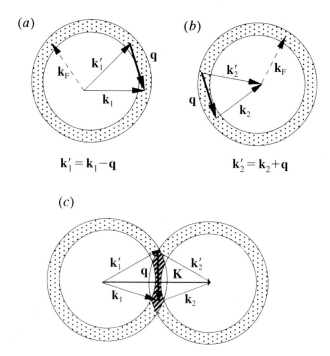

Figure 12.7. Formation of Cooper pair. (a) Emission and (b) absorption of a virtual phonon. The dotted shell across the Fermi level represents Δk in equation (12.29). (c) Total process combining (a) and (b). Only in the region where the two shells overlap can scatterings (a) and (b) occur simultaneously for a phonon of an arbitrary wave vector $\mathbf{q}$.

Following the symbols used in Fig. 12.6, we denote the wave vectors of a pair of electrons before scattering as $\mathbf{k}_1$ and $\mathbf{k}_2$, and those after scattering as $\mathbf{k}'_1$ and $\mathbf{k}'_2$. We have two relations $\mathbf{k}'_1 = \mathbf{k}_1 - \mathbf{q}$ and $\mathbf{k}'_2 = \mathbf{k}_2 + \mathbf{q}$. If we write $\mathbf{K} = \mathbf{k}_1 + \mathbf{k}_2$, then the relation $\mathbf{K} = \mathbf{k}'_1 + \mathbf{k}'_2$ holds. Furthermore, all these wave vectors must be found within the shell between $\mathbf{k}_F$ and $\mathbf{k}_F + \Delta \mathbf{k}$. The phonon emission and absorption processes are shown in Fig. 12.7(a) and (b), respectively. They are combined in a single diagram by sharing both $\mathbf{q}$ and $\mathbf{K}$, as shown in (c).

It is now clear that the scattering satisfying both conditions $\mathbf{K} = \mathbf{k}_1 + \mathbf{k}_2$ and $\mathbf{K} = \mathbf{k}'_1 + \mathbf{k}'_2$ is limited to the hatched area formed by the overlap of the two spherical shells. The shaded area must be maximized in order to cause such attractive electron–electron scattering processes to occur as frequently as possible. This is realized, when $\mathbf{K} = 0$ or the two Fermi spheres completely overlap each other. In other words, the electron of wave vector $\mathbf{k}$ should be paired with an electron of wave vector $-\mathbf{k}$. In addition, the lowest state is realized when the two electrons have opposite spins ↑ and ↓. The entity formed by such an interaction is called a Cooper pair. Both the spin quantum number S and its z-component S_z of the resultant spin **S** of a Cooper pair are therefore zero. This is called the singlet state.[6]

The triplet states $|k\uparrow, -k\uparrow\rangle$, $|k\downarrow, -k\downarrow\rangle$ and $(|k\uparrow, -k\downarrow\rangle + |k\downarrow, -k\uparrow\rangle)/\sqrt{2}$ characterized by the spin quantum number $S=1$ and $S_z = 1, -1$ and 0, respectively, may also be taken as possible candidates. However, they are theoretically proved not to form the bound state. To summarize, the total energy of the conduction electron system can be most effectively lowered when the spin-up electron of wave vector **k** is paired with the spin-down electron of wave vector $-\mathbf{k}$. A Cooper pair is hereafter denoted as $(\mathbf{k}_i\uparrow, -\mathbf{k}_i\downarrow)$.

12.12 The superconducting ground state and excited states in the BCS theory

Bardeen, Cooper and Schrieffer had to struggle further to relate a single "Cooper pair" with the many-electron theory and to cope with the overlap of many pairs. Schrieffer portrayed the problem in an analogy with couples dancing on a crowded floor [9]: "Even though partners dance apart for considerable periods and even though other dancers come between, each pair remains a couple. The problem was to represent this situation mathematically."

BCS successfully constructed the wave function by taking a linear combination of many normal-state configurations in which the Bloch states are occupied by a pair of opposite momenta and spins:

$$\Phi(\mathbf{r}_1, \mathbf{r}_2) = \sum_i a_i \phi(\mathbf{k}_i\uparrow, -\mathbf{k}_i\downarrow), \tag{12.30}$$

where the sum extends over all possible pair configurations within a range $\Delta k = m\omega_q/\hbar k_F$ about k_F and $|a_i|^2$ represents the probability of finding a pair of electrons with the states $\mathbf{k}_i\uparrow$ and $\mathbf{k}_i\downarrow$. The wave function given by equation (12.30) is called a Cooper pair. Therefore, it is important to realize that each Cooper pair is composed of the Bloch states with all possible wave vectors $\mathbf{k}_i$.

Once the attractive interaction between two electrons dominates over the repulsive screened Coulomb interaction, the system would produce as many Cooper pairs as possible to lower its energy. BCS [3] showed how this attractive

---

[6] Consider the wave function $\psi_{\sigma\sigma'}(\mathbf{r}_1, \mathbf{r}_2)$ of a pair of electrons, positioned at $\mathbf{r}_1$ and $\mathbf{r}_2$, respectively. This function depends only on the relative coordinate $\mathbf{r}_1 - \mathbf{r}_2 \equiv \mathbf{r}$ and is given by the product of the orbital and spin wave functions: $\psi_{\sigma\sigma'}(\mathbf{r}) = \phi(\mathbf{r}) \cdot (1/\sqrt{2})(|\uparrow\downarrow\rangle - |\downarrow\uparrow\rangle)$ (see Section 14.3). Suppose the orbital function $\phi(\mathbf{r})$ to be the eigen function of the resultant orbital angular momentum **L** (see Section 13.3). Then, its eigen value can be specified in terms of the quantum number $L = 0, 1, 2, 3 \cdots$ or s, p, d, f, $\cdots$ symmetries. Since the spin function in the BCS ground state is characterized by the singlet state, the orbital wave function must be symmetric, i.e., $\phi(\mathbf{r}) = \phi(-\mathbf{r})$. Hence, only s, d, $\cdots$ waves are permissible. In the BCS theory, a constant attractive interaction is assumed so that an energy gap opens in any direction in reciprocal space (see equation (12.33)). This is consistent with the s-type orbital wave function for the paired electrons. In the case of high-T_c superconductors, a number of experimental studies indicate the existence of anisotropy in the energy gap and are consistent with the possession of d-wave symmetry.

interaction can give rise to a cooperative many-particle state which is lower in energy than the normal state by an amount proportional to $(\hbar\omega_D)^2$ in agreement with the isotope effect. They constructed the ground-state wave function in the superconducting state by using a Hartree-like approximation and expressed it as a product of the individual Cooper pair wave functions given by equation (12.30):

$$\psi_0(\mathbf{r}_1,\mathbf{r}_2, \ldots , \mathbf{r}_{n_0}) = \Phi(\mathbf{r}_1,\mathbf{r}_2)\Phi(\mathbf{r}_3,\mathbf{r}_4) \ldots ,\Phi(\mathbf{r}_{n-1},\mathbf{r}_n) \qquad (12.31)$$

where n is the total number of electrons participating in the superconductivity, $\mathbf{r}_n$ is the position coordinate of the n-th electron and the Φs on the right-hand side are the same for all pairs. The square of the many-electron wave function (12.31) gives the probability of finding superconducting electrons at $\mathbf{r}_1, \mathbf{r}_2, \ldots , \mathbf{r}_{n_0}$ regardless of their momenta. The Cooper pair $\Phi(\mathbf{r}_{n-1},\mathbf{r}_n)$ involved in equation (12.31) can be regarded as a single particle obeying the Bose–Einstein statistics.

The ground-state energy per electron in the superconducting state is shown to be lower than that in the normal state by the amount:

$$\frac{\Delta\varepsilon}{n} \cong \left(\frac{1}{2}\right)\frac{N(\varepsilon_F)\Delta^2}{n} \approx \frac{\Delta^2}{\varepsilon_F}, \qquad (12.32)$$

where $N(\varepsilon_F)$ is the density of states at the Fermi level in the normal state and Δ is one-half the energy gap characteristic of the superconducting state, as already mentioned in Sections 12.7 and 12.8. In the BCS theory, the net attractive interaction is assumed to be constant, regardless of the direction in reciprocal space. Then, the energy gap 2Δ is deduced as

$$\Delta \cong 2\hbar\omega_D \exp\left[-\frac{1}{N(\varepsilon_F)V}\right], \qquad (12.33)$$

where V is a constant matrix element representing the strength of the net attractive interaction and $\hbar\omega_D$ is the mean phonon energy.

We noted in Section 12.10 that the Fröhlich theory predicted the energy gain per electron to be proportional to $(k_B\Theta_D)^2/\varepsilon_F$ in equation (12.28) and that it is too large to account for the experimental data. In the BCS theory, $k_B\Theta_D$ is replaced by Δ, which is smaller by the exponential factor than $\hbar\omega_D$. The number of Cooper pairs is gradually destroyed with increasing temperature and, accordingly, the energy gap 2Δ decreases and vanishes at the superconducting transition temperature T_c. As shown in Fig. 12.8, a good agreement with the BCS theory is found in the temperature dependence of the energy gap determined from tunneling measurements. According to the BCS theory, the superconducting transition temperature T_c is approximated as

$$k_B T_c \cong \Delta(0). \qquad (12.34)$$

12.12 Superconducting ground state and excited states in BCS theory

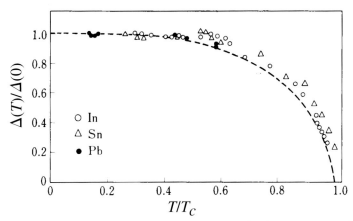

Figure 12.8. Normalized energy gap versus normalized temperature derived from tunneling experiments for superconducting In, Sn and Pb. The dashed curve is derived from the BCS theory. [I. Giaever and K. Megerle, *Phys. Rev.* **122** (1961) 1101]

This relation immediately leads to the isotope effect in equation (12.26), since the relation $k_B T_c \cong \Delta(0) \propto \hbar \omega_D \propto 1/\sqrt{M}$ holds.

BCS also evaluated the wave function and energy in the excited states and calculated the free energy, from which the specific heat and critical magnetic field were calculated. Their results can explain these experimental data very well. The Meissner effect is also well interpreted. In this way, the BCS theory could successfully provide the theoretical basis for various properties inherent in the superconducting state.

When the net electron–electron interaction becomes attractive, all electrons within the range $\Delta k = m\omega_q/\hbar k_F$ about k_F are coupled to form Cooper pairs in the ground state. The paired electrons described by the wave function (12.31) are repeatedly scattered between single-electron states so that their motions can no longer be distinguished as a result of so frequent scattering events. This means that the wave vector **k** specifying a single-electron state in the vicinity of the Fermi level is no longer a good quantum number and that the Fermi sphere corresponding to the BCS ground state is blurred within the range $\hbar \omega_D$ across the Fermi level. This is illustrated in Fig. 12.9. Thus, the electronic states in the superconducting ground state cannot be uniquely described in reciprocal space.

Let us discuss the excited states of a superconductor. We have already shown in Sections 12.7 and 12.8 the presence of a small energy gap, as evidenced from various experiments like infrared absorption, specific heat and tunneling experiments. As is clear from equation (12.30), the wave function of the Cooper pair is composed of all possible single-electron states having a set of the wave vectors ($\mathbf{k}_i \uparrow$ and $-\mathbf{k}_i \downarrow$) within the range $\Delta k = m\omega_q/\hbar k_F$ about k_F. Thus, all

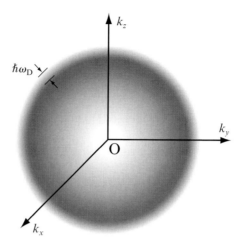

Figure 12.9. The Fermi sphere in the superconducting state at 0 K. The Fermi sphere within $\hbar\omega_D$ is smeared as a result of the attractive electron–electron interaction mediated by phonons. The electronic structure in the superconducting state is no longer well described in reciprocal space.

allowed wave vectors are used up to form a Cooper pair while the electrons in each pair should always have equal but opposite momenta. Hence, the momenta of the paired electrons cannot be freely increased when energy is imparted to a superconductor. The only possible way to expend the energy is to break up a Cooper pair. According to the BCS theory, the minimum amount of energy needed is 2Δ given by equation (12.33) and is able to produce two electrons with well-defined wave vectors.

As is clear from the argument above, both ground and excited states in a superconductor can be more precisely described by an energy spectrum rather than in reciprocal space. As shown in Fig. 12.10(a), all Cooper pairs are condensed into a single energy level ε_0 in the ground state at absolute zero. If the Cooper pair receives energies higher than the energy gap 2Δ, then the pair is broken into two independent electrons. The resulting electrons, which are called quasiparticles, are subjected to the Fermi statistics. This is schematically illustrated in Fig. 12.10(a). The energy spectrum of the quasiparticles, therefore, represents the excited states of a superconductor.

As has been discussed in Section 10.5, the excitation of electrons into the conduction band conversely results in the excitation of holes in the valence band. Such representation can be used also in a superconductor. An energy spectrum for the quasiparticles in a superconductor at a finite temperature below T_c is expressed as shown in Fig. 12.10(b). One can see that the quasiparticles cannot occupy states in the energy range 2Δ across the Fermi level. This

12.12 Superconducting ground state and excited states in BCS theory 357

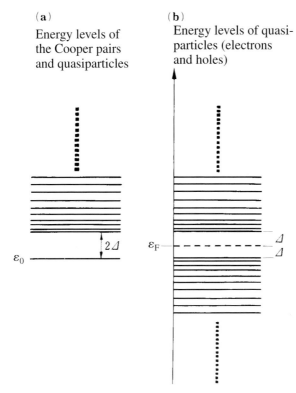

Figure 12.10. (a) All Cooper pairs in the ground state are in the energy level ε_0. The lowest excited energy of the two quasiparticles (electrons) is 2Δ higher than ε_0. Each level is filled by two electrons with spin-up and spin-down, since the quasiparticle obeys the Fermi statistics. (b) Energy levels for the excited states of a superconductor. The Fermi level for electrons is raised by Δ relative to that in the normal state. Likewise, the Fermi level for holes is lowered by Δ. Accordingly, the energy gap 2Δ is opened.

is the forbidden energy gap of a superconductor. It is, therefore, important to note that a well-defined energy gap exists, though the Fermi sphere is blurred in reciprocal space even at absolute zero. The density of states near the Fermi level in a superconductor is enhanced as a result of the congestion of the energy levels, as shown in Fig. 12.10.

Now it is obvious that the entity responsible for the long-range order of the momentum London proposed is the Cooper pair. Because of their resultant zero spin, Cooper pairs can behave as Bose particles and, because of the resultant zero momentum, the system is in an ordered state. In the BCS theory, the energy gap 2Δ is shown to remain unchanged even when a magnetic field is applied. This means that the density of the Cooper pairs and, hence, the ordered state, remains unchanged. This leads to the possession of a uniform order parameter in a superconductor, being consistent with the

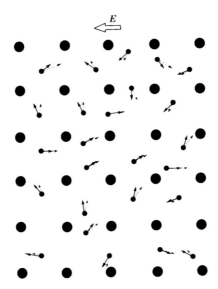

Figure 12.11. Motion of conduction electrons in the normal state. Large and small circles represent ions and conduction electrons, respectively. The momentum of each conduction electron is shown by a small arrow. When an electric field (marked by the white arrow) is applied, all conduction electrons gain additional momentum and the drift velocity defined by equation (10.2).

Ginzburg–Landau theory discussed in Section 12.6 and is indeed the characteristic feature of the Meissner effect.

12.13 Secret of zero resistance

Now a naive question is addressed as to why the resistance becomes zero in the superconducting state. Before answering this, we will briefly review the electron conduction in the normal state. We have shown in Section 10.2 that the Fermi surface moves as a whole to the direction parallel to the electric field **E**. The direction of the wave vector randomly changes on the Fermi surface, as soon as the conduction electron is scattered by imperfections like phonons and impurities. The Fermi sphere is displaced only by a certain distance in the presence of a constant field and a steady state is established within the relaxation time τ, as is illustrated in Fig. 10.1. This is Ohm's law and gives rise to a finite resistivity equal to $\rho = E/J$. Electron conduction in the normal state is illustrated schematically in Fig. 12.11.

As emphasized in the preceding section, the superconducting state is in a constrained condition such that the momentum of the paired electrons cannot be altered at will. Indeed, the energy $2\Delta(0)$ is needed to destroy the Cooper pair at absolute zero. As a consequence, the scattering which changes the direction

12.14 Magnetic flux quantization in a superconducting cylinder

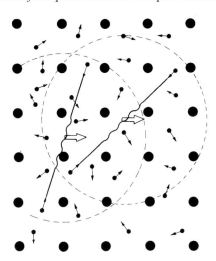

Figure 12.12. Motion of electrons in the superconducting state. When a superconductor is connected to a current source, all Cooper pairs receive the same drift velocity and, hence, the same momentum, in a direction (indicated by the white arrows) opposite to the applied field E. The effective range where the attractive interaction extends is shown by a dashed circle. This is the coherence length. Its diameter is about 100 nm for pure metals like Pb.

of the wave vector is prohibited for the paired electrons. This situation is shown schematically in Fig. 12.12. Once a current is induced, each Cooper pair acquires the same velocity vector **v** in parallel to the applied field. Thus, the drift velocity of all Cooper pairs must be **v**. Thus, we see that all the Cooper pairs acquire the same momentum and shift the Fermi surface endlessly in a direction parallel to the field. A current flowing without disturbing the ordered state is indeed a resistanceless conduction. Therefore, once a current is induced by applying the magnetic field to a superconducting ring, it persists forever as long as the Cooper pairs remain stable.

12.14 Magnetic flux quantization in a superconducting cylinder

F. London predicted that the long-range order of the momentum would result in a specific quantum effect [6]. If a Cooper pair gains a finite momentum **P** as discussed in the preceding section, its wave function is presented by

$$\Phi_\mathbf{P} = \Phi(\mathbf{r}_1,\mathbf{r}_2)e^{i\mathbf{P}\cdot\mathbf{r}/\hbar}, \qquad (12.35)$$

where $\Phi(\mathbf{r}_1,\mathbf{r}_2)$ is the wave function of the Cooper pair given by equation (12.30). Equation (12.35) obviously indicates that the pair travels with the momentum **P** without being scattered throughout the whole volume of a superconductor. This means that the phase coherence of the traveling wave is infinitely long and that the quantum effect should be manifested on a macroscopic scale. For example, we consider a superconducting cylinder and apply a magnetic field parallel to its axis. As discussed in Section 12.2, a surface current is induced such that the resultant field in the cylinder is zero, i.e., $\mathbf{B}=0$. This is the Meissner effect and the manifestation of a perfect diamagnetism. The circulating surface current induces a magnetic flux inside the cylinder. London suggested that the magnetic flux trapped in a superconducting cylinder would be quantized in the units of h/e, since an orbiting electron forms a stationary standing wave in the same manner as an electron in a free atom.

We know that the orbit of a core electron around a nucleus is quantized and takes only discrete values. This is the phenomenon in a microscopic atom. The orbit of the superconducting electrons circulating in the cylinder is really on a macroscopic scale and the phase coherence extends over centimeters. London's prediction was proved experimentally in 1961 [10]. The trapped magnetic fluxes are not only quantized but also the units are $h/2e$, which is just one-half the value predicted by London. This can be easily understood, since the superconducting electron has a charge $(-2e)$ instead of $(-e)$ as a result of the formation of the Cooper pair. This experiment demonstrated the great intuition of London and the validity of the BCS theory. In Section 12.2, the charge of the superconducting electron is intentionally expressed as q_s but now it is proved to be $q_s = -2e$.

12.15 Type-I and type-II superconductors

Upon the exposure to magnetic fields below H_c, a superconductor exhibits a perfect diamagnetism and the magnetic susceptibility is given by $\chi = \mathbf{M}/\mu_0\mathbf{H}_{ext} = -1$ ($\chi = -1/4\pi$ in CGS units). Once the magnetic field exceeds H_c, the superconducting state is transformed into the normal state, the magnetization of which becomes practically zero or, more precisely, of the order of 10^{-5} mol as listed in Table 3.2 (in CGS units). Therefore, the magnetic field dependence of the magnetization forms a triangle as shown in Fig. 12.13(a). The superconducting pure elements like Pb, Sn and Hg exhibit a magnetization curve like that shown in Fig. 12.13(a). They are called type-I superconductors.

All superconductors do not always exhibit such behavior. The magnetization curve shown by Fig. 12.13(b) is observed in type-II superconductors. There are two critical magnetic fields H_{c1} and H_{c2}, which are called the lower and upper critical fields, respectively. As will be discussed below, the magnetic fluxes begin to penetrate into the superconductor in a quantized form and the superconducting state is gradually destroyed, once the applied field exceeds H_{c1}. It is completely transformed to the normal state, only when the applied

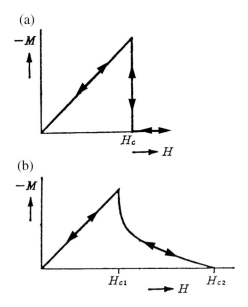

Figure 12.13. Magnetization curve for (a) ideal type-I and (b) ideal type-II superconductors. The magnetization curves shown are reversible in both cases.

field exceeds the value of H_{c2}. Both superconducting and normal states coexist in magnetic fields between H_{c1} and H_{c2}. This is called the mixed state. The resistivity remains zero up to H_{c2}.

Prior to a discussion of a type-II superconductor, it may be worthwhile discussing the intermediate state in a type-I superconductor. For example, we apply a magnetic field to a type-I superconducting sphere. The density of the magnetic lines becomes the thickest along the equator perpendicular to the magnetic field. This can be seen in Fig. 12.2(d). When the applied field reaches $\frac{2}{3}$ of the critical field H_c, the magnetic field at the equator has already reached H_c and the magnetic field begins to penetrate into the superconductor. The whole sphere becomes normal when the applied field reaches H_c. The state in the fields $\frac{2}{3}H_c \leq H \leq H_c$ is called the intermediate state and is a mixture of the superconducting and normal states.[7]

As is clear from the argument above, the boundary energy plays a crucial role in determining if the intermediate state or the mixed state is formed upon application of the magnetic field. The magnetic flux penetrates into the superconductor over the distance of the penetration depth λ to cancel the flux density

---

[7] An intermediate state is realized when the boundary energy is positive, whereas the mixed state is realized when it is negative. Normal and superconducting phases in the intermediate state are fairly large in size and can be seen with the naked eye. In the mixed state, the magnetic fluxes enter into the superconducting phase in a quantized form and are distributed over the superconducting phase with a size less than 10^{-8}m. Note that only its core is in the normal state, as is illustrated schematically in Fig. 12.14.

inside. A positive magnetic energy is built up in the superconductor and its value increases from zero to $\frac{1}{2}\mu_0 H^2$ per unit volume, as given by equation (12.15), over the distance λ in the superconductor. This positive magnetic energy competes with the negative contribution to the free energy due to electron ordering in the superconducting state. The value of the order parameter $|\psi(\mathbf{r})|^2$, which represents the density of the superconducting electrons, must be finite and uniform well inside the superconducting region. The ordering energy certainly lowers the free energy of the superconducting region relative to the normal region. However, if the order parameter changes abruptly at the boundary, the kinetic energy goes to infinity there. Therefore, the order parameter $|\psi(\mathbf{r})|^2$ has to increase gradually from zero to a finite value over the distance ξ in order to minimize an increase in the kinetic energy. The distance ξ is called the coherence length and is known to be essentially equivalent to the coherence length introduced in the BCS theory to characterize the spatial correlation of the paired electrons. The coherence length in type-I superconductors is known to be about $\xi \approx 10^{-6}$m ($=10^3$nm) but that in type-II superconductors is much shorter because of the decreasing mean free path of electrons.[8] The competing magnetic and ordering energies cancel near the boundary only if $\xi \approx \lambda$.

Ginzburg and Landau [7] showed that the boundary energy becomes positive and the intermediate state is stabilized, if the ratio $\kappa = \lambda/\xi$ is lower than $1/\sqrt{2} \cong 0.707$. For instance, the Ginzburg–Landau constant κ for pure Pb is 0.4 and satisfies this condition. On the other hand, the boundary energy becomes negative, if $\kappa > 1/\sqrt{2}$. Now the magnetic field can penetrate into a superconductor by dividing it into as many thin normal states as we wish. Abrikosov [11] later extended the Ginzburg and Landau theory and showed theoretically that, when the magnetic field exceeds H_{c1}, the magnetic fluxes begin to penetrate into the superconductor in a quantized form and the mixed state persists up to the upper critical field H_{c2}. The Abrikosov theory provided the theoretical basis for a superconductor with $\kappa > 1/\sqrt{2}$ and established a very important ground for the application-oriented research on superconductors. In summary, we see that type-II superconductors are designated as those with $\kappa > 1/\sqrt{2}$ while type-I superconductors as those with $\kappa < 1/\sqrt{2}$.

12.16 Ideal type-II superconductors

The magnetization curve observed in a real superconductor of either type-I or type-II is always more or less irreversible due to the presence of unavoidable

---

[8] The coherence length ξ_0 in an ideally pure metal is decided by its intrinsic property, while that in an impure metal or alloy decreases with a decrease in the electron mean free path Λ_e. In an impure specimen, the coherence length ξ is approximated as $\frac{1}{\xi} \cong \frac{1}{\xi_0} + \frac{1}{\Lambda_e}$. [A. B. Pippard, *Physica* **19** (1953) 765.]

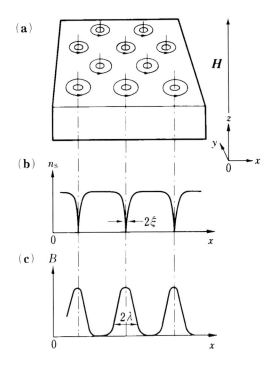

Figure 12.14. Schematic illustration of the mixed state in an ideal type-II superconductor. (a) Magnetic flux distribution penetrating into the superconductor. A circle with an arrow indicates the screening current flowing around a quantized flux line. The magnetic fluxes are distributed so as to form a close-packed regular triangular lattice. (b) Number density distribution of the superconducting electrons, where 2ξ represents the coherence length. (c) Magnetic flux density distribution, where 2λ represents the penetration depth. [From ref. 12.]

impurities and defects. However, we consider in this section the mixed state of an "ideal" type-II superconductor, which is defined as that whose magnetization curve is reversible in both increasing and decreasing magnetic fields, as illustrated in Fig. 12.13(b). The mixed state is achieved by applying a magnetic field in the range $H_{c1} \leq H \leq H_{c2}$. The resulting quantized magnetic flux distribution is illustrated schematically in Fig. 12.14(a). It is seen that the flux lines are distributed uniformly in the superconducting matrix. Its core, through which the magnetic flux penetrates, is in the normal state and the superconducting current is circulating around it. The magnetic flux lines repel each other and form a close-packed regular triangular lattice to minimize the repulsive energy. Figure 12.14(b) illustrates how the order parameter or the number density of superconducting electrons approaches zero towards the center of the magnetic flux lines in the region 2ξ, whereas (c) shows how the magnetic flux penetrates into the superconducting region within the range 2λ. As the applied magnetic field H is increased beyond H_{c1}, the density of the quantized fluxes increases in the superconductor and the average distance between neighboring fluxes is gradually reduced. At the upper critical field H_{c2}, the magnetization **M** becomes essentially zero as shown in Fig. 12.13(b), and the magnetic flux density **B** equals $\mu_0 \mathbf{H}_{c2}$. The superconductor is completely transformed into the normal state. In ideal type-II superconductors, the superconducting current is hardly conveyed in contrast to the non-ideal one containing various types of imperfections, which will be discussed in the following section.

12.17 Critical current density in type-II superconductors

Type-II superconductors, which exhibit a large hysteresis in the magnetization curve, are practically of great importance, since they are capable of carrying a large amount of superconducting current. To begin with, we discuss the two different types of currents flowing through a superconductor. One is the transport current supplied from an external source and the other is the screening current induced by applying a magnetic field to a superconductor. Hence, the total current density is given by

$$\mathbf{J} = \mathbf{J}_i + \mathbf{J}_H, \tag{12.36}$$

where $\mathbf{J}_i$ is the transport current and $\mathbf{J}_H$ is the screening current [12]. There is a maximum superconducting current density above which a finite resistivity appears. This is defined as the critical current density J_c. In a type-I superconductor, a finite resistance appears when the sum of the applied magnetic field and the magnetic field caused by the transport current exceeds the critical field H_c at any point on the surface of the superconductor. Thus, the stronger the external field, the smaller the value of the J_c is. In other words, the critical current density for type-I superconductors is simply decided by the value of the critical magnetic field.

The behavior of J_c is more complex in type-II superconductors, because the superconducting current flows not only just at the surface but also around the magnetic fluxes distributed inside the superconductor. First, we consider an ideal type-II superconductor, to which a magnetic field exceeding the lower critical field H_{c1} is applied along its z-axis, as illustrated in Fig. 12.14(a). We know that the screening current J_H flows around each quantized magnetic flux. Since the distribution of magnetic fluxes is everywhere uniform in the ideal type-II superconductor, the x-components of $\mathbf{J}_H$ circulating around the adjacent magnetic fluxes are equal in magnitude but opposite in direction to each other so that they cancel out and give rise to no net current. This is indeed an obvious

12.17 Critical current density in type-II superconductors

solution from the Maxwell equation $\text{rot}\mathbf{B} = \mu_0 \mathbf{J}_s$ ($\text{rot}\mathbf{B} = (4\pi/c)\mathbf{J}_s$ in CGS units), since $\text{rot}\mathbf{B} = 0$ holds in the case of a spatially uniform magnetic flux distribution.

In a "non-ideal" type-II superconductor, defects or non-superconducting precipitates are present as pinning centers and, because of this, the magnetization curve becomes irreversible. As emphasized in Fig. 12.14(b), the core of the magnetic flux is in the normal state. Thus, the magnetic flux tends to be trapped or pinned by these non-superconducting regions, because, otherwise, the magnetic flux has to expend some extra energy to destroy the superconducting state. Clearly, the flux pinning effect must be responsible for the manifestation of the irreversible magnetization curve.

In type-II superconductors containing pinning centers, the distribution of magnetic fluxes is no longer uniform. Consider such a type-II superconductor having an infinite xz-plane and a finite length L in the y-direction with the magnetic field applied along the z-axis. The flux distribution is not uniform, as illustrated in Fig. 12.15(a). We see clearly that the magnetic fluxes are more concentrated near the surface than at the center because of the pinning effect, giving rise to a gradient in the magnetic flux distribution along the y-direction. This is shown in Fig. 12.15(b). Obviously, the cancellation of the screening current $\mathbf{J}_H$ no longer occurs when the magnetic flux distribution has a finite gradient. Indeed, the Maxwell equation above results in a net critical current flowing along the x-direction, as shown in Fig. 12.15(c).

When the current density $\mathbf{J}$ appears in the presence of the magnetic flux density $\mathbf{B}$, the Lorentz force $\mathbf{F} = \mathbf{J} \times \mathbf{B}$ is generated. If the pinning force $\mathbf{F}_p$ per unit length of core is stronger than the Lorentz force, the pinned magnetic fluxes will not move. Thus, the superconducting current density $\mathbf{J}_s$ can flow along the x-direction in the presence of the magnetic field. However, when the Lorentz force exceeds the pinning force, the magnetic flux begins to move. A finite $\partial \mathbf{B}/\partial t$ gives rise to a voltage through equation (12.5), resulting in a finite resistance. Thus, it is critically important to introduce intentionally effective pinning centers into type-II superconductors so as to achieve as high a critical current density as possible. This is equivalent to making the hysteresis in the irreversible magnetization curve as large as possible (see Fig. 12.23).

Type-II superconductors have received much attention from the point of view of their practical applications. For example, a very high magnetic field can be produced by winding a solenoid coil with wires of a type-II superconducting material and feeding a large superconducting current through it. This is a superconducting magnet. A superconducting magnet capable of producing magnetic fields up to 18 tesla is commercially available. The superconducting properties of representative type-II superconducting materials are listed in Table 12.2.

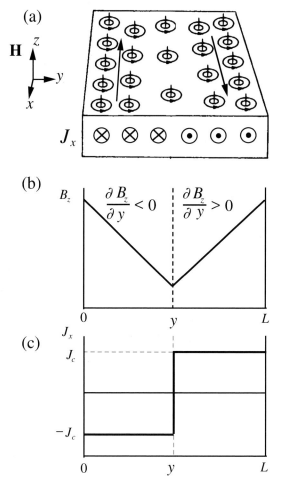

Figure 12.15. (a) Distribution of the magnetic fluxes in a type-II superconductor containing pinning centers. A magnetic field is applied parallel to the z-axis. The specimen is assumed to be infinitely long in both x- and z-directions but L in the y-direction. (b) The derivative of the z-component flux density, $\partial B_z/\partial y$, is assumed to be constant. (c) The critical current density J_c flows along the x-direction.

The Nb–Ti and Nb$_3$Sn superconductors have been widely used as superconducting wires in a superconducting magnet.[9] Figure 12.16 shows the magnetic field dependence of the critical current density J_c at 4.2 K for these

---

[9] A complete solid solution is formed in the Nb–Ti system. Superconducting Nb–Ti composite wire consists of a larger number of Nb–Ti filaments of the composition near Nb$_{40}$Ti$_{60}$ embedded in a Cu matrix. This is achieved by repeated cold elongation with subsequent annealing in the range 300–500 °C to precipitate alpha-Ti fine particles which act as pinning centers in the superconducting matrix. The superconducting characteristics in the Nb–Zr system are poorer than those in Nb–Ti alloys. Because of its brittle nature, the Nb–Zr is not commercially produced as superconducting wire.

12.17 Critical current density in type-II superconductors

Table 12.2. *Superconducting characteristics of representative type-II superconducting materials*

superconducting material	T_c (K)	H_{c2} (T) at 4.2 K	structure
V_3Ga	14.8	23.6	A15
Nb_3Sn	18.0	26.0	A15
Nb_3Al	18.7	29.5	A15
Nb_3Ga	20.2	33.0	A15
Nb_3Ge	22.5	37.0	A15
Nb–Ti[a]	10	12	bcc
$PbMo_6S_8$	15	54	hexagonal[b]

Note:
[a] See footnote 9, p. 366, concerning the Nb–Ti system.
[b] The compound crystallizes into the hexagonal structure at ambient temperatures but transforms into a triclinic structure at low temperatures. There exist a number of isomorphous compounds given by the chemical formula $M_xMo_6X_8$ ($1<x<4$) with M = Pb, Sn, In, Zn, Cd, Al, Cu, Ag or any 3d-transition metal from Cr to Ni and X = chalcogen (O, S, Se, Te) or halogen (F, Cl, Br, I). They were studied extensively by a French chemist, R. Chevrel, and christened "Chevrel compounds". Most of them undergo a transition to the superconducting state at low temperatures.

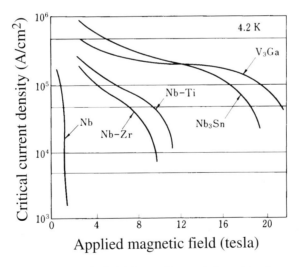

Figure 12.16. Applied magnetic field dependence of the critical current density at 4.2 K for various superconducting materials. (See footnote 9, concerning Nb–Ti and Nb–Zr.) [Courtesy of Dr Y. Tanaka]

superconducting materials. It is clear that J_c values exceeding 50 000 A/cm² have been achieved at 4.2 K in a magnetic field of 10 tesla for Nb–Ti and of 19 tesla for Nb_3Sn.

12.18 Josephson effect

We have shown in Section 12.8 that the energy gap of a superconductor can be deduced by measuring the tunneling current passing through a very thin insulating layer sandwiched between a superconductor and a normal metal. In this experiment, the tunneling current begins to flow only when the electron at the Fermi level in the normal metal can tunnel through the insulating layer. This condition is fulfilled when the applied voltage exceeds the energy gap. What happens if two superconductors are separated by a very thin insulating layer? Now we have to study the tunneling condition for the Cooper pair superconducting electrons.

A macroscopic wave function is formed in a superconductor, since a large number of Cooper pairs constitute a coherent state as described in Sections 12.12–12.14. Let us write the macroscopic wave function as $\psi(\mathbf{r}) = |\psi| e^{i\eta}$ with its phase shift η. Suppose we have two superconductors A and B. If A and B are independent, then their phase shifts η_A and η_B are also independent of each other. However, the situation changes if A and B are separated by a very thin insulating layer, the thickness of which is narrower than the coherence length of the Cooper pair electrons. Now the Cooper pair electrons in superconductor A can tunnel into superconductor B across the insulating layer. As a result of the tunneling effect, η_A and η_B become no longer independent. Josephson [13] proved that the tunneling current i_s crossing the insulating layer is given by

$$i_s = i_c \sin(\eta_B - \eta_A) = i_c \sin\Delta\eta, \qquad (12.37)$$

where i_c is the maximum superconducting tunneling current obtained when $\Delta\eta = \pi/2$. Here i_c is often called the Josephson critical current. According to equation (12.37), the phase difference $\Delta\eta$ is not uniquely determined from the measured i_s but takes either $2\pi n + \Delta\eta$ or $2\pi n + (\pi - \Delta\eta)$.

The device consisting of two superconductors separated by a very thin insulating layer is called the Josephson device and its junction the Josephson junction. For example, the Pb–PbO–Pb Josephson device is made of two Pb metal layers having $T_c = 7.2$ K and a PbO insulating layer of about 1 nm. The I–V characteristics shown in Fig. 12.17(a) can be obtained if the current I is fed to the Josephson device and the voltage across it is measured. No voltage is generated, as long as the feeding current is lower than i_c. This implies that the superconducting tunneling current flows without any voltage drop between the

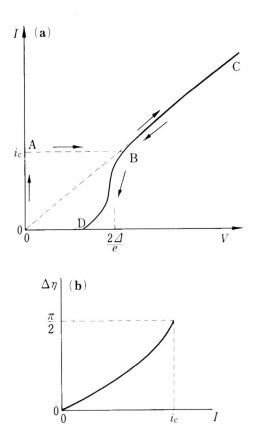

Figure 12.17. (a) I–V characteristics for the Josephson device. 0A represents the DC Josephson effect. (b) Phase difference versus DC Josephson current in region 0A in (a).

two superconductors. The I–V characteristic along the line 0A in Fig. 12.17(a) is often called the DC Josephson effect.

The value of the phase difference $\Delta\eta$ increases with increasing current I and reaches its critical value $I = i_c$ at $\Delta\eta = \pi/2$. Once the current I exceeds i_c, the voltage jumps from point A to point B in the I–V characteristics. At this instant, the device gives rise to a voltage equal to $2\Delta/e$, where 2Δ refers to the energy gap of the superconductor deduced from equation (12.33). By further increase in the current I, we obtain a more or less linearly changing I–V characteristic shown as the line BC. Hysteresis appears and follows the path C–B–D, when the current I is decreased from point C. The I–V characteristic corresponding to C–B–D is due to the tunneling of the quasiparticles produced by the destruction of the Cooper pairs and is different from the Josephson effect.

Let us discuss the AC Josephson effect which appears when a DC voltage V is applied across the junction. Under this condition, the superconducting tunneling current passing the junction changes with time and behaves as an alternating current. This implies that the phase difference $\Delta\eta$ becomes time dependent. Josephson [13] derived theoretically that the time-dependent phase difference is related to the DC voltage V through the relation:

$$2eV = \hbar \frac{d\Delta\eta}{dt}. \tag{12.38}$$

By integrating equation (12.38) with respect to time, we obtain

$$\Delta\eta = \left(\frac{2e}{\hbar}\right)Vt + (\Delta\eta)_0 = \omega t + (\Delta\eta)_0, \tag{12.39}$$

where $\omega = (2e/\hbar)V$ is the angular frequency. An insertion of this relation into equation (12.37) yields

$$i_s = i_c \sin[\omega t + (\Delta\eta)_0], \tag{12.40}$$

showing that an alternating current component appears in the Josephson current. The AC Josephson effect has been confirmed by experiments and has contributed to the development of a technique to measure a DC voltage with a very high precision and to determine the universal constant e/h very accurately.

The understanding of the DC and AC Josephson effects may be facilitated by making use of a close analogy between a Josephson junction and a pendulum [12]. Let us consider the general case in which the superconducting tunneling currents are time dependent. Then, equation (12.38) assures the presence of a DC voltage in the circuit. Because of the presence of the DC voltage across the junction, quasiparticles cross the junction through the normal tunneling process. This must be resistive and can be represented by a resistance R across the junction. In addition, the insulating junction layer would possess a capacitance C, since the two superconducting metal surfaces are parallel and in close proximity to one another. Summing up these contributions, we can draw an equivalent circuit as shown in Fig. 12.18(a). If the current I is supplied from a constant current source, then the following equation holds:

$$I = C\frac{dV}{dt} + \frac{V}{R} + i_c \sin\Delta\eta. \tag{12.41}$$

The relation $V = 0$ should hold when the current I is increased from zero to i_c. Here the first and second terms in the right-hand side are absent and only the third term remains. This corresponds to the line 0A in the I–V characteristics

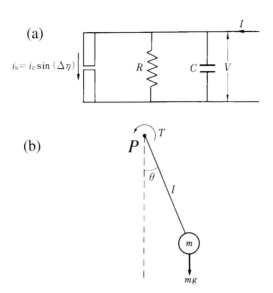

Figure 12.18. (a) Equivalent circuit to the Josephson junction and (b) the rigid pendulum.

shown in Fig. 12.17(a). Once I exceeds i_c, the DC voltage appears and equation (12.38) holds. Its insertion into equation (12.41) yields

$$I = \frac{C\hbar}{2e}\frac{d^2\Delta\eta}{dt^2} + \frac{\hbar}{2eR}\frac{d\Delta\eta}{dt} + i_c \sin\Delta\eta. \qquad (12.42)$$

We see from equation (12.42) that the total current I is now expressed in terms of the phase difference associated with the superconducting tunneling current.

Equation (12.42) represents the differential equation involving the first and second derivatives and sine function of the phase difference. One can easily show that the motion of the rigid pendulum shown in Fig. 12.18(b) can be described in terms of the same differential equation as equation (12.42). The rigid pendulum has its arm of length l with a bob of mass m at its lower end. The pendulum can rotate freely about the pivot P. Let us apply a torque T to the pendulum. The Newton equation of motion is given by

$$M\frac{d^2\theta}{dt^2} = T - mgl\sin\theta - \kappa\frac{d\theta}{dt}, \qquad (12.43)$$

where θ is the deflection angle and M is the moment of inertia of the pendulum about P. Here the first term in the right-hand side represents the external torque, the second term the contribution due to the weight of the bob and the third term that due to the viscous force proportional to angular velocity $d\theta/dt$, which arises from the movement of the pendulum in air. This equation is easily rearranged to the same form as equation (12.42):

$$T = M\frac{d^2\theta}{dt^2} + \kappa\frac{d\theta}{dt} + mgl\sin\theta. \qquad (12.44)$$

A comparison of equation (12.44) with equation (12.42) immediately leads to the following one-to-one correspondence. First, the phase difference $\Delta\eta$ plays the same role as the deflection angle θ of the pendulum. The total current I fed from the current source is equivalent to the external torque T. The capacitance C and the inverse of the resistance $1/R$ correspond to the moment of inertia M and the viscous damping κ, respectively. We also see that the superconducting tunneling current $i_s = i_c \sin\Delta\eta$ corresponds to the horizontal displacement $x = l\sin\theta$ of the bob and the voltage $V = (\hbar/2e)(d/dt)\Delta\eta$ across the junction to the angular velocity $d\theta/dt$ of the pendulum.

Obviously, an increase in the current I is equivalent to an increase in the torque T. When the torque is small, the pendulum stops at a constant deflection angle θ. Thus $d\theta/dt$ is zero. This explains why there is no voltage at the Josephson junction when a current is small. However, when the deflection angle reaches $\theta = \pi/2$ corresponding to the horizontal position of the pendulum, the restoring torque due to gravity reaches its maximum value of mgl. This certainly corresponds to the situation where the superconducting tunneling current reaches its critical value of i_c.

If we further increase the torque T, the condition $T > mgl\sin\theta$ holds so that the pendulum rises and passes through the vertical position. Once the deflection angle exceeds the vertical position, then both applied and restoring torques act in the same direction. Now the pendulum continues to rotate around its axis, so long as the torque continues to be applied. Here the angular velocity $d\theta/dt$ is no longer zero but becomes finite. We immediately see that a finite $d\theta/dt$ corresponds to the appearance of a DC voltage across the Josephson junction. Once the current I exceeds the critical value i_c, the I–V characteristic jumps from point A to point B in Fig. 12.17(a). This explains the AC Josephson effect on the basis of the analogy with the motion of a rigid pendulum.

Let us once again consider the situation where the pendulum continues to rotate. By projecting the rotational motion of the pendulum onto a horizontal plane, we can describe its motion as an oscillator with amplitude $2l$. As is clear from Fig. 12.18(b), the position of the projected bob is given by $l\sin\theta$, whereas the corresponding quantity in the Josephson junction is $i_c\sin\Delta\eta$. We see, therefore, that the oscillation of the pendulum with amplitude $2l$ is equivalent to that of the superconducting tunneling current with amplitude $2i_c$.

The angular velocity is not constant but varies during each revolution. The frequency of the rotation in the pendulum is defined as $(1/2\pi)\langle d\theta/dt\rangle$, where $\langle d\theta/dt\rangle$ is the time average of the angular velocity over one cycle. Similarly, the superconducting tunneling current oscillates across the junction with the frequency ν expressed as $\nu = (1/2\pi)\langle d(\Delta\eta)/dt\rangle$. The frequency of the superconducting tunneling current can be calculated by inserting equation (12.38) into this relation:

$$\nu = \frac{1}{2\pi}\left\langle\frac{d\Delta\eta}{dt}\right\rangle = \frac{1}{2\pi}\frac{2e}{\hbar}\langle V\rangle = \frac{2e}{h}V_{DC}, \qquad (12.45)$$

where the DC component V_{DC} denotes the time average of the voltage. Hence, we see that the frequency ν of the oscillation is related to the DC voltage across the junction.

The AC Josephson tunneling effect is observed as a ripple of frequency $(2e/h)V_{DC}$ superimposed onto the DC voltage V_{DC}. The coefficient $2e/h$ is equal to 4.836×10^{14} [hertz/volt] and, hence, a DC voltage of $1\,\mu V$ gives rise to a frequency of 483.6 MHz. We have thus established a technique of measuring the DC voltage very accurately by making use of the relation (12.45), since the frequency of the electromagnetic wave can be measured with a much higher precision than the DC voltage. The DC voltage determined by this technique has been adopted as the standard voltage. The presence of the AC component in the AC Josephson effect can also be confirmed by measuring the emission of the electromagnetic wave with the frequency ν from the junction, though its intensity is very weak.

12.19 Superconducting quantum interference device (SQUID) magnetometer

A SQUID magnetometer utilizing the Josephson effect has been developed and is widely used in many laboratories because of its high sensitivity to extremely weak magnetic fields. It consists of a superconducting ring containing one or more Josephson junctions. Here we will discuss the superconducting ring with two junctions [12], as depicted in Fig. 12.19.

A measuring current I is supplied to this circuit. Since the ring is symmetrical, it is precisely divided into two and $I/2$ flows through each junction. Now a magnetic field of gradually increasing flux density B is applied perpendicular to the plane of the ring and a circular current i is induced within the ring. Thus, the current $i+(I/2)$ flows through the right-hand side of the ring, whereas the current $i-(I/2)$ flows through the left-hand side. The phase change around any closed superconducting circuit must be equal to an integral multiple of 2π in order to assure the coherence of the Cooper pair wave function throughout the superconducting ring.

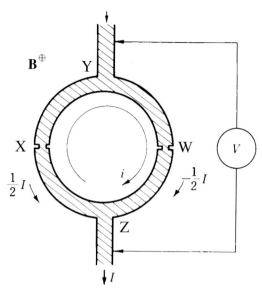

Figure 12.19. Principle of a SQUID magnetometer. X and W represent Josephson junctions. The current I is fed through the junctions and the magnetic field of flux density B is applied in a direction perpendicular to the plane of the ring. [From ref. 12.]

Let us assume the phase difference of the tunneling current passing through the Josephson junctions in the right- and left-hand sides of the ring to be α and β, respectively, and the phase difference caused by the magnetic field to be $\Delta\eta(B)$. If the ring is superconducting everywhere, we should have the relation:

$$\alpha + \beta + \Delta\eta(B) = 2\pi n, \tag{12.46}$$

where n is an integer. The phase difference due to the magnetic field can be expressed as a function of the applied field or the magnetic flux ϕ_a passing through the ring in the following form [12]:

$$\Delta\eta(B) = \left(\frac{2\pi\phi_a}{\phi_0}\right), \tag{12.47}$$

where ϕ_0 is a quantum magnetic flux equal to $\phi_0 = h/2e = 2.0678 \times 10^{-15}$ tesla m² mentioned in Section 12.14.[10]

The relation $\alpha = \beta = \pi[n - (\phi_a/\phi_0)]$ holds in equation (12.46), when the measuring current I is turned off. The parameter α is no longer equal to β, when the current I is on. Since $\alpha + \beta$ remains constant, we can write

$$\alpha = \pi\left[n - \left(\frac{\phi_a}{\phi_0}\right)\right] - \delta$$

---

[10] $\phi_0 = hc/2e = 2.0678 \times 10^{-7}$ gauss cm² in CGS units. A quantized flux unit is called the flux quantum or fluxon or fluxoid.

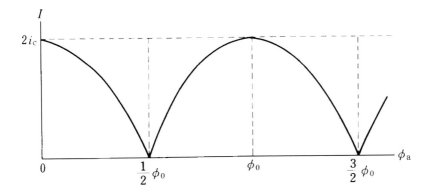

Figure 12.20. Oscillating critical measuring current in presence of applied magnetic field in SQUID magnetometer. [From ref. 12.]

and

$$\beta = \pi\left[n - \left(\frac{\phi_a}{\phi_0}\right)\right] + \delta, \quad (12.48)$$

where δ depends on the magnitude of the measuring current I. Now equation (12.37) is explicitly written as

$$i - \frac{I}{2} = i_c \sin\left[\pi\left(n - \frac{\phi_a}{\phi_0}\right) - \delta\right]$$

and

$$i + \frac{I}{2} = i_c \sin\left[\pi\left(n - \frac{\phi_a}{\phi_0}\right) + \delta\right] \quad (12.49)$$

An elimination of i from these two equations immediately leads to

$$I = 2i_c \cos\left[\pi\left(n - \frac{\phi_a}{\phi_0}\right)\right] \sin\delta. \quad (12.50)$$

Since $\sin\delta \leq 1$, we have the relation

$$I \leq 2i_c \cos\left|\pi\frac{\phi_a}{\phi_0}\right|. \quad (12.51)$$

Equation (12.51) indicates that the critical measuring current is given by $I_c = 2i_c \cos|\pi(\phi_a/\phi_0)|$, which is shown in Fig. 12.20 as a function of the applied field ϕ_a. It exhibits oscillations with a period of ϕ_0 and takes its maxima whenever the magnetic flux ϕ_a becomes a multiple of ϕ_0. Hence, the magnetic fluxes passing through the ring can be measured as multiples of ϕ_0. The measurement of such a digitized quantity can be made very accurately, particularly since the magnitude of ϕ_0 is extremely small. This is the reason why a very weak magnetic field can be measured by using a SQUID magnetometer. For example, magnetic fields as small as 10^{-10} Oe, like a brain wave, can be measured. The SQUID magnetometer is being very widely used not only in the fields of science and technology but also in medical science.

12.20 High-T_c superconductors

Until 1986, the Nb_3Ge compound with T_c equal to 23 K had been recognized as the superconductor having the highest superconducting transition temperature. Bednorz and Müller [2] observed in 1986 a sharp drop in resistivity below 35 K in a La–Ba–Cu–O compound and suggested the possibility of synthesizing a high-T_c superconducting oxide. Immediately after their report, Tanaka's group in the University of Tokyo confirmed that the $(La_{1-x}Sr_x)_2CuO_4$ compounds exhibit both a resistivity zero and the Meissner effect below 33 K and took this as evidence for the onset of superconductivity. Since then, researches to synthesize new superconducting ceramic oxides have been intensively carried out all over the world.

In 1987, Wu et al. discovered the $YBa_2Cu_3O_7$ compound, which undergoes a superconducting transition at 90 K. Its crystal structure was identified later, and is shown in Fig. 12.21. In 1988, Maeda et al. synthesized a Bi–Sr–Ca–Cu–O compound with the onset T_c value of 110 K and Sheng and Hermann a Tl–Ba–Ca–Cu-O compound with a T_c of 110 K. The highest T_c value reported so far is 164 K, which was revealed in $HgBa_2Ca_2Cu_3O_8$ under a pressure of 31 GPa. All these high-T_c superconductors possess CuO_2 planes in a layered structure, as shown in Fig. 12.21, which are now known to be responsible for conveying superconducting currents. Representative high-T_c superconductors are listed in Table 12.3. The electronic structure and electron transport properties of high-T_c cuprate superconductors have been extensively studied in the last decade in both superconducting and normal states with the aim of clarifying the mechanism of superconductivity. Its development will be briefly outlined in Chapter 14.

High-T_c superconducting materials have also received strong attention from the point of view of practical applications. Though a number of superconductors with T_c exceeding 77 K, the boiling point of liquid nitrogen, have been synthesized, it does not necessarily mean that they are immediately ready for practical use. As mentioned in Section 12.17, the development of a type-II superconductor having a high J_c value in the presence of magnetic fields is critically important. Indeed, we have shown in Fig. 12.16 that ordinary low-T_c

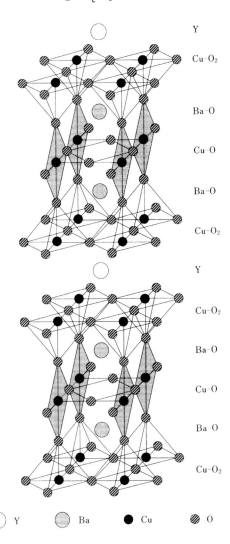

Figure 12.21. Crystal structure of $YBa_2Cu_3O_{7-\delta}$. The parameter δ ($0 \leq \delta \leq 1$) indicates the deficiency of oxygen atoms in the Cu–O chain. $YBa_2Cu_3O_7$ with $\delta = 0$ corresponds to 100 % occupation of oxygen atoms and exhibits superconductivity with $T_c = 92$ K. [See, for example, J. D. Jorgensen et al., Phys. Rev. **B 41** (1990) 1863]

superconductors like Nb–Ti and Nb_3Sn exhibit J_c values exceeding 50 000 A/cm² in a magnetic field of 10 tesla at 4.2 K and have already been employed as superconducting wires in commercially available superconducting magnets.

A large number of data has been reported concerning the critical current densities of high-T_c superconductors. For example, a Bi2223-type superconductor with the chemical formula $(BiPb)_2Sr_2Ca_2Cu_3O_x$ has been synthesized as tapes. Figure 12.22 shows the magnetic field dependence of the critical current

Table 12.3. Representative high-T_c superconductors

	T_c (K)	References
La$_{2-x}$Ba$_x$CuO$_4$	30	J. G. Bednorz and K. A. Müller, *Z. Physik* **B64** (1986) 189
YBa$_2$Cu$_3$O$_7$ (Y123)	90	M. K. Wu *et al.*, *Phys. Rev. Lett.* **58** (1987) 908
YBa$_2$Cu$_4$O$_8$ (Y124)	80	J. Karpinski *et al.*, *Nature* **336** (1988) 660
Bi$_2$Sr$_2$CaCu$_2$O$_8$ (Bi2212)	80	H. Maeda *et al.*, *Japan. J. Appl. Phys.* **27** (1988) L209
Bi$_2$Sr$_2$Ca$_2$Cu$_3$O$_{10}$ (Bi2223)	110	H. Maeda *et al.*, *Japan. J. Appl. Phys.* **27** (1988) L209
Tl$_2$Ba$_2$CaCu$_2$O$_8$ (Tl2212)	110	Z. Z. Sheng and A. M. Hermann, *Nature* **332** (1988) 138
Tl$_2$Ba$_2$Ca$_2$Cu$_3$O$_{10}$ (Tl2223)	125	M. A. Subramanian *et al.*, *Nature* **332** (1988) 631
HgBa$_2$Ca$_2$Cu$_3$O$_9$ (Hg1223)	134	A. Schilling *et al.*, *Nature* **363** (1993) 56

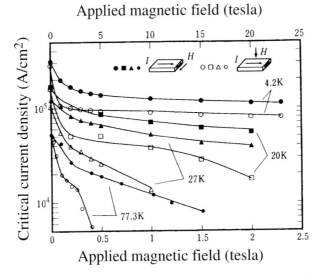

Figure 12.22. Magnetic field dependence of the critical current density of a Ag-sheathed Bi2223 superconducting tape made by packing Bi2223 powder into an Ag tube followed by plastic deformation. The bottom abscissa scale applies only to data at 77.3 K, the top to data at 4.2, 20 and 27 K. The c-axis of the Bi2223 phase is oriented preferentially perpendicular to the surface due to the mechanical deformation. [M. Ueyama et al., Adv. in Superconductivity VII, *Proc. ISS '94* (Springer, Tokyo 1995)]

density for this material reported in 1995. At 4.2 K a critical current density exeeding 10^5 A/cm² can be achieved even in the presence of a magnetic field of 20 tesla. Unfortunately, however, it drops substantially at 77 K, being 12 000 A/cm² in a magnetic field of 1 tesla and dropping to the order of only 10^3 A/cm² when a magnetic field of only 0.5 tesla is applied perpendicular to the plane in which the current flows (open circles). This clearly indicates that the pinning effect is severely weakened at 77 K in Bi2223-type superconductors. A similar situation exists in the case of Bi2212-type superconductors.

As mentioned in Section 12.16, the critical factor determining achievement of high critical current densities is whether or not effective pinning centers can be introduced in the superconducting matrix. The introduction of pinning centers effective even at 77 K is much easier in rare earth–123 type superconductors than in Bi2223-type superconductors.

The magnetization curve shown in Fig. 12.23 is for the Nd123 superconductor synthesized by the melt-processing technique [14], through which non-superconducting $Nd_4Ba_2Cu_2O_{10}$ fine particles can be homogeneously dispersed throughout the superconducting matrix. The critical current density J_c can be determined from the current above which a finite resistance appears. We call the value thus determined the transport J_c. Alternatively, the value of J_c can be

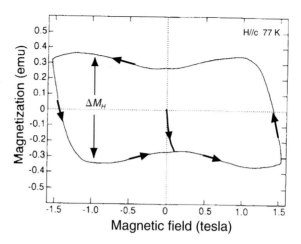

Figure 12.23. Magnetization curve at 77 K for a NdBa$_2$Cu$_3$O$_{7-\delta}$ superconducting quasi-single crystal grown by the melt-processing technique. The sample was 1.8 mm × 1.75 mm × 0.87 mm in size and the magnetic field applied parallel to the c-axis. [A. Takagi et al., Physica **C250** (1995) 222]

evaluated from the magnetization curve by applying the Bean critical-state model [15]. For the sake of simplicity, consider a rectangular specimen with the dimensions $L \times L \times 2d$ with $L \gg 2d$ and apply the magnetic field parallel to the $L \times L$ plane, as in Fig. 12.15(a). This may be approximated as an infinite plane with a thickness $2d$ with the magnetic field parallel to the plane. According to the Bean model, the critical current density J_c at a given applied field is easily calculated by inserting the measured width ΔM_H in the hysteresis curve, as marked in Fig. 12.23, into the relation:

$$J_c = \frac{\Delta M_H}{\mu_0 d} = 20 \Delta M_H / (2d), \qquad (12.52)$$

where J_c, ΔM_H and d are in the practical units of A/cm^2, emu/cm^3 and cm, respectively [15]. The critical current density J_c thus obtained for the data in Fig. 12.23 is deduced to be 45 000 A/cm^2 in a magnetic field of 1 tesla at 77 K.

Once the magnetic fluxes are pinned in a superconductor, they essentially remain there after the magnetic field is removed, so long as the specimen is in the superconducting state. As a result, the superconductor behaves as a permanent magnet. Recently, the fabrication of an extremely strong superconducting permanent magnet has been reported [16]. A c-axis oriented SmBa$_2$Cu$_3$O$_{7-\delta}$ superconductor, 36 mm in diameter, was grown by the melt-processing technique. Non-superconducting Sm$_2$BaCuO$_5$ fine particles a few μm in diameter and Ag particles 20–50 μm in size were homogeneously distributed in the superconducting 123-phase matrix. The Ag particles dispersed in the matrix

12.20 High-T_c superconductors

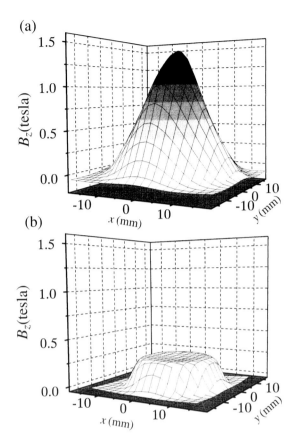

Figure 12.24. (a) Trapped flux distribution at 77 K for a $SmBa_2Cu_3O_{7-\delta}$ superconductor, 36 mm in diameter. The axial component B_z was measured by scanning the Hall sensor 1.1 mm above the surface of the sample. [From ref.16] (b) Trapped flux distribution 0.5 mm above the surface of a commercially available Nd–Fe–B permanent magnet, 22 mm in diameter. (See footnote 11, below, concerning the Nd–Fe–B system.)

enhanced the mechanical strength against fracture during magnetization and thermal cycling. The sample was then magnetized by cooling it below T_c in the presence of a magnetic field. The trapped field distribution, which was measured after the removal of the applied magnetic field, is shown in Fig. 12.24(a). It can be seen that the trapped magnetic flux density reaches 1.7 tesla at the center of the superconductor at 77 K. Its value could be increased to 9 tesla by magnetizing it at 25 K. The $Nd_2Fe_{14}B$ magnet, known as the most powerful permanent magnet available at present, possesses a magnetic flux density of only 0.25 tesla 0.5 mm above its surface, as shown in Fig. 12.24(b).[11] This clearly

---

[11] Alloys with the formulae $Nd_2Fe_{14-x}Co_xB$ and $Nd_{2-y}Dy_yFe_{14}B$ are often employed in the fabrication of Nd–Fe–B permanent magnets.

demonstrates the strength of a superconducting permanent magnet, which is known as a "Superconducting Bulk Magnet (SBM)". Its development is very promising and, hopefully SBMs will be used in various applications, such as a superconducting motor, by taking full advantage of the extremely high trapped fields.

Exercises

12.1 Show that $\nabla^2 \mathbf{J}_s = (\mu_0 n_s q_s^2/m_s)\mathbf{J}_s$ by using the London equation $\nabla^2 \mathbf{B} = (\mu_0 n_s q_s^2/m_s)\mathbf{B}$ in equation (12.12). This indicates that the screening current $\mathbf{J}_s$ obeys the same spatial distribution as the penetrated magnetic flux density.

12.2 A permanent current is induced in a superconducting ring with circumference L by approaching a permanent magnet. Apply the Sommerfeld condition (7.4) and show that the magnetic flux penetrating the ring is quantized.

Chapter Thirteen

Magnetism, electronic structure and electron transport properties in magnetic metals

13.1 Prologue

In Section 3.6, we explained the magnetism of simple metals, such as Na, Zn and Al, as arising from contributions due to both conduction electrons and ion cores. The former gives rise to Pauli paramagnetism and Landau diamagnetism whereas the latter the diamagnetism associated with the orbital motion of the core electrons. The magnetic susceptibility of these metals is essentially temperature independent and is only of the order of 10^{-5} mol, as listed in Table 3.2. This weak magnetism scarcely affects the electron transport properties. For this reason, these metals and their alloys are classified as non-magnetic metals. In this chapter, we discuss the magnetism and its effect on the electron transport of magnetic metals. Here magnetic metals include: ferromagnetic and antiferromagnetic ones possessing spontaneous magnetization below the ordering temperatures called the Curie temperature and the Néel temperature, respectively; spin-glasses; and paramagnetic ones whose magnetic susceptibility obeys the Curie–Weiss law down to low temperatures. In all cases, a magnetic moment is present and substantially affects the electron transport properties.

There are a number of excellent textbooks concerning both basic and application-oriented magnetism [1–5]. The emphasis in the present chapter is placed on the interrelationship between magnetism and the electronic structure and electron transport properties of magnetic metals.

13.2 Classification of crystalline metals in terms of magnetism

Crystalline metals can be divided into five different groups in terms of magnetism, as listed in Table 13.1. Metals in groups (I) to (IV) are characterized by

Table 13.1. Classification of crystalline metals and alloys in terms of magnetism

group	magnetic system			non-magnetic system	
	I ferromagnetism	II weak ferromagnetism	III spin-glass or magnetically dilute alloys	IV paramagnetism	V weak paramagnetism or diamagnetism
representative metals and alloys	Fe, Co, Ni	ZrZn$_2$, Sc$_3$In	Au–(Fe), Cu–(Fe)	Ti, Zr, Nb	Na, Mg, Al, Cu, Ag, Au, Zn, Pb
characteristic features of magnetism	($T_C > 300$ K) Fe 1043 K Co 1400 K Ni 631 K	($T_C \ll 300$ K) ZrZn$_2$ 21K Sc$_3$In 5.5 K	the spin freezing temperature in spin-glass or Curie–Weiss type magnetic susceptibility in dilute alloys (Kondo effect)	weak temperature dependence of magnetic susceptibility due to Pauli paramagnetism	temperature independent magnetic susceptibility
resistivity at 300 K ($\mu\Omega$-cm)	Fe 9.7 Co 6.2 Ni 6.8			Nb 14.5 Mo 5.7 Pd 10.8	Al 2.69 Cu 1.67 Na 4.6

the possession of a Fermi level either in the d band or the d states.[1] Among them, those in groups (I) to (III) are magnetic whereas those in group (IV) are non-magnetic. Metals in group (V) possess a Fermi level in the sp band and are certainly non-magnetic.[2]

Group (I) contains ferromagnetic metals and alloys with a Curie temperature well above room temperature. Here the Curie temperature T_C refers to the temperature above which ferromagnetism becomes unstable and is taken over by paramagnetism. The temperature dependence of magnetization at low temperatures in group (I) metals may be well described in terms of spin wave excitations, which will be discussed in Section 13.4. The metals Fe, Co, Ni and their alloys belong to this group. Antiferromagnetic metals and alloys with a Néel temperature T_N above 300 K are also included in group (I).

Metals and alloys in group (II) also exhibit spontaneous magnetization only at low temperatures. However, the temperature dependence of magnetization is no longer described in terms of the spin wave approximation. The $ZrZn_2$ and Sc_3In intermetallic compounds are included in this group. The details of the weak ferromagnetism will be found in the literature [6].

Metals and alloys in group (III) carry a localized moment but exhibit no spontaneous magnetization down to the lowest temperature available. Spin-glass is classified within this group. It is defined as a substance obeying the Curie–Weiss law down to the spin freezing temperature T_f, below which the randomly oriented magnetic moments are "frozen" in motion without resulting in any spontaneous magnetization. Paramagnetic metals and alloys obeying the Curie law down to the lowest temperature available are also included in group (III). Among these are dilute alloys, in which very small amounts of impurity atoms carrying finite magnetic moments are dissolved. As a typical example, we cite a Cu metal containing only a few ppm of Fe atoms. A resistivity minimum phenomenon is often observed at low temperatures in these magnetically dilute alloys and is known as the Kondo effect. Here the s–d interaction plays a critical role and has been discussed as one of the most exciting topics in the electron theory of metals in 1970s and 80s.

Metals and alloys in group (IV) carry no localized magnetic moments and are non-magnetic, though the Fermi level is situated in the middle of the d band. Hence, a relatively large Pauli paramagnetism is observed. The magnetic susceptibility shows only a weak temperature dependence of the Pauli

---

[1] There are other series of magnetic metals containing the rare-earth and actinide elements, in which 4f and 5f electrons form an incomplete shell, respectively, and are responsible for yielding the magnetic moment.
[2] Note that the p-like states always mix with the s-like states in a solid, even though only s electrons exist as outermost electrons in a free atom. In particular, electronic states near the Fermi level are dominated by the p-like states in typical mono- and divalent metals like Na, K, Mg, Zn, etc. Hence, the valence electrons in a solid are simply referred to as sp electrons.

paramagnetism (see Exercise 3.1). Typical transition metals like Ti, V, Zr, Nb, Mo, Pd, Pt and their alloys are included in this group. Metals and alloys possessing a Fermi level in the sp band are classified within group (V). All non-magnetic metals like Na, Cu, Ag, Au, Mg, Zn, Al, Pb and their alloys belong to this group. The electronic properties of metals and alloys in group (V) have already been discussed in previous chapters.

As will be discussed in Chapter 15, amorphous alloys are also classified into five groups in the same way as in Table 13.1: groups (I) to (IV) include amorphous alloys possessing a Fermi level in the d band, whereas group (V) include those in the sp band. In the former, both sp and d electrons coexist at the Fermi level and, in principle, may equally contribute to the electron conduction. This is indeed true for high-resistivity amorphous alloys, where the mean free path of the sp-electrons is shortened and is comparable to an average atomic distance (see Section 15.5). In the case of crystals, however, the Bloch theorem holds and, hence, the mean free path of the sp electrons is much longer than an average atomic distance so that only sp electrons are responsible for the electrical conduction. Instead, the d electrons at the Fermi level in group (I) to (IV) crystal metals are assumed to be immobile.

13.3 Orbital and spin angular momenta of a free atom and of atoms in a solid

Our aim in this section is to study the origin of the magnetic moment of the free atom or ion and those in a solid. It is well known that the 3d-transition metals Fe, Co and Ni are ferromagnetic at room temperature. The electronic configurations in the corresponding free atom are composed of $(1s)^2$, $(2s)^2$, $(2p)^6$, $(3s)^2$, $(3p)^6$ core electrons plus the $(3d)^n$ and $(4s)^2$ outer electrons, where the integer n is 6, 7 and 8 for Fe, Co and Ni, respectively. Since the Ar core consisting of 18 inner electrons forms a closed shell and each orbital is shared by equal numbers of spin-up and spin-down electrons, its total orbital and spin angular momenta are reduced to zero. Thus, the Ar core bears no magnetic moment and can be ignored in the rest of the discussion.

The electron configurations outside the Ar core determine the magnetic structure in a free atom. The 3d orbitals can accommodate a total of 10 electrons per atom. However, the energy difference between the 4s and 3d orbitals is so small that the 4s orbitals are often occupied prior to the occupation of the 3d orbitals, and the 3d orbitals form an incomplete shell. As noted in footnote 1, p. 385, 4f and 5f orbitals also form an incomplete shell in the rare-earth and actinide metals and their alloys, respectively. In the present section, we discuss exclusively the magnetism involving either 3d or 4f electrons.

13.3 Orbital and spin angular momenta: free atom and atoms in a solid

When more than two electrons occupy the incomplete shell, the quantum number ℓ of the orbital angular momentum and its z-component m_ℓ of each electron no longer act as good quantum numbers owing to the presence of the electron–electron Coulomb interaction.[3] It is necessary to consider how the orbital and spin momenta of the 3d or 4f electrons are combined to form the momentum of the atom. The orbital angular momentum $\mathbf{l}_i$ of the i-th electron in an incomplete shell is added vectorially to form a resultant orbital momentum $\mathbf{L}$ for the atom and the spin angular momentum $\mathbf{s}_i$ is added likewise to form a resultant spin angular momentum $\mathbf{S}$.

The quantum number L of the resultant angular momentum $\mathbf{L}$ for an atom having two electrons specified by orbital angular momenta $\mathbf{l}_1$ and $\mathbf{l}_2$ is allowed to take the values:

$$L = (\ell_1 + \ell_2), (\ell_1 + \ell_2 - 1), \ldots, |\ell_1 - \ell_2| \tag{13.1}$$

where ℓ_1 and ℓ_2 are their respective quantum numbers. The same rule is applied to S. Both L and S and their z-components M_L and M_S are conserved and employed as good quantum numbers in a free atom or ion.[4] A magnetic moment is associated with finite angular momenta $\mathbf{L}$ and $\mathbf{S}$.[5]

Let us consider the spin configuration of the V^{+3} free ion having two 3d electrons outside the Ar core. These two 3d electrons possess the same quantum numbers $\ell_1 = \ell_2 = 2$. Hence, allowed values of L for the V^{+3} ion are 4, 3, 2, 1 and 0 whereas those of S are 1 and 0. Among various combinations of L and S, the electronic states incompatible with the Pauli exclusion principle must be excluded. For example, $L = 4$ can take nine different M_L values in the range 4 to -4 and the $(L=4, M_L=4)$ state arises when $m_{\ell_1} = m_{\ell_2} = 2$. Similarly, $S = 1$ can take three different M_S values equal to 1, 0 and -1 and the $(S=1, M_S=1)$ state arises when $m_{s_1} = m_{s_2} = \frac{1}{2}$. Thus, we see that the two 3d electrons occupy the same electronic state $\ell = 2$, $m_\ell = 2$ and $m_s = \frac{1}{2}$ when the quantum number of the ion is assigned to $L = 4$ and $S = 1$. This is in conflict with the Pauli exclusion principle and must be excluded. Further simple manipulations lead to the following five states being allowed: $(L=4, S=0)$, $(L=3, S=1)$, $(L=2, S=0)$,

---

[3] The quantum number of the orbital angular momentum ℓ is alternatively called the azimuthal quantum number and allowed to take the values $\ell = 0, 1, 2, \cdots (n-1)$, where n is the principal quantum number. The magnetic quantum number m_ℓ determines the component of the orbital angular momentum in the direction of the applied field and takes $(2\ell+1)$ values given by $m_\ell = \ell, (\ell-1), \cdots, -(\ell-1), -\ell$. The spin quantum number s represents an intrinsic angular momentum about an internal axis and its z-component m_s can take either 1/2 or $-1/2$. The spin angular momentum and its z-components are $\hbar/2$ and $\pm \hbar/2$, respectively.
[4] The eigenvalue of the square of the orbital angular momentum $\mathbf{L}^2$ for a free atom is given by $L(L+1)\hbar^2$ or $|\mathbf{L}| = \sqrt{L(L+1)}\hbar$. Similarly, we have $|\mathbf{S}| = \sqrt{S(S+1)}\hbar$.
[5] The magnetic moment $\mathbf{m}$ is related to both orbital and spin angular momenta through the relation $|\mathbf{m}| = g\mu_B \sqrt{J(J+1)}$, where g is unity for $J = L$ and 2 for $J = S$. The parameter g is called the Landé g-factor and is given by $g = (3/2) + \{S(S+1) - L(L+1)\}/\{2J(J+1)\}$ [refs. 1, 2].

($L=1$, $S=1$) and ($L=0$, $S=0$).[6] Among them, we must know which spin configuration results in the ground state with the lowest energy.

The spin configuration of the free atom or ion in its ground state is determined by the Hund rule, which is stated as a combination of the following rules:

1. The quantum number S of the resultant spin angular momentum is decided by maximizing $\sum_i m_{s_i}$ consistent with the Pauli exclusion principle (see Section 14.3).

2. The quantum number L of the resultant orbital angular momentum is decided by maximizing $\sum_i m_{\ell_i}$ consistent with rule (1).

3. J for a shell less than half occupied is given by $J=|L-S|$, while that for a shell more than half occupied is given by $J=L+S$.

When more than two electrons enter the 3d or 4f orbitals, their spins are aligned in parallel to each other to maximize the resultant spin angular momentum S. Among those satisfying rule (1), the spin configuration with the maximum resultant orbital angular momentum minimizes the total energy of the free atom or ion. The application of the Hund rule to the V^{3+} free ion immediately leads to the combination of $S=\frac{1}{2}+\frac{1}{2}=1$ and $L=2+1=3$ or $^3F_2$ in its ground state. The ground state of a free atom or ion may be conveniently assigned by positioning electrons in a matrix of column m_s and row m_ℓ so as to be consistent with the Hund rule above. As an example, the determination of the ground state of the Dy^{3+} ion possessing $(4f)^9$ electrons in its 4f shell is illustrated in Fig. 13.1.

We have shown above that the spin configuration of the free atom can be specified in terms of a combination of the resultant orbital and spin angular momentum quantum numbers L and S. The 3d orbitals in the free atom are degenerate corresponding to the different z-components M_L of the quantum number L. Because of this, the expectation value of the orbital angular momentum remains finite (see Exercise 13.1). Note that both **L** and **S** in the free atom or ion are not separately observed but what we observe is the total angular momentum **J** given by their vector sum $\mathbf{L} \pm \mathbf{S}$.

Let us now consider 3d-transition metal ions in a solid. The degeneracy of the 3d orbitals is lifted owing to exposure to the crystalline electric field induced by surrounding ions, and the expectation value of the orbital angular momentum **L** for any non-degenerate state vanishes [1–3] (see Exercise 13.1). This is

---

[6] The spin configuration of an atom is often denoted as $^{(2S+1)}L_J$ following the Russel–Saunders nomenclature. Capital letters S,P,D,F,G,H, ⋯ are used in place of $L=0, 1, 2, 3, 4, 5, \cdots$, respectively. For example, ($L=3$, $S=1$ and $J=2$) is expressed as $^3F_2$.

13.3 Orbital and spin angular momenta: free atom and atoms in a solid 389

z-component of orbital angular momentum	spin-up ($\uparrow$) $m_s = \frac{1}{2}$	spin-down ($\downarrow$) $m_s = -\frac{1}{2}$
$m_\ell = 3$	①	⑧
2	②	⑨
1	③	
0	④	
−1	⑤	
−2	⑥	
−3	⑦	

Figure 13.1. Spin configuration of $(4f)^9$ electrons in an incomplete shell. The nine 4f electrons in the Dy^{3+} ion are filled in the shell so as to be consistent with the Hund rule. Its ground state is $L = 5$, $S = 5/2$ and $J = 15/2$ otherwise expressed as $^6H_{15/2}$.

known as quenching of the 3d orbital angular momentum. On the other hand, the spin state does not depend on the orbital motion of the 3d electrons and, hence, is not influenced by the crystalline electric field. Thus, the magnetic moment associated with 3d-transition metal ions arises almost exclusively from their resultant spin angular momentum **S**. The L, S and J values of the free atom in the 3d-transition metal series consistent with the Hund rule are shown in Fig. 13.2(a), and the observed magnetic moment for the 3d-transition metal ions in a solid is shown in Fig. 13.2(b). It is clear that the measured moment arises essentially from the spin angular momentum **S**.

The rare-earth element generally exists as a trivalent ion in metals and compounds by releasing the outermost $(5d)^1$ and $(6s)^2$ electrons as conduction electrons. It is also important to note that the orbitals of 4f electrons are closer to the nucleus than those of the $(5s)^2$ and $(5p)^6$ electrons, which form a closed shell so that the 4f electrons are almost completely screened from the crystalline electric field in a solid. Therefore, the orbital angular momentum in a free atom remains conserved in a solid [1–5]. This is the reason why the magnetic moment associated with the total angular momentum **J** is observed in the rare-earth metals, as shown in Fig. 13.3.

In the following sections, we discuss ferromagnetism in the 3d-transition metals. It is noted that the 3d electrons in a metal are not moving as freely as the sp electrons but are not so tightly bound to each atom as core electrons.

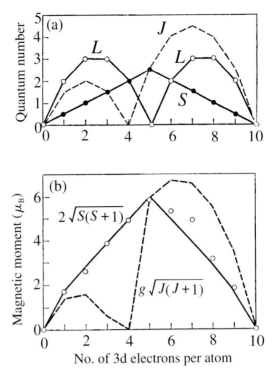

Figure 13.2. (a) L, S and J of free atom in the 3d-transition metal series. (b) Observed magnetic moment (open circles) together with the calculated values of $2\sqrt{S(S+1)}$ and $g\sqrt{J(J+1)}$ of the 3d-transition metal ion in a solid.

This intermediate character of the 3d electrons cannot be rigorously treated in the electron theory of metals. The magnetism associated with the 3d electrons has been discussed in terms of the two different approaches: the localized electron model versus the itinerant electron model.

13.4 Localized electron model and spin wave theory

In Section 3.6, we derived the Curie law by assuming the conduction electron as a classical particle carrying a magnetic moment. As a matter of fact, Langevin assumed each atom in a metal to bear a localized magnetic moment and derived the Curie law by calculating the distribution of the component of the localized moment parallel to the external field in the thermal equilibrium with its surroundings at temperature T. In 1907, Weiss proposed a new theory concerning the origin of ferromagnetism by introducing the concept of an internal magnetic field into the Langevin theory [1, 2]. Here the internal magnetic field is assumed to be proportional to the spontaneous magnetization M.

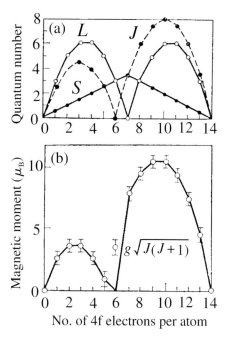

Figure 13.3. (a) L, S and J of free atom in the rare-earth series. $J = L - S$ holds for less than half-filled shells and otherwise $J = L + S$. (b) Observed magnetic moment (open circles) and the calculated values of $g\sqrt{J(J+1)}$ of the rare-earth ions in a solid.

He obtained not only the condition for the onset of ferromagnetism but also the Curie–Weiss law, which describes the temperature dependence of the magnetic susceptibility in the paramagnetic state above the Curie temperature.

The energy of the atomic spins aligned by the internal field can be equated with the thermal energy at the Curie temperature where the ferromagnetism collapses. The internal field H_i, therefore, satisfies the relation

$$n\mu_B H_i \cong k_B T_C, \quad (13.2)$$

where T_C is the Curie temperature and μ_B is the Bohr magneton (see Section 3.6). Here each atom is assumed to possess the magnetic moment $n\mu_B$. As listed in Table 13.2, we have $n = 2.2$ and $T_C = 1043\,\text{K}$ for pure Fe metal. The internal field turns out to be an extremely large value of $H_i = 7 \times 10^6$ Oe or 700 tesla for Fe. Since a static magnetic field available in a laboratory is at most 2×10^5 Oe or 20 tesla, we realize how strong is the internal field. The origin of such an extremely large magnetic field cannot be explained within the framework of classical theory.

The very strong internal field originates from the exchange interaction of the 3d electrons. Let us first study the exchange interaction of two electrons in the

Table 13.2. *Magnetic properties of several ferromagnetic metals and intermetallic compounds*

substance	spontaneous magnetization extrapolated to 0 K		magnetic moment extrapolated to 0 K $(\mu_B)^a$	Curie temperature (K)
	(gauss/cm^3)	(Wb/m^2)		
Fe	1744	0.175	2.216	1043
Co	1435	0.145	1.72	1400
Ni	512	0.051	0.616	631
Gd	1980	0.198	7.0	293
Dy	3030	0.292	10.2	85
MnBi	675	0.068	3.52	630
Cu$_2$MnAl	580	0.058	4.0	603

Note:
$\mathbf{B} = \mu_0 \mathbf{H} + \mathbf{M}$ (SI) and $\mathbf{B} = \mathbf{H} + 4\pi\mathbf{M}$ (CGS)
^{a} See footnote 6 on p. 46 in Chapter 3.

hydrogen molecule. According to the Pauli principle, the total wave function given by the product of the orbital and spin wave functions must be antisymmetric. The orbital wave function in a hydrogen molecule is antisymmetric if the spin configuration forms the triplet state or $S = 1$.[7] Conversely, it is symmetric if the singlet state or $S = 0$ is formed. We can calculate the expectation value for the ground state of a hydrogen molecule by using symmetric and antisymmetric orbital wave functions.

The difference in the ground-state energy is explicitly written as

$$J = \left(\frac{\varepsilon_s - \varepsilon_t}{2}\right)$$

$$= \iint \phi_1^*(\mathbf{r}_1)\phi_2^*(\mathbf{r}_2)\left(\frac{e^2}{|\mathbf{r}_1 - \mathbf{r}_2|} + \frac{e^2}{|\mathbf{R}_1 - \mathbf{R}_2|} - \frac{e^2}{|\mathbf{r}_1 - \mathbf{R}_2|} - \frac{e^2}{|\mathbf{r}_2 - \mathbf{R}_1|}\right)\phi_1(\mathbf{r}_2)\phi_2(\mathbf{r}_1)d\mathbf{r}_1 d\mathbf{r}_2,$$

(13.3)

where ε_s and ε_t are the ground-state energies in the singlet and triplet states, respectively (see Section 14.3). The energy difference given by equation (13.3) is called the exchange integral J, since the integrand involves the various Coulomb energy terms multiplied by the orbital wave functions $\phi_i(\mathbf{r}_1)$ and $\phi_i(\mathbf{r}_2)$ with their coordinates exchanged. As is clear from equation (13.3), the

---

[7] There are four possible spin configurations for the two electrons in a hydrogen molecule. The triplet state refers to (↑↑), (↑↓+↓↑) and (↓↓) configurations or $S = 1$ with $M_s = 1, 0$ and -1. All of them are symmetric with respect to an interchange of spin variables. The singlet state refers to (↑↓−↓↑) or $S = 0$ with $M_s = 0$. It is antisymmetric with respect to spin variables (see Sections 12.11 and 14.3).

13.4 Localized electron model and spin wave theory

exchange integral J can be equivalently viewed as arising from the difference in alignments of spins, i.e., the triplet or singlet states, though it originates from the electrostatic Coulomb energy.

As is clear from the argument above, the Hamiltonian for the exchange energy between the two atom spins located at the lattice sites i and j can be formulated as the scalar product of the spin operators $\mathbf{S}_i$ and $\mathbf{S}_j$:

$$H^{\mathrm{spin}} = -\sum_i \sum_{j \neq i} J_{ij} \mathbf{S}_i \cdot \mathbf{S}_j, \tag{13.4}$$

where J_{ij} is an exchange integral given by equation (13.3). This is called the Heisenberg model. The exchange integral is related to the overlap of the charge distributions of the atoms i and j. Hence, it is often a good approximation to consider the exchange interaction of a given atom only with its nearest neighbor atoms and write J_{ij} simply as J in the right-hand side of equation (13.4). When $J > 0$, the lowest ground-state energy is achieved by making the two atom spins $\mathbf{S}_i$ and $\mathbf{S}_j$ parallel to each other. This leads to the ferromagnetic coupling of the two spins and the triplet state is formed. On the other hand, when $J < 0$, the two spins are antiferromagnetically coupled and the singlet state is formed.

At absolute zero, all spins are oriented in one direction in a ferromagnetic metal.[8] As discussed in Section 3.6, the magnetization is defined as the vector sum of the magnetic moments per unit volume. The value at a maximum, where all relevant magnetic moments are aligned in one direction, is termed saturation magnetization. For the moment, we treat the spin operator of the j-th atom given by equation (13.4) as a classical spin angular momentum $\mathbf{S}_j$ and assume N identical spins to be equally spaced on a circle. The total exchange energy in this system is written as

$$U = -2J \sum_{j=1}^{N} \mathbf{S}_j \cdot \mathbf{S}_{j+1}. \tag{13.5}$$

Its value is obviously equal to $U_0 = -2NJS^2$, since $\mathbf{S}_j \cdot \mathbf{S}_{j+1} = S^2$ holds. Let us consider the first excited state by reversing one spin in this system. The energy of the system is increased to $U_1 = U_0 + 8JS^2$.[9] This is, however, energetically unfavorable. Instead, we can construct an energetically more favorable excited state with

---

[8] A ferromagnet is composed of a number of regions called magnetic domains, inside which the magnetic moments are fully aligned in one direction due to the exchange interaction. The directions of magnetizations in different domains need not be parallel to each other. Hence, approximately zero resultant magnetization, the state of which is called "demagnetized", can be realized, for example, by heating a ferromagnet above the Curie temperature with subsequent cooling to room temperature in the absence of a magnetic field. The size of a magnetic domain is widely ranged over 10–1000 μm and can be larger than or smaller than a crystal grain. See more details in references [1, 2].

[9] Suppose the j-th spin to be reversed in equation (13.5). The total energy is expressed as $U = -2J\{\cdots + \mathbf{S}_{j-1} \cdot \mathbf{S}_j + \mathbf{S}_j \cdot \mathbf{S}_{j+1} + \cdots\} = -2J\{(N-2)S^2\} - 2J\{\mathbf{S}_{j-1} \cdot \mathbf{S}_j + \mathbf{S}_j \cdot \mathbf{S}_{j+1}\}$. The relation $-2J(\mathbf{S}_{j-1} \cdot \mathbf{S}_j + \mathbf{S}_j \cdot \mathbf{S}_{j+1}) = -(2J) \times (-2S^2)$ holds, since the spin at the j-th site is antiferromagnetically coupled with its neighboring spins. This leads to $U = U_0 + 8JS^2$.

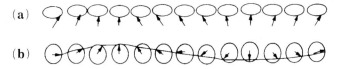

Figure 13.4. Excitations of spin waves viewed from (a) side and (b) top.

one spin effectively reversed. This is realized by sharing the reversal of one spin by all N spins in the lattice, as illustrated in Fig. 13.4. This is called a spin wave excitation. Magnons are created as quasiparticles by quantization of the spin waves. The spin wave theory was initially proposed by Bloch in 1930 analogous to the concept of phonons for the collective motion of lattice waves. The lattice wave describes oscillations in the relative positions of atoms on a lattice whereas the spin wave describes those in the relative orientations of atom spins on a lattice.

The dispersion relation of the spin waves for a ferromagnetic simple cubic metal with lattice constant a is given by

$$\hbar\omega = 2JS\left[z - \sum \cos(\mathbf{k}\cdot\boldsymbol{\delta})\right], \tag{13.6}$$

where z is the number of the nearest neighbor atoms, $\boldsymbol{\delta}$ is the vector pointing to the nearest neighbor atom with $|\boldsymbol{\delta}| = a$, ω is the angular frequency and $\mathbf{k}$ is the wave vector of the spin waves [1–5] (see Exercise 13.2). The summation in equation (13.6) is taken over the z nearest neighbor atoms.

The distribution function of magnons at a finite temperature T is given by the Planck distribution function in the same way as that of phonons discussed in Section 4.7:

$$n(\omega, T) = \frac{1}{\exp(\hbar\omega/k_\mathrm{B}T) - 1}. \tag{13.7}$$

The magnetization at temperature T is easily calculated from equations (13.6) and (13.7):

$$\frac{M(T)}{M(0)} = \frac{\int_0^\infty D(\omega)n(\omega,T)d\omega}{NS} = 1 - \frac{0.0587}{SQ}\left(\frac{k_\mathrm{B}T}{2JS}\right)^{3/2}, \tag{13.8}$$

where $D(\omega)$ is the magnon density of states and Q is the number of atoms in a unit cell, which is equal to 1, 2 and 4 for a simple cubic, body-centered cubic and face-centered cubic lattice, respectively. Equation (13.8) is called the Bloch $T^{3/2}$-law and can describe well the temperature dependence of magnetization of ferromagnetic metals and alloys in group (I) at low temperatures well below the Curie temperature T_C.

13.5 Itinerant electron model

The metals Fe, Co and Ni are typical of ferromagnetic metals, since their atom spins are ferromagnetically coupled with the exchange integral $J>0$. If all 3d electrons are localized at each atom, the magnetic moment per atom must be an integer multiple of the Bohr magneton μ_B. However, as listed in Table 13.2, the observed magnetic moments of Fe, Co and Ni metals are non-integers and are equal to $2.2\mu_B$, $1.7\mu_B$ and $0.6\mu_B$, respectively. This means that the 3d electrons are not completely localized at each atom.

The itinerant electron model assumes that the 3d electron propagates in the lattice as a Bloch wave. Let us assume the numbers of spin-up and spin-down electrons per unit volume to be $N\uparrow$ and $N\downarrow$, respectively. The total number of electrons N per unit volume and the quantity M proportional to the magnetization, are given by

$$N = N\uparrow + N\downarrow \tag{13.9a}$$

and

$$M = N\downarrow - N\uparrow. \tag{13.9b}$$

By using the parameters N and M, we can write the exchange energy as

$$\varepsilon_X = JN\downarrow \cdot N\uparrow = \frac{1}{4}J(N^2 - M^2). \tag{13.10}$$

Stoner [7] employed the molecular field approximation to treat the exchange interaction between the 3d Bloch electrons and assumed the molecular field or exchange integral J in equation (13.10) to be independent of the wave vector of the Bloch wave. According to equation (13.10), the exchange energy can be lowered by generating a finite magnetization. The magnetization arises by shifting the spin-up band relative to the spin-down band, as shown in Fig. 13.5. Electrons must be transferred from the spin-down band to the spin-up band so as to coincide the Fermi level between the two sub-bands. The transfer of electrons results in an increase in the kinetic energy.

Let us assume that the exchange interaction causes a shift of the spin-up band relative to the spin-down band by the amount equal to 2Δ. An increase in kinetic energy is then approximated as

$$\Delta\varepsilon_{kin} \cong N(\varepsilon_F)\Delta^2$$

$$= \left[\frac{1}{4N(\varepsilon_F)}\right]M^2, \tag{13.11}$$

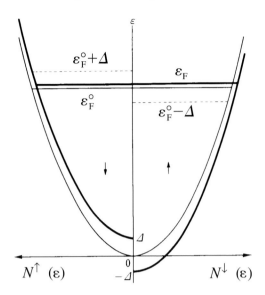

Figure 13.5. Splitting of spin-up and spin-down sub-bands due to the exchange energy. The Fermi level ε_F^o without the exchange energy is shown by the thin line. The Fermi level is displaced to $\varepsilon_F^o + \Delta$ and $\varepsilon_F^o - \Delta$ in the respective bands due to the exchange energy. A new Fermi energy ε_F is formed after charge transfer.

where $N(\varepsilon_F)$ is the density of states of 3d electrons at the Fermi level (see Exercise 13.3).[10] The total energy change associated with the generation of magnetization is given by the sum of equations (13.10) and (13.11):

$$\varepsilon = \frac{1}{4}\left[\frac{1}{N(\varepsilon_F)} - J\right]M^2 + \cdots. \tag{13.12}$$

The energy is lowered if the coefficient of M^2 is negative. In other words, the ferromagnetic state is stabilized if the relation holds:

$$JN(\varepsilon_F) > 1. \tag{13.13}$$

Equation (13.13) is known as the Stoner condition for stabilizing the ferromagnetic state in the itinerant electron model. It is clear from equation (13.13) that the ferromagnetic state is certainly favored when the exchange integral J is large but that it is also favored if the density of states at the Fermi level is high. The reason for the need of a high density of states at the Fermi level is to

---

[10] The magnetization induced by the applied magnetic field in a non-magnetic metal was given by equation (3.41). In a ferromagnet, spontaneous magnetization appears at absolute zero. We obtain from Fig. 13.5 the following relation:

$$M = \int_0^{\varepsilon_F} [N(\varepsilon+\Delta) - N(\varepsilon-\Delta)]d\varepsilon \cong 2\Delta \int_0^{\varepsilon_F} \left[\frac{dN(\varepsilon)}{d\varepsilon}\right]d\varepsilon = 2\Delta \cdot N(\varepsilon_F).$$

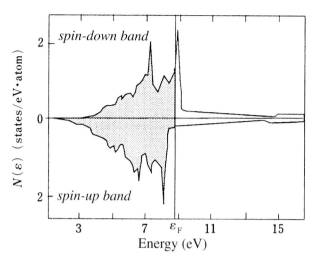

Figure 13.6. Valence band in the ferromagnetic state of pure Ni. The spin-up band is shifted to lower binding energies relative to the spin-down band due to the exchange energy. [D.A. Papaconstantopoulos, *Handbook of the Band Structure of Elemental Solids* (Plenum Press, 1986)]

minimize an increase in the kinetic energy upon the transfer of electrons from the spin-down band to the spin-up band.

Figure 13.6 shows the calculated valence band structure for the spin-up and spin-down electrons in ferromagnetic Ni metal. It can be seen that the spin-up band is shifted to higher binding energies relative to the spin-down band because of the exchange energy. The Fermi level is determined by fitting a total of 10 electrons per atom into the hybridized 3d and 4s bands. It is clear that the 3d spin-up band is fully filled by electrons while the Fermi level falls in the middle of the 3d spin-down band, leaving holes in this sub-band. The number of holes per atom is calculated to be 0.6. This is obviously proportional to the magnetization of pure Ni metal and is consistent with the observed non-integer value listed in Table 13.2.

Figure 13.7 shows the solute concentration dependence of the saturation magnetic moment for various Ni-based fcc alloys. It is clear that magnetization disappears at 60 at.%Cu or $x=0.6$ in the $Ni_{1-x}Cu_x$ alloys and that the concentration at which magnetization vanishes decreases with increasing valency of the solute atom. This is easily understood from the electronic structure of pure Ni discussed above. In the case of the $Ni_{1-x}Cu_x$ alloys, the Cu atom has a total of 11 electrons whereas the Ni atom has 10 electrons outside the Ar core electrons. Since an average electron concentration for the $Ni_{1-x}Cu_x$ alloy is given by $11x+10(1-x)=10+x$, the Ni 3d spin-down band possessing 0.6 holes per atom will be filled when $x=0.6$ or 60 at.%Cu. If the Zn atom is dissolved into Ni metal, an average electron concentration for the $Ni_{1-x}Zn_x$ alloy is given by

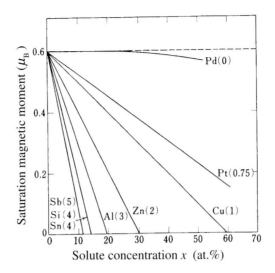

Figure 13.7. Solute concentration dependence of saturation magnetic moment in $Ni_{100-x}M_x$ alloys. The number in parentheses indicates the number of valence electrons per atom. [A. H. Morrish, *The Physical Principles of Magnetism* John Wiley & Sons, Inc., 1965]

$12x + 10(1 - x) = 10 + 2x$. Hence, the Ni 3d spin-down band will be filled at $x = 0.3$. Similarly, one can easily confirm that the addition of trivalent Al and quadravalent Sn atoms to Ni metal would fill the Ni sub-band when their concentrations reach 20 and 15 at.%, respectively. This is clearly seen in Fig. 13.7.

Figure 13.8 shows the observed saturation magnetization of various 3d-transition metal alloys as a function of the number of electrons per atom e/a outside the Ar core electrons. This is known as the Slater–Pauling curve. As discussed above, magnetization disappears at an average electron concentration of 10.6 when the Ni metal is alloyed with an element having more electrons than Ni. When Ni is alloyed with an element having less electrons than Ni, the magnetic moment increases by $1\mu_B$ as the number of electrons per atom decreases by unity. In other words, the data obtained for the fcc Fe–Ni, Fe–Co, Ni–Co and Ni–Cu alloys fall on a straight line with a slope of -1 which intercepts the horizontal axis at $e/a = 10.6$. A maximum magnetization of $2.5\mu_B$ is obtained at 70 at.%Fe–Co. The Slater–Pauling curve corresponding to the straight line with a slope of -1 on the right-hand side is readily explained within the framework of the itinerant electron model.

The magnetization always decreases when the early transition metal elements like V, Cr and Mn are alloyed with Fe, Co and Ni metals. The data fall on more or less straight lines with a slope of $+1$, each starting from pure Fe, Co and Ni. The nuclear charge for an early transition metal ion is smaller than those of Fe, Co and Ni because of their location to the left in the periodic table. This

13.5 Itinerant electron model

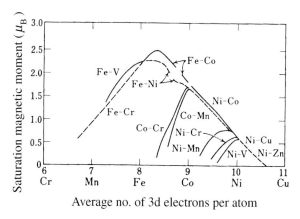

Figure 13.8. The Slater–Pauling curve representing the dependence of the saturation magnetization on the number of 3d electrons per atom in 3d-transition metal alloys.

indicates that the ionic potential fastening the 3d electrons to the nucleus is weaker so that the 3d electrons around an early transition metal atom would be more easily released and occupy the 3d orbitals of the neighboring late transition metal atoms like Fe, Co and Ni. Friedel explained in this way why the addition of the early transition metals to Fe, Co and Ni always decreases magnetization. Indeed, neutron diffraction experiments reveal that the early transition metal atoms like Cr and V carry no magnetic moment when dissolved in Fe, Co and Ni metals. The interpretation due to Friedel takes into account the localized nature of the magnetic moment while relying on the itinerant electron model.

The itinerant electron model was successful in explaining not only the fact that the magnetic moment in 3d-transition metals and alloys is in non-integer multiples of the Bohr magneton but also the electron concentration dependence of magnetization in the Slater–Pauling curve. However, there exist other phenomena which cannot be well accounted for in terms of the itinerant electron model. For example, the itinerant model is unable to explain the $T^{3/2}$-law for the temperature dependence of magnetization in ferromagnetic metals at low temperatures and the Curie–Weiss law of the magnetic susceptibility at high temperatures above the Curie temperature. Both are deeply related to the phenomenon called spin fluctuation [6]. The failure of the itinerant electron model stems from the fact that the average field approximation employed is limited within the second-order perturbation theory and that the spin fluctuations or electron correlation effects have to go beyond the second-order perturbation theory.

Although both localized and itinerant electron models are limited in their applicability, the 3d electrons were identified experimentally as itinerant electrons in the 1960s by the de Haas–van Alphen effect for the ferromagnetic metals Fe and Ni. In the localized model, the 3d electrons are assumed to be localized in real space, whereas, in the itinerant model, the Bloch state is localized in reciprocal space. More recently, various efforts have been directed to construct more self-consistent models by introducing spin fluctuations into the itinerant electron model [6]. The weak ferromagnetic metals in group (II) may be characterized by the excitation of modes only with small wave vectors, whereas ordinary ferromagnets in group (I) by the excitation of all modes up to large wave vectors. The latter, when it is Fourier-transformed, gives rise to the localized electron picture in real space.

13.6 Electron transport in ferromagnetic metals

As shown in Fig. 13.6, the valence band of the 3d-transition metals like Ni consists of the superposition of a narrow 3d band over a wide 4s band and the density of states at the Fermi level is shared by both 3d electrons and 4s or 4p electrons. Our first objective in this section is to discuss the role of the sp and d electrons at the Fermi level in the electron transport of the 3d-transition metals. As listed in Table 10.1, the resistivity of transition metals like Ti, Cr and Fe is always higher than that of simple sp-electron metals like Al and Cu.

Mott [8] tried to explain why a transition metal always possesses a larger resistivity than a simple metal in group (V). He assumed that sp electrons at the Fermi level exclusively convey the electrical current and placed an emphasis on the fact that the density of states at the Fermi level in the transition metal is very high because of its location in the middle of the d band. As is clear from equations (10.81) and (10.82) coupled with equation (10.53), the scattering probability $1/\tau$ of the conduction electron is proportional to the final density of states at the Fermi level. This means that the higher the density of states at the Fermi level, the more frequently sp conduction electrons are scattered into it. A higher final density of states at the Fermi level would result in a shorter relaxation time τ and, in turn, a higher resistivity in transition metals. This is the Mott s–d scattering model (see Section 15.8.3).

The temperature dependence of the normalized electrical resistivity of pure Ni and Pd is plotted in Fig. 13.9. Pure Ni is ferromagnetic below 631 K while pure Pd remains paramagnetic over the whole temperature range. We can immediately see from Fig. 13.9 that the resistivity in the ferromagnetic state is lower than that in the paramagnetic state and that spin ordering apparently causes a reduction in resistivity.

Mott argued the lowering of the resistivity in the ferromagnetic state of pure Ni in terms of his s–d scattering model. As shown in Fig. 13.6, the spin-up band in pure Ni is displaced relative to the spin-down band owing to the exchange interaction in the ferromagnetic state, resulting in a complete filling of the spin-up band. Mott pointed out that, at low temperatures, spin-orientation of the conduction electron must be unchanged upon scattering so that spin-up

13.6 Electron transport in ferromagnetic metals

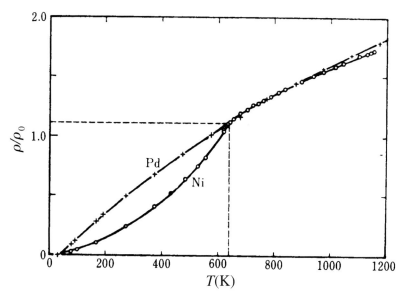

Figure 13.9. Temperature dependence of the normalized electrical resistivity in Ni and Pd. The data are shown such that both sets of data coincide with each other at the Curie point of pure Ni. [J. M. Ziman, *Electrons and Phonons* (Clarendon Press, Oxford 1962) p. 380]

conduction electrons in Ni cannot make transitions to the spin-up d band because it is full. This implies that these electrons would have a longer mean free path than those with the opposite spin. Above the Curie temperature, however, s–d transitions should occur equally for both spin-up and spin-down electrons, since Ni becomes paramagnetic. Thus, nickel above the Curie temperature behaves in the same manner as Pd. However, the interpretation based on the Mott s–d scattering model is not universal. For example, the Mott model encounters difficulty in interpreting a decrease in resistivity upon the ferromagnetic transition of Gd [4], where the 4f electrons are responsible for the onset of ferromagnetism but the 4f band is located about 8 eV below the Fermi level ε_F so that there is no chance for sp conduction electrons at ε_F to be scattered into the 4f states (see Fig. 7.18 and Section 13.11).

The temperature dependence of the electrical resistivity in ferromagnetic metals at low temperatures has been often discussed in terms of the electron–magnon interaction. It was emphasized in Section 10.8 that the electrical resistivity arises from the disruption of a periodicity of the lattice. At absolute zero, all the magnetic moments in a ferromagnetic metal (strictly speaking, in a magnetic domain) are completely oriented in one direction so that the Bloch state would not be disturbed at all. At finite temperatures, however, the localized moment begins to be thermally agitated, the motion of which is well described in terms of the excitation of spin waves or magnons. Therefore, both phonons

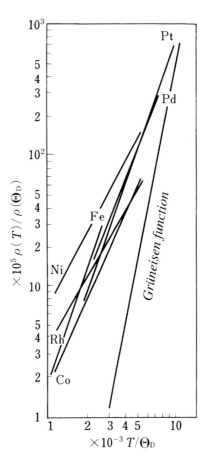

Figure 13.10. Normalized electrical resistivity versus normalized temperature for several transition metals at low temperatures. [K. Mendelssohn, *Can. J. Phys.* **34** (1956) 1315]

and magnons are excited at finite temperatures and interact with the conduction electrons. The resistivity below the Curie temperature may be explained by taking into account both electron–phonon and electron–magnon interactions.

Because of the coexistence of scattering due to phonons and magnons, the temperature dependence of the electrical resistivity in ferromagnetic metals in group (I) is expected to differ from that derived from the Bloch–Grüneisen law, which leads to an exponent $n = 5$ in $\rho_0 + AT^n$ at low temperatures, as discussed in Section 10.12. The temperature dependence of the resistivity for typical transition metals at low temperatures is shown in Fig. 13.10 on a normalized log–log scale. In contrast to non-magnetic metals like pure Na with $n = 5$, the exponent n for ferromagnetic metals is distributed in the range 2–3. A smaller exponent is believed to originate from the electron–magnon interaction in ferromagnetic metals.

13.7 Electronic structure of magnetically dilute alloys

The electronic structure and electron transport properties of magnetic metals and alloys in group (III) will be discussed in Sections 13.7–13.11. First, non-magnetic metals containing magnetic impurities will be discussed. By magnetic impurities we mean impurities that contribute a Curie–Weiss term to the susceptibility. In Section 9.3, we discussed the increment of resistivity upon the addition of 1 at.% polyvalent metals like Zn, Ga, Ge, etc., to the noble metals Cu and Ag. There the impurity potential was approximated by the screened Coulomb potential given by equation (9.1), from which the differential cross-section of equation (9.9) is calculated. Though the Linde rule can be explained from equation (9.10), it overestimates the experimental value, indicating the limit of applicability of the Thomas–Fermi approximation employed in Section 9.3. In this section, we focus on the scattering phenomenon in a system where a single magnetic impurity, Fe, is embedded in a non-magnetic metal, Cu.

Let us locate a single Fe atom at the center of a Cu metal sphere with the radius $\mathfrak{R}$. Since the impurity potential can be assumed to be spherically symmetric, the wave function of the scattered wave is expressed in terms of spherical coordinates as $\psi(r,\theta,\varphi) = R(r)Y_{\ell m}(\theta,\varphi)$, where $R(r)$ is the radial wave function and $Y_{\ell m}(\theta,\varphi)$ is the ℓ-th-order spherical harmonic function [9]. The radial wave function $R(r)$ satisfies the following Schrödinger equation:

$$\frac{1}{r^2}\frac{d}{dr}\left[r^2\frac{dR(r)}{dr}\right] + \left\{\frac{2m}{\hbar^2}[E - V(r)] - \frac{\ell(\ell+1)}{r^2}\right\}R(r) = 0,$$

which is rewritten as

$$\frac{1}{r^2}\frac{d}{dr}\left[r^2\frac{dR(r)}{dr}\right] + \left\{k^2 - \left[U(r) + \frac{\ell(\ell+1)}{r^2}\right]\right\}R(r) = 0, \qquad (13.14)$$

where $k^2 = 2mE/\hbar^2$ and $U(r) = (2m/\hbar^2)V(r)$. It can be seen that the potential field is given by the sum of the attractive Coulomb potential $U(r)$ and the repulsive centrifugal potential $\ell(\ell+1)/r^2$ arising from the orbital angular momentum of the electron, and that the centrifugal potential increases as the quantum number ℓ increases.

For the sake of simplicity, the ion potential $U(r)$ of the Fe atom is truncated at the distance $r = r_0$ and set to zero in the range $r > r_0$. As illustrated in Fig. 13.11, a thick curve represents the potential with $\ell = 0$ experienced by the 3s electron. Owing to the presence of the centrifugal potential, a potential barrier is formed for electrons with $\ell \geq 1$ in the vicinity of $r = r_0$. The total potential after adding the centrifugal potential $\ell(\ell+1)/r^2$ for electrons with $\ell \geq 1$ is shown by a thin curve. Let us consider the scattering of an electron with a positive energy by this potential. Then the motion of the electron may be treated

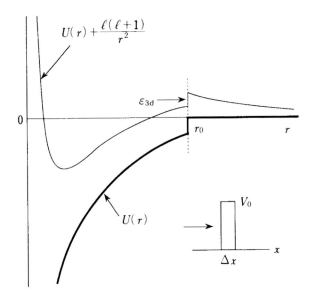

Figure 13.11. Impurity potential due to an Fe atom embedded in pure Cu. Its potential $U(r)$ is approximated as the Coulomb potential in the range $r \leq r_0$ and zero in $r > r_0$. A potential barrier is formed for electrons with $\ell \geq 1$ as a result of the centrifugal potential. Its role may be approximated by the positive square-well potential shown inset.

in the same manner as that in the tunneling phenomenon for the square-potential shown in the inset to Fig. 13.11.

The ionic potential of the Fe atom is shallower than that of Cu because of its smaller atomic number. Hence, the 3d level of the Fe atom would appear between the top of the Cu 3d band and the Fermi level. Since $U(r)=0$ in the range $r > r_0$, we take this as the origin of the energy axis and adjust the bottom of the Cu 4s band to coincide with this energy. This is illustrated in Fig. 13.12. The 3s and 3p electrons of the Fe atom will form core levels with negative energies but the energy level of the 3d electrons will be formed at a positive energy. If the potential barrier were absent, the conduction electron with a positive energy would have been only weakly scattered by the impurity potential. Instead, the existing potential barrier would enhance the tendency of the incident electron to localize in the range $r \leq r_0$.

As shown in Fig. 13.12, the incident electron with a positive energy interacts strongly with the Fe 3d states because of the presence of the potential barrier. However, some portion of the incident electron can escape from the impurity potential and mix with the wave function of the electrons forming the valence band of pure Cu. Thus the conduction electron coupled with the 3d electron is extended over a certain range in both real space and energies and results in a narrow energy band near the Fermi level. The amplitude of the wave function with $\ell = 2$ is certainly enhanced in the range $r \leq r_0$. The presence

13.8 Scattering in a magnetically dilute alloy – "partial wave method"

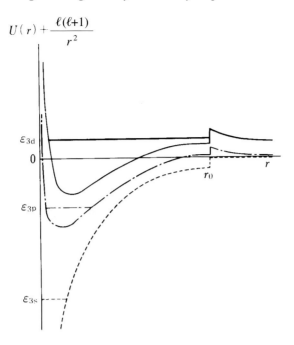

Figure 13.12. Impurity potential and 3s, 3p and 3d energy levels of an Fe atom embedded in pure Cu.

of such electronic states was pointed out for the first time by Friedel in 1956, who named it the virtual bound state [10]. As will be discussed in the next section, the formation of the virtual bound state is believed to be responsible for an increase in the residual resistivity when small amounts of 3d-transition metal atoms are uniformly distributed within the matrix of a non-magnetic metal like pure Cu.

13.8 Scattering of electrons in a magnetically dilute alloy – "partial wave method"

We discuss in this section the use of the partial wave method to treat the scattering phenomenon of the conduction electron due to a magnetic impurity atom. The impurity atom is again located at the center of a metal sphere with radius $\mathfrak{R}$ and the wave function of the conduction electron is expressed as $\psi(r,\theta,\varphi) = R(r) Y_{\ell m}(\theta,\varphi)$. At a position far from the impurity atom, the scattered wave function may be well approximated as a superposition of the wave function propagating radially outward from the center of the impurity atom and the incident plane wave along the z-direction:

$$\psi(r,\theta,\varphi) \underset{r \to \infty}{\to} A\left[e^{ikz} + \frac{1}{r}f(\theta,\varphi)e^{ikr}\right]. \qquad (13.15)$$

We calculate the radial wave function in a system having the impurity potential shown in Fig. 13.11. First consider the region $r > r_0$, where the potential $U(r)$ is zero. A general solution of equation (13.14) is obviously given by a linear combination of the spherical Bessel function $j_\ell(r)$ and the spherical Neumann function $n_\ell(r)$.[11] Since their coefficients can be taken as $\cos\eta_\ell$ and $\sin\eta_\ell$, we immediately obtain its general solution as

$$R_\ell(r) = A_\ell[\cos\eta_\ell j_\ell(kr) - \sin\eta_\ell n_\ell(kr)], \tag{13.16}$$

where η_ℓ is real.[12] At a position far from the potential, the spherical Bessel and spherical Neumann functions can be replaced by their asymptotic forms and equation (13.16) is approximated as

$$R_\ell(r) \underset{r\to\infty}{\to} (kr)^{-1} A_\ell \sin(kr - \tfrac{1}{2}\ell\pi + \eta_\ell). \tag{13.17}$$

In the absence of the impurity potential, the phase shift η_ℓ must be zero and the wave function should be of the form:

$$R_\ell(r) = A_\ell j_\ell(kr) \underset{r\to\infty}{\to} (kr)^{-1} A_\ell \sin(kr - \tfrac{1}{2}\ell\pi), \tag{13.18}$$

where $n_\ell(r)$ is excluded because of its divergence at $r=0$. We see that the scattering effect due to the potential $U(r)$ in the range $r \leq r_0$ is represented by the phase shift η_ℓ. The scattering effect vanishes, when the phase shift is equal to either 0 or π.

The radial wave function in the range $r \leq r_0$ has to be smoothly connected to equation (13.16). From this, the phase shift is uniquely determined. Practically, the logarithmic derivative of R_ℓ, $d\ln R_\ell/dr = (1/R_\ell)(dR_\ell/dr)$, obtained from inside and outside of r_0 must be equated at $r = r_0$. The value can be easily calculated in the range $r > r_0$ from equation (13.16). But the value in the range

---

[11] The spherical Bessel function $j_\ell(r)$ and spherical Neumann function $n_\ell(r)$ are defined in terms of the Bessel function $J_\ell(r)$:

$$j_\ell(r) = \left(\frac{\pi}{2r}\right)^{1/2} J_{\ell+(1/2)}(r) \text{ and } n_\ell(r) = (-1)^{\ell+1}\left(\frac{\pi}{2r}\right)^{1/2} J_{-\ell-(1/2)}(r).$$

They are approximated as

$$j_\ell(r) \underset{r\to\infty}{\to} \frac{1}{r}\cos\left[r - \frac{1}{2}(\ell+1)\pi\right] \text{ and } n_\ell(r) \underset{r\to\infty}{\to} \frac{1}{r}\sin\left[r - \frac{1}{2}(\ell+1)\pi\right]$$

$$j_\ell(r) \underset{r\to 0}{\to} \frac{r^\ell}{1\cdot 3\cdot 5\cdots(2\ell+1)} \text{ and } n_\ell(r) \underset{r\to 0}{\to} -\frac{1\cdot 1\cdot 3\cdot 5\cdots(2\ell-1)}{r^{\ell+1}}.$$

The spherical Neumann function goes to infinity at $r = 0$.

[12] A choice of $\cos\eta_\ell$ and $\sin\eta_\ell$ as the coefficients automatically satisfies the normalization condition, since the probability densities are proportional to $\cos^2\eta_\ell$: $\sin^2\eta_\ell$.

13.8 Scattering in a magnetically dilute alloy – "partial wave method"

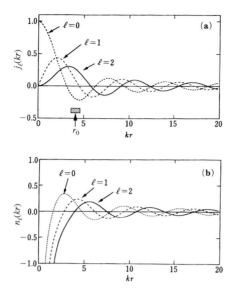

Figure 13.13. (a) Spherical Bessel functions for three values of ℓ. The range of r_0 for typical metals like Cu is shown by hatches. (b) Spherical Neumann functions for the same three values of ℓ. See footnote 11, p. 406.

$r \leq r_0$ depends on the choice of the potential $U(r)$ and, in general, cannot be analytically solved. If we denote it as γ_ℓ, then we have the following relation as the boundary condition:

$$\frac{k[j'_\ell(kr_0)\cos\eta_\ell - n'_\ell(kr_0)\sin\eta_\ell]}{j_\ell(kr_0)\cos\eta_\ell - n_\ell(kr_0)\sin\eta_\ell} = \gamma_\ell, \qquad (13.19)$$

which is rewritten as

$$\tan\eta_\ell = \frac{kj'_\ell(kr_0) - \gamma_\ell j_\ell(kr_0)}{kn'_\ell(kr_0) - \gamma_\ell n_\ell(kr_0)}. \qquad (13.20)$$

Any scattered wave function can be expanded into a linear combination of the spherical functions with all possible quantum numbers ℓ. As shown in Fig. 13.13, the spherical Bessel function $j_\ell(kr)$ takes its maximum at $r_{max} \approx 2\ell/k$ and is approximated as $j_\ell(r) \propto r^\ell$ at $r \to 0$. Hence, only partial waves with small ℓs satisfying the condition $r_{max} < r_0$ need be considered.[13] This is because all higher partial waves satisfying $r_{max} > r_0$ are small in the range $r \leq r_0$ where the potential $U(r)$ is substantial.

The phase shift disappears even for partial waves having such small ℓs, if the numerator of equation (13.20) happens to be zero. The Thomas–Fermi

---

[13] The magnitude of r_0 is of the order of a few Å, while the Fermi wave vector is about 1.35 Å^{-1} in pure Cu. Thus, the relation $r_0 \approx (4\sim 5)/k_F$ roughly holds and the components ℓ higher than 3 may be neglected.

approximation discussed in Section 9.2 overestimates the increment of resistivity upon the addition of a small amount of a non-magnetic impurity like Zn to pure Cu. The reason for this is that the scattering of the 4s wave with $\ell = 0$ and the 4p wave with $\ell = 1$ causes the numerator of equation (13.20) to be small enough to weaken the scattering. The situation is different when a transition metal element like Fe and Cr is added to Cu. Now the denominator of equation (13.20) becomes close to zero for the 3d wave with $\ell = 2$. As will be discussed in the next section, the phase shift η_2 passes $\pi/2$ at the energy ε_0 and results in strong scattering when ε_0 coincides with the Fermi level.

The differential scattering cross-section $\sigma(\theta)$ is expressed as

$$\sigma(\theta) = |f(\theta)|^2 = \frac{1}{k^2}\left|\sum_{\ell=0}^{\infty}(2\ell+1)e^{i\eta_\ell}\sin\eta_\ell P_\ell(\cos\theta)\right|^2, \qquad (13.21)$$

where $P_\ell(\cos\theta)$ is the Legendre polynomial [8] (see footnote 4, p. 247, in Section 9.8). The total scattering cross-section is obtained by integrating equation (13.21) over a whole solid angle:

$$\sigma = 2\pi\int_0^\pi \sigma(\theta)\sin\theta d\theta = \frac{4\pi}{k^2}\sum_{\ell=0}^{\infty}(2\ell+1)\sin^2\eta_\ell. \qquad (13.22)$$

The factor $(1-\cos\theta)$ must be incorporated in the scattering cross-section contributing to the electrical resistivity (see Section 10.10). The transport cross-section σ_{tr} is then calculated as

$$\sigma_{tr} = 2\pi\int_0^\pi \sigma(\theta)(1-\cos\theta)\sin\theta d\theta = \frac{4\pi}{k^2}\sum_{\ell=0}^{\infty}(\ell+1)\sin^2(\eta_\ell - \eta_{\ell+1}) \qquad (13.23)$$

(see Exercise 13.4). The electrical resistivity containing N_{imp} impurities in unit volume is now given by

$$\rho = \frac{mv_F}{ne^2\Lambda} = \frac{N_{imp}mv_F}{ne^2}\sigma_{tr} = \frac{4\pi N_{imp}\hbar}{ne^2 k_F}\sum_{\ell=0}^{\infty}(\ell+1)\sin^2(\eta_\ell - \eta_{\ell+1}), \qquad (13.24)$$

where n is the number of conduction electrons in unit volume and η_ℓ is the phase shift of the ℓ-th partial wave at the Fermi level. This is the resistivity formula in the partial wave method.

At the end of this section, we derive the Friedel sum rule which is often employed as an additional constraint to be satisfied in the partial wave method. In Section 2.6, the wave vector of the conduction electron was quantized by confining the electron in a cube with edge length L. Instead, the conduction electron is now confined in a sphere with radius $\mathfrak{R}$. The boundary condition is imposed such that the spherical wave function (13.17) vanishes at the boundary $r = \Re$. Then the wave vector is quantized in the form:

$$k_n \Re - \frac{\ell \pi}{2} + \eta_\ell(k_n) = n\pi \qquad (n = \pm 1, \pm 2, \pm 3, \cdots). \qquad (13.25)$$

The differentiation of n in equation (13.25) with respect to the wave vector k_n yields the relation $dn/dk = (\Re/\pi)[1 + \Re^{-1}(d\eta_\ell/dk)]$. Since each ℓ is $(2\ell+1)$-fold degenerate, the density of states in the range k to $k + \delta k$ turns out to be

$$n(k)\delta k = \sum_\ell (2\ell + 1)\left(\frac{\Re}{\pi}\right)\left(1 + \frac{1}{\Re}\frac{d\eta_\ell}{dk}\right)\delta k. \qquad (13.26)$$

We are interested in a change in the density of states induced by an impurity atom located at the center of a metal sphere. The impurity atom introduces an excess nuclear charge $e\Delta Z$, which must be screened by the same amount of the electronic charge, as required from the charge neutrality condition. Here ΔZ represents the difference in the valency between the impurity and matrix, as was introduced in Section 9.3. Thus, a redistribution of conduction electrons occurs and results in a change in the density of states in the energy range ε to $\varepsilon + d\varepsilon$. This is easily calculated from equation (13.26):

$$\Delta n(k)\delta k = \frac{1}{\pi}\sum_\ell (2\ell+1)\left(\frac{d\eta_\ell}{d\varepsilon}\right)\left(\frac{d\varepsilon}{dk}\right)\delta k = \frac{1}{\pi}\sum_\ell (2\ell+1)\left(\frac{d\eta_\ell}{d\varepsilon}\right)\delta\varepsilon \equiv \Delta n(\varepsilon)\delta\varepsilon$$

and, hence,

$$\Delta n(\varepsilon)\delta\varepsilon = \frac{1}{\pi}\sum_\ell (2\ell+1)\left(\frac{d\eta_\ell}{d\varepsilon}\right)\delta\varepsilon. \qquad (13.27)$$

The total excess electronic charge is obtained by integrating equation (13.27) up to the Fermi level. This must be equal to the excess nuclear charge $e\Delta Z$:

$$\Delta Z = 2\int_0^{\varepsilon_F} \Delta n(\varepsilon)d\varepsilon = \frac{2}{\pi}\sum_\ell (2\ell+1)\eta_\ell(\varepsilon_F), \qquad (13.28)$$

where the factor 2 is due to the spin degeneracy. Equation (13.28) is known as the Friedel sum rule [10]. The Friedel sum rule is very useful when one intends to construct an impurity potential better than the Thomas–Fermi one. The phase shift is calculated from equation (13.20) but the resulting phase shift must be determined so as to satisfy the Friedel sum rule.

We showed above that the excess nuclear charge around the impurity atom is screened by conduction electrons on an atomic scale. However, there are oscillations in the screening charge distribution that fall off rapidly with

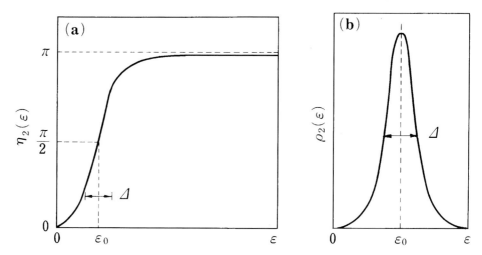

Figure 13.14. Energy dependence of (a) phase shift and (b) partial density of states for $\ell = 2$. [J. Kondo, *Kinzoku Densi Ron* (Shokabo, Tokyo 1983) (in Japanese)]

distance. The screened potential of a point charge contains a term proportional to

$$\phi(r) \approx \frac{1}{r^3}\cos 2k_F r \qquad (13.29)$$

at distance r from an impurity atom. The oscillations are known as the Friedel oscillations [11]. It is shown that the presence of a sharp Fermi edge is responsible for oscillations of the screened potential with a wavelength of $1/2k_F$ (see Section 15.6).

13.9 Scattering of electrons by magnetic impurities

We noted in Section 13.8 that there is an energy ε_0 at which the $\ell = 2$ partial wave causes the denominator in equation (13.20) to be zero. This occurs when the incident electron has an energy near ε_{3d} in Fig. 13.12 and the phase shift η_2 becomes equal to $\pi/2$. This is called resonant scattering and enhances the tendency for these electrons to localize inside the potential barrier as a result of the centrifugal potential.

The energy dependence of the phase shift η_2 for the $\ell = 2$ partial wave is depicted in Fig. 13.14(a). It passes through $\pi/2$ within a narrow energy range centered at ε_0, which is close to ε_{3d} in Fig. 13.12. As is clear from Fig. 13.14(a), the derivative $d\eta/d\varepsilon$ is largest at ε_0 and so is Δn from equation (13.27), resulting in a sharp peak in the density of states at $\varepsilon = \varepsilon_0$ with the width Δ. This is the virtual bound state and is illustrated in Fig. 13.14(b).

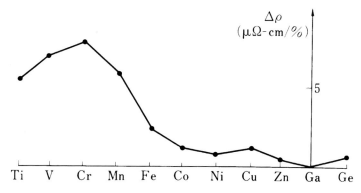

Figure 13.15. Residual resistivity due to addition of 1 at.% transition metal impurity to pure Al. [From ref. 10.]

The electronic specific heat coefficient is naturally enhanced when the virtual bound state is formed across the Fermi level. We expect from equation (13.24) that the electrical resistivity is also enhanced. Figure 13.15 shows the increment in the electrical resistivity induced when 1 at.% of the 3d-transition metal element M is dissolved in fcc Al. Clearly, the resistivity increment is largest for $M=$ Cr. Friedel interpreted this unique behavior in such a way that, within the context of the rigid-band model, the peak of the virtual bound state moves to higher binding energy with increasing atomic number in the series of the 3d-transition metals in the periodic table and happens to pass across the Fermi level of pure Al when $M=$ Cr. Indeed, he noted in his paper [10] that the resonant effect is intuitively realized from the experimental data shown in Fig. 13.15, in which the M dependence of $\Delta\rho$ is clearly not monotonic but exhibits a maximum at Cr.[14]

There is another important effect associated with a magnetic impurity embedded in a non-magnetic metal [12]. As was already noted, magnetic impurities would possibly exhibit a Curie–Weiss-type magnetic susceptibility $\chi = C/(T-T_C)$, where T is the temperature and T_C is the Curie temperature. These impurities are either from the 3d-transition element series or the 4f rare-earth series in the periodic table. Unfortunately, the Curie–Weiss behavior cannot be explained within an average field approximation, as discussed in Section 13.5.

---

[14] The Friedel model is employed to explain transport properties in dilute alloys dissolving only less than 1 at.% of the transition metal element in simple metals like Al. As shown in Fig. 13.14(b), the height of the virtual bound state is at most two times that of the sp band and its width is generally less than 1 eV. Therefore, the electronic structure of a magnetically dilute alloy is entirely different from that to which the Mott s–d scattering model is applied. The Mott s–d scattering model is applied to an alloy composed of transition metal elements as major components, where the d band is 2–10 times that of the sp band and extends over several eV across the Fermi level. See also Section 15.8.

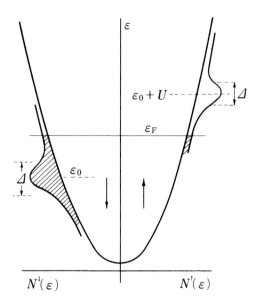

Figure 13.16. Anderson model for the formation of localized moment. [From ref. 13.]

Let us consider again a system where a single 3d-transition metal atom is embedded in a non-magnetic metal like Al and assume that this impurity atom possesses only one spin-up 3d electron. Suppose that a spin-down 3d electron is added to the same orbit as the already existing spin-up 3d electron. The addition of this electron increases the energy of the system by U as a result of the Coulomb repulsive interaction. In the atomic 3d orbitals, the Coulomb interaction is very large and is of the order of 30 eV. Its magnitude is reduced to 1–7 eV for 3d electrons in a metallic environment due partly to the delocalization of the orbital and due partly to screening by the other electrons. This is still fairly large and plays a key role in the following discussions.

In 1961, Anderson [13] discussed the criterion for the appearance of the local magnetic moment under the simplest condition of non-degenerate 3d orbitals. He assumed that the energy of a second spin-up electron increases by the Coulomb energy U if the first spin-down electron has already occupied the lowest 3d orbital with the energy ε_0. Furthermore, the matrix element V between the conduction electron and the 3d electron is introduced to allow the conduction electron to enter into and/or escape from the 3d orbital. Anderson predicted the local moment to appear, provided that the conditions $\varepsilon_0 + U \gg \varepsilon_F$ and $\varepsilon_0 \ll \varepsilon_F$, together with $|\varepsilon_0 + U - \varepsilon_F| \gg \Delta$ and $|\varepsilon_F - \varepsilon_0| \gg \Delta$, are satisfied. Since the energy of the second spin-up electron is $\varepsilon_0 + U$, this state will not be occupied, as indicated in Fig. 13.16. Therefore, the 3d orbital is filled only by the spin-down 3d electron, resulting in the appearance of the magnetic moment. In contrast, the magnetic moment will not appear, if both energy levels ε_0 and $\varepsilon_0 + U$ appear above or below the Fermi level. The effective exchange coupling J between the localized spin and conduction electron is expressed as a function of the parameters U and V and is shown to be negative for scattering of conduction electrons near the Fermi level.

The width Δ of the virtual bound state increases in proportion to the density of states at the Fermi level of the host metal and the magnitude of the matrix element V representing the degree of the mixing of the localized wave function inside the barrier and the spherical wave function outside [14,15]. Its width in Al-based alloys is believed to be greater than that in Cu-based alloys, since the density of states at the Fermi level in pure Al is definitely higher than that in pure Cu (see Table 3.1, in which their electronic specific heat coefficients, proportional to their density of states at the Fermi level, are listed). The ratio $\pi\Delta/U$ is often used as a measure for the occurrence of the localized moment: its formation is unfavorable when the ratio is larger than unity. Instead, the localized moment will be formed, when the ratio is lower than unity, as in the case where impurities such as Mn and Fe are dissolved in pure Cu.

In the Anderson model, the interaction of the conduction electron with the 3d electrons was treated within an average field approximation and, hence, the dynamical effect associated with the flipping of the localized spin was not included. Let us consider the 3d spin-up electron at the impurity site. There is a chance of the spin-down conduction electron jumping into the impurity potential to cause the direction of the 3d spin-up electron to reverse after scattering. The phenomenon associated with the exchange of the spin orientation between the magnetic moment of the impurity atom and that of the conduction electron is called spin fluctuation. The lifetime of localized spin fluctuations is denoted as τ_{sf}, which is meaningful only when the virtual bound state is well defined. A moment can exist if $\tau_\Delta \ll \tau_{sf}$, where $\tau_\Delta = \hbar/\Delta$ describes the relaxation time for an electron from the virtual bound state to the continuum.[15]

The question of the existence of a localized moment is, however, a matter of time or temperature [15]. It depends on whether or not the spin fluctuations can be sufficiently slow that there appears to be a moment on the time-scale of our experimental probe. The relaxation time associated with thermal fluctuations at temperature T is defined as $\tau = \hbar/k_B T$. For example, a moment may be observed even in a simple metal at temperatures $T > T_F$, where the measuring time-scale τ is shorter than $\tau_F = \hbar/E_F$. The reason why the moment is absent in simple metals at ordinary temperatures arises from the fact that the measuring

---

[15] The width Δ is generally of the order of 0.5 eV or 5000 K on the temperature scale, equivalent to a lifetime of $\tau_\Delta \approx 1.3 \times 10^{-15}$ s. The value of τ_{sf} is of the order of τ_Δ in the non-magnetic limit but increases with increasing U or decreasing ratio $\pi\Delta/U$ [15].

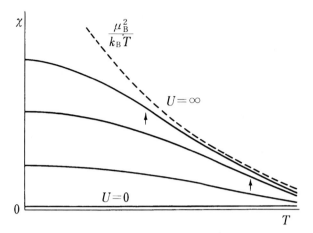

Figure 13.17. Temperature dependence of magnetic susceptibility with Coulomb energy U as a parameter. The arrows indicate the temperatures above which spinfluctuations dominate. [J. Kondo, *Kinzoku Densi Ron* (Shokabo, Tokyo 1983) (in Japanese)]

temperature is so low or the measuring time-scale so long that we observe only an average over repeated flips of spins. In the present case, a localized moment will not be observed at low temperatures, where the condition $\tau_\Delta \ll \tau_{sf} < \tau$ or $T < T_{sf} \ll T_\Delta$ is satisfied. Here the frequency of spin fluctuations is so high over the thermal fluctuation time-scale τ that only an average of up- and down-spins is observed. However, a localized moment will become observable at high temperatures $T > T_{sf}$ and the magnetic susceptibility will obey the Curie–Weiss-type temperature dependence.

Let us summarize the magnetic properties of a magnetically dilute alloy as a function of temperature and Coulomb energy U. Obviously, a localized moment will not appear and the temperature-independent Pauli paramagnetic susceptibility will dominate, when U is zero. Its value is given by $N(\varepsilon_F)\mu_B^2$. In contrast, a localized moment appears, when U becomes large and satisfies the condition $\pi\Delta/U < 1$. Thus, the magnetic susceptibility obeys the Curie law owing to the moment localized at the impurity atom. However, the moment will apparently disappear at low temperatures where the condition $T < T_{sf} \ll T_\Delta$ is satisfied. This behavior is schematically illustrated in Fig. 13.17.

13.10 s–d interaction and Kondo effect

When a small amount of Fe impurities are added to pure Cu, a resistivity minimum appears at low temperatures and the resistivity increases logarithmically with further decrease in temperature, as is shown in Fig. 13.18. This

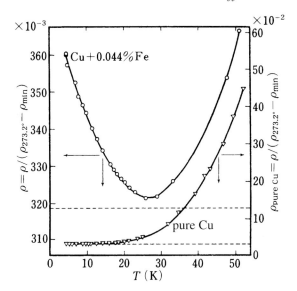

Figure 13.18. Temperature dependence of the electrical resistivity for Cu containing 0.044 at.%Fe. Data for pure Cu are also shown. [J. M. Ziman, *Electrons and Phonons* (Clarendon Press, Oxford Press, 1962) p.344]

unique electron transport behavior was a theoretical puzzle for a long time. The experimental data showed that there exists a correlation between the presence of the Curie–Weiss term due to the local magnetic moment and the occurrence of the resistance minimum and that the resistivity increment observed below the resistivity minimum temperature is linearly proportional to the concentration of magnetic impurities. The latter fact apparently ruled out the possibility of an impurity–impurity interaction and, instead, suggested an isolated impurity phenomenon.

Even though the magnetic moments of the 3d-transition metal impurities dissolved in a non-magnetic host metal are isolated from each other, an interaction between them can be mediated through the conduction electrons. This is called the s–d interaction. In 1964, Kondo made the first important step toward the understanding of this unique phenomenon and employed the s–d exchange Hamiltonian

$$H_{sd} = -2J(\mathbf{r} - \mathbf{R}_n)(\mathbf{s} \cdot \mathbf{S}_n), \tag{13.30}$$

where $\mathbf{s}$ is the spin operator of the conduction electron at the position $\mathbf{r}$, $\mathbf{S}_n$ is that of the 3d impurity atom at the position $\mathbf{R}_n$, and $J(\mathbf{r} - \mathbf{R}_n)$ is the exchange integral between $\mathbf{s}$ and $\mathbf{S}_n$ [14]. Kondo treated the s–d interaction H_{sd} as a perturbation and calculated the scattering probability of the conduction electron by taking its effect up to the third-order. The resistivity due to the s–d interaction is given by

$$\rho_{\text{spin}} = c\rho_0 \pi^2 [N(\varepsilon_F)]^2 J^2 S(S+1) \left[1 + 4JN(\varepsilon_F) \log \frac{k_B T}{D} \right], \quad (13.31)$$

where c is the impurity concentration, and $N(\varepsilon_F)$ is the density of states at ε_F in a flat band of width $2D$ [12, 14, 15].

The total resistivity is expressed as the sum of the Bloch–Grüneisen contribution due to the electron–phonon interaction and the s–d interaction given by equation (13.31). We have already shown in Section 10.12 that the former, say in pure Cu, is well approximated as $\rho_0 + AT^5$ at low temperatures. Hence, the total resistivity in a dilute magnetic alloy is expressed as

$$\rho = \rho_0 + AT^5 + B\log T, \quad (13.32)$$

where ρ_0 is the residual resistivity and A and B are numerical constants. Note that the resistivity minimum occurs when the exchange constant J and, hence, the coefficient B is negative. Equation (13.32) yields a resistivity minimum at $T_{\min} = (|B|/5A)^{1/5}$ and the resistivity increases logarithmically with decreasing temperature below $T_{\min}$, consistent with the experimental data shown in Fig. 13.18.

The second term in equation (13.31), derived for the first time by Kondo, represents a many-body effect arising from the scattering process in which the spins of the conduction electron and localized electron are flipped [12,14,15]. According to equation (13.31), the dimensionless parameter $JN(\varepsilon_F)\log(k_B T/D)$ eventually leads to a divergence with decreasing temperature, no matter how small is the quantity $JN(\varepsilon_F)$. This indicates the breakdown of the perturbation theory at temperatures below the so-called Kondo temperature defined as

$$T_K = \frac{D}{k_B} \exp[-1/|J|N(\varepsilon_F)]. \quad (13.33)$$

A more comprehensive theory was needed to explain the difficulty involved in the logarithmic divergence as well as the actual ground state of the conduction electron–magnetic impurity system. The search for such a theory became known as the "Kondo problem" and attracted much theoretical interest in the late 1960s and early 1970s [12]. Briefly, an antiferromagnetic coupling between the localized spin and the spin of the conduction electron is essential in the Kondo effect. It induces spin polarization of the conduction electron in a direction opposite to that of the localized moment and cancels completely the localized moment, leading to the formation of a singlet state $S=0$ at absolute zero.

We can say that the Curie law holds in the temperature range above the

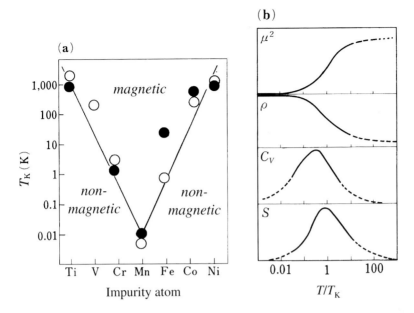

Figure 13.19. (a) Kondo temperature for Cu (●) and Au (○) alloyed with various 3d-transition metal impurities. (b) Temperature dependence of the magnetic moment μ^2, electrical resistivity ρ, specific heat C_V and thermoelectric power S. Temperature is normalized with respect to the Kondo temperature. Solid curves are based on experimental behavior with the properties of the pure host material subtracted. The dashed portions have not been examined experimentally and represent predictions based on theory. [M. D. Daybell, *Magnetism* edited by H.Suhl, (Academic Press, 1973) vol.5, pp. 121–147]

Kondo temperature T_K but that spin fluctuations become substantial in the temperature range below T_K and gradually approaches the singlet state at absolute zero. Figure 13.19(a) shows the Kondo temperature T_K for the metals Cu and Au containing various 3d-transition metal impurities. It has also been established that various physical properties exhibit unique temperature dependences across the Kondo temperature. The magnetic moment, electrical resistivity, specific heat and thermoelectric power are shown schematically in Fig. 13.19(b) as a function of temperature normalized with respect to the Kondo temperature T_K.

As is seen from Fig. 13.19(a), the Kondo temperature varies over a wide range, depending on the combination of magnetic impurity atom and non-magnetic host metal. For example, the Kondo temperature T_K is only 0.01 K in the Cu–(Mn) system and the logarithmic temperature dependence of resistivity is observed below about 10 K. In contrast, T_K is 300 K in the Au–(V) system, where the logarithmic temperature dependence is observed in the range

300–1000 K. Figure 13.19(b) clearly shows that the localized moment tends to zero while the resistivity is saturated and becomes temperature independent at temperatures well below T_K.

The Kondo effect is observed over the whole concentration range $0 \leq x \leq 1$ in some rare-earth compounds $RE_xM_{1-x}X$, where the rare-earth element RE such as Ce and Yb is substituted by an element M such as Y or La and X stands for simple metals like Cu, Al and so on. For example, in $Ce_xLa_{1-x}Cu_6$ alloys, the localized magnetic moment due to 4f-electrons resides at each Ce^{3+} ion and forms a periodic lattice. With increasing Ce concentration, the f electrons begin to move together with the conduction electrons [16]. Hence, the electronic structure and electron transport properties are dominated by conduction electrons with an extremely heavy effective mass. This is often referred to as the "heavy fermion" system and various-rare earth and actinide alloys based on Ce, Yb and U are classified into the family of dense Kondo systems or Kondo lattices [12].

13.11 RKKY interaction and spin-glass

In order to discuss the line width of the nuclear spin resonance, Ruderman and Kittel [17] showed in 1954 that the two nuclear spins in nearest neighbor atoms interact via the conduction electron through the relation

$$-\frac{2\pi}{9}n^2\frac{A^2}{\varepsilon_F}F(2k_FR_{nm})(\mathbf{I}_n\cdot\mathbf{I}_m), \qquad (13.34)$$

where $\mathbf{I}_n$ and $\mathbf{I}_m$ are nuclear spins of the neighboring atoms, n is the number of conduction electrons per atom, ε_F is the Fermi energy, k_F is the Fermi radius, A is the hyperfine coupling constant between the nuclear spins, R_{nm} is the distance between the two nuclei and the function $F(x)$ is given by

$$F(x) = \frac{-x\cos x + \sin x}{x^4}. \qquad (13.35)$$

The oscillatory function $F(x)$ decreases in an inverse proportion to the third power of the distance.[16]

Kasuya in 1956 applied equation (13.34) to the interaction of the conduction electron with the localized magnetic moment, instead of the nuclear spin, and discussed the temperature dependence of the electrical resistivity in ferromagnetic metals like Gd [18]. Here the nuclear spins $\mathbf{I}_n$ and $\mathbf{I}_m$ in equation

---

[16] The second term is generally negligible relative to the first term. For instance, take pure Cu. Following the discussion in Section 2.7, the Fermi diameter $2k_F$ of pure Cu is 2.7 Å^{-1}. Since the nearest neighbor distance is 2.56 Å, the variable x defined as $x = 2k_FR_{nm}$ is found to be 6.9. The ratio of the second term over the first one is of the order of 10^{-2}.

(13.34) are replaced by the localized magnetic spins $\mathbf{S}_n$ and $\mathbf{S}_m$ and the hyperfine coupling constant A by the s–d exchange constant J. Equation (13.34) is rewritten as

$$-\frac{2\pi}{9}n^2\frac{J^2}{\varepsilon_\mathrm{F}}\sum_{n,m}F(2k_\mathrm{F}R_{nm})(\mathbf{S}_n\cdot\mathbf{S}_m), \tag{13.36}$$

where the function $F(x)$ is the same as in equation (13.35). Yosida in 1957 considered that the interaction given by equation (13.36) causes spin polarization of the conduction electron around the localized magnetic moment and showed that the spatial distribution of the spin polarization of the conduction electron is described by equation (13.35) [19]. We see that the spin of the conduction electron mediates two localized moments at m-th and n-th sites, their positions being not necessarily an immediate neighbor. Equation (13.36) has been referred to as the RKKY interaction after Ruderman, Kittel, Kasuya and Yosida.

The heavy rare-earth metals Gd, Tb, Dy, Ho, Er and Tm undergo ferromagnetic transitions at 293, 219, 89, 20, 20 and 32 K, respectively, but their spin configurations are complex. As has been noted in Section 13.3, the 4f electrons bear a localized moment but direct interaction between them is weak because they are almost fully screened by the 5s and 5p outer electrons. The interaction between the 4f magnetic moments must be brought about by the spin of the conduction electrons. The formation of complex magnetic structures in heavy rare-earth metals has been discussed in terms of the RKKY interaction [2–4].

Figure 13.20 shows the temperature dependence of the magnetic susceptibility measured under an alternating magnetic field for Au containing 1 and 2 at.%Fe. The smaller the amplitude of the alternating magnetic field, the sharper is the peak of the magnetic susceptibility. The temperature corresponding to the peak is called the spin freezing temperature T_f. Below T_f, the motion of the spins in the paramagnetic state freezes while keeping their directions at random. This unique magnetic state is called a spin-glass. Its origin has also been discussed in terms of the RKKY interaction given by equations (13.35) and (13.36), which gives rise to either ferromagnetic or antiferroagnetic coupling as a function of the distance x or R_{nm} between two neighboring spins.

The temperature dependence of the electrical impurity resistivity for Au alloys containing differing amounts of Fe is shown in Fig. 13.21. A resistivity minimum typical of the Kondo effect is observed at 11 K when the Fe concentration is only 0.01 at.%. An increase in the Fe concentration induces the RKKY interaction among the magnetic moments of the Fe atoms and yields a resistivity maximum. The magnetic state is characterized by a spin-glass when the Fe concentration exceeds about 0.15 at.%, where the resistivity minimum disappears. Note also the broad maximum observed over the concentration

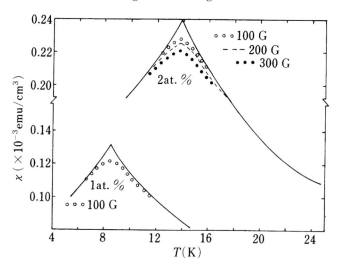

Figure 13.20. Temperature dependence of AC magnetic susceptibility for Au containing 1 and 2 at.%Fe. A solid curve refers to the data where the magnetic field is extrapolated to zero. [V. Cannella and J. A. Mydosh, *Phys. Rev.* **B6** (1972) 4220]

range 0.06 to 0.25 at.%. The theoretical treatment for the electron transport mechanism in a spin-glass is complex and further details may be found in the literature [4].

13.12 Magnetoresistance in ferromagnetic metals

Fe–Co and Fe–Ni ferromagnetic alloys are used as magnetic field sensors because they possess a large magnetoresistance. The magnetoresistance can be longitudinal or transverse, depending on whether the magnetic field is applied in a direction parallel to or perpendicular to the direction of the electrical current, respectively. Figure 13.22 shows the change in the normalized resistivity upon the application of a magnetic field, either in the longitudinal or in the transverse direction, to demagnetized pure Ni. These are called the longitudinal and transverse magnetoresistance, respectively.[17] An initial large change in resistivity is accompanied by growth of magnetic domains parallel to the direction of the magnetic field. Once it is saturated, the resistivity changes more or less linearly with increasing magnetic field. The intercepts obtained by back-extrapolating the linearly dependent behavior may be denoted as $\rho_{//}$ and $\rho_{\perp}$ in

---

[17] The term "magnetoresistance", meaning the change in resistance or resistivity upon the application of a magnetic field, is expressed in different ways: (1) $\Delta R = R(H) - R(0)$; (2) $\Delta R/R(0)$; (3) $R(H)/R(0)$; (4) $\Delta \rho = \rho(H) - \rho(0)$; (5) $\Delta \rho/\rho(0)$; and (6) $\rho(H)/\rho(0)$, where $R(H)$ and $\rho(H)$ are the resistance and resistivity, respectively, in an applied magnetic field H.

13.12 Magnetoresistance in ferromagnetic metals

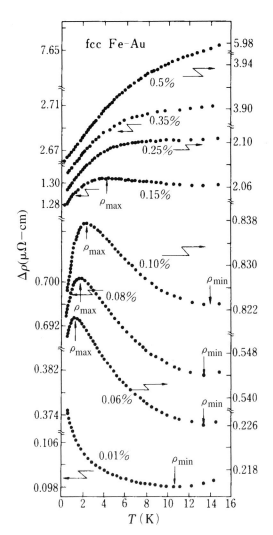

Figure 13.21. Temperature dependence of electrical resistivity for fcc Au–Fe alloys. Concentrations are given in at.%Fe. $\Delta\rho$ is the impurity resistivity, defined as $\rho_{alloy} - \rho_{Au}$. [P. J. Ford, T. E. Whall and J. W. Loram, *Phys. Rev.* **B2** (1970) 1547]

the longitudinal and transverse configurations, respectively. Now the anisotropy in magnetoresistance is defined as

$$\frac{\Delta\rho}{\rho_{//}} = (\rho_{//} - \rho_{\perp})/\rho_{//}. \tag{13.37}$$

The ratio $\Delta\rho/\rho_{//}$ is often called FAR or ferromagnetic anisotropy of resistivity. The largest value of FAR at room temperature so far reported in the literature

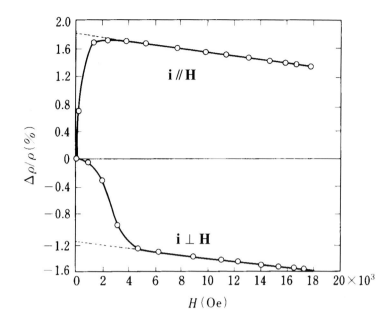

Figure 13.22. Longitudinal and transverse magnetoresistance of pure Ni. [E. Englert, *Ann. Phys.* **14** (1932) 589]

is 6.5% for the $Ni_{70}Co_{30}$ alloy. The value increases with decreasing temperature and exceeds 10% at the liquid nitrogen temperature of 77 K.

The FAR effect has been discussed in terms of the two-current model originally proposed by Mott [8]. Magnetization in ferromagnetic metals arises as a result of the splitting of the spin-up or majority-spin band relative to the spin-down or minority-spin band, as shown in Fig. 13.6. Mott suggested that conduction electrons in ferromagnetic metals can propagate by repeating scattering events without changing spin orientations at temperatures well below the Curie temperature (see Section 13.6).[18] This implies that the spin-up and spin-down conduction electrons can be treated independently. Electron conduction in ferromagnetic metals may be conveniently discussed by using a parallel circuit due to the currents of spin-up and spin-down conduction electrons. This is the Mott two-current model [8]. As can be recognized from Fig. 13.6, the valence band structure differs substantially, depending on the spin orientation. This suggests that the spin-up and spin-down conduction electrons would possess different relaxation times.

---

[18] In Sections 13.9–13.11, we were considering metals in group (III) of Table 13.1, where the spin fluctuations play a dominant role. In this section, we are discussing the electron transport properties of ferromagnetic metals in group (I), where the spin fluctuations may be ignored.

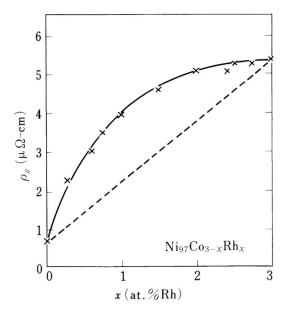

Figure 13.23. Longitudinal resistivity at 4.2 K for $Ni_{97}Co_{3-x}Rh_x$ alloys. [J. W. F. Dorleijn and A. R. Miedema, *J. Phys. F: Metal Phys.* **5** (1975) 1543]

In order to discuss the electron transport on the basis of the two-current model, one must find a way to separate the spin-up electron conduction from the spin-down one. This is made possible by measuring the resistivity for pseudo-binary dilute alloys. For instance, Ni-based ferromagnetic alloys in the form of $Ni_{97}A_{3-x}B_x$ have been studied [20]. The concentration dependence of the resistivity is plotted in Fig. 13.23 for the $Ni_{97}Co_{3-x}Rh_x$ alloys. According to the two-current model, the electron conduction of this ternary alloy is described by using the equivalent circuit shown in Fig. 13.24 and the resultant resistivity is given by

$$\rho = \frac{(c_A \rho_A^\uparrow + c_B \rho_B^\uparrow)(c_A \rho_A^\downarrow + c_B \rho_B^\downarrow)}{c_A \rho_A^\uparrow + c_B \rho_B^\uparrow + c_A \rho_A^\downarrow + c_B \rho_B^\downarrow}, \quad (13.38)$$

where c_A and c_B are the concentrations of the elements A and B and $c_A + c_B = 3$ holds in the present case. $\rho_A^\uparrow$ and $\rho_B^\uparrow$ represent the residual resistivity caused by the scattering of the spin-up electrons by A and B atoms, respectively. Four unknown parameters $\rho_A^\uparrow, \rho_B^\uparrow, \rho_A^\downarrow$ and $\rho_B^\downarrow$ are involved in this equation. We can determine them by measuring the resistivities for more than four samples with different concentrations.

Such experiments have been carried out for a large number of pseudo-binary dilute alloys [20]. For instance, $Ni_{97}Cr_{3-x}M_x$ alloys with $M = $ Al, Fe, Mn, Ti, etc. were prepared to study the effect of the addition of Cr on Ni. In this way,

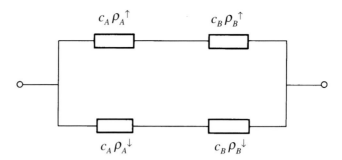

Figure 13.24. Equivalent circuit based on the two-current model of the residual resistivity at 4.2 K for a ferromagnetic metal containing c_A % of A atoms and c_B % of B atoms. [J. W. F. Dorleijn and A. R. Miedema, J. Phys. F: Metal Phys. **5** (1975) 487]

the values of $\rho_{Cr}^{\uparrow}$ and $\rho_{Cr}^{\downarrow}$ are deduced. The validity of the model may be judged by checking whether or not the resulting value is independent of the atomic species M. Indeed, the values of $\rho_{Cr}^{\uparrow}$ and $\rho_{Cr}^{\downarrow}$ turned out to be 25 and 6 $\mu\Omega$-cm, respectively, without depending seriously on the atomic species M. The resistivity increment due to spin-up and spin-down conduction electrons per 1 at.% addition of various transition metal elements to Ni is plotted in Fig. 13.25 for 3d-, 4d- and 5d-transition metal series. It can be seen that the resistivity increment $\rho^{\downarrow}$ due to the spin-down electron is always larger for the lighter elements Ti, Zr and Hf and decreases with increasing atomic number in the respective series.

The FAR can be calculated within the two-current model in the following form [20]:

$$\frac{\Delta\rho}{\rho_{//}} = \left(\frac{\alpha}{1+\alpha}\right)\left(\frac{\Delta\rho}{\rho_{//}}\right)^{\uparrow} + \left(\frac{1}{1+\alpha}\right)\left(\frac{\Delta\rho}{\rho_{//}}\right)^{\downarrow}, \qquad (13.39)$$

where the ratio $\Delta\rho/\rho_{//}$ in the left-hand side represents the FAR and can be determined from experiments, and the parameter α, defined as $\alpha = \rho_{//}^{\downarrow}/\rho_{//}^{\uparrow}$, can be read off from Fig. 13.25. The resulting FAR measured at 4.2 K is plotted in Fig. 13.26 as a function of α for a number of Ni-based alloys. The data fall on a hyperbolic curve shown by the solid line, indicating the coefficients $(\Delta\rho/\rho_{//})^{\uparrow}$ and $(\Delta\rho/\rho_{//})^{\downarrow}$ in equation (13.39) to be uniquely determined for Ni-based alloys. Indeed, the least square fitting of the data points to equation (13.39) results in $(\Delta\rho/\rho_{//})^{\uparrow} = +10\%$ and $(\Delta\rho/\rho_{//})^{\downarrow} = -2\%$. The parameters thus obtained may be considered to reflect the electronic structure of the host metal Ni. Then we can say that the spin-up conduction electrons in Ni possess a larger scattering cross-section in the longitudinal configuration than in the transverse configuration whereas the anisotropy is small for the spin-down electrons.

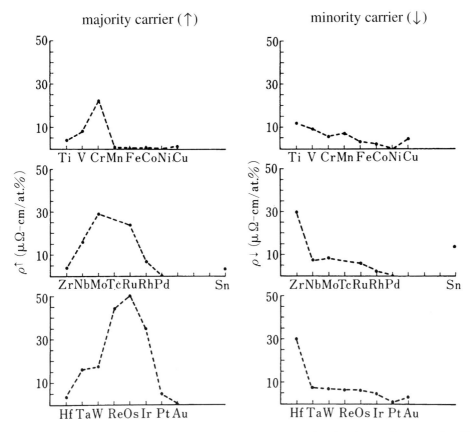

Figure 13.25. Residual resistivity due to spin-up and spin-down electrons for Ni containing various transition metals. [J. W. F. Dorleijn and A. R. Miedema, *J. Phys. F: Metal Phys.* **5** (1975) 487]

It is noted from equation (13.39) that the magnitude of FAR in a dilute alloy is mainly decided by the resistivity ratio α of the spin-down conduction electrons over the spin-up electrons. This tendency holds true even in concentrated alloys, though the theoretical understanding of the FAR effect in concentrated alloys is still beyond our reach. Aside from the theoretical analysis, the largest FAR has been obtained at 30 at.%Co in the Ni–Co system and the second largest at 15 at.%Fe in the Ni–Fe system. Magnetic field sensors have been commercially manufactured using these alloys.

Before ending this section, we briefly discuss the giant magnetoresistance effect (abbreviated as GMR) discovered in Fe/Cr multilayered films in 1988 [21]. The Fe/Cr multilayered film consists of alternate stacks of the ferromagnetic Fe and non-magnetic Cr layers. As is shown in Fig. 13.27, the resistivity is found to decrease substantially upon application of magnetic fields, resulting in the ratio $[\rho(H=0) - \rho(H=20\text{ kOe})]/\rho(H=0)$ amounting to 50% at 4.2 K.

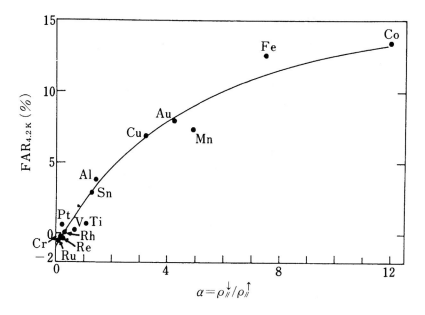

Figure 13.26. α dependence of FAR at 4.2 K for Ni-based alloys. The solid curve refers to equation (13.39) with the coefficients $(\Delta\rho/\rho_{-11})\uparrow$ and $(\Delta\rho/\rho_{-11})\downarrow$ set equal to 10 and -2%, respectively. [J. W. F. Dorleijn and A. R. Miedema, *J. Phys. F: Metal Phys.* **5** (1975) 1543]

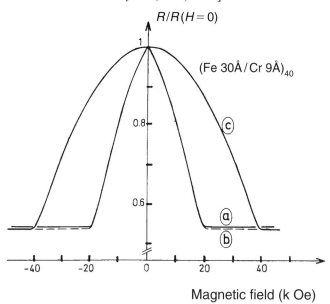

Figure 13.27. Magnetoresistance of [Fe 30 Å/Cr 9 Å]$_{40}$ superlattice at 4.2 K. The current is along [110] and the magnetic field is in the layer plane ⓐ parallel to the current direction, ⓑ perpendicular to the current and ⓒ perpendicular to the layer plane. [From ref. 21.]

13.12 Magnetoresistance in ferromagnetic metals   427

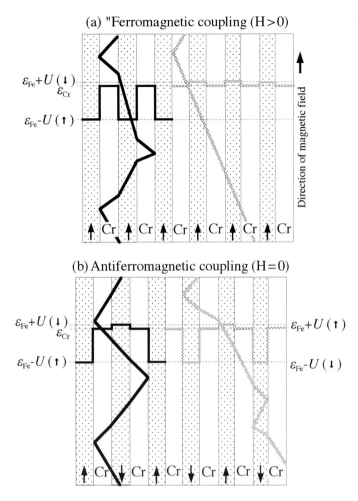

Figure 13.28. GMR mechanism in Fe/Cr multilayered film. The electron scattering path of spin-up (solid line) and spin-down (hatched line) electrons (a) in the presence and (b) in the absence of a magnetic field. The horizontal dotted lines indicate the average potentials $\varepsilon_{Fe} \pm U$ and ε_{Cr} associated with Fe-3d and Cr-3d states. The potentials experienced by spin-up and spin-down electrons across the layers are marked by solid and hatched lines, respectively.

The occurrence of the GMR effect has been interpreted as arising from spin-dependent electron scattering in the film.

Figure 13.28(a) represents the situation where magnetizations of the ferromagnetic layers, as marked by arrows, are forced to be aligned parallel to the direction of the external magnetic field. The 3d band in the Fe layer is split into spin-up and spin-down sub-bands due to the exchange energy U. We consider spin-up and spin-down electrons to experience different average potentials $(\varepsilon_{Fe} - U)$ and $(\varepsilon_{Fe} + U)$, respectively, where ε_{Fe} is the mean energy of the Fe-3d band in the absence of U. Such band splitting does not occur in the non-magnetic Cr layers. But its ionic potential is shallower than that of Fe, since Cr has a smaller atomic number than Fe. Thus, $\varepsilon_{Fe} < \varepsilon_{Cr}$ holds, where ε_{Cr} is the mean energy of the Cr-3d band. The detailed calculations show that $\varepsilon_{Fe} + U \approx \varepsilon_{Cr}$. This means that spin-down electrons can propagate through the boundary between the layers without being heavily scattered, whereas spin-up electrons are heavily scattered there due to the large potential barrier. This is illustrated schematically in Fig. 13.28(a).

In the absence of a magnetic field, the magnetizations in neighboring Fe layers couple antiferromagnetically, as shown in Fig. 13.28(b). Now the potential barrier at every boundary acts almost equally for both spin-up and spin-down electrons as a result of the reversal of the magnetization in every other Fe layer. Hence, both spin-up and spin-down electrons are equally scattered by the disorder at the boundary, resulting in a larger resistance. This is believed to be responsible for the sharp decrease in resistance upon application of a magnetic field and to explain the GMR effect in the Fe/Cr multilayered film [22]. As is clear from Fig. 13.27, an extremely large applied magnetic field of 20–40 kOe is needed to extract the maximum GMR effect in multilayered films like Fe/Cr. This is certainly not favorable in practical applications. Readers may find recent progress in the development of GMR devices in the literature [22].

13.13 Hall effect in magnetic metals

The Hall effect in magnetic metals is measured using the same geometrical configuration as shown in Fig. 11.11. The magnetic field is applied along the z-direction while the electrical current is fed along the x-direction of a specimen and the resulting Hall voltage generated in the y-direction is measured. As mentioned in Section 11.8, a conduction electron subject to the Lorentz force contributes to the Hall effect. If the Hall resistivity ρ_H is defined as the ratio of the transverse field E_y over the current density J_x, then we immediately have the relation $\rho_H = R_H B_z$ from equation (11.34) for non-magnetic metals. Obviously, measured values of ρ_H, when plotted against the applied magnetic induction B_z, fall on a straight line passing through the origin, since R_H is independent of B_z. This is, however, no longer true in magnetic metals because of the additional contribution arising from the localized magnetic moment.

Figure 13.29(a) shows the Hall resistivity of the ferromagnetic $Ni_{97}Al_3$ alloy as a function of the magnetic field applied along the z-direction. It is seen that the Hall resistivity initially increases very rapidly but its slope becomes small with increasing magnetic field. As is illustrated schematically in (b), the

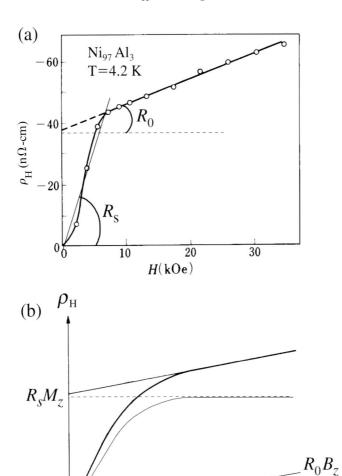

Figure 13.29. (a) Magnetic field dependence of the Hall resistivity for Ni$_{97}$Al$_3$ alloy at 4.2 K. (b) The Hall resistivity is decomposed into the normal and anomalous contributions. [From refs. 20 and 23.]

non-linear behavior of the ρ_H–B curve is decomposed into the normal Hall effect due to the Lorentz force and a strongly temperature-dependent component proportional to magnetization M_z, and is explicitly written as

$$\rho_H = \frac{E_y}{J_x} = R_0 B_z + R_s M_z, \tag{13.40}$$

where R_0 is the normal Hall coefficient and R_s is called the anomalous Hall coefficient. The second term can be present in a ferromagnetic domain even in

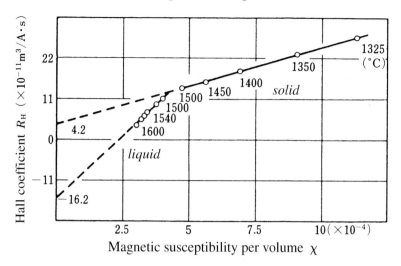

Figure 13.30. Hall coefficient versus magnetic susceptibility per volume for liquid and solid Co. In each state, the normal and anomalous Hall coefficients are derived from the intercept and slope, respectively, of the straight line drawn through the data points. [H.-J. Güntherodt *et al.*, *Liquid Metals*, (The Institute of Physics, 1977) p. 342]

the absence of an applied field. Thus, it is a spontaneous contribution to ρ_H in this case. The normal Hall coefficient is deduced from a slope of the ρ_H–B curve at high magnetic fields while the anomalous Hall coefficient can be roughly estimated from the initial slope. More details about the Hall coefficient in ferromagnetic metals will be found in the literature [23].

Finally, we consider the Hall effect in the paramagnetic state above the Curie temperature, where the magnetization M_z can be expressed as $M_z = \chi \mu_0 H$. Now equation (13.40) is rewritten as

$$R_H \equiv \frac{E_y}{J_x B_z} = R_0 + \frac{\chi}{1+\chi} R_S \approx R_0 + \chi R_S, \qquad (13.41)$$

where χ in the denominator can be ignored relative to unity. For example, the magnetic susceptibility of Co above its Curie temperature ($T_C = 1400\,\text{K}$) follows well the Curie–Weiss law. The data shown in Fig. 13.30 were obtained by measuring the magnetic susceptibility and Hall coefficient for pure Co above its Curie temperature. The data in the solid and liquid states fall on respective lines with different slopes and intercepts. The normal Hall coefficient R_0 and the anomalous Hall coefficient R_S are derived from the intercept and slope, respectively. One can clearly see that the normal Hall coefficient in the paramagnetic state in pure Co is positive.

Exercises

13.1 The expectation value of the z-component of the orbital angular momentum, $\langle L_z \rangle$, vanishes when the degeneracy is lifted and a single level is the lowest in energy. We study this by considering one electron in a p-state ($L=1$). The three degenerate wave functions in the free atom are denoted by p_{+1}, p_0 and p_{-1} where the subscript refers to M_L. Suppose that the atom is exposed to the crystalline field of rhombic symmetry given by $V(x, y, z) = Ax^2 + By^2 + Cz^2$. By taking linear combinations of p_{+1}, p_0 and p_{-1}, one can construct non-degenerate wave functions so as to make each of them to be real. They are given by $\psi_1 = (p_{+1} + p_{-1})/\sqrt{2} = xf(r)$, $\psi_2 = -i(p_{+1} - p_{-1})/\sqrt{2} = yf(r)$ and $\psi_3 = p_0 = zf(r)$.

(a) Show that all non-diagonal elements $\int \psi_i^* V \psi_j d\tau$ ($i \neq j$) vanish but that three diagonal elements are finite and result in the splitting of the degenerate energy level into three.

(b) Show that the expectation value of L_z given by $\int \psi_i^* L_z \psi_i d\tau$ for $i = 1, 2, 3$ is zero. This is called the quenching of the orbital angular momentum.

13.2 Equation (13.6) is reduced to the form $\hbar\omega = (2JSa^2)k^2$ at the long-wavelength limit $ka \ll 1$. Show that the density of states for magnons is given by $D(\omega) = (1/4\pi^2)(\hbar/2JSa^2)^{3/2}\omega^{1/2}$. Calculate the total number of magnons at temperature T by inserting the density of states into the expression $\Sigma_k n_k = \int_0^\infty n(\omega, T) D(\omega) d\omega$ and then derive equation (13.8) by using the relation $\Sigma_k n_k / NS = \int_0^\infty = \Delta M / M(0)$.

13.3 Derive equation (13.11) by using Figure 13.5.

13.4 Derive equations (13.22) and (13.23) from equation (13.21).

Chapter Fourteen

Electronic structure of strongly correlated electron systems

14.1 Prologue

There exists a family of solids, in which the electron–electron interaction plays so substantial a role that the one-electron approximation fails. This is known as the strongly correlated electron system. Historically, De Boer and Verwey were the first to point out, as early as in 1937, that NiO in the NaCl structure should be metallic, since the Fermi level falls in the middle of the Ni-3d band. This already posed serious difficulty in the one-electron band calculations at that time, since NiO is known to exist as a transparent insulator having a band gap of a few eV. Peierls noted in the same year that this difficulty stemmed from the neglect of the repulsive interaction between the electrons and that the electron–electron interaction must be treated beyond the Hartree–Fock one-electron approximation [1].

Various transition metal oxides, including NiO and various layered perovskite cuprates, the latter being known to undergo a transition to the superconducting state upon carrier doping, have now been recognized as solids typical of a strongly correlated electron system. Their electronic structures and electron transport properties have been extensively studied in the last ten years, i.e., the 1990s. In this chapter, we introduce first the concept of the Fermi liquid theory, which justified the one-electron approximation for electrons near the Fermi level in ordinary metals and alloys, and then extend our discussion to cases where the one-electron approximation fails because of the electron–electron interaction. The Hubbard model is introduced as a model appropriate to describe the short-range motion of electrons transferring from one atomic site to another in competion with the on-site repulsive Coulomb interaction. Finally, the electronic structure of the cuprate compounds is briefly discussed on the basis of the Hubbard model.

14.2 Fermi liquid theory and quasiparticle

Landau considered that any homogeneous system composed of a large number of particles has low-lying excited states of waves and introduced the concept of the quasiparticle to describe the waves.[1] A typical example is an assembly of atoms, from which lattice waves are excited and phonons are created as quasiparticles. He extended the idea to describe the low-lying excited states of the interacting electron system in terms of quasiparticles. The interacting electron system is called the electron liquid or the Fermi liquid in contrast to an ideal electron gas or the Fermi gas for the case of non-interacting electrons. The Coulomb interaction between electrons is included in the energy dispersion of the quasiparticle. As described below, the Fermi liquid theory developed by Landau has justified the one-electron approximation for electrons near the Fermi level in interacting electron systems. Indeed, all discussions so far made are based on the one-electron approximation and the Schrödinger equation is constructed by assuming each conduction electron experiences an average Coulomb field created by other electrons.

The repulsive Coulomb force is exerted between any two electrons and the Coulomb potential energy is inversely proportional to their distance apart, regardless of the quantum states involved. As noted in Section 8.2, the Coulomb interaction between the conduction electrons in a metal is reduced to the screened short-range potential $\exp(-\lambda r)/r$, because the Fourier components of the Coulomb potential in the range of short wave vectors are separated as collective excitations, called the plasmon. As a result, the Coulomb interaction is screened at large distances and each electron effectively keeps other electrons away from its neighbors and behaves as if it carries a "positively charged" cloud along with it. This effect is reflected in the energy dispersion relation of the conduction electron, which may be written as

$$\varepsilon(k) = \hbar^2 k^2 / 2m^*, \qquad (14.1)$$

where m^* is called the effective mass of the quasiparticle or quasielectron and may be slightly different from that of the non-interacting free-electron.

The Fermi liquid theory developed by Landau assumes a one-to-one correspondence between the interacting and non-interacting electron systems. The quasiparticle is defined in such a way that the energy of each electron is modified when the interaction between electrons is gradually turned on. Hence, the quantum state of the quasiparticle is still uniquely assigned in terms of its

---

[1] The word "quasiparticle" was also used in Section 12.12 to describe two independent electrons generated by the breaking of a Cooper pair.

momentum **p** and position vector **r** and the number of quasiparticles is defined as $f(\mathbf{p},\mathbf{r})d\mathbf{p}d\mathbf{r}/h^3$ in the phase space $d\mathbf{p}d\mathbf{r}$, where $f(\mathbf{p},\mathbf{r})$ is the distribution function. The quasiparticle is assumed to carry the electronic charge $(-e)$ in the same way as the non-interacting free-electron and to interact with the ionic potentials. In this way, the energy dispersion for the quasiparticle can be uniquely defined.

The energy of a quasiparticle with the wave vector **k** is expressed in terms of equation (14.1) by using the effective mass m^*. These quasiparticles are shown to be stable at energies close to the Fermi level but to dissipate with time due to transitions to other states at energies far from the Fermi level [2]. Thus, the Fermi liquid at absolute zero possesses a well-defined Fermi surface. The Fermi liquid theory has provided a firm basis for the notion of a Fermi surface, electron transport theories and electronic properties dominated by electrons near the Fermi level even in the presence of the electron–electron interaction. For example, the electronic specific heat in the interacting electron system is proved to be linearly proportional to the absolute temperature while the Pauli paramagnetic susceptibility is shown to be independent of temperature. The Fermi liquid theory can be applied successfully to a system forming an extended band in ordinary metals and alloys. The aim of this chapter is, however, to discuss the electronic structure of strongly correlated electron systems, which goes beyond the framework of the Fermi liquid theory.

14.3 Electronic states of hydrogen molecule and the Heitler–London approximation

In order to recognize the limit of the one-electron approximation, we try to solve the equation of motion for the two-electron system of a hydrogen molecule while intentionally ignoring the electron–electron interaction between the two electrons [3]. Since each electron interacts only with two protons 1 and 2, the Schrödinger equation (8.1) is explicitly written as

$$\left[\sum_{i=1,2}\left(-\frac{\hbar^2}{2m}\nabla_i^2 - \frac{e^2}{|\mathbf{r}_i - \mathbf{R}_1|} - \frac{e^2}{|\mathbf{r}_i - \mathbf{R}_2|}\right)\right]\psi(\mathbf{r}_1,\mathbf{r}_2) = E\psi(\mathbf{r}_1,\mathbf{r}_2), \quad (14.2)$$

where $\mathbf{r}_i$ is the position vector of the i-th electron, $\mathbf{R}_i$ is that of the i-th proton and $\psi(\mathbf{r}_1,\mathbf{r}_2)$ is the molecular orbital wave function of the two electrons. Let us rewrite equation (14.2) as

$$(H_1 + H_2)\psi(\mathbf{r}_1,\mathbf{r}_2) = E\psi(\mathbf{r}_1,\mathbf{r}_2), \quad (14.3)$$

where

$$H_i = -\frac{\hbar^2}{2m}\nabla_i^2 - \frac{e^2}{|\mathbf{r}_i - \mathbf{R}_1|} - \frac{e^2}{|\mathbf{r}_i - \mathbf{R}_2|} \quad (i=1,2). \quad (14.4)$$

14.3 Electronic states of hydrogen molecule; Heitler–London approximation

Since H_i represents the one-electron Hamiltonian, we drop the subscript i from equation (14.4) and write the Schrödinger equation as

$$H\psi(\mathbf{r}) = \varepsilon\psi(\mathbf{r}), \tag{14.5}$$

where $\psi(\mathbf{r})$ is the one-electron wave function and ε is its energy in the presence of the two protons. The solutions of equation (14.5) with the lowest energy ε_0 and the second-lowest energy ε_1 are denoted as $\psi_0(\mathbf{r})$ and $\psi_1(\mathbf{r})$, respectively.

Now we consider the total orbital wave function $\psi(\mathbf{r}_1,\mathbf{r}_2)$ for two electrons in a hydrogen molecule within the one-electron approximation. Equation (14.3) at the lowest energy $E_s = 2\varepsilon_0$ is simply given by the product of $\psi_0(\mathbf{r}_1)$ and $\psi_0(\mathbf{r}_2)$:

$$\psi_s(\mathbf{r}_1,\mathbf{r}_2) = \psi_0(\mathbf{r}_1)\psi_0(\mathbf{r}_2). \tag{14.6}$$

Equation (14.6) is the symmetric orbital wave function, since it is unchanged upon the interchange of the coordinates 1 and 2. Similarly, the state with the next-lowest energy $E_t = \varepsilon_0 + \varepsilon_1$ can be written as

$$\psi_t(\mathbf{r}_1,\mathbf{r}_2) = \psi_0(\mathbf{r}_1)\psi_1(\mathbf{r}_2) - \psi_0(\mathbf{r}_2)\psi_1(\mathbf{r}_1), \tag{14.7}$$

which is antisymmetric with respect to the interchange in the coordinates. Here the relation $E_s < E_t$ holds, since $E_s - E_t = \varepsilon_0 - \varepsilon_1 < 0$.

Prior to the discussion of the failure of the one-electron approximation, we consider the singlet and triplet states of the hydrogen molecule. As noted in Section 8.2, the total wave function given by the product of the orbital and spin wave functions must be antisymmetric with respect to an interchange in the space and spin coordinates in accordance with the Pauli exclusion principle. The spin wave function must be antisymmetric (or symmetric), if the orbital wave function is symmetric (or antisymmetric). The spin states of a two-electron system like the hydrogen molecule are given by a linear combination of the four different states $|\uparrow\uparrow\rangle$, $|\uparrow\downarrow\rangle$, $|\downarrow\uparrow\rangle$ and $|\downarrow\downarrow\rangle$:

$$\frac{1}{\sqrt{2}}(|\uparrow\downarrow\rangle - |\downarrow\uparrow\rangle), |\uparrow\uparrow\rangle, \frac{1}{\sqrt{2}}(|\uparrow\downarrow\rangle + |\downarrow\uparrow\rangle), |\downarrow\downarrow\rangle, \tag{14.8}$$

Obviously, the first spin state is antisymmetric while the latter three are symmetric with respect to an interchange of spins. The quantum number of the resultant spins and their z-components for the two-electron system is easily found to be $S = 0, 1, 1$ and 1 and $S_z = 0, 1, 0$ and -1, respectively (see Exercise 14.1). Thus, we see that the singlet state ($S = 0$ and $S_z = 0$) must accompany the symmetric orbital wave function given by equation (14.6), whereas the triplet state ($S = 1$ and $S_z = 1, 0$ and -1) must accompany the antisymmetric orbital wave function given by equation (14.7).

As pointed out in Section 8.5, the extended wave function is constructed

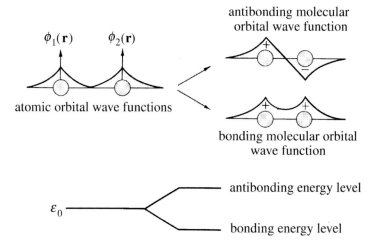

Figure 14.1. (*Left*) Atomic orbital wave functions of a diatomic molecule and (*right*) antibonding and bonding molecular orbital wave functions; with corresponding energy levels.

from a linear combination of atomic orbitals in the tight-binding method. This method is preferably called the molecular orbital method, when applied to a molecule like the hydrogen molecule. It may be appropriate to approximate $\psi(\mathbf{r}_1,\mathbf{r}_2)$ in equation (14.3) particularly in the immediate vicinity of each nucleus, where the molecular orbital should resemble an atomic orbital. Since $N=2$, there are only two different linear combinations of the wave functions:

$$\psi_0(\mathbf{r}) = \phi_1(\mathbf{r}) + \phi_2(\mathbf{r}) \tag{14.9a}$$

and

$$\psi_1(\mathbf{r}) = \phi_1(\mathbf{r}) - \phi_2(\mathbf{r}), \tag{14.9b}$$

where $\phi_i(\mathbf{r})$ is the 1s atomic orbital wave function of the hydrogen atom at the positions $i=1$ and 2. Equation (14.9a) represents the bonding molecular orbital of a given electron, since the probability density $|\psi_0(\mathbf{r})|^2$ is finite at a position intermediate between two protons. On the other hand, equation (14.9b) represents the antibonding molecular orbital, since the probability density becomes zero there. This is illustrated schematically in Fig. 14.1.

The symmetric and antisymmetric total orbital wave functions for the hydrogen molecule containing two electrons at positions $\mathbf{r}_1$ and $\mathbf{r}_2$ are now constructed by inserting equation (14.9) into equations (14.6) and (14.7), respectively:

$$\psi_s(\mathbf{r}_1,\mathbf{r}_2) = \psi_0(\mathbf{r}_1)\psi_0(\mathbf{r}_2)$$
$$= \phi_1(\mathbf{r}_1)\phi_2(\mathbf{r}_2) + \phi_2(\mathbf{r}_1)\phi_1(\mathbf{r}_2) + \phi_1(\mathbf{r}_1)\phi_1(\mathbf{r}_2) + \phi_2(\mathbf{r}_1)\phi_2(\mathbf{r}_2) \tag{14.10}$$

14.3 Electronic states of hydrogen molecule; Heitler–London approximation

and

$$\psi_t(\mathbf{r}_1,\mathbf{r}_2) = \psi_0(\mathbf{r}_1)\psi_1(\mathbf{r}_2) - \psi_0(\mathbf{r}_2)\psi_1(\mathbf{r}_1)$$
$$= 2[\phi_2(\mathbf{r}_1)\phi_1(\mathbf{r}_2) - \phi_1(\mathbf{r}_1)\phi_2(\mathbf{r}_2)]. \quad (14.11)$$

The first and second terms in equation (14.10) describe the motion of each electron in orbit around the respective protons $i=1$ and 2, whereas the third and fourth ones describe the electronic state of the H^- ion plus the bare proton, since two electrons simultaneously occupy the same orbit in one of the two protons. If we were to include the electron–electron interaction in equation (14.2), then the third and fourth terms should not have occurred because of an increase in the Coulomb repulsion between the two electrons. Now it is clear that equation (14.10) cannot be taken as the ground state in the hydrogen molecule, since the electron–electron interaction cannot be neglected.

In order to discuss properly the ground state of the hydrogen molecule, one must retain only the first and second terms of equation (14.10) but drop the third and fourth terms:

$$\psi_s(\mathbf{r}_1,\mathbf{r}_2) = \phi_1(\mathbf{r}_1)\phi_2(\mathbf{r}_2) + \phi_2(\mathbf{r}_1)\phi_1(\mathbf{r}_2). \quad (14.12)$$

Indeed, the Heitler–London model chooses equations (14.11) and (14.12) as the wave functions appropriate to the triplet and singlet states of the hydrogen molecule, respectively. As is clear from the argument above, the tight-binding approximation automatically includes terms in which the Coulomb repulsive energy is inevitably increased as a result of the simultaneous occupation of two electrons in the same orbit around a given atom.

Before ending this section, we can calculate the energy difference between the triplet and singlet states of the hydrogen molecule from equations (14.11) and (14.12). It is easily obtained as

$$\Delta E = E_s - E_t$$
$$= 2\iint \phi_1^*(\mathbf{r}_1)\phi_2^*(\mathbf{r}_2)\left(\frac{e^2}{|\mathbf{r}_1-\mathbf{r}_2|} + \frac{e^2}{|\mathbf{R}_1-\mathbf{R}_2|} - \frac{e^2}{|\mathbf{r}_1-\mathbf{R}_1|} - \frac{e^2}{|\mathbf{r}_2-\mathbf{R}_2|}\right)\phi_1(\mathbf{r}_2)\phi_2(\mathbf{r}_1)d\mathbf{r}_1 d\mathbf{r}_2. \quad (14.13)$$

The integral in equation (14.13) was already referred to as an exchange integral in Section 13.4. The first and second terms in the bracket always give rise to positive contributions whereas the third and fourth terms negative contributions. Since the third and fourth terms can be rewritten as

$$I_{12} = \int \phi_1^*(\mathbf{r}_1)\left(\frac{e^2}{|\mathbf{r}_1-\mathbf{R}_1|}\right)\phi_2(\mathbf{r}_1)d\mathbf{r}_1 \int \phi_2^*(\mathbf{r}_2)\phi_1(\mathbf{r}_2)d\mathbf{r}_2$$
$$= \int \phi_1^*(\mathbf{r}_1)\phi_2(\mathbf{r}_1)d\mathbf{r}_1 \int \phi_2^*(\mathbf{r}_2)\left(\frac{e^2}{|\mathbf{r}_2-\mathbf{R}_2|}\right)\phi_1(\mathbf{r}_2)d\mathbf{r}_2,$$

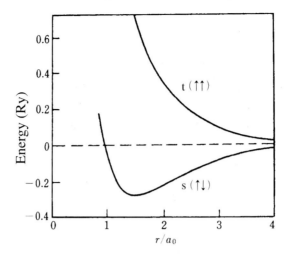

Figure 14.2. Energy of the hydrogen molecule in the Heitler–London model. The symbols s(↑↓) and t(↑↑) refer to the symmetric and antisymmetric orbital wave functions, respectively. The energy is taken to be zero, when two hydrogen atoms are infinitely apart. The distance is normalized with respect to the Bohr radius $a_0 = 0.053$ nm.

we find both terms to be equal to each other. If the two electrons enter different orbits in a given atom, the integral $\int \phi_1^*(\mathbf{r})\phi_2(\mathbf{r})d\mathbf{r}$ in equation (14.13) vanishes because of the orthogonality of the wave functions. The second term also vanishes for the same reason. Then, we obtain $E_s > E_t$, since the remaining first term is always positive. This means that the energy of the system can be lowered by aligning two spins in different orbits parallel to each other. This leads to the Hund rule discussed in Section 13.3. On the other hand, the integral I_{12} becomes finite when the orthogonality condition no longer holds, which occurs when electrons occupy orbits around different atoms as in the hydrogen molecule. In this case, equation (14.13) generally becomes negative and the singlet state becomes more stable. This is the case for the hydrogen molecule, as shown in Fig. 14.2.

14.4 Failure of the one-electron approximation in a strongly correlated electron system

In this section, we extend our discussion of the diatomic molecule with $N=2$ to that of the system with $N=N$. The tight-binding method is employed, in which the Bloch wave is constructed from a linear combination of the wave functions of the free atom. One can immediately realize that equation (8.31) is derived as an extension of equation (14.9) with $N=2$ to a crystal with $N=N$; the wave functions with $k=0$ and π/a in equation (8.31) for $N=2$ lead to the bonding and

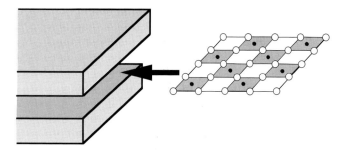

Figure 14.3. Alternate stacks of block layer and CuO_2 plane of the high-T_c cuprate superconductors.

antibonding molecular orbital wave functions of equation (14.9), respectively.[2] The antisymmetric total wave function can be constructed from equation (8.31) in the determinantal form (8.10) in the same way as the derivation of equations (14.10) and (14.11) with $N=2$. Obviously, the same difficulty associated with the third and fourth terms in equation (14.10), i.e., the simultaneous occupation of two electrons in the same orbit around a given atom occurs, unless some means like the Heitler–London approximation is incorporated into it to circumvent the occurence of such unfavorable electronic configurations.

We now consider a cuprate compound typical of a strongly correlated electron system, which undergoes a transition to the superconducting state by carrier doping, and show how the one-electron approximation fails when applied to it. A large number of high-T_c cuprate compounds crystallize into the layered perovskite structure, in which the CuO_2 plane and the so-called "block layer" are stacked, as sketched in Fig. 14.3 [4]. The block layer represents an intervening layer composed of metal ions and oxygen ions.

As a typical example, we show the crystal structure of the $La_{2-x}Sr_xCuO_4$ compound in Fig. 14.4. It consists of a repetition of the rock salt-type La_2O_2 block layer and the CuO_2 plane. The CuO_2 plane per cell is charged to -2, since Cu and O atoms are ionized to $+2$ and -2, respectively. The average charge of the block layer must be $+2$ in order to maintain charge neutrality in the whole crystal. The replacement of La^{3+} by Sr^{2+} ions in the $La_{2-x}Sr_xCuO_4$ compound causes the charge of the block layer to decrease to $3(2-x)+2x-4=2-x$ and the difference x must be compensated for in the CuO_2 plane to maintain charge neutrality. This means that the excess positive charge enters into the CuO_2 plane and makes it to be conductive. This is called hole doping.

---

[2] Equation (8.31) is reduced to $\psi_k(r) = \sum_{l=0,a} e^{ik \cdot l} \phi_l(r) = \phi(r) + e^{ik \cdot a} \phi(r-a)$ for the $N=2$ system. We have $\psi_0(r) = \phi_1(r) + \phi_2(r)$ or $\psi_1(r) = \phi_1(r) - \phi_2(r)$, since $k = n\pi/a$ ($n=0$ and 1).

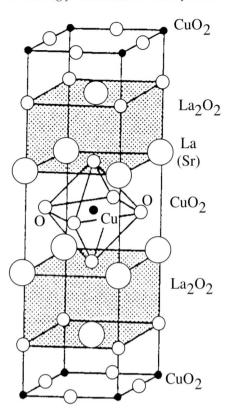

Figure 14.4.   Structure of $La_{2-x}Sr_xCuO_4$. [Y. Tokura, H. Takagi and S. Uchida, *Nature* **337** (1989) 345]

The one-electron band calculations have been performed for such oxides. For example the electronic structure of the undoped La_2CuO_4 compound has been calculated by the linear augmented plane wave (LAPW) method in combination with the local density functional method (see Section 8.3). The results are reproduced in Fig. 14.5 [5]. A fairly large hole Fermi surface is centered at X. This clearly indicates the formation of a metal, though this undoped compound is an antiferromagnetic insulator. This is indeed a demonstration of the failure of the ordinary band calculations, even though the electron–electron interaction is taken into account through the use of the local density functional method.

In spite of the apparent failure of the band calculations, we can identify essential features from the calculated electronic structure. First, bands associated with La appear above the Fermi level E_F in good agreement with a full ionization of the La ion: La-5d states appear at about $+2$ eV and La-4f orbitals form flat bands at about $+4$ eV. The bands labeled A and B in Fig. 14.5 are confirmed to originate from hybridization of the $3d_{x^2-y^2}$ orbitals of Cu sites with $2p_\sigma$ orbitals of the neighboring oxygen atoms of the CuO_2 plane, which is

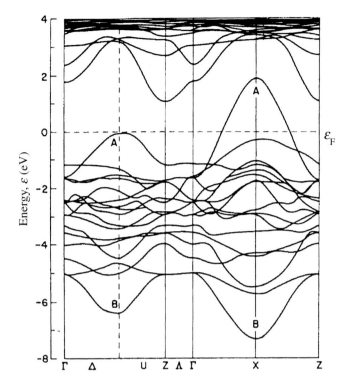

Figure 14.5. Energy bands for La_2CuO_4 in the bct Brillouin zone [5]. A and B represent the antibonding and bonding states between Cu-$3d_{x^2-y^2}$ and O-$2p_\sigma$ orbitals, respectively.

illustrated schematically in Fig. 14.6. This hybridization results in one hole on each Cu site, which gives rise to a large Fermi surface centered at the point X in Fig. 14.5. As will be discussed in the next section, we see that the failure of the one-electron band calculations stems from the difficulty in evaluating properly the Coulomb repulsive energy associated with the Cu-3d orbital.

14.5 Hubbard model and electronic structure of a strongly correlated electron system

Hubbard [6] proposed a model which is capable of describing the extended electronic states expected from the band theories on the one hand and the localized states dominated by the on-site Coulomb energy on the other. The Hubbard Hamiltonian is expressed as

$$H = -t\sum_{i,j}(c^+_{i\uparrow}c_{j\uparrow} + c^+_{i\downarrow}c_{j\downarrow}) + U\sum_i n_{i\uparrow}n_{i\downarrow}, \qquad (14.14)$$

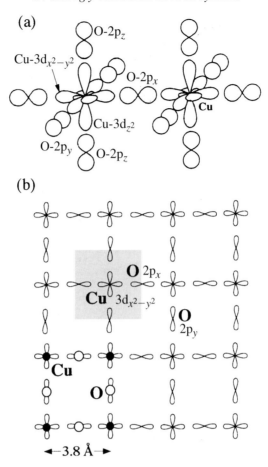

Figure 14.6. (a) Three-dimensional electronic configurations of the CuO_2 plane and (b) that within the CuO_2 plane. The square unit cell contains one Cu atom (●) and two oxygen atoms (○).

where $c^+_{i\uparrow}$ creates a spin-up electron at site i, $c_{j\downarrow}$ annihilates a spin-down electron at the site j and $n_{i\uparrow} = c^+_{i\uparrow} c_{i\uparrow}$ represents the number of spin-up electrons at site i and takes either 0 or unity. The second term indicates that the energy of the system increases by U when an spin-down electron is added to an orbit where a spin-up electron already exists.[3] The parameter U is called the on-site Coulomb energy. The parameter t in the first term is called the transfer integral or the hopping matrix element and represents the kinetic energy involved upon the transfer of a spin-up electron at site j to a neighboring site i without changing its spin orientation. The band becomes more extended, as the transfer integral t increases. Thus, we see that the Hubbard model treats the

---

[3] The electron with spin-up need not be considered because of the Pauli exclusion principle.

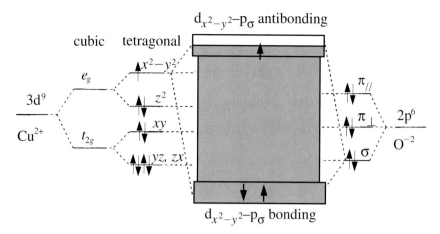

Figure 14.7. Electronic structure of CuO_2 plane in a crystalline field of tetragonal symmetry. The $d_{x^2-y^2}$ orbital is the only relevant one in the plane. The essential features are well reflected in the $\varepsilon - \mathbf{k}$ relation shown in Fig. 14.5.

Hamiltonian as a short-range phenomenon in contrast to the ordinary band calculations.

Let us consider again the CuO_2 plane in the layered perovskite cuprate compounds and apply the Hubbard model to it by focusing on the local electronic structure in the CuO_2 plane. There exist two oxygen atoms per Cu atom in the CuO_2 plane, and hence, the molecular wave function is constructed by a linear combination of six 2p and five 3d orbitals. The electronic structure of the CuO_2 plane exposed to a three-dimensional octahedral crystalline field of tetragonal symmetry is depicted in Fig. 14.7. As emphasized in Section 14.4, the hybridization of the $3d_{x^2-y^2}$ orbitals of Cu sites with $2p_\sigma$ orbitals of the neighboring oxygen atoms in the CuO_2 plane is critically important and gives rise to bonding and antibonding states. This is shown schematically in Fig. 14.7. Here it is important to remember that the oxygen atom has a very high electronegativity in comparison to other atoms.[4] Because of this nature, two electrons, one spin-up and the other spin-down, in the $d_{x^2-y^2}-p_\sigma$ orbital reside exclusively on the oxygen site and only one spin-up electron is left in the $d_{x^2-y^2}$ orbital on the Cu site.

Since all $d_{x^2-y^2}$ orbitals on Cu^{2+} sites are half-filled, we can in principle add

---

[4] Pauling [7] introduced a quantity termed electronegativity χ to discuss the nature of bonds between unlike atoms. It represents the power of an atom in a molecule to attract an electron to itself. Electronegativity values for relevant elements are as follows: Ti (1.5), V (1.6), Cr (1.6), Mn (1.5), Fe (1.8), Co (1.8), Ni (1.8), Cu (1.9), Zn (1.6) and O (3.5). Hence, the oxygen atom attracts more electrons than transition metal atoms. The square of the electronegativity difference of the atom pair is called the ionic resonance energy Δ. The larger the electronegativity difference or the ionic resonance energy, the stronger is the ionic character of the bond. Conversely, a bond with a small value of Δ will be more covalent.

a spin-down electron without any violation of the Pauli exclusion principle. This is indeed the case in metals, as shown in Fig. 14.5. However, when the condition $U \gg t$ holds in equation (14.14), the addition of the spin-down electron raises the on-site Coulomb energy by U. This leads to the formation of a new 3d band of the spin-down electrons at the energy U above the 3d band of the spin-up electrons. These are called the upper and lower Hubbard bands, which are separated by the energy U. Note that the band calculations based on the one-electron approximation are unable to produce this unique band structure. Hindrance of the electron transfer from site i to its neighbors means that the system becomes an insulator in spite of the fact that the band is only half-filled. An insulator thus obtained is called a Mott–Hubbard insulator.

The energy ε_i of the i-th electron in a metal crystal is obtained as the solution of equation (8.8) in the one-electron approximation. According to Koopmans theorem [8], the difference in the total energy of an N- and an $(N-1)$-electron system is equal to the energy of the electron that has been omitted. This theorem is based on the assumption that the individual one-electron wave functions are the same in both N- and $(N-1)$-electron systems. Koopmans theorem obviously fails for the Mott–Hubbard insulator. As is clear from the argument above, the one-electron band picture breaks down when the Coulomb energy U plays a critical role for a half-filled band. This is a characteristic feature of a strongly correlated electron system.

Research on the metal–insulator transition induced by strong electron–electron interaction dates back many years. For instance, data of the metal–insulator transition for the V_2O_3 system, which is known as a typical Mott–Hubbard insulator, are depicted in Fig. 14.8 [9]. It can be seen that the metal–insulator transition temperature systematically changes with either varying pressure or substitution of Cr or Ti for V and that the antiferromagnetic insulating phase is stable at low temperatures but transforms into the metallic phase at high temperatures.

14.6 Electronic structure of 3d-transition metal oxides

We have emphasized so far the importance of the on-site repulsive Coulomb energy associated with the 3d orbital. As noted above, the p orbitals of the neighboring oxygen atoms also play a crucial role in the electronic structure near the Fermi level. The energy needed for the transfer of 3d electrons to the oxygen atom is important. The charge transfer energy or resonance energy Δ, which is essentially proportional to the square of the electronegativity difference [7], is employed as its measure (see footnote 4). The value of Δ between the 3d-transition metal atom and a neighboring oxygen atom tends to decrease

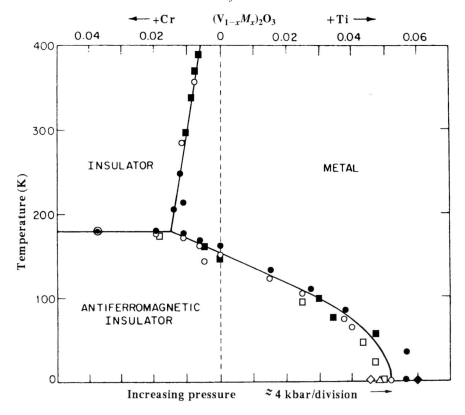

Figure 14.8. Generalized phase diagram of transition temperature as a function of pressure and as a function of doping with $M=$Cr or Ti in $(V_{1-x}M_x)_2O_3$ system. Solid and open circles are obtained from the temperature dependence of resistivity at 1 atm for mixed oxides on increasing and decreasing temperature, respectively. Triangles, squares and diamonds are obtained for $x=0$; $x=0.04$, $M=$Cr; and $x=0.04$, $M=$Ti, respectively, for increasing (*solid*) and decreasing (*open*) pressure (or temperature), respectively. The pressure was scaled to the composition, i.e., 4 kbar/division, using the difference in the critical pressure for the "V_2O_3" and "$x=0.04$, $M=$Cr" samples (for more details, see ref. [9]).

with increasing atomic number and, as a result, the positions of both upper and lower Hubbard bands of the 3d orbitals come closer to the O-2p band. The interplay between the two parameters U and Δ determines the essential feature of the electronic structure of a strongly correlated electron system [10].

Figure 14.9(a) illustrates the electronic structure of a system with $\Delta > U$, where the p band of a non-metallic element such as oxygen is located far below the lower Hubbard band. When the band width W is wider than the on-site Coulomb energy U, the upper and lower Hubbard bands are combined into a single band into which the Fermi level falls. This is the case for an ordinary metal. However, if $U>W$ holds, the upper and lower Hubbard bands are

(a) Δ is large

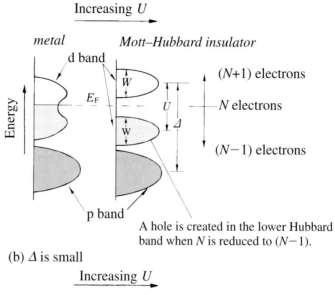

(b) Δ is small

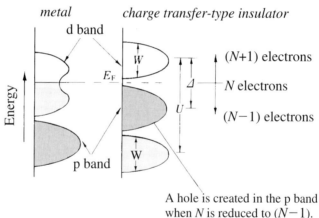

Figure 14.9. Electronic structure of (a) Mott–Hubbard insulator and (b) charge transfer-type insulator in comparison with the corresponding metals. The parameter Δ is large in (a) but small in (b). The ground state refers to the system with N electrons. The system is excited to those with $(N-1)$ or $(N+1)$ electrons either by ejecting one photoelectron from the occupied states or by absorbing one electron in the unoccupied state. Both $(N-1)$ or $(N+1)$ states can be observed by photoemission and inverse photoemission experiments but the N-electron states cannot be observed. Contrary to normal metals, the electronic structure in a strongly correlated system depends on how many electrons are involved in the system.

Table 14.1. Coulomb energy U and charge transfer energy Δ in strongly correlated systems [11]

system	U (eV)	Δ (eV)
CuO	7.5	3
NiO	7.5	4
FeO	7.3	6.5
MnO	7.0	7.5
VO	(~2)	~9
Fe_2O_3	8	3

separated by an energy gap, resulting in the Mott–Hubbard insulator. Here the p band is unimportant. Typical examples are the early transition metal oxides like Cr_2O_3, Ti_2O_3 and V_2O_3, the data for the last one being shown in Fig. 14.8.

The electronic structure for a system with $\Delta < U$ is shown in Fig. 14.9(b). Here the p band of the non-metallic element appears at an energy between the lower and upper Hubbard bands and strongly hybridizes with the d orbitals. An energy gap opens between the p band and the upper Hubbard band. This is typical of the electronic structure of a charge transfer-type insulator. Indeed, the formation of such an electronic structure has been experimentally confirmed by means of photoemission spectroscopy for late transition metal oxides like NiO and many cuprates. Table 14.1 lists the parameters U and Δ deduced from photoemission spectroscopy measurements [11]. It is clear that CuO and NiO are charge transfer-type insulators while VO is a Mott–Hubbard insulator. Insulators of both types are often collectively called Mott insulators.

14.7 High-T_c cuprate superconductors

A high-T_c cuprate superconductor is obtained near the composition where a charge transfer-type insulator is stabilized. As discussed in Section 14.4, high-T_c cuprate superconductors such as $La_{2-x}Sr_xCuO_4$ ($0.05 < x < 0.26$) (Fig. 14.4) and $YBa_2Cu_3O_{7-\delta}$ ($0 \leq \delta < 0.5$) (Fig. 12.21) are characterized by the possession of a two-dimensional layered perovskite structure formed by stacking CuO_2 planes separated by block layers. The electronic configuration of the Cu atom in the CuO_2 plane is $3d^9$, thereby each Cu atom being occupied by a hole with

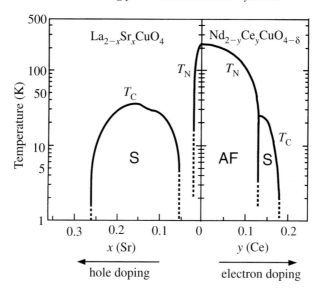

Figure 14.10. Phase diagram of $La_{2-x}Sr_xCuO_4$ and $Nd_{2-y}Ce_yCuO_{4-\delta}$ systems. S and AF stand for superconducting and antiferromagnetic phases, respectively. Curves labeled T_c and T_N show changes in the superconducting transition and Néel temperature, respectively.

the $d_{x^2-y^2}$ symmetry. The resulting $3d_{x^2-y^2}$ band is half-filled and is split into the upper and lower Hubbard bands. The oxygen $2p_\sigma$ band appears in between them, as indicated in Fig. 14.9(b). Hence, their undoped parent compounds are typical of a charge transfer-type insulator. The hole with the $d_{x^2-y^2}$ symmetry resides on each Cu atom in the CuO_2 plane and carries a magnetic moment of $S=\frac{1}{2}$, which aligns antiferromagnetically at low temperatures.

Mobile holes can be doped into the CuO_2 plane by substituting a Sr^{2+} ion for a La^{3+} ion in the block layers in $La_{2-x}Sr_xCuO_4$.[5] This means that electrons in the oxygen $2p_\sigma$ band are depleted or, equivalently, holes are introduced into the oxygen p_σ band. Now holes in the CuO_2 plane are responsible for the electron conduction and superconductivity as well. Electron doping is also made possible, for example, by substituting a Ce^{4+} ion for a Nd^{3+} ion in the $Nd_{2-y}Ce_yCuO_{4-\delta}$ system. As shown in Fig. 14.10, the superconducting phase appears next to the antiferromagnetic insulating phase with an increase in either electron or hole concentrations. It is also clear that an excessive increase in the carrier concentration suppresses the superconducting phase and,

---

[5] By analogy with semiconductors, the word "doping" is often employed to describe the introduction of carriers to the CuO_2 plane. Note, however, that the concentration of dopant is only in the range of a few ppm up to 0.1 at.% in semiconductors but is ranged over 5–10 at.% in high-T_c superconductors. Thus, the effect of the distortion of the crystal on the electronic structure must be significant in the latter.

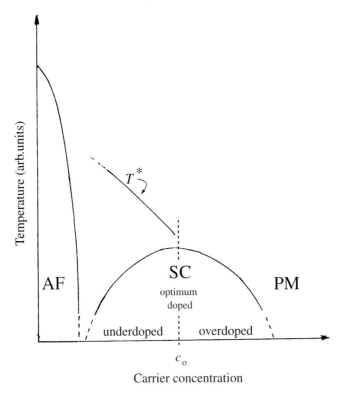

Figure 14.11. Typical phase diagram of high-T_c cuprates. AF, SC and PM stand for the antiferromagnetic phase, superconducting phase and paramagnetic metallic phase, respectively, and c_o is the optimum concentration, which divides the SC into underdoped and overdoped regions. T^* is called the spin-gap or pseudogap temperature observed in the underdoped region.

instead, stabilizes the non-superconducting metallic phase. This suggests that superconductivity appears only when the metallic nature is still insufficient.

A successive appearance of phases is quite commonly observed for a large number of high-T_c cuprates and, hence, a universal phase diagram can be drawn as a function of the concentration of carriers doped into the antiferromagnetic insulating phase of the parent compound. This is shown in Fig. 14.11. The superconducting transition temperature T_c increases up to the so-called optimum concentration c_o. But it begins to decrease with further increase in carrier concentration beyond c_o and the superconducting phase is finally replaced by a paramagnetic non-superconducting metallic phase. The superconducting region is often divided into three: the optimum doped region centered at c_o; the underdoped region in the range $c < c_o$; and the overdoped region in the range $c > c_o$. Observed physical properties above T_c are known to be strongly dependent on the three distinctive regions.

The electrical resistivity in the *ab*-plane of various cuprate compounds in the optimum doped region obeys surprisingly well a linear temperature dependence over a very wide temperature range covering from just above T_c up to 1000 K, in contrast to the behavior expected from the Bloch–Grüneisen law shown in Fig. 10.9. Deviation from a linear temperature dependence is, however, evident in both overdoped and underdoped regions. The electronic properties observed at temperatures above T_c are unique, in the sense that these behaviors cannot be consistently interpreted in terms of the ordinary transport theories applicable to normal metals. The presence of a "pseudogap" below T^* in the underdoped region is indicated from measurements such as neutron inelastic scattering [12].[6] Extensive studies have been carried out to clarify if the observed gap in the range $T_c < T < T^*$ is related to the signature of the residual energy gap inherent in either superconductivity or spin fluctuations associated with the $S = \frac{1}{2}$ spin on the Cu site. A more detailed discussion of the mechanism of superconductivity for the cuprate compounds goes beyond the level of the present book. References are listed for further studies of this topic [11, 12].

Exercise

14.1 As shown in equation (14.8), there are four spin functions for two electrons in the hydrogen molecule. One can easily assign the quantum numbers with $S=0, 1, 1$ and $S_z = 0, 1, -1$ to the $(1/\sqrt{2})(|\uparrow\downarrow\rangle - |\downarrow\uparrow\rangle)$, $|\uparrow\uparrow\rangle$ and $|\downarrow\downarrow\rangle$ states, respectively. In contrast, it is not so obvious why the quantum numbers $S=1$ and $S_z = 0$ are assigned to the $(1/\sqrt{2})(|\uparrow\downarrow\rangle + |\downarrow\uparrow\rangle)$ state. Show this by using the following relations for the resultant spin operator $\mathbf{S}$ and spin operators $\mathbf{S}_1$ and $\mathbf{S}_2$ for electrons 1 and 2:

$$\mathbf{S}^2 = (\mathbf{S}_1 + \mathbf{S}_2)^2 = S_1^2 + S_2^2 + 2(S_{x1}S_{x2} + S_{y1}S_{y2} + S_{z1}S_{z2}), \quad (14Q.1)$$

where spin operators are explicitly written as

$$S^2 = \frac{3}{4}\hbar^2 \begin{pmatrix} 1 & 0 \\ 0 & 1 \end{pmatrix}, \quad S_x = \frac{\hbar}{2}\begin{pmatrix} 0 & 1 \\ 1 & 0 \end{pmatrix}, \quad S_y = \frac{\hbar}{2}\begin{pmatrix} 0 & -i \\ i & 0 \end{pmatrix}, \quad S_z = \frac{\hbar}{2}\begin{pmatrix} 1 & 0 \\ 0 & -1 \end{pmatrix}. \quad (14Q.2)$$

---

[6] The term "pseudogap" was originally introduced by Mott for the dip in the density of states across the Fermi level, which grows upon the expansion of liquid mercury (see Sections 15.9 and 15.14). The origin of the pseudogap in this section is obviously different.

Chapter Fifteen

Electronic structure and electron transport properties of liquid metals, amorphous metals and quasicrystals

15.1 Prologue

When a crystal is melted, the periodic lattice is destroyed and the atomic distribution is randomized. This causes the Bloch theorem to fail in liquid metals. The discussion of the conduction electron in liquid metals dates back as early as 1936 when the Mott and Jones book [1] on the electron theory of metals was first published. Since the 1970s, new melt-quenching techniques have been developed and amorphous alloys stable at room temperature have become available in ribbon form. Abundant production of thermally stable amorphous alloys has enabled us to study their electron transport properties down to very low temperatures and has certainly widened the research field in the electron theory of a non-periodic system.

In 1984, Shechtman *et al.* [2] revealed that the electron diffraction pattern of the melt-quenched $Al_{86}Mn_{14}$ alloy exhibits two-, three- and five-fold symmetries incompatible with the translational symmetry of a crystal and suggested that this material belongs to a new family of substances different from crystals. Since then a number of solids of this class have been discovered along with progress in theoretical studies [3]. They are indeed new solids classified, in crystallographic terms, as quasicrystals. Because of the possession of five-fold symmetry incompatible with the translational symmetry, the Bloch theorem breaks down. Hence, they are grouped together with liquid metals and amorphous alloys into the category of non-periodic systems in spite of the possession of a high degree of ordering, evidenced from very sharp Bragg reflections.

The fundamental understanding of the atomic structure of quasicrystals has been deepened through the construction of a quasiperiodic Penrose lattice obtained by projecting lattice points in a selected region of a six-dimensional lattice onto a three-dimensional physical space.[1] The concept of the periodic

---

[1] Penrose is a mathematician famous for his remarkable work on a plane tiling with five-fold symmetry by building up two rhombs with a specific quasiperiodicity and long-range order (see Section 15.11).

Penrose lattice, which carries the same local structure as that of the quasiperiodic one, was also established. Complex compounds like the Frank–Kasper compound (see Section 15.11) have now been found to possess such a lattice. Such compounds are called approximants. The atomic structure of the periodic Penrose lattice gradually converges to a quasiperiodic one as the degree of "approximation" becomes less and less. Hence, studies of various approximants together with quasicrystals and amorphous alloys are important to bridge the gap between the periodic and non-periodic solids. In this chapter, we discuss the electronic structure and electron transport properties of liquid metals, amorphous alloys and quasicrystals as non-periodic systems, and approximants as periodic systems having large lattice constants, in the range 1–2.5 nm.

15.2 Atomic structure of liquid and amorphous metals

The atomic structure of a crystal can be uniquely determined, once its unit cell is defined. This is possible because of its possession of translational symmetry. However, the unit cell cannot be defined in liquid and amorphous metals because of the disordered distribution of atoms. Figure 15.1 illustrates the atom distribution in a liquid metal and an amorphous solid in comparison with that in a gas and in a crystal. Let us take one of the atoms as an origin and plot the number of atoms found between the two radii r and $r+dr$ from the origin. The results are shown in the right-hand side of Fig. 15.1. Obviously, the distribution of atoms in a crystal is described by a series of delta functions, since a definite number of atoms are found at a definite distance.

In contrast, the density of atoms in a gas is so low that their distribution is independent of the distance r except for the region below the diameter of the atom a, where two neighboring atoms cannot be geometrically overlapped. This is no longer true in the case of liquid metals and amorphous solids because the density is high. Indeed, some correlation in the atom distribution is clearly visible in both cases, as shown in Fig. 15.1. The distribution is again zero in the region $r \leq a$ because of the geometrical restriction mentioned above. There is a prominent peak at the nearest neighbor distance in both liquids and amorphous solids, indicating that the probability of finding atoms at this distance is high. This is called the first peak. Further peaks corresponding to the second, third, . . . nearest neighbor distances appear with increasing distance r but their intensities are rapidly weakened and eventually converge into an average density. The oscillatory function shown in Figs. 15.1(b) and (c) is a characteristic feature of liquid metals and amorphous solids and is referred to as the pair distribution function.

15.2 Atomic structure of liquid and amorphous metals

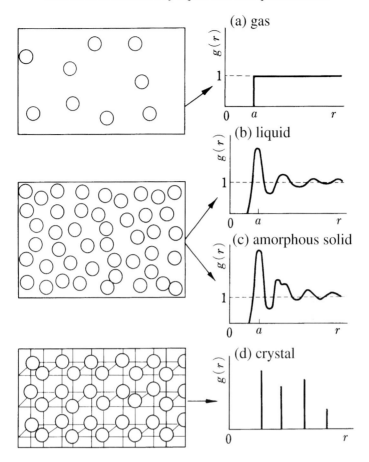

Figure 15.1. Distributions of atoms with radius $a/2$ and the corresponding pair distribution functions in (a) gas, (b) liquid, (c) amorphous solid and (d) crystal. (b) and (c) are drawn on a plane projected from the three-dimensional space so that atoms are positioned as if they are overlapped.

The pair distribution function $g(r)$ is mathematically defined as follows. The number of atoms found in the volume element $d\mathbf{r}(=4\pi r^2 dr)$ at the radius r from an atom at $r=0$ is expressed as

$$\rho_0 g(r) 4\pi r^2 dr, \tag{15.1}$$

where $\rho_0 = N/V$ is an average number density given by the ratio of N atoms over the volume V. Since the volume of the spherical shell, $4\pi r^2 dr$, increases with increasing radius r, the number of atoms involved in the shell increases and the quantity in equation (15.1) divided by $4\pi r^2 dr$ naturally approaches the average number density ρ_0. Thus, we obtain the relation

$$\lim_{r \to \infty} g(r) = 1. \tag{15.2}$$

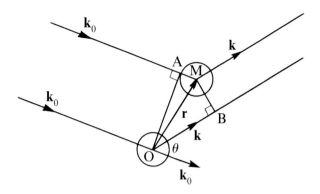

Figure 15.2. X-ray scattering due to atoms at O and M.

The pair distribution function $g(r)$ can be determined experimentally by using x-rays, electrons or neutrons. We consider below the x-ray diffraction technique for the structural analysis of a disordered system. Figure 15.2 illustrates the situation where the incident x-ray of the wave vector $\mathbf{k}_0$ is scattered into the wave of the wave vector $\mathbf{k}$ through an angle θ by two atoms located at the positions O and M. From a simple geometrical consideration, we can easily find that the difference in the optical path of the two scattered waves is given by

$$\overrightarrow{OB} - \overrightarrow{AM} = \mathbf{r} \cdot (\mathbf{k} - \mathbf{k}_0), \qquad (15.3)$$

where $\mathbf{r}$ is the distance between the two atoms. The scattering vector $\mathbf{K}$ is defined as $\mathbf{K} = \mathbf{k} - \mathbf{k}_0$. It is clear from Fig. 15.2 that the relation $|\mathbf{K}| = (4\pi/\lambda)\sin(\theta/2)$ holds, where λ is the wavelength of the incident x-rays.

The amplitude $A_M(\mathbf{K})$ of the x-ray scattered by the atom at the position M is related to $A_o(\mathbf{K})$ by the atom at the position O through the phase shift:

$$A_M(\mathbf{K}) = A_o(\mathbf{K}) \exp(-i\mathbf{K} \cdot \mathbf{r}). \qquad (15.4)$$

The intensity $I(\mathbf{K})$ of the x-rays scattered from N identical atoms is then calculated as

$$I(\mathbf{K}) = |A(\mathbf{K})|^2 = |A_o(\mathbf{K})|^2 \sum_{i=1}^{N} \sum_{j=1}^{N} \exp(-i\mathbf{K} \cdot \mathbf{r}_i) \exp(i\mathbf{K} \cdot \mathbf{r}_j)$$

$$= f^2 \left\{ N + \sum_{i \neq j}^{N} \sum_{j=1}^{N} \exp[-i\mathbf{K} \cdot (\mathbf{r}_i - \mathbf{r}_j)] \right\}, \qquad (15.5)$$

15.2 Atomic structure of liquid and amorphous metals

where $f = A_0(K)$, called the atomic scattering factor or scattering amplitude, depends on the atomic species involved and the magnitude of K or the scattering angle in the case of x-rays.[2]

The atom distribution in liquid and amorphous metals is assumed to be isotropic or spherically symmetric. Hence, both the intensity $I(\mathbf{K})$ and the atomic scattering factor f would solely depend on the magnitude of the scattering wave vector $\mathbf{K}$. Equation (15.5) is now further rewritten in the form:

$$S(\mathbf{K}) \equiv \frac{I(\mathbf{K})}{Nf^2} = 1 + \left(\frac{1}{N}\right) \sum_{\substack{i \neq j}}^{N} \sum_{j=1}^{N} \exp[-i\mathbf{K} \cdot (\mathbf{r}_i - \mathbf{r}_j)], \qquad (15.6)$$

where $S(\mathbf{K})$ is called the structure factor defined by the ratio of the coherent scattering intensity $I(\mathbf{K})$ over the non-interfering scattering intensity Nf^2 due to N atoms.

The pair distribution function $g(\mathbf{r})$ is defined as

$$\rho(\mathbf{r}) = \rho_0 g(\mathbf{r}) = \left(\frac{1}{N}\right) \left\langle \sum_{j=1}^{N} \sum_{i=1}^{N} \delta(\mathbf{r} - (\mathbf{r}_j - \mathbf{r}_i)) \right\rangle - \delta(\mathbf{r}), \qquad (15.7)$$

where $\rho(\mathbf{r})$ is the number-density function, ρ_0 is the average number density already defined, $\delta(\mathbf{r})$ is the delta function and $\langle \cdots \rangle_V$ is to take the sum over atom pairs i–j with a distance $\mathbf{r}$ apart in the volume V [4]. Since the relation $\lim_{\mathbf{r} \to \infty}(1/N)\langle \sum_{j=1}^{N} \sum_{i=1}^{N} \delta(\mathbf{r} - (\mathbf{r}_j - \mathbf{r}_i)) \rangle_V = \rho_0$ holds, equation (15.7) is consistent with equation (15.2). Furthermore, the delta function in the second term of equation (15.7) assures $g(0) = 0$.[3] Now let us multiply both sides of equation (15.7) by $\exp(-i\mathbf{K} \cdot \mathbf{r})d\mathbf{r}$ and integrate over the volume V:

$$\left(\frac{1}{N}\right) \sum_{j=1}^{N} \sum_{i=1}^{N} \exp[-i\mathbf{K} \cdot (\mathbf{r}_i - \mathbf{r}_j)] = 1 + \rho_0 \int g(\mathbf{r}) \exp(-i\mathbf{K} \cdot \mathbf{r}) d\mathbf{r},$$

where the relation $\int \delta(\mathbf{r} - \mathbf{r}') \exp(-i\mathbf{K} \cdot \mathbf{r}) d\mathbf{r} = \exp(-i\mathbf{K} \cdot \mathbf{r}')$ is used. Here the left-hand side of this equation is nothing but the structure factor. Thus we find that the structure factor is linked with the pair distribution function through the relation:

$$S(\mathbf{K}) = 1 + \rho_0 \int g(\mathbf{r}) \exp(-i\mathbf{K} \cdot \mathbf{r}) d\mathbf{r}. \qquad (15.8)$$

---

[2] The x-ray scattering amplitude depends on K because the dimensions of the core electron distribution are comparable to the x-ray wavelength. In the case of neutron scattering, the amplitude is independent of K, since the size of the nucleus, of the order of 10^{-4} nm, is small in comparison with the neutron wavelength of 10 nm.

[3] Note that the delta function is in the units of [length]$^{-3}$, since the relation $\int_V \delta(\mathbf{r}) d\mathbf{r} = 1$ holds. Thus, the right-hand side of equation (15.7) has the dimension of [length]$^{-3}$ and represents the number density in the volume V.

Since an isotropic atom distribution is assumed for liquid and amorphous metals, we have $g(\mathbf{r}) = g(r)$. The volume integral in equation (15.8) is explicitly rewritten in polar coordinates as

$$S(K) = 1 + 4\pi\rho_0 \int_0^\infty [g(r) - 1] r^2 \left(\frac{\sin Kr}{Kr}\right) dr + 4\pi\rho_0 \int_0^\infty r^2 \left(\frac{\sin Kr}{Kr}\right) dr, \quad (15.9)$$

where the third term contributes to the structure factor only through the forward scattering in the region $K \approx 0$.[4] Since this term corresponds to the case with $g(r) = 1$, it represents the diffraction effect due to a body having a rigorously uniform density and cannot be distinguished from a direct beam. The measurements are always carried out to avoid the effect due to the direct beam. Hence, we will ignore the third term in the rest of our discussion.

The structure factor is now written as

$$S(K) = 1 + 4\pi\rho_0 \int_0^\infty [g(r) - 1] r^2 \left(\frac{\sin Kr}{Kr}\right) dr. \quad (15.10)$$

The pair distribution function $g(r)$ is obtained by the Fourier transformation of equation (15.10):

$$g(r) = 1 + \left(\frac{1}{2\pi^2 \rho_0}\right) \int_0^\infty [S(K) - 1] K^2 \left(\frac{\sin Kr}{Kr}\right) dK. \quad (15.11)$$

The pair distribution function is experimentally determined as follows: the coherent scattering intensity $I(K)$ is first derived by subtracting incoherent contributions such as inelastic scattering from the diffraction spectrum. The structure factor $S(K)$ is then calculated by taking the ratio of $I(K)$ over the intensity Nf^2 of non-interfering scattering due to N atoms. The pair distribution function is finally obtained by inserting the resulting $S(K)$ into equation (15.11).

Figure 15.3 shows the structure factor $S(K)$ and the pair distribution function $g(r)$ for liquid and amorphous Ni. It is clear that the overall atomic structures of liquid and amorphous Ni resemble each other, unless a detailed comparison is made. It is also noted that both $S(K)$ and $g(r)$ oscillations rapidly decrease their intensity with increasing K and r, respectively. This is a characteristic feature observed in both liquid and amorphous metals. The first peak in $S(K)$ appears at the wave number K_p, which is roughly equal to the inverse of the average atomic distance r_0 corresponding to the first peak in $g(r)$. There

---

[4] When K is sizeable, the function $(\sin Kr)/Kr$ oscillates around the r-axis and converges to zero with increasing r. Its positive and negative contributions to the integral are largely cancelled and, hence, the third term can be neglected, so long as K is not too close to zero.

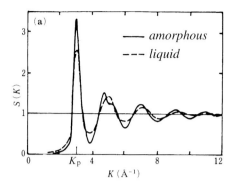 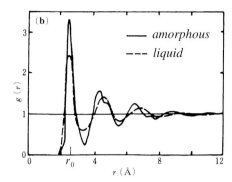

Figure 15.3. (a) Structure factor and (b) its pair distribution function of liquid and amorphous Ni. K_p refers to the wave number corresponding to the first peak of the structure factor. [Y. Waseda et al., J. Mat. Sci. **12** (1977) 1927]

exists no measurable difference in the position of the first peak between the liquid and the amorphous metal. However, a close inspection of the spectra reveals the following differences between them:

1. The intensity of the first peak in both $S(K)$ and $g(r)$ of the amorphous phase is stronger than that of the liquid phase.
2. The second peak of $S(K)$ and $g(r)$ is often split into two components in the amorphous phase but generally not in the liquid phase.
3. The oscillatory behavior in the amorphous phase persists up to a greater distance than that in the liquid phase.

The differences mentioned above originate from the fact that the atom distribution in the amorphous phase is more dense than that in the liquid phase. Indeed, the packing density of atoms in a liquid metal is about 0.5, whereas that in an amorphous metal is generally about 0.7 and is closer to that in a crystal metal.

We have so far compared the atomic structure of liquid and amorphous phases in a single-component system. However, an amorphous phase is most frequently obtained in an alloy system. The atomic structure of a multi-component liquid or amorphous alloy depends not only on the geometrical arrangements of the atoms but also on the number of ways of distributing the different chemical species over short and medium ranges.[5] Let us consider a binary alloy system composed of atom A and atom B. There are three different

---

[5] The short-range structure in an amorphous alloy refers to that of nearest neighboring atoms around a given atom and is often discussed in terms of the coordination number and average atomic distance in comparison with those of its parent crystal. The medium-range structure generally refers to the connectivity of the short-range one and extends to the range of second or third nearest neighbor atoms.

atom pairs A–A, B–B and A–B. If the number of the relevant atom pair is larger than its statistical average, then we say that there exists a favorable bonding in this atom pair. Conversely, a tendency to repulsive bonding appears if it is smaller than the statistical average. The deviation from the statistical average is often discussed in terms of a chemical short-range order. If attractive bonding dominates, the atomic distance would be shortened relative to the sum of the Goldschmidt radii of the two atoms.[6] There always exists some short-range order in amorphous alloys, though its degree depends on the system chosen. We will learn later that the short-range order affects substantially the electronic properties of amorphous alloys.

Let us denote the atomic scattering factors of atoms A and B to be f_A and f_B, respectively, in an A–B alloy. The amplitude of the scattered x-rays is given by

$$A(\mathbf{K}) = f_A \sum_{i=1}^{N_A} \exp(-i\mathbf{K}\cdot\mathbf{r}_{i(A)}) + f_B \sum_{i=1}^{N_B} \exp(-i\mathbf{K}\cdot\mathbf{r}_{i(B)}), \qquad (15.12)$$

where the subscript $i(\alpha)$ means that the i-th atom is $\alpha = A$ or B and $N = N_A + N_B$ is the total number of atoms in a volume V. The coherent scattering intensity is obtained by taking the product of equation (15.12) with its complex conjugate:

$$I(K) = |A(\mathbf{K})|^2 = \sum_{\alpha,\beta} f_\alpha^* f_\beta \sum_{j(\beta)} \sum_{i(\alpha)} \exp[-i\mathbf{K}\cdot(\mathbf{r}_{i(\alpha)} - \mathbf{r}_{j(\beta)})]$$

$$= f_A^2 \sum_{j=1}^{N_A} \sum_{i=1}^{N_A} \exp[-i\mathbf{K}\cdot(\mathbf{r}_{i(A)} - \mathbf{r}_{j(A)})] + f_B^2 \sum_{j=1}^{N_B} \sum_{i=1}^{N_B} \exp[-i\mathbf{K}\cdot(\mathbf{r}_{i(B)} - \mathbf{r}_{j(B)})]$$

$$+ 2 f_A f_B \sum_{j=1}^{N_B} \sum_{i=1}^{N_A} \exp[-i\mathbf{K}\cdot(\mathbf{r}_{i(A)} - \mathbf{r}_{j(B)})], \qquad (15.13)$$

where sums appearing in the first, second and third terms correspond to the partial structure factors of the atom pairs A–A, B–B and A–B, respectively, in the same manner as equation (15.6) for the single-component system.

Ashcroft and Langreth [5] defined the partial structure factor for the atom pair α–β as

---

[6] The traditional sets of ionic radii of each element determined by Goldschmidt *et al.* (1926) and Pauling (1927) have been used with considerable success. More comprehensive values of ionic radii have been compiled by Shannon and Prewitt [*Acta Cryst.* **B25** (1969) 925]. They are all referred to as the Goldschmidt radius.

15.2 Atomic structure of liquid and amorphous metals

$$S_{\alpha\beta}(\mathbf{K}) = (N_\alpha N_\beta)^{-1/2} \left\{ \sum_{j=1}^{N_\beta} \sum_{i=1}^{N_\alpha} \exp[-i\mathbf{K}\cdot(\mathbf{r}_{i(\alpha)} - \mathbf{r}_{j(\beta)})] - \delta_{\mathbf{K},0} \right\}, \quad (15.14)$$

where the second term $\delta_{\mathbf{K},0}$ in the curly bracket appears to exclude the forward scattering. By using the Ashcroft–Langreth structure factor in equation (15.14), we can rewrite equation (15.13) as

$$I(\mathbf{K}) = N \sum_\alpha \sum_\beta (c_\alpha c_\beta)^{1/2} f_\alpha f_\beta S_{\alpha\beta}(\mathbf{K}), \quad (15.15)$$

where $c_\alpha = N_\alpha/N$ represents the concentration of α atoms, where α is A or B.

The number-density function $\rho_{\alpha\beta}(\mathbf{r})$ is introduced to represent the number of β atoms found at a radius $\mathbf{r}$ from the atom α at the origin. The partial pair distribution function $g_{\alpha\beta}(\mathbf{r})$ is defined as

$$\rho_{\alpha\beta}(\mathbf{r}) = c_\beta \rho_0 g_{\alpha\beta}(\mathbf{r})$$
$$= N_\alpha^{-1} \left\langle \sum_i \sum_j \delta[\mathbf{r} - (\mathbf{r}_{i(\alpha)} - \mathbf{r}_{j(\beta)})] \right\rangle_V - \delta_{\alpha\beta}\delta(\mathbf{r}), \quad (15.16)$$

where the symbol $\delta_{\alpha\beta}$ in the second term of the right-hand side indicates that it is unity for the like atom pair α–α while it is zero for the unlike atom pair α–β. This assures the condition $g_{\alpha\beta}(0) = 0$. As has already been defined in equation (15.7), the bracket $\langle \cdots \rangle_V$ is to take the sum over atom pairs with a distance $\mathbf{r}$ apart in the volume V. We multiply both sides of equation (15.16) by $\exp(-i\mathbf{k}\cdot\mathbf{r})d\mathbf{r}$ and integrate over the volume V in the same way as in the single-component system. The partial structure factor $S_{\alpha\beta}(\mathbf{K})$ defined by equation (15.14) is now expressed in terms of $g_{\alpha\beta}(\mathbf{r})$ as follows:

$$S_{\alpha\beta}(\mathbf{K}) = \delta_{\alpha\beta} + (c_\alpha c_\beta)^{1/2} \rho_0 \int [g_{\alpha\beta}(\mathbf{r}) - 1]\exp(-i\mathbf{K}\cdot\mathbf{r})d\mathbf{r}. \quad (15.17)$$

This definition of the partial structure factor is, however, not unique. Faber and Ziman [6] employed a different expression for the partial structure factor in their discussion of the electron transport properties of liquid binary alloys. They define the partial structure factor $a_{\alpha\beta}(K)$ in terms of the partial pair distribution function $g_{\alpha\beta}(\mathbf{r})$ in the form:

$$a_{\alpha\beta}(\mathbf{K}) = 1 + \rho_0 \int [g_{\alpha\beta}(\mathbf{r}) - 1]\exp(-i\mathbf{K}\cdot\mathbf{r})d\mathbf{r}. \quad (15.18)$$

By using the Faber–Ziman partial structure factor, one can write the coherent scattering intensity of equation (15.13) as

$$I(\mathbf{K}) = \sum_{\alpha,\beta} f_\alpha^* f_\beta \sum_{j(\beta)} \sum_{i(\alpha)} \exp[-i\mathbf{K}\cdot(\mathbf{r}_{i(\alpha)} - \mathbf{r}_{j(\beta)})]$$

$$= \sum_{\alpha,\beta} f_\alpha^* f_\beta N c_\alpha \left\{ \delta_{\alpha\beta} + \sum_{i \neq j} \exp[-i\mathbf{K}\cdot(\mathbf{r}_{i(\alpha)} - \mathbf{r}_{j(\beta)})] \right\}$$

$$= \sum_\alpha \sum_\beta f_\alpha^* f_\beta N c_\alpha \{\delta_{\alpha\beta} + c_\beta [a_{\alpha\beta}(K) - 1]\}, \qquad (15.19)$$

which is further rewritten as

$$I(\mathbf{K}) = N\{c_A c_B [f_A(\mathbf{K}) - f_B(\mathbf{K})]^2 + \sum_\alpha \sum_\beta f_\alpha(\mathbf{K}) f_\beta(\mathbf{K}) a_{\alpha\beta}(\mathbf{K})\} \qquad (15.20)$$

for a binary A–B alloy system. Here the $\mathbf{K}$ dependence of f_A and f_B is explicitly indicated (see footnote 2, p. 455).

The partial pair distribution function $g_{\alpha\beta}(r)$ for an isotropic system is obtained by the Fourier transformation of equation (15.18):

$$g_{\alpha\beta}(r) = 1 + \left(\frac{1}{2\pi^2 \rho_0}\right) \int_0^\infty [a_{\alpha\beta}(K) - 1] K^2 \left(\frac{\sin Kr}{Kr}\right) dK. \qquad (15.21)$$

The number of the structure factors is increased to three in a binary system: a_{AA}, a_{BB} and a_{AB}. Thus, we need three independent equations (15.15) or (15.20) to determine uniquely the three partial structure factors. Once they are obtained, the local atomic structure can be calculated from equation (15.21) or the Fourier transformation of equation (15.17).

Figure 15.4 shows the local atomic structure of the amorphous $Ni_{81}B_{19}$ alloy derived from the neutron diffraction technique. In the case of neutrons, the atomic scattering factor in equation (15.15) or (15.20) is replaced by the neutron scattering amplitude b. By making full use of the fact that the isotope ^{62}Ni possesses a negative scattering amplitude, one can intentionally prepare a so-called "neutron zero-alloy" having zero scattering amplitude by mixing an appropriate amount of ^{62}Ni and natural Ni which has a positive scattering amplitude.[7] The scattering from the Ni atoms is cancelled to zero in the zero-alloy. Hence, we can accurately determine the partial structure factor S_{B-B} associated with the boron B–B atom pair as if it were in a single-component system.

---

[7] The neutron wave function scattered from a nucleus is expressed as the sum of the incident plane wave and the scattered spherical wave: $\psi = e^{-ikx} - (b/r)e^{-ikr}$, where b is the scattering amplitude. The phase shift between incident wave and scattered wave at the nucleus site is π or $b > 0$ in many nuclei but is zero or $b < 0$ in some nuclei like ^{62}Ni. Scattering amplitudes in the istopes of Ni are as follows. ^{58}Ni: $b = 1.44 \times 10^{-12}$ cm (68%), ^{60}Ni: $b = 0.28 \times 10^{-12}$ cm (26 %), ^{62}Ni: $b = -0.87 \times 10^{-12}$ cm (3.6%). The values in parentheses indicate the natural abundances.

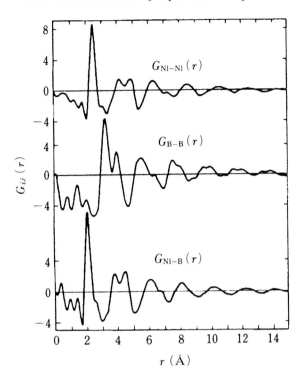

Figure 15.4. Partial pair distribution functions of the amorphous $Ni_{81}B_{19}$ alloy. Partial pair distribution function is defined as $G_{ij}(r) = 4\pi r_0[g_{ij}(r) - 1]$. [P. Lamparter *et al.*, *Z. Naturforsch.*, **37a** (1982) 1223]

We still need information about the local atomic structure around the Ni atom. Hence, two other samples with the same composition must be prepared, one using only ^{62}Ni and the other only natural Ni. By solving the three independent equations thus obtained, one can determine the remaining partial structure factors associated with the Ni–Ni and Ni–B atom pairs. As shown in Fig. 15.4, the B–B atom pair exhibits the largest separation among the three different atom pairs and the B atom is always surrounded by Ni atoms without allowing other B atoms as its nearest neighbors. In contrast, the Ni–B atom pair is formed at the shortest distance, thereby lending support to the formation of a strong covalent bonding between Ni and B atoms. As is clearly understood from the argument above, the atomic structure of a binary amorphous alloy is by no means simply given by a random mixture of two different atom species but is characterized by a short-range order unique to a given alloy system.

15.3 Preparation of amorphous alloys

Unique preparation techniques must be devised to synthesize amorphous alloys, since they exist only as a metastable phase. The currently available techniques may be divided into three: gas quenching, melt quenching and solid state reaction. In the gas quenching technique, an amorphous film can be prepared by depositing gas atoms or gas molecules onto a substrate. Gas evaporation and sputtering methods are included in this group. The melt-quenching technique has been widely used as a powerful tool to produce an amorphous alloy in a ribbon form since the early 1970s when a single-roll spinning wheel apparatus was developed. The mechanical alloying technique has received much attention since 1980s and has been recognized as a method to produce amorphous powders through a solid state reaction without involving a melting process. For example, an amorphous phase can be formed even in the immiscible Cu–Ta system by mechanical alloying [7].

In the gas evaporation method, atoms are evaporated in a vacuum by heating a metal or an alloy and are deposited onto a substrate. The atom arriving at the substrate carries a kinetic energy of, at most, 0.1 eV, which is of the order of the thermal energy at the melting point. Thus, an amorphous phase would be formed, if the diffusion of atoms after the deposition can be adequately suppressed. For this reason, the substrate is often kept at low temperatures. In 1954, Buckel and Hilsch were able to prepare amorphous thin films by evaporating pure elements such as Bi onto a substrate at 4.2 K cooled by liquid helium [8]. They discovered that amorphous Bi becomes metallic and undergoes a superconducting transition at about 6 K. As discussed in Section 6.8, Bi possesses the electronic structure typical of a semimetal as a result of the Fermi surface–Brillouin zone interaction. The disappearance of the Brillouin zone upon amorphization must be responsible for the onset of a metallic state and superconductivity. However, these amorphous pure metals are immediately crystallized when the temperature is raised to around 20 K. Hence, studies of the atomic structure and electronic properties were fairly limited.

In the DC sputtering technique, argon gas is ionized by applying several hundreds to thousands of volts between the anode and cathode plates, which face each other in the Ar gas atmosphere of 10^{-1}–10^{-2} Torr. The target is bombarded by accelerated ionized Ar particles and a large momentum is transferred from the bombarding ion to the atoms in the target. The energy of the ejected target atoms during sputtering reaches about 10 eV so that the sputtered film is generally more firmly adhered to the substrate than that prepared by gas evaporation in a vacuum. However, the temperature of the substrate is inevita-

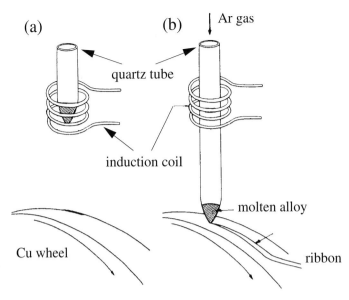

Figure 15.5. Single-roll melt spinning apparatus. (a) An alloy is inductively melted in a quartz tube with an orifice at its lower end. (b) After complete melting, the quartz tube is lowered immediately above the rotating Cu wheel and molten alloy is ejected onto it through the orifice by applying pressurized Ar gas to it.

bly increased during sputtering and, hence, the substrate is often cooled by water. An amorphous alloy film having a crystallization temperature well above room temperature can be relatively easily prepared by the DC sputtering method. The sputtering rate is practically very important and is now increased to as high as 1 μm/min so that an amorphous film of the order of 0.1 mm thickness can be prepared by continuous sputtering over 2 days.

In 1960 Duwez and his group [9] developed the gun technique of splat quenching, which enables fine molten droplets produced by a shock wave to be impinged onto a water-cooled Cu substrate with a cooling rate reaching 10^8–10^{10} K/s. For example, the amorphous $Au_{70}Si_{30}$ alloy could be formed in the form of tiny flakes about $0.2 mm^2$ in area and 10 μm in thickness. They were still far from practical use at that time. In 1969, Pond and Maddin [10] succeeded in producing amorphous ribbons by ejecting molten alloy onto the inner wall of a rotating metal drum. In 1970, Chen and Miller [11] developed the twin-roll quenching method in which molten alloy was ejected into a narrow gap between two rotating metal rolls so that an amorphous ribbon with a uniform thickness could be produced. The single-roll quenching method being currently widely used was established in 1971–2. Its principle is illustrated schematically in Fig. 15.5. The control of the gap is very delicate in the twin-roll method but the single-roll method is free from this difficulty.

Amorphous ribbons can be produced with a cooling rate of about 10^6 K/s in a large quantity within a short time.

One may wonder if the atomic structure and physical properties of an amorphous alloy would depend on the preparation technique and/or on the cooling rate. However, they generally reflect well its intrinsic nature, irrespective of the preparation methods. But care must always be exercised to check the quality of the amorphous phase.

15.4 Thermal properties of amorphous alloys

An amorphous phase is metastable and does not exist in the equilibrium phase diagram. Hence, it crystallizes upon heating. The temperature at which crystallization occurs is referred to as the crystallization temperature T_x and is generally several degrees higher than the glass transition temperature discussed below. Figure 15.6 illustrates the temperature dependence of the free volume of a given substance. The volume of a liquid decreases with decreasing temperature and discontinuously drops upon solidification at the melting point T_m, as can be seen in Fig. 15.6 (line AB). During the cooling process, the supercooling phenomenon may occur. This is apt to proceed, particularly if the cooling rate is very high. The motion of atoms in a supercooled liquid becomes sluggish as the supercooling proceeds. If its viscosity exceeds some critical value of about 10^{13} poise, the motion of the atoms is practically frozen.[8] This results in the formation of an amorphous phase. The temperature at which the viscosity reaches this critical value is called the glass transition temperature T_g.

Upon heating, the amorphous phase may enter the supercooled liquid state, if crystallization is somehow suppressed across the glass transition temperature, but eventually crystallizes, say, at a temperature at point F below the melting point T_m. The volume corresponding to the first derivative of the free energy is continuous but its slope changes across the glass transition temperature marked as the point D. The same is true in the temperature dependence of the entropy. Thus, the specific heat corresponding to the second derivative of the free energy shows a discontinuous jump at T_g.

Both the crystallization temperature T_x and glass transition temperature T_g can be experimentally determined using differential thermal analysis (DTA) or differential scanning calorimetry (DSC). In both DTA and DSC measure-

---

[8] A tangential stress proportional to the velocity of the fluid appears, if the velocity is position dependent. Its proportional coefficient is called the viscosity and expressed in CGS units of poise. 1 poise = 1 dyne·s/cm^2 = 0.1 Pa. The viscosity of water at 0 °C is 1.79×10^{-2} poise.

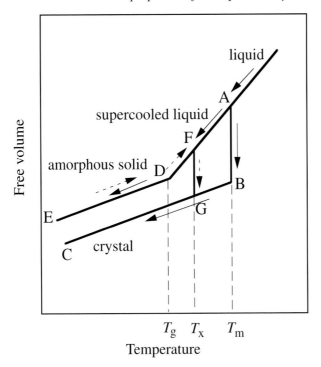

Figure 15.6. Temperature dependence of free volume. In thermal equilibrium, liquid solidifies at A and its volume is decreased to B. This is the melting point T_m. Liquid, however, solidifies at temperature lower than T_m under rapid cooling. Supercooled liquid is formed in the region A–D. It solidifies when its viscosity exceeds some critical value at D. The temperature at D is called the glass transition temperature T_g. The crystal is formed in the region B–C. When an amorphous solid is heated, it becomes supercooled liquid at D and crystallizes at F. The temperature at F is called the crystallization temperature T_x.

ments, an amorphous alloy sample is heated together with a reference material possessing approximately the same heat capacity as the sample. DTA records the temperature difference between them as a function of temperature whereas DSC measures the heat supplied by the heater so as to make the temperature difference zero. The relation $T_g < T_x < T_m$ generally holds. We can determine both T_g and T_x from the measured spectra, since a small endothermic reaction occurs at T_g but the exthothermic reaction occurs at T_x in the heating process. It is to be noted that they are not properties inherent in a given amorphous alloy but depend on the heating rate. It may also be noted that the DSC measurement allows us to determine the enthalpy difference between an amorphous phase and the crystallized phases by integrating the total area of the exothermic peak upon crystallization.

15.5 Classification of amorphous alloys

In Section 13.2, we classified crystalline metals into five different groups in terms of the magnetic state involved. It is also convenient to classify in a similar manner non-periodic metals including amorphous alloys, liquid alloys and quasicrystals into five groups to facilitate the discussion of the electron transport properties of these non-periodic substances [12]. The Bloch theorem fails in a non-periodic system and, hence, the mean free path of sp conduction electrons is certainly shorter than that in a crystal. Thus, the d electrons at the Fermi level E_F having more localized character can contribute to the electron conduction, together with sp electrons. The behavior of the conduction electrons becomes more complex in magnetic systems because of their interaction with magnetic moments. Here only amorphous alloys are classified into five different groups depending on their magnetic state, though a similar classification is also possible for quasicrystals and liquid alloys.[9]

Group (I): ferromagnetic, having a Curie temperature well above 300 K.
Group (II): weakly ferromagnetic, having a Curie temperature well below 300 K.
Group (III): having no spontaneous magnetization over the whole temperature range. But they exhibit a strongly temperature-dependent magnetic susceptibility. Those exhibiting spin-glass behavior and the Kondo effect are included in this group.
Group (IV): non-magnetic, with a fairly large Pauli paramagnetism.
Group (V): non-magnetic, with a small Pauli paramagnetism.

Typical amorphous alloys in the respective groups are listed in Table 15.1. An emphasis is laid on the difference in the electronic structure between the group (IV) and (V) alloys. They are all non-magnetic but those in group (IV) possess E_F in the d band, whereas those in group (V) in the sp band. Therefore, amorphous alloys in which sp electrons dominate the electron transport properties, are limited only to those in group (V). As will be described later, the Ziman theory based on the nearly-free-electron model can be applied only to liquid metals in group (V). Liquid metals in group (V) are sometimes referred to as simple liquid metals. In order to grasp the essence of the electron theory of a non-periodic system, we limit our discussion only to the electronic structure and electron transport properties of non-magnetic liquid metals, amorphous alloys and quasicrystals in groups (IV) and (V).

---

[9] Many quasicrystals are non-magnetic and belong either to group (IV) or group (V). However, the Mg–Zn–Ho quasicrystal has been identified as a spin-glass [T. J. Sato *et al.*, *Phys. Rev. Lett.* **81** (1998) 2364] and, hence, is classified in group (III).

15.6 Electronic structure of amorphous alloys

The electronic structure of liquid and amorphous alloys has been studied both experimentally and theoretically. Measurements such as the de Haas–van Alphen effect rely on the mean free path, which must be long enough to allow the conduction electron to make a closed orbit in the presence of a magnetic field (see Section 7.2). Hence, the mean free path-dependent probes cannot be applied to non-periodic systems. Instead, measurements of the electronic specific heat coefficient (Sections 3.4–3.5), magnetic susceptibility (Section 3.6), positron annihilation (Section 7.3), photoemission spectroscopy (Sections 7.5–7.7), soft x-ray spectroscopy (Section 7.8) and optical properties (Sections 11.9–11.12) have been employed as powerful tools to investigate the electronic structure of non-periodic systems.

Let us first discuss the main feature of the band structure of liquid and amorphous alloys. In a crystalline metal, the band structure is uniquely decided by the interaction of the Fermi surface with the Brillouin zone derived from the Fourier transformation of the periodic lattice in real space. Even in a disordered system like liquid and amorphous alloys, we pointed out the presence of correlations in the atomic arrangements in Section 15.2. In contrast to Bragg peaks in a crystal, the structure of liquid or amorphous alloys can be characterized by a main peak at the wave number K_p with subsequent diminishing oscillations in the structure factor, as indicated in Fig. 15.3(a). The presence of the main peak in the structure factor is taken as evidence for the existence of a long-range correlation in the assembly of atoms in liquid metals and amorphous alloys. In other words, the conduction electron in such non-periodic systems would experience the effect of a "weak periodicity" when its wave number coincides with $K_p/2$ or one-half of the reciprocal lattice vector (see equation (5.42) in Section 5.6). Indeed, the "smeared Fermi surface–Brillouin zone" effect gives rise to a faint van Hove singularity across E_F in the valence band spectra of the amorphous Mg–Zn alloy in group (V) [13]. As will be described below, however, the development of a short-range order or covalent bonding affects more substantially the electronic structures near E_F in an amorphous alloy and, in turn, electron transport properties, than the long-range correlation discussed above.

Let us begin with reviewing how the short-range structure in a disordered system can be incorporated into the theoretical band calculations. Because of the failure of the Bloch theorem, conventional band calculations which need information only about the atomic arrangements in a unit cell, have to be abandoned. In calculations of the electronic structure of an amorphous A–B alloy, the interatomic pair potentials of all the atom pairs A–A, B–B and A–B are

Table 15.1. *Classification of amorphous alloys in terms of magnetic states*

group	I ferromagnetism	II weak ferromagnetism	III spin-glass or Kondo state	IV paramagnetism	V weak paramagnetism or diamagnetism
characteristic features	Curie temperature $T_C > 300$ K	Curie temperature $T_C \ll 300$ K	presence of the spin freezing temperature (spin-glass) Curie–Weiss-type temperature dependence of magnetic susceptibility (Kondo state)	$\chi \simeq 10^{-4}$/mol temperature dependence of the Pauli paramagnetism $\gamma > 3$ mJ/mol·K^2	$\chi \simeq 10^{-5}$/mol negligible temperature dependence of magnetic susceptibility $1.5 < \gamma < 3$ mJ/mol·K^2	$\chi < 10^{-6}$/mol negligible temperature dependence of magnetic susceptibility $\gamma < 1.5$ mJ/mol·K^2
main carrier at the Fermi level	d electrons	d electrons	d electrons or (sp+d) electrons	d electrons	(sp+d) electrons	sp electrons
typical amorphous alloys	Fe–Co–Zr Co–B Fe–Co–B–Si	Fe–Zr Fe–Hf	Pd–Si–Mn Fe–Mn–B–Si Co–Mn–B–Si	Cu–Zr Cu–Ti Ni–Zr Y–Al La–Al	Ca–Al Ca–Mg Ni–P Mo–Ru–P	Mg–Zn–Ga Ag–Cu–Mg Ag–Cu–Ge Mg–Cu

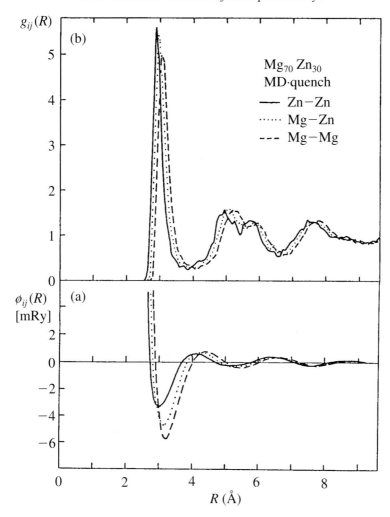

Figure 15.7. (a) Interatomic pair potentials $\phi_{ij}(R)$ and (b) partial pair distribution function $g_{ij}(R)$ of amorphous $Mg_{70}Zn_{30}$ alloy. [From J. Hafner, *From Hamiltonians to Phase Diagrams* Springer Series in Solid State Sciences 70, (Springer-Verlag, 1987)]

constructed first and the atomic structure is calculated via molecular dynamics simulations. The interatomic pair potentials calculated for the amorphous $Mg_{70}Zn_{30}$ alloy are shown in Fig. 15.7(a). Each pair potential has a deep minimum at the nearest neighbor distance with subsequent Friedel oscillations due to the screening charges (see Section 13.8). Once the interatomic pair potentials are assigned to all atom pairs involved, a high kinetic energy is fed equally to them to produce a liquid-like atomic structure in the computer. In the next step, the kinetic energy is rapidly deprived from each atom. All atoms will try to find their own stable positions and eventually settle down in the local minima in the matrix. The resulting partial pair distribution functions for the amorphous $Mg_{70}Zn_{30}$ alloy are shown in Fig. 15.7(b). In principle, the atomic structure thus obtained can reproduce the short-range order characterized by each interatomic pair potential. Once the atomic structure is constructed in the computer, the electronic structure is then calculated by using techniques such as the LMTO-recursion method [13, 14].

The calculations of both the atomic and electronic structures must be continued until a good agreement is achieved with the radial distribution functions deduced from neutron and/or x-ray diffraction measurements as well as valence band structures deduced from measurements such as photoemission spectroscopy and electronic specific heat coefficient. The effect of the short-range order on the electronic structure may be clearly singled out in a ternary amorphous A–(B–C) alloy by choosing different elements as the third element A.

We show a typical example in Fig. 15.8. The resistivity increases and the Hall coefficient changes its sign, when trivalent Al atoms are added to the amorphous $Cu_{40}Y_{60}$ alloy, whereas the resistivity monotonically decreases and the Hall coefficient remains negative when the divalent Mg atom is chosen as the third element [15]. Photoemission spectroscopy in combination with neutron diffraction experiments clearly reveals the formation of a strong covalent bonding in the Al–Y and Al–Cu atom pairs in the amorphous Al–(Cu–Y) alloys while the Cu–Y bonding is always dominant in the amorphous Mg–(Cu–Y) alloys. Both atomic and electronic structures have been calculated via molecular dynamics simulations in combination with the LMTO-recursion method for both the amorphous Al–(Cu–Y) and Mg–(Cu–Y) alloys [14]. The parameters involved in the relevant interatomic potentials were optimized until the results became consistent with the experimentally derived partial radial distribution functions and valence band structure. A combination of experimental and theoretical studies confirmed the formation of bonding states associated with Al–Y and Al–Cu atom pairs across E_F in the amorphous Al–(Cu–Y) alloy. The observed increase in resistivity in the ternary Al–(Cu–Y) amorphous alloys was attributed to the unique band structure brought about by the development of the short-range order upon the addition of Al to the amorphous Cu–Y matrix. The data discussed above emphasize the importance of the short-range order in interpreting the electronic structure and electron transport properties of amorphous alloys. The importance of the short-range order will be further emphasized in the discussion of quasicrystals (see Section 15.13).

The self-consistent determination of both atomic and electronic structures is essential in a disordered system, where a unit cell cannot be defined. Within a limited computing time and the limited memory size of a computer, a

15.6 Electronic structure of amorphous alloys

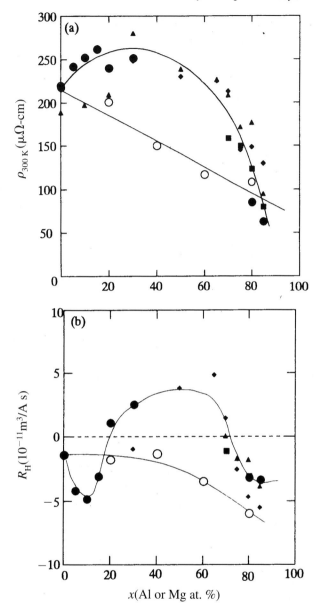

Figure 15.8. Al or Mg concentration dependence of the electrical resistivity at 300 K and the Hall coefficient for amorphous $Al_x(Cu_{0.4}Y_{0.6})_{100-x}$ and $Mg_x(Cu_{0.4}Y_{0.6})_{100-x}$ alloys. [From ref. 15.]

construction of a reliable but efficient scheme to calculate the electronic structure and electron transport properties of a disordered system is critically important. It has long been a puzzle why some liquid and amorphous alloys, like the Al–(Cu–Y) amorphous alloys mentioned above, exhibit a positive Hall coefficient in spite of the absence of the hole Fermi surface. Tanaka and Itoh [16] could explain soundly the mechanism of the occurrence of a positive Hall coefficient in liquid Fe by applying the tight-binding linear muffin-tin orbital-particle source method developed by Tanaka to the Itoh formula of the Hall conductivity. It was shown that p–d hybridization near E_F must be responsible for the observed positive Hall coefficient in liquid Fe. It is of great interest to examine whether their model can be extended to explain the observed positive Hall coefficient in the amorphous Al–(Cu–Y) alloys, in which bonding states are formed across E_F as a result of the hybridization of the Al-3p states with the Y-4d and Cu-3d states.

15.7 Electron transport properties of liquid and amorphous metals

One of the most challenging objectives in the studies of liquid and amorphous alloys is to gain a deeper insight into the electron conduction mechanism in non-periodic systems. Electron conduction in liquid and amorphous alloys is certainly due to electrons at E_F, whose number density generally amounts to the order of $10^{22}/cm^3$. Electron transport in liquid metals was discussed in the well-known textbook by Mott and Jones first published in 1936 [1]. However, a breakthrough was brought about by the Ziman theory put forward in 1961, which successfully explained the electrical resistivity behavior of simple liquid metals like pure Na and Zn [17]. At the International Conference on Liquid Metals held in Tokyo in 1972 [18], Ziman reported that the magnitude and temperature dependence of the electrical resistivity of simple liquid metals could be well accounted for at a quantitative level but that there still existed great difficulties in the treatment of the electron transport of liquid transition metals. This implies that the Ziman theory discussed in Section 10.10 had been widely accepted as a valid model for liquid metals in group (V) but that the role of the d electrons in the electron conduction of liquid transition metals in group (IV) still remained unsettled at that time.

The development of a single-roll melt-spinning wheel apparatus in the early 1970s concomitant with the Tokyo conference on liquid metals mentioned above certainly stimulated studies of various physical properties in amorphous alloys, since a large number of amorphous ribbon samples could now be relatively easily produced in many laboratories. But almost all amorphous ribbons thus produced at that time contained transition metals like Fe and Ni as major

15.7 Electron transport properties of liquid and amorphous metals

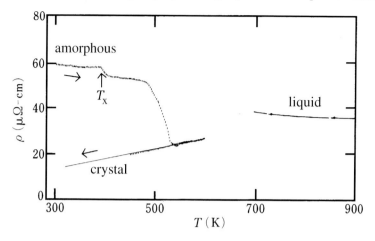

Figure 15.9. Temperature dependence of electrical resistivity of amorphous and liquid $Mg_{70}Zn_{30}$ alloy. The amorphous phase is crystallized at T_x. [Amorphous phase: T. Matsuda *et al.*, *J. Phys. F: Met. Phys.* **14** (1984) 1193; liquid phase: S. Steeb *et al.*, *Z. Metallkunde* **61** (1970) 637]

constituent elements. Unfortunately, however, the measured resistivity and its temperature dependence had been frequently discussed on the basis of the Ziman theory for simple liquid metals without seriously considering the limit of its applicability.

We review, at this stage, characteristic features of the electrical resistivity behavior of liquid and amorphous phases in comparison with those in the crystalline phase. To avoid complications due to possible d electron contributions, we choose the data in group (V) to which the nearly-free-electron model can be applied. Figure 15.9 shows the temperature dependence of the electrical resistivity of the amorphous $Mg_{70}Zn_{30}$ alloy in group (V). The data in the liquid phase are also incorporated. The resistivity decreases slightly with increasing temperature in the amorphous phase. Thus, the temperature coefficient of resistivity or TCR $(=(1/\rho)(d\rho/dT))$ is negative in the amorphous phase. Crystallization proceeds in two steps at about 400 and 500 K for this sample and is accompanied by a larger drop in resistivity at the second crystallization. The TCR becomes positive after complete crystallization. The room temperature resistivity is reduced to only $15\,\mu\Omega$-cm in contrast to $60\,\mu\Omega$-cm in the amorphous phase. It is clear that not only the magnitude of the resistivity is substantially different but also the sign of TCR changes between amorphous and crystalline phases. In contrast, the resistivity in the liquid phase is close to the value obtained by extrapolating the temperature dependence of the resistivity in the amorphous phase, and its TCR is negative in agreement with that in the amorphous phase.

Characteristic features of the electrical resistivities of liquid and amorphous alloys are summarized below.

1. The residual resistivity of amorphous alloys is generally in the range $20 \sim 1000\,\mu\Omega$-cm (see Fig. 15.16) and is 5–100 times that in the crystallized phase. The resistivity values in amorphous alloys are, instead, comparable to those in liquid metals.
2. The resistivity of amorphous alloys changes with temperature only by a few % over the temperature range 2–300 K. The change in resistivity with temperature is also fairly small in liquid metals.
3. The sign of TCR in amorphous alloys and liquid metals is either positive or negative and is sometimes extremely close to zero, depending on the composition of a given alloy system.

As is clear from the discussion above, electron transport properties of amorphous alloys are definitely different from those in crystals but are seemingly similar to those of the corresponding liquid phase. However, we will show below that there exists a distinctive difference in the electron transport mechanism between liquid and amorphous phases: elastic scattering dominates in the liquid phase whereas inelastic scattering and/or the quantum interference effect play a key role in the amorphous phase at temperatures below 300 K.

15.8 Electron transport theories in a disordered system

The negative TCR phenomenon is a characteristic feature of a semiconductor with a well-defined energy gap. Thus, one may address a naive question as to why a negative TCR, though its magnitude is small, appears for many liquid and amorphous alloys, in which the carrier concentration is as high as $10^{22}/\text{cm}^3$ and a well-defined Fermi edge exists without any energy gap. Various theories have been proposed to shed more light on the electron transport mechanism including the origin of a negative TCR in a disordered metallic system. As noted in the preceding section, the Ziman theory developed for simple liquid metals was frequently employed without much success to explain the occurrence of a negative TCR observed in various amorphous alloys in groups (I) to (IV) in the late 1970s to early 1980s. Apart from the Ziman theory, the Anderson localization theory in 1958 [19] has played a crucial role in the progress of the electron theory of a disordered system. Mott [20] has made further important contributions to this field and laid the basis for the concept of the weak localization effect in a series of his papers over the period 1966–90. We discuss below the more fundamental theories and related topics on disordered systems, including those in both marginally metallic and insulating regimes.

15.8.1 Ziman theory for simple liquid metals in group (V)

We derived in Section 10.10 the resistivity formula of equation (10.82) by assuming only elastic scattering in a perfect crystal at finite temperatures. However, it was emphasized that equation (10.82) is not adequate to describe the temperature dependence of the resistivity of a crystal because of the neglect of the inelastic electron–phonon interaction. Its rigorous treatment for a crystal has been described in Sections 10.11 and 10.12. This does not necessarily mean that equation (10.82) is meaningless.

As a matter of fact, equation (10.82) has more often been referred to as the Ziman resistivity formula for simple liquid metals, because he derived it for the first time to describe their resistivity behavior [17]. In place of equation (10.73), Ziman expressed the total ionic potential for a simple liquid metal as $U(\mathbf{r}) = \Sigma_i U_p(\mathbf{r} - \mathbf{R}_i)$, where the vector $\mathbf{R}_i$ denotes the position of the i-th ion, the vector $\mathbf{r}$ the position of the conduction electron and $U_p(\mathbf{r} - \mathbf{R}_i)$ is its pseudopotential. Here, the position vector $\mathbf{R}_i$ of the ion is fixed in a liquid metal and the displacement vector $\mathbf{u}_l$ in equation (10.73) need not be considered. In other words, we limit ourselves to the interaction of conduction electrons with the *static* distribution of ions in liquid metals. This is certainly equivalent to the assumption of elastic scattering of the conduction electron by ions.

As has been described in Section 10.10, the Ziman resistivity formula is constructed on the basis of the following three assumptions. Firstly, the linearized Boltzmann transport equation is assumed. It implies that the mean free path of the conduction electron must be longer than an average atomic distance and, hence, the Ziman theory will work only for systems with resistivities less than about 150 μΩ-cm.[10] Secondly, the Born approximation is assumed to calculate the transition probability. This is justified by the pseudopotential approach, which holds only for systems in group (V) but fails for systems in group (IV). Thirdly, elastic scattering is assumed. This limits the applicability of the Ziman formula only to liquid metals, because elastic scattering dominates either at very low temperatures or at temperatures well above the Debye temperature

---

[10] The mean free path of the conduction electron for crystals and amorphous alloys in types (a) to (c) (see Fig. 15.12), to which the nearly-free-electron model is applicable, is roughly estimated from the relation

$$\Lambda_F = \frac{m v_F}{n e^2 \rho} = \frac{9.1 \times 10^{-28} \times 10^8 \times 9 \times 10^{11}}{6 \times 10^{23} \times (4.8 \times 10^{-10})^2} \cdot \frac{A}{d\rho} = 5.92 \times 10^{-5} \frac{A}{d\rho},$$

where A is the atomic weight [g], d is the density [g/cm³], ρ is the resistivity [Ω-cm] and Λ_F is the mean free path [Å]. For example, the mean free path in pure Cu is deduced to be about 260 Å by inserting a resistivity of 1.6 μΩ-cm at 300 K, whereas that in an amorphous alloy with $\rho = 150$ μΩ-cm and $A/d \approx 10$ is deduced to be about 4 Å, comparable to an average atomic distance.

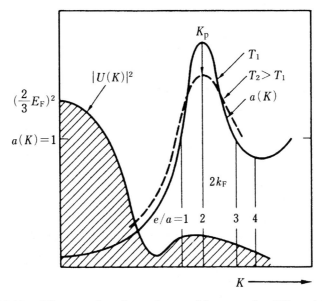

Figure 15.10. Wave number dependence of integrands $a(K)$ and $|U(K)|^2$ appearing in the Ziman theory. The interference function $a(K)$ is spread as shown by the dashed curve when temperature T_1 is raised to T_2. The positions of $2k_F$ corresponding to various e/a values are shown by vertical lines. $|U(K)|^2$ is obtained by taking the square of the pseudopotential such as that shown in Fig. 8.5. [U. Mizutani, *Prog. Mat. Sci.* **28** (1983) 97]

(see Section 11.4). Therefore, we see that the Ziman resistivity formula should be applicable only to low-resistivity liquid metals and alloys in group (V).

Let us study how the Ziman resistivity formula has been applied to liquid metals and alloys in group (V), for which a spherical Fermi surface can be reasonably assumed. The elastic scattering of conduction electrons always takes place on the Fermi surface, as shown in Fig. 10.6. The magnitude K of the scattering vector is then related to the scattering angle θ through the relation:

$$\sin\frac{\theta}{2} = \frac{K}{2k_F}. \tag{15.22}$$

The allowed range of θ is obviously $0 \leq \theta \leq \pi$ and, hence, $0 \leq K \leq 2k_F$, indicating that the scattering vector is limited by the diameter of the Fermi sphere. The integrand in equation (10.82) consists of the interference function $a(K)$, the square of the Fourier component of the ion potential, $|U(K)|^2$, and the weighted factor K^3. Because of the presence of the weighted factor, the contribution of $a(K)$ and $|U(K)|^2$ to the integral becomes substantial in the K region only near the upper limit $2k_F$.

Figure 15.10 illustrates the K dependence of both $a(K)$ and $|U(K)|^2$ for a

15.8 Electron transport theories in a disordered system

typical liquid metal. The wave number dependence of the pseudopotential was already shown for pure Al in Fig. 8.5. Generally speaking, the K dependence of $|U(K)|^2$ is fairly small and flat in the region of our interest ($1.0 \leq e/a \leq 4.0$ in Fig. 15.10). Thus, more important is the contribution from the interference function $a(K)$, which is shown schematically in Fig. 15.10. It is characterized by the first main peak centered at the wave number K_p, the inverse of which roughly corresponds to an average atomic distance of a liquid metal. As described below, the position of the upper limit $2k_F$ relative to K_p plays a key role in the Ziman theory.

In monovalent liquid metals like Na, the $2k_F$ value corresponding to $e/a = 1.0$ is relatively small and the condition $2k_F < K_p$ holds. In divalent liquid metals like Zn and Cd, the $2k_F$ value is increased and the condition $2k_F \approx K_p$ holds. The value of $2k_F$ is further increased in polyvalent liquid metals like Al and Sn, leading to $2k > K_p$. As shown in Fig. 15.10, the integration is carried out only up to the rising slope of the main peak for monovalent metals. The resistivity value is found to be fairly small, since $a(K)$ remains relatively small (~ 1) up to the upper limit $2k_F$. Similarly, $a(K)$ becomes relatively small near $2k_F$ for polyvalent metals with $e/a = 3.0$–4.0 and, thus, the resistivity becomes low again. The largest resistivity is expected to occur for divalent metals where the upper limit $2k_F$ coincides with the main peak in $a(K)$.

The sign of TCR can also be predicted from equation (10.82). When the temperature of a liquid metal is increased, the main peak will be broadened, as shown by a dashed curve in Fig. 15.10, as a result of an increasing free volume and structural disorder. The resistivity will increase with increasing temperature for monovalent metals, since the integration is terminated at the enhanced outskirts of the main peak, where $a(K)_{T=T_2} > a(K)_{T=T_1}$ holds. This explains well the occurrence of a positive TCR in monovalent and polyvalent metals. However, when $2k \approx K_p$ is satisfied, a decrease in the height of the main peak with increasing temperature contributes to reduce the resistivity, thereby leading to a negative TCR for divalent metals.

The validity of the Ziman theory has been tested for a series of liquid alloys obtained by adding polyvalent metals like Sn and In to monovalent noble metals like Cu and Ag, all of which are typical of group (V). The value of $2k_F$ can be continuously increased with increasing concentration of the polyvalent element. Both the value of resistivity and the sign of TCR for noble metal alloys are plotted in Fig. 15.11(a) as a function of $2k_F/K_p$ rather than the concentration of the polyvalent element. It is clear that both the maximum resistivity and a negative TCR concurrently occur at $2k_F/K_p = 1$, in good agreement with the Ziman theory. In this way, the Ziman theory has been proved to be successful in interpreting the resistivity behavior of simple liquid metals and alloys in group (V).

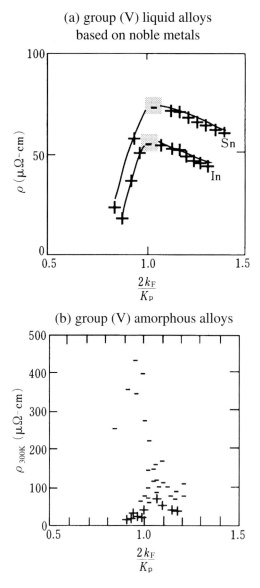

Figure 15.11. Resistivity as a function of $2k_F/K_p$ for (a) group (V) liquid alloys based on noble metals [reproduced from G. Busch and H.-J. Güntherodt, *Solid State Physics*, vol. 29, edited by H. Ehrenreich, F. Seitz and D. Turnbull, (Academic Press, New York, 1974)] and (b) group (V) amorphous alloys. The sign of the TCR is shown by symbols + or −. [U. Mizutani, *Materials Science and Technology* edited by R. W. Cahn, P. Haasen and E. J. Kramer, (VCH, Germany, 1994), vol.3B: *Electronic and Magnetic Properties of Metals and Ceramics, Part II*, volume editor K. H. J. Buschow, pp. 97–157]

15.8.2 Baym–Meisel–Cote theory for amorphous alloys in group (V)

The resistivity at 300 K for a large number of amorphous alloys in group (V) is plotted in Fig. 15.11(b) as a function of $2k_F/K_p$. Its behavior is obviously different from that in Fig. 15.11(a) and no clear $2k_F/K_p$ dependence is observed in amorphous alloys. Regardless of the magnitude of $2k_F/K_p$, a positive TCR appears when the resistivity is lower than about 50–60 $\mu\Omega$-cm and, otherwise, a negative TCR dominates.[11] This clearly demonstrates the failure of the Ziman theory to explain the resistivity behavior of amorphous alloys even in group (V).

Among the three underlying assumptions in the Ziman theory, the linearized Boltzmann transport equation and the Born approximation are equally justified for amorphous alloys in group (V). However, the assumption of the elastic scattering collapses in amorphous alloys, since their electron transport properties are discussed at temperatures well below the Debye temperature Θ_D or below 300 K in most cases. At such low temperatures, ions can no longer be treated as independent particles but the concept of collective excitations of phonons must be introduced to treat the thermal vibrations of ions.[12] A proper evaluation of the inelastic electron-phonon interaction is of prime importance in dealing with the electron transport properties of amorphous alloys in group (V) at temperatures below Θ_D. We abandon the Ziman theory of equation (10.82) and, instead, employ the Baym resistivity formula of equation (10.96) as a starting equation, through which the inelastic electron–phonon interaction is incorporated.

The Baym resistivity formula of equation (10.96) is more explicitly written as

$$\rho = \frac{3\pi\Omega_0}{4e^2\hbar v_F^2 k_F^4} \int_0^{2k_F} \left[\int_{-\infty}^{\infty} \frac{(\hbar\omega/k_B T)}{\exp(\hbar\omega/k_B T) - 1} a(K,\omega) d\omega \right] |U(K)|^2 K^3 dK, \quad (15.23)$$

where $a(K,\omega)$ is the dynamical structure factor at frequency ω and wave number K. As noted in Section 10.11, the ω-dependent integral in equation (15.23) is reduced at high temperatures $T > \Theta_D$ to

---

[11] We have pointed out in Fig. 15.9 the similarity of the resistivity behavior in both amorphous and liquid phases of the $Mg_{70}Zn_{30}$ alloy. The condition $2k_F = K_p$ is satisfied for this alloy, since both Mg and Zn are divalent. But the temperature dependence of the resistivity for the amorphous phase in the range 2–300 K can be interpreted only in the context of the Baym–Meisel–Cote theory [21].

[12] The lattice specific heat at low temperatures exemplifies the need for introducing the concept of phonons. At temperatures well above Θ_D, the lattice specific heat of $3R$ is well explained in terms of the Boltzmann equipartition law by treating ions as classical individual particles. However, a rapid decrease in the lattice specific heat below Θ_D can be explained only by taking into account the collective motions of ions (see Section 4.8).

$$\int_0^\infty \frac{(\hbar\omega/k_B T)}{\exp(\hbar\omega/k_B T)-1} a(K,\omega)d\omega \approx \int_0^\infty a(K,\omega)d\omega = a(K), \qquad (15.24)$$

indicating that the dynamical structure factor is replaced by the static structure factor at $T > \Theta_D$. We see, therefore, that the Ziman equation (10.82) is a high temperature limit of the Baym equation (15.23).[13]

In Section 10.12, the Bloch–Grüneisen law is derived by applying the Baym equation to a crystal metal. Now equation (15.23) is applied to amorphous alloys in group (V). Its details have been described in successive papers by Meisel and Cote [21]. They showed that the temperature dependence of the resistivity for amorphous alloys in group (V) is well approximated as

$$\rho = [\rho_0 + \Delta\rho(T)]\exp[-2W(T)], \qquad (15.25)$$

where ρ_0 is the residual resistivity, $\Delta\rho(T)$ is the term arising from the inelastic electron–phonon interaction and the exponential term $\exp[-2W(T)]$ represents the Debye–Waller factor (see Section 10.12 and Exercise 10.8 for the definition of $W(T)$) [21]. The term $\Delta\rho(T)$ is shown to exhibit $+T^2$ dependence at low temperatures, say, below 20 K, and $+T$ at higher temperatures. This is compared with the Bloch–Grüneisen law for a crystal metal, where $+T^5$ dependence holds at low temperatures and $+T$ at higher temperatures. The Debye–Waller factor yields a $(1-\alpha T^2)$ dependence at low temperatures and $(1-\beta T)$ above about several tens of degrees K (see Exercise 15.1). The validity of equation (15.25) has been experimentally confirmed by Mizutani [12], as shown below.

In the case of group (V) amorphous alloys with residual resistivities lower than about 50–60 $\mu\Omega$-cm, the term $\Delta\rho(T)$ dominates and $+T^2$ dependence is observed at temperatures below about 20 K whereas $+T$ at temperatures above about 30 K. Hence, the TCR is positive over the whole temperature range 2–300 K. This ρ–T behavior is hereafter called type (a) and is illustrated in Fig. 15.12. As ρ_0 gradually increases beyond 60 $\mu\Omega$-cm, the Debye–Waller factor begins to play a more important role at higher temperatures. Type (b) is assigned to the ρ–T behavior characterized by $+T^2$ dependence at low temperatures but $(1-\beta T)$ at higher temperatures. This naturally results in a shallow

---

[13] As noted in footnote 12, p. 284, in Chapter 10, the reduction of the dynamical structure factor to the static structure factor at high temperatures is brought about as a result of the smearing of the Fermi distribution of the order of $k_B T$ which exceeds the maximum transfer energy of $k_B \Theta_D$ between electrons and phonons. Strictly speaking, this is not equivalent to the assumption of elastic scattering and is often referred to as quasi-elastic scattering (see Section 11.4).

15.8 Electron transport theories in a disordered system

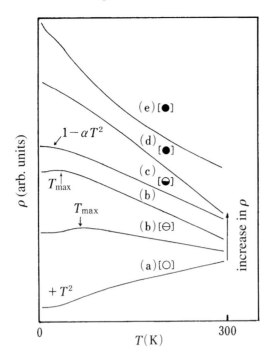

Figure 15.12. Temperature dependence of the electrical resistivity of non-magnetic amorphous alloys in the temperature range 2–300 K. Symbols representing the ρ–T types are used in Figs. 15.16 and 15.17. [U. Mizutani, *Phys. Stat. Sol.* (b) **176** (1993)9]

resistivity maximum at an intermediate temperature, as shown in Fig. 15.12. With increasing ρ_0, the resistivity maximum is shifted to lower temperatures and finally vanishes [12]. Now the ρ–T dependence is dominated only by the Debye–Waller factor over the whole temperature range: $(1 - \alpha T^2)$ at low temperatures and $(1 - \beta T)$ at high temperatures. This is type (c), which is obviously characterized by a negative TCR over the whole temperature range. Type (c) is generally observed in group (V) amorphous alloys with resistivities in the range 100–150 $\mu\Omega$-cm.

The temperature dependence of the electrical resistivities in the range 2–300 K for amorphous alloys in both groups (IV) and (V) can be systematically classified into five different ρ–T types. They always appear in the sequence (a)→(b)→(c)→(d)→(e) with increasing resistivities, as illustrated in Fig. 15.12. The gradual change in types (a)→(b)→(c) is accompanied by shortening of the mean free path down to an average atomic distance [12]. We say, therefore, that an increase in resistivity in this regime is due entirely to the mean free path effect and is free from the band structure effect. Indeed, the data are well explained in terms of equation (15.25) within the framework of the Baym–Meisel–Cote theory based on the Boltzmann transport equation. The sign of TCR shown in Fig. 15.11(b) refers to that near room temperature and, hence, a positive TCR corresponds to type (a) whereas a negative TCR to the remaining types (b) to (e). This means that the sign of TCR in amorphous alloys is determined by the interplay between $\Delta\rho(T)\exp[-2W(T)]$ and $\rho_0\exp[-2W(T)]$ in equation (15.25) instead of $2k_F/K_p$ in liquid metals.

The ρ–T types (d) and (e) are exclusively observed in high-resistivity amorphous alloys with resistivities exceeding about 200 $\mu\Omega$-cm. This indicates that the mean free path is shortened to an average atomic distance of a few Å. An increase in resistivity in this regime should be caused by a change in the electronic structure, since the mean free path can no longer be decreased. The ρ–T types (d) and (e) are observed in some group (V) and many group (IV) amorphous alloys. They cannot be explained in terms of the Baym–Meisel–Cote theory but must be treated by theories beyond those based on the Boltzmann transport equation.

15.8.3 Mott s–d scattering model

Both sp and d electrons coexist at E_F in group (IV) amorphous alloys. Since the electron transport properties are exclusively determined by electrons at E_F, it is a crucial matter to clarify how sp and d electrons share the electron conduction. Mott [22] explained the reason for the possession of a relatively large resistivity in the transition liquid metals in terms of the so called s–d scattering model (see Section 13.6). He assumed that the sp electrons are exclusively responsible for electron conduction in a transition liquid metal. According to equations (10.80) and (10.81) coupled with (10.54), the scattering probability $1/\tau$ of the conduction electron is found to be proportional to the final density of states at E_F, $N(E_F)$. Since E_F is located in the middle of the d band in the transition metal, Mott attributed the observed large resistivity to its large final density of states.

Mott's s–d scattering model presumes that the sp conduction electron possesses a mean free path much longer than that of the d electrons at E_F.[14] This condition is certainly better satisfied in a crystal where the Bloch theorem holds. Its failure in liquid and amorphous alloys causes the mean free path of sp electrons to be shortened. In particular, we will show in Section 15.9 that both sp and d electrons would equally contribute to the electron conduction in amorphous alloys characterized by ρ–T types (d) or (e).

---

[14] If the Mott s–d scattering model worked for amorphous alloys in group (IV), then the residual resistivity ρ_0 would increase in proportion to $N(E_F)$ or the electronic specific heat coefficient γ. However, as indicated in Fig. 15.16, this is generally not observed.

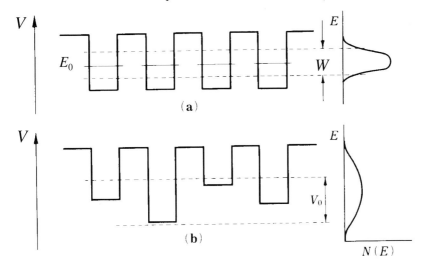

Figure 15.13. Potential arrays in periodic and non-periodic lattices. (a) Electrons with an average energy E_0 form the Bloch state with band width W in the periodic potential. (b) Electrons are localized if V_0/W exceeds some critical value, where V_0 represents the degree of disorder in the potential. [N. F. Mott, *Metal–Insulator Transitions* (Taylor & Francis Ltd, 1990)]

15.8.4 Anderson localization theory

As shown in Fig. 15.13(a), we consider the conduction electron with energy E_0 to be loosely bound in the periodic potential. The Bloch wave derived from the tight-binding approximation will form a relatively narrow band with a width W. Now some disorder is introduced into the potential distribution. Its amplitude is assumed to vary irregularly in the range $-V_0/2 < V < -V_0/2$, as shown in Fig. 15.13(b). Anderson [19] proved that all electrons within the band cannot form the Bloch wave but localize in real space, if the ratio V_0/W exceeds some critical value. At finite temperatures, localized electrons will be able to exchange their energy with phonons and to hop from one site to another. This is termed hopping conduction. The Anderson localization theory has been further elucidated by Mott and others and concepts such as weak localization, mobility edge, minimum metallic conductivity, scaling law and metal–insulator transition have been established. More details are to be found in the literature [20].

As is inferred from equation (15.25), the inelastic electron–phonon interaction plays a key role in determining the temperature dependence of the electrical resistivity in amorphous alloys but its contribution becomes less important as the residual resistivity increases. The residual resistivity arises from the elastic scattering of conduction electrons due to random distributions of ions

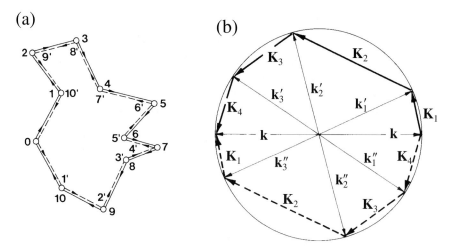

Figure 15.14. Repeated elastic scattering events of conduction electron and resulting localization effect in (a) real and (b) reciprocal space. [G. Bergman, *Phys. Rev.* **B28** (1983) 2914]

at absolute zero. The elastic scattering is indeed essential in the Anderson localization theory.

Following Bergman [23], we consider why the repetition of the elastic scattering leads to an enhancement of the localization. He introduced two different lifetimes of the conduction electron at low temperatures: one the elastic life time τ_o representing that of the electron in an eigenstate of momentum and the other the inelastic life time τ_i in an eigenstate of energy. At low temperatures below, say, 4 K, elastic scattering dominates and the inelastic lifetime becomes longer than the elastic one by several orders of magnitude. As a result, an electron of wave vector **k** is scattered by impurities without losing its phase coherence.

Figure 15.14(a) illustrates the situation where the conduction electron of wave vector **k** is scattered into the state $-\mathbf{k}$ by repeated elastic scattering events with impurities. This implies that a series of elastic scatterings result in back-scattering. An exactly opposite passage is also equally possible and is shown by a dashed line. The same elastic processes can be considered on the Fermi surface in the reciprocal space. As shown in Fig. 5.14(b), two passages refer to the multiple scatterings: one goes via $\mathbf{k}'_1$, $\mathbf{k}'_2$, $\mathbf{k}'_3$ and the other via $\mathbf{k}''_1$, $\mathbf{k}''_2$ and $\mathbf{k}''_3$ before reaching the state $-\mathbf{k}$. The repeated elastic scatterings yield the momentum transfers $\mathbf{K}_1$, $\mathbf{K}_2$, $\mathbf{K}_3$ and $\mathbf{K}_4$ in the former whereas they yield $\mathbf{K}_4$, $\mathbf{K}_3$, $\mathbf{K}_2$ and $\mathbf{K}_1$ in the latter. Among numerous possible scattering processes, we can show that the processes mentioned above tend to occur more frequently than others under the condition of the elastic scattering.

15.8 Electron transport theories in a disordered system

We can show that the scattering amplitudes $A' = |A'|e^{i\theta'}$ and $A'' = |A''|e^{i\theta''}$ upon multiple scattering from state $\mathbf{k}$ to state $-\mathbf{k}$ in these two complementary passages are the same. Firstly, the scattering probability is proportional to the product of the Fourier components of the scattering potential, $U(\mathbf{K}_1)U(\mathbf{K}_2)U(\mathbf{K}_3)U(\mathbf{K}_4)$, and, hence, is the same for the two passages. Secondly, the phase remains unchanged and $\theta' = \theta''$, since all scatterings involved are elastic. Thus, we have the relations $|A'| = |A''| = |A|$, $A'^*A'' = |A|^2$ and $A'A''^* = |A|^2$. The probability density of the electron state of $-\mathbf{k}$ is calculated as

$$|A' + A''|^2 = |A'|^2 + |A''|^2 + A'^*A'' + A'A''^* = 4|A|^2. \tag{15.26}$$

However, the third and fourth terms will disappear and the probability density is reduced to one-half or $2|A|^2$, if the scattering is inelastic and the two phases involved are incoherent, namely, $\theta' \neq \theta''$. We see, therefore, that the multiple elastic scattering accompanying the momentum transfer $2k_F$ is more frequent than others and that the probability density of electrons at the position before scattering is doubly enhanced. This is indeed the localization effect. The scattering described above is often called the quantum interference effect or $2k_F$ scattering.

Only elastic scattering survives at absolute zero. Therefore, Anderson localization is prone to occur in systems possessing a high residual resistivity. At finite temperatures, electrons begin to be scattered inelastically with phonons so that the phase coherence will be gradually lost during the successive scatterings, thereby leading to delocalization of electrons. Obviously, this yields a negative TCR in the temperature dependence of the electrical resistivity.

Altshuler and Aronov [24] noted that the electron–electron interaction is enhanced when the electron tends to be localized. As emphasized in Sections 8.2 and 14.2, the success of the one-electron approximation in normal metals and alloys is due to the fact that conduction electrons are so mobile that the Coulomb field is screened by other electrons. However, when electron localization sets in and the screening effect is weakened, the electron–electron Coulomb interaction tends to be enhanced. The enhanced electron–electron interaction coupled with the weak localization effect causes both the electrical conductivity σ and the Hall coefficient R_H to obey a $\sqrt{T}$-dependence at temperatures below about 20 K in a three-dimensional system [24]:

$$\sigma(T) = \sigma_0(1 + \alpha\sqrt{T}) \quad (T \lesssim 20\,K) \tag{15.27}$$

and

$$R_H(T) = R_H(0)(1 + \beta\sqrt{T}) \quad (T \lesssim 20\,K). \tag{15.28}$$

Furthermore, the relation $\beta = 2\alpha$ is theoretically predicted [24] and experimentally confirmed [12]. At temperatures above about 30 K, the conductivity follows $\sigma \propto T$ whereas the Hall coefficient becomes temperature independent [12]. Here it should be emphasized that the ρ–T type (d) or (e) (see Fig. 15.12) is converted to the temperature dependence of the conductivity ($\sqrt{T}$ below 20 K and $+T$ above 30 K). This unique feature observed in high-resistivity amorphous alloys is in sharp contrast to the temperature independent Hall coefficient and successive change in the ρ–T types from (a) to (c) for low-resistivity amorphous alloys in group (V), to which the Baym–Meisel–Cote theory is applicable.

15.8.5 Variable-range hopping model

A system of interest is obviously a metal if the conductivity extrapolated to 0 K is finite and an insulator if it is zero. However, the extrapolation of the measured conductivity to 0 K is generally very delicate even when the measurement is extended down to, say, 0.5 K. It is often indispensable to measure some other properties to judge if the system is a metal or an insulator. At first sight, one may consider the measurement of the electronic specific heat coefficient γ_{exp} to be decisive. However, this is not true. As will be shown below, a finite γ_{exp} value remains even in an insulator and plays a key role in the temperature dependence of its conductivity. Hence, a decisive test is often made by analyzing the temperature dependence of conductivity at low temperatures. We have already discussed above that the ρ–T types (d) and (e) appear on the metallic side of the metal–insulator transition.

The conductivity on the insulating side becomes finite at finite temperatures through phonon-assisted hopping of electrons. The variable-range hopping model proposed by Mott [25] assumes that the electronic states at E_F are finite but are localized at 0 K. The corresponding wave function generally decays exponentially at large distances in a spherically symmetrical potential and, hence, we can write

$$\psi(r) \approx \exp(-r/a), \qquad (15.29)$$

where a is its characteristic radius. The two localized states centered at $\mathbf{R}_i$ and $\mathbf{R}_j$ can interact through the overlap of the wave functions:

$$\int \psi^*(\mathbf{r} - \mathbf{R}_i) \psi(\mathbf{r} - \mathbf{R}_j) d\mathbf{r} \approx \exp(-R/a), \qquad (15.30)$$

where $R = |\mathbf{R}_i - \mathbf{R}_j|$. Thus, electrons can hop from site i to site j, only if the overlap of wave functions is finite. Since the probability P of a transition from $\mathbf{R}_i$ to $\mathbf{R}_j$ is proportional to the square of the overlap integral, we have the relation $P \propto \exp(-2R/a)$.

15.8 Electron transport theories in a disordered system

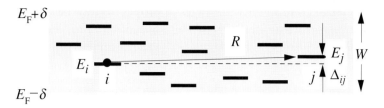

Figure 15.15. Variable-range hopping model. Localized electron can hop from site i with energy E_i to site j with energy E_j separated by the distance R.

The energy eigenvalue of each localized state will depend sensitively on the degree of overlap from wave functions of other centers distributed at random in the matrix and spread into a band with a characteristic width W at E_F. This is illustrated in Fig. 15.15. Even when R is large enough to make the overlap integral very small, hopping from site i to site j will still occur if the energy difference $E_i - E_j$ is compensated for by the absorption or emission of a phonon. Thus, the probability for hopping will be modulated with the rate of the excitation of a phonon and is expressed as

$$P = \nu_{ph} \exp(-|\Delta_{ij}|/k_B T) \exp(-2R/a) \tag{15.31}$$

where $\Delta_{ij} = E_i - E_j$ and ν_{ph} is the characteristic frequency of phonons.

Let us assume that only a single site j is available within the distance R from the site i, provided that their energy difference is within Δ_{ij}. This implies that the number of electrons in the energy range Δ_{ij} across E_F must be unity in a spherical volume with the radius R. This leads to the condition:

$$\frac{4\pi}{3} R^3 N(E_F) \Delta_{ij} \approx 1, \tag{15.32}$$

where $N(E_F)$ is the density of states per unit volume at E_F. By inserting equation (15.32) into equation (15.31), we obtain the transition probability as a function of the distance R:

$$P = \nu_{ph} \exp\left[-\frac{3}{4\pi R^3 N(E_F) k_B T} - \frac{2R}{a}\right]. \tag{15.33}$$

The most probable hopping will be realized by maximizing equation (15.33). The condition $dP/dR = 0$ yields

$$R = \left[\frac{9a}{8\pi N(E_F) k_B}\right]^{1/4} T^{-1/4}. \tag{15.34}$$

Since the electrical conductivity increases in proportion to the transition probability, the temperature dependence of the conductivity $\sigma(T)$ can be expressed as

$$\sigma(T) \propto v_{\text{ph}} \exp\left[-\frac{B}{T^{1/4}}\right], \tag{15.35}$$

where $B = (8/3)(9/8\pi)^{1/4}[N(E_F)k_B a^3]^{-1/4} = 2.062[N(E_F)k_B a^3]^{-1/4}$. This is known as the $T^{-1/4}$ law derived from the variable-range hopping model proposed by Mott [25]. The exponentially dependent $T^{-1/4}$ behavior has often been observed on the insulating side of the metal–insulator transition (see Fig. 15.29).

15.9 Electron conduction mechanism in amorphous alloys

We assume that, on the metallic side of the metal–insulator transition, all electrons at E_F, including both sp and d electrons contribute equally to the electron conduction and that the conductivity formula $\sigma = (e^2/3)\Lambda_F v_F N(E_F)$ in equation (10.54) is still applicable to them. Furthermore, we assume in this limit that conduction electrons take a minimum diffusion coefficient $D = \Lambda_F v_F/3$ in the conductivity formula.[15] For the moment, it is set equal to 0.25 cm²/s, which is simply deduced by using Λ_F equal to an average atomic distance of 3 Å and v_F equal to one-fourth of the free-electron Fermi velocity, namely, 0.25×10^8 cm/s. The reason for this choice will be discussed later.

Figure 15.16 shows sets of measured residual resistivity ρ_0 versus the electronic specific heat coefficient for a large number of amorphous alloys in groups (IV) and (V). Here the experimental electronic specific heat coefficient is assumed to represent $N(E_F)$. The dashed curve represents a hyperbolic curve $\rho_0 N(E_F) = e^{-2} D_{\text{min}}^{-1}$ with the minimum diffusion coefficient of $D_{\text{min}} = 0.25$ cm²/s mentioned above. This is our high-resistivity limiting curve. The curve indicates that the larger the density of states at E_F, the lower is the resistivity.

Four different symbols are used in Fig. 15.16 to distinguish the ρ-T types: (○) for type (a), (⊖) for type (b), (⊙) for type (c), (●) for types (d) and (e). The data of types (a)→(b)→(c) always appear in this sequence with increasing resistivity and all these data fall far below the high-resistivity limiting curve.

---

[15] Particles can diffuse from a higher to a lower concentration region by repeating random motions, when a concentration gradient exists in the particle distribution. Here the relation $J = -D\text{grad}n$ holds, where J is its flow, $\text{grad}n$ is its concentration gradient and D is the diffusion coefficient (see Section 11.13). No activation energy is involved in the diffusion process. In a disordered system such as an amorphous alloy, the conduction electrons do not form the Bloch wave but flow by repeated random scattering with ions. Hence, motion of electrons can be described in terms of the diffusion process. The diffusion coefficient defined as equation (11.82) is deduced to be $D = (1/3)v\Lambda$ in units of cm²/s. This relation appeared in equation (11.4) in the derivation of the specific heat and thermal conductivity in the kinetic theory of gases.

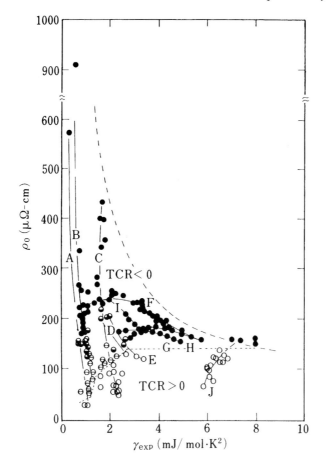

Figure 15.16. Residual resistivity–electronic specific heat coefficient diagram for groups (I), (IV) and (V) amorphous alloys. The ρ–T types are shown by the symbols used in Fig. 15.12. Letters refer to alloy systems: (A) Ag–Cu–X (X=Ge, Si), (B) Al–Ni–X (X=Si, Ge), (C) Ca–Mg–Al, Ca–Zn, (D) Ni–P, (E) Mo–Ru–P, (F) Ni–Zr–X (X=Cu, Al, Si, B), (G) Cu–Zr, (H) Ni–Zr, (I) Cu–Zr–Al, (J) Co–B–X (X=Si, Al, Ni). The dotted curve shows the boundary, across which the TCR near 300 K changes its sign. The dashed line is the high-resistivity limiting curve. [U. Mizutani, *Phys. Stat. Sol.* (b) **176** (1993) 9]

Instead, the data of types (d) and (e) appear immediately below the high-resistivity limiting curve. Obviously, all amorphous alloys in the metallic regime fall below this limiting curve. This means that amorphous alloys of types (d) or (e) essentially possess a very low diffusion coefficient of the order of 0.2–0.3 cm²/s and that all electrons at E_F, whether sp or d electrons, contribute equally to electron conduction in accordance with the conductivity formula.

Let us consider further the physical meaning of the high-resistivity limiting curve in Fig. 15.16. In order to discuss electron transport phenomena in the high-resistivity metallic regime, Mott [20] introduced the g-parameter defined as

$$g = N(E_F)/N(E_F)^{free}, \qquad (15.36)$$

where $N(E_F)$ is the density of states at E_F and $N(E_F)^{free}$ is the corresponding free-electron value. The condition $g<1$ refers to the situation where E_F is located within the valley in the density of states or on a declining slope of the density of states peak. Mott [20] christened such a valley in the density of states the "pseudogap". Obviously, the location of E_F within the pseudogap results in $g<1$. When the Fermi level is located within the pseudogap, Mott proposed the electrical conductivity formula of equation (10.52) to be modified in the following form:

$$\sigma = g^2 S_F^{free} e^2 a / 12\pi^3 \hbar, \qquad (15.37)$$

where S_F^{free} is the area of the free-electron Fermi sphere, a is an average atomic distance and the g-parameter is varied over the range $0 \leq g \leq 1$.[16] It should be noted that Mott ingeniously replaced the mean free path Λ_F in equation (10.52) by an average atomic distance a in equation (15.37) in the spirit of the Ioffe–Regel criterion discussed in Section 10.10.[17] This means that the mean free path effect is not expected in the high-resistivity limit but only the electronic structure effect through g^2 needs to be considered.

Equation (15.37) may be alternatively expressed in the form of equation (10.54) where the area of the free-electron Fermi surface is replaced by the product of $N(E_F)$ and the Fermi velocity v_F. The g^2-dependence of equation (15.37) is then rewritten as

$$\sigma = g^2 (e^2/3) a v_F^{free} N(E_F)^{free} = (e^2/3) a g v_F^{free} N(E_F) = (e^2/3) a v_F N(E_F), \qquad (15.38)$$

where $v_F = g v_F^{free}$ is assumed in order to reconcile equation (15.38) with equation (10.54).[18]

---

[16] What happens when the g-parameter exceeds unity? This corresponds to a Fermi surface positioned just prior to contact with the Brillouin zone and a Fermi level located prior to the van Hove singularity peak. In such circumstances, the Fermi velocity, proportional to $\hbar^{-1} \partial E/\partial k$ is lowered relative to that of the free-electron value and is approximated as $v_F = v_F^{free}/g$. Note that this differs from $v_F = g v_F^{free}$ for $g<1$. On the other hand, the scattering probability $1/\tau$ is proportional to the final density of states at the Fermi level and, hence, the g-parameter. The mean free path is then deduced to be $\Lambda_F = v_F \tau = (v_F^{free}/g)(\tau^{free}/g) = \Lambda_F^{free}/g^2$. Insertion of this relation into equation (15.37) restores the free-electron expression $\sigma_0 = e^2 \Lambda_F^{free} S_F^{free}/12\pi^3 \hbar$ in the case of $g>1$.

[17] Mott [20] referred to the conductivity obtained by inserting $g=1$, $S_F^{free} = 4\pi k^2$ and $k = \pi/a$ for a half-filled band into equation (15.37) as the Ioffe-Regel conductivity $\sigma_{IR} = e^2/3\hbar a$. The corresponding resistivity is $360\,\mu\Omega$-cm if $a = 3$ Å. This may be used as a rough guide for a critical resistivity, above which the mean free path effect is lost and, instead, the quantum interference effect dominates (see also footnote 10, p. 475).

[18] This does not hold unconditionally. Strictly speaking, the decomposition of the diffusion coefficient into the mean free path and the velocity would not be justified in the diffusional motion of electrons.

15.9 Electron conduction mechanism in amorphous alloys

Mott [20] claimed that the Ziman theory based on the Boltzmann transport equation is valid when $g \geq 1$ but that the electron localization effect sets in as g is lowered below unity. He conjectured the minimum metallic conductivity to occur at $g \approx 0.2$–0.3. A system in the range $0.2 < g < 1$ is metallic but the localization effect dominates at low temperatures. The conduction electron in this metallic regime is said to be weakly localized. As noted in Section 10.2, simple metals like pure Na, Zn and Al possess a free-electron-like Fermi velocity almost equal to $v_F^{free} = 10^8$ cm/s. The Fermi velocity for the minimum metallic conductivity is then roughly estimated to be $v_F = gv_F^{free} = (0.2$–$0.3) \times 10^8$ cm/s. This was chosen earlier as a possible minimum Fermi velocity in the evaluation of the high-resistivity limiting curve.

The g-parameter can be determined experimentally by taking the ratio of the measured electronic specific heat coefficient γ_{exp} over the corresponding free-electron value γ_{free} for amorphous alloys in group (V). The parameter can also be determined from the ratio of the free-electron value over the measured Hall coefficient [12]. Figure 15.17 shows logarithmic plots of the electrical conductivity at 300 K as a function of the measured g-parameter for amorphous (Ag–Cu)–Ge and (Ca–Mg)–Al alloys in group (V). Included are the data for liquid mercury reported by Even and Jortner [26]. They employed an apparatus capable of increasing the pressure from 1 to 1600 atm while increasing temperature from 20 to 1475 °C and measured the electrical conductivity, the Hall coefficient and density. It was revealed that the system gradually loses metallic conduction and approaches an insulating state when the density is decreased from 13.6 to 8.5 g/cm^3.

It is clear from Fig. 15.17 that both amorphous alloys and liquid mercury exhibit essentially the same behavior. The electron conduction is controlled by the pressure and temperature in liquid mercury and by the concentration of the non-metallic element like Ge in the amorphous (Ag–Cu)–Ge alloys. The ρ–T types for the amorphous alloys are again shown in Fig. 15.17, using the same symbols as those in Fig. 15.16. It is found that types (a), (b) and (c) fall on the vertical line with $g = 1$ in good agreement with the assumption in the Baym–Meisel–Cote model, whereas types (d) and (e) fall on a straight line with a slope of $+2$. This means that the ρ–T type (d) or (e) is uniquely observed in systems characterized by the pseudogap at E_F and $\Lambda_F \approx a$, where the Mott relation $\sigma \propto g^2$ holds well.

It is worthwhile commenting further on the ρ–T type (d) or (e). As shown in Fig. 15.12, a more or less linear temperature dependence persists down to about 10 K for type (d) whereas the concave curvature dominates over the temperature range 10–300 K for type (e). Hence, there is apparently a clear difference in this characteristic feature between types (d) and (e). However, we have

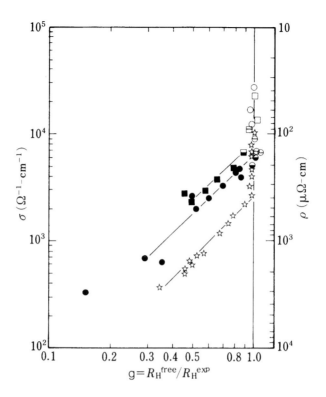

Figure 15.17. g-parameter dependence of conductivity in (●) amorphous $(Ag_{0.5}Cu_{0.5})_{100-x}Ge_x$ $(0 \leq x \leq 90)$ alloys, (■) amorphous Ca–Mg–Al alloys and (☆) liquid mercury. The ρ–T types are shown by symbols used in Fig. 15.12. This also applies to the square symbols. The corresponding ρ–T data cannot be measured for liquid mercury. [U. Mizutani, *Phys. Stat. Sol* (b) **176** (1993) 9]

noted that types (d) and (e) are equally characterized by a $\sqrt{T}$-dependence of conductivity below about 20 K, as indicated in equation (15.27), and $\sigma \propto T$ above about 30 K. Typical data are shown in Fig. 15.18. It was shown that the Hall coefficient also obeys $R_H \propto \sqrt{T}$ below about 20 K [12], being consistent with the theory proposed by Altshuler and Aronov [24] (see Section 15.8). All these results can be taken as a clear demonstration of the manifestation of the weak localization effect in amorphous alloys of types (d) and (e) in groups (IV) and (V). It is important to remind ourselves that the weak localization effect begins to participate as soon as the g-parameter is lowered below unity in amorphous alloys. This is because the mean free path Λ_F has already been decreased to an average atomic distance a prior to the growth of the pseudo-gap. This is different from the situation in quasicrystals and approximants, as will be discussed in Section 15.13.

In summary, electron transport phenomena of non-magnetic amorphous

15.9 Electron conduction mechanism in amorphous alloys

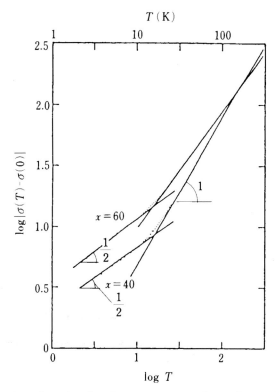

Figure 15.18. Log–log plot showing temperature dependence of the conductivity for the amorphous $(Ag_{0.5}Cu_{0.5})_{100-x}Ge_x$ ($x=40$, 60) alloys. Square-root temperature dependence is observed below about 20 K, while T-linear dependence above about 30 K. [U. Mizutani, *Phys. Stat. Sol* (b) **176** (1993) 9]

alloys can be understood in a unified picture. They are divided into two classes: one is described by ρ–T types (a), (b) and (c), the other by type (d) or (e).

1. The ρ–T types change in accordance with (a)→(b)→(c) with increasing resistivity in the low-resistivity regime of 20–200 $\mu\Omega$-cm and the successive changes in the ρ–T type reflect a decreasing mean free path Λ_F down to an average atomic distance [12]. The electronic structure is approximated by the nearly-free-electron model and the g-parameter remains essentially equal to unity. The behavior is well explained within the framework of the Baym–Meisel–Cote theory based on the Boltzmann transport equation.

2. The ρ–T type of either (d) or (e) occurs, as soon as the g-parameter is lowered below unity and the mean free path is constrained by an average atomic distance, and evidences the participation of the weak localization effect, particularly at low temperatures. The weak localization effect enhances the electron–electron interaction and a $\sqrt{T}$ dependence appears in the temperature dependence of both electrical conductivity and the Hall coefficient below about 20 K. The conductivity at 300 K is proportional to g^2, in good agreement with the Mott prediction.

15.10 Structure and preparation method of quasicrystals

A quasicrystal is defined as a solid satisfying the following conditions: (1) the diffraction intensities consist of an infinite number of δ-functions, (2) the number of basic vectors is larger than that of its dimension,[19] and (3) rotational symmetries forbidden in crystals exist. Icosahedral and decagonal quasicrystals may be taken as representative. An icosahedral quasicrystal possesses a three-dimensional quasiperiodicity with rotational symmetries characteristic of an icosahedron.[20] In contrast, a decagonal quasicrystal possesses a two-dimensional quasiperiodicity with five-fold symmetry in one plane but periodicity along the direction perpendicular to it.

There exist both thermally stable and metastable quasicrystals. As in the preparation of amorphous ribbons, a single-roll melt-spinning apparatus has been frequently employed to produce quasicrystalline ribbon samples by liquid quenching. Thermally stable single-grained quasicrystals can be grown by slow-cooling of the liquid phase in several alloy systems, such as Al–Pd–Mn and Mg–Zn–Ho [27].

Figure 15.19(a) is a scanning electron micrograph of the thermally stable icosahedral Al–Cu–Fe quasicrystal. The crystal habit showing a regular pentagon is a clear manifestation of the possession of the five-fold symmetry of its atomic arrangements. An electron diffraction pattern taken with the incident electron beam parallel to the five-fold axis is shown in Fig. 15.19(b). The diffraction spot is very sharp and the five-fold symmetry incompatible with a crystal is clearly seen. Diffraction patterns with two- and three-fold rotational symmetries are also observed by rotating the specimen in a manner consistent with the symmetries in an icosahedron. Therefore, we learn that the atomic structure of the quasicrystal appearing in both real and reciprocal spaces is entirely different from that in liquid and amorphous alloys and is characterized by highly ordered atomic arrangements. However, the five-fold rotational

---

[19] As discussed in Sections 5.8 and 5.9, we employed three basic vectors to describe the bcc, fcc and hcp structures. Hence, the number of basic vectors coincides with the dimensionality in a crystal. However, six and five basic vectors are needed for the description of the icosahedral and decagonal quasicrystals, respectively.

[20] Icosahedral quasicrystals are divided into two groups in terms of the cluster unit building up its structure: one is described by the Mackay icosahedron containing 54 atoms and the other by the rhombic triacontahedron containing 45 atoms. The former is abbreviated as the MI-type quasicrystal and is suited to describe the atomic structure of the Al–Mn-type quasicrystal. Typical examples are Al–Pd–Re, Al–Pd–Mn and Al–Cu–Fe, in which the hybridization effect between Al-3p and transition metal d-states is substantial. They are classified in group (IV) in Table 15.1. The latter is abbreviated as the RT-type quasicrystal. Typical examples are Al–Mg–Zn, Al–Li–Cu and Al–Mg–Ag, in which the DOS at the Fermi level is dominated by the free electron-like sp electrons and the hybridization effect is rather weak. They are classified in group (V).

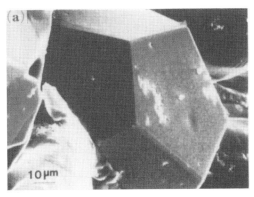 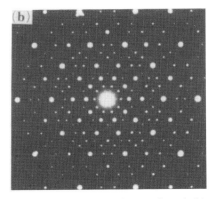

Figure 15.19. (a) Scanning electron micrograph of Al–Cu–Fe quasicrystal and (b) electron diffraction pattern showing five-fold rotational symmetry of the same sample. [A. P. Tsai *et al.*, *Jpn. J. Appl. Phys.*, **26** (1987) L1505]

symmetry breaks the translational symmetry of atomic arrangements and, hence, quasicrystals are classified as non-periodic systems.

15.11 Quasicrystals and approximants

A regular pentagon can be defined only if four vectors from a center to the vertices are specified in a plane. Hence, a four-dimensional space is required to accommodate periodic tilings (see below) having the rotational symmetries of a pentagon. Similarly, a regular icosahedron is defined by six vectors from the center to the vertices in three-dimensional space and, hence, a six-dimensional space is required to accommodate a periodic lattice with icosahedral symmetries. A three-dimensional quasiperiodic lattice can be constructed by merging our three-dimensional physical space into an n-dimensional hyperspace [27, 28]. In the case of icosahedral symmetries, we take a six-dimensional space, which is tilted relative to the six-dimensional cubic Bravais lattice so that six edge directions are projected onto the six-fold axes of a regular icosahedron in the three-dimensional physical space (see footnote 22, p. 497). For the sake of simplicity, we show below how a one-dimensional quasilattice is constructed from a two-dimensional hyperspace.

Figure 15.20 illustrates the situation such that the vertical and horizontal edges of the square lattice in xy-coordinates are projected onto the x'-axis of the $x'y'$-coordinates, which are tilted by $\tan\theta = \tau^{-1}$ relative to the xy-coordinates. Here τ is the golden ratio given by $\tau = (1+\sqrt{5})/2$. The x'-axis is called the physical space or parallel space, since atoms are projected onto it. Only lattice points which fall within a band or "window" parallel to the x'-axis are

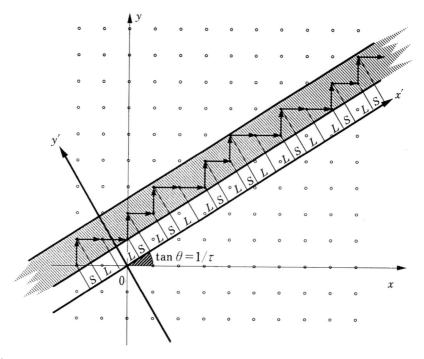

Figure 15.20. Construction of the one-dimensional Fibonacci lattice by projection from a two-dimensional square lattice. The $x'y'$-coordinate system is rotated around the origin by $\tan\theta = 1/\tau$ relative to the xy-coordinate system. Only the lattice points in the shaded region, called the "window", are projected onto the x'-axis.

projected. The width of the window is selected so as to be equal to the projection of the unit cell of the square lattice onto the y'-axis or the perpendicular space, as shown in Fig. 15.20. Accordingly, two different basic vectors or tiles marked as L (long) and S (short) are created in the one-dimensional physical space.

The two tiles L and S are not randomly distributed but are subjected to the following rule, when the tilt angle is chosen as the golden ratio τ. Its sequence is obtained by repeating the substitution of L→L+S and S→L, resulting in the so-called Fibonacci chain.[21] The Fibonacci chain is not periodic but has some order in the appearance of the two different tiles L and S. For example, L may appear twice successively but is never repeated successively three times.

---

[21] Fibonacci numbers are generated by the recurrence relation $F_{n+1} = F_n + F_{n-1}$, where $F_1 = 1$ and $F_2 = 1$. A simple manipulation leads to the Fibonacci sequence ($F_i = 1, 1, 2, 3, 5, 8, 13, 21, 34, \cdots$). The ratio F_{n+1}/F_n approximates the golden ratio τ and thus Fibonacci approximants are close to quasicrystals in their local structure.

S never appears twice successively. The local structure over several tiles is always found elsewhere in the sequence. This implies that the Fibonacci chain possesses self-similarity or scale invariance in such a way that it transforms into another Fibonacci chain with a different size through the substitution rule mentioned above. We call a non-periodic but ordered sequence like the Fibonacci chain "quasiperiodic".

Let us construct the Fibonacci lattice by starting from the tile L. The ratio N_L/N_S of the long tile L over the short tile S is obviously 1/0 in the zero-th generation, where N_L and N_S are the numbers of generated respective tiles. In the first generation, L and S are created from the parent L. Hence, its sequence is LS and the ratio $N_L/N_S = 1/1$ is obtained. The second and third generations yield the sequence LSL with the ratio $N_L/N_S = 2/1$ and LSLLS with $N_L/N_S = 3/2$, respectively. A repetition of the operation results in the ratio $N_L/N_S = 1/0, 1/1, 2/1, 3/2, 5/3, 8/5, \ldots$ and the ratio eventually approaches the golden ratio τ. The lattice formed on the one-dimensional physical space in Fig. 15.20 constitutes the Fibonacci chain and the ratio N_L/N_S is proved to be τ.

By terminating the operation at the n-th generation, we can construct a periodic structure with a large unit cell consisting of L and S tiles. This is equivalent to projecting the lattice points in the strip, which is tilted to the slope of an integer ratio N_S/N_L, onto the x'-axis of the $x'y'$ coordinates in Fig. 15.20. The periodic lattice thus obtained is called an approximant, since it closely approximates the quasiperiodic structure. The ratio N_S/N_L begins from 0/1 and continues with the sequence of $1/1=1$, $1/2=0.5$, $2/3=0.66 \cdots$ and approaches the inverse of the golden ratio $\tau^{-1} = 0.618 \cdots$ with an increasing number of operations.

Any realistic quasicrystal exists in the three-dimensional physical space. We consider a simple hypercubic lattice in six-dimensional space with coordinates $X_1 X_2 \cdots X_6$ and the edges of the square lattice are projected onto the three-dimensional physical space $X'_1 X'_2 X'_3$ of the $X'_1 X'_2 \cdots X'_6$ coordinates, which are tilted by the angle $\theta_i = \tan^{-1} 1/\tau$ ($i = 1, 2$ and 3) relative to the original space $X_1 X_2 X_3$.[22] The subspace $X'_1 X'_2 X'_3$ and its complementary subspace $X'_4 X'_5 X'_6$ are called the parallel and perpendicular spaces, respectively. Further, we choose a rhombic triacontahederon as the window or domain in the perpendicular space $X'_4 X'_5 X'_6$ and only lattice edges projected onto it are allowed. Now the icosahedral quasilattice is constructed by projecting only the allowed lattice

---

[22] The tilting angles are generally chosen as $\theta_i = \tan^{-1}(q_i/p_i)$ ($i=1, 2, 3$). The approximants are obtained by choosing $(q_1/p_1, q_2/p_2, q_3/p_3)$ equal to the inverse, N_S/N_L, of the Fibonacci ratio. For example, the 1/1-1/1-1/1 approximant is cubic, whereas the 3/2-2/1-2/1 approximant is tetragonal.

edges onto the parallel space $X'_1 X'_2 X'_3$.[23] This operation creates the three-dimensional Penrose quasilattice, which is composed of the two different rhombohedra without any gap or overlap. The structure thus obtained has the quasiperiodicity, rotational symmetries characteristic of an icosahedron and self-similarity discussed above.

A realistic icosahedral quasicrystal is obtained by decorating the Penrose quasilattice with atoms. Quasicrystals are found in binary or more frequently ternary alloy systems. Decoration of the quasilattice by two or three differing atomic species is indeed a formidable task because of the lack of lattice periodicity. The atomic structure of a quasicrystal has been conjectured from that of its approximant. For instance, Elser and Henley [29] pointed out in 1985 that the α-phase Al–Mn–Si compound with the lattice constant of 12.68 Å contains 138 atoms in its unit cell and corresponds to the 1/1-1/1-1/1 approximant to the MI-type Al–Mn quasicrystal.

The $Al_x Mg_{39.5} Zn_{60.5-x}$ ($20.5 \leq x \leq 50.5$) compound known as the Frank–Kasper phase contains 160 atoms in its unit cell with the lattice constant of 14.2Å [30, 31].[24] This compound is now established as the 1/1-1/1-1/1 approximant to the RT-type Al–Mg–Zn, Al–Mg–Cu, Al–Mg–Ag and Al–Mg–Pd quasicrystals. In addition to the 1/1-1/1-1/1 approximant, the 2/1-2/1-2/1 approximant with the lattice constant of 22.9Å has been discovered in the Al–Mg–Zn system [31]. The x-ray diffraction spectra for three different

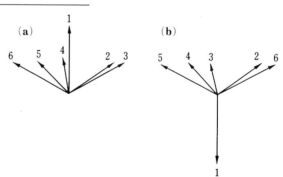

Six independent basic vectors for an icosahedral quasicrystal in (a) parallel and (b) perpendicular spaces.

[23] By applying the cut-projection method, one can produce an icosahedron which subtends the six basic vectors in the three-dimensional physical and perpendicular spaces, as shown in (a) and (b), respectively. A simple cubic lattice is formed also in the six-dimensional reciprocal space. Thus, the six reciprocal basic vectors can be constructed in the three-dimensional physical reciprocal space in the same way as (a). If these basic vectors are written as $\mathbf{b}_i$ ($i = 1-6$), then any reciprocal lattice vector takes the form of $\mathbf{g}_{n_1 n_2 n_3 n_4 n_5 n_6} = (n_1 \mathbf{b}_1 + n_2 \mathbf{b}_2 + \cdots n_6 \mathbf{b}_6)$, where ($n_1 n_2 \cdots n_6$) denotes a set of integers, called the six-dimensional Miller indices. For example, (211111) represents the diffraction line obtained when the electron or x-ray beam is incident parallel to the five-fold axis of an icosahedral quasicrystal. The Brillouin zone is constructed by planes formed by bisecting perpendicularly the reciprocal lattice vectors.

[24] Bergman et al. [Acta Cryst. **10** (1957) 254] originally reported 162 atoms in the unit cell. However, the x-ray Rietveld analysis for a series of the Al–Mg–Zn 1/1-cubic approximants revealed that 160 atoms exist in its unit cell [30].

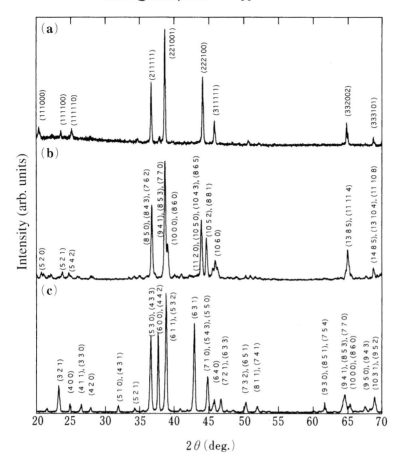

Figure 15.21. X-ray diffraction patterns of three different compounds in the Al–Mg–Zn system. (a) Quasicrystal, (b) 2/1-2/1-2/1 approximant and (c) 1/1-1/1-1/1 approximant. [From ref. 31.]

compounds are shown in Fig. 15.21. The diffraction lines of the quasicrystal are split into multiple lines in the approximant because of the lowering of the symmetry. For example, the (222100) diffraction lines in Fig. 15.21(a) for the quasicrystal are 60-fold degenerate but should decompose into (710), (543) and (631) lines in the 1/1-1/1-1/1 approximant and (10 5 2), (865) and (11 2 0) lines in the 2/1-2/1-2/1 approximant. It is clear from Fig. 15.21 that the diffraction spectrum of the 2/1-2/1-2/1 approximant more resembles that of the quasicrystal than does that of the 1/1-1/1-1/1 approximant. From this, we can say that the 2/1-2/1-2/1 approximant already has an atomic structure fairly close to that of the quasicrystal. So far approximants up to 3/2-2/1-2/1 have been synthesized experimentally in a bulk form. Further increase in the degree of the approximant would make the differentiation from the structure of a quasicrystal difficult.

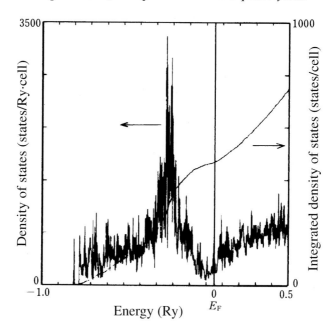

Figure 15.22. Calculated density of states for the 1/1-1/1-1/1 Al–Li–Cu approximant. [T. Fujiwara and T. Yokokawa, *Phys. Rev. Letters* **66** (1991) 333]

15.12 Electronic structure of quasicrystals

Since its unit cell is infinitely large, one cannot perform band calculations for a quasicrystal. A periodic boundary condition is imposed in the ordinary k-space band calculations. This means that band calculations are possible for approximants. Figure 15.22 shows the density of states calculated for the Al–Li–Cu 1/1-1/1-1/1 approximant [27, 32]. One can clearly see a V-shaped dip (valley) immediately below E_F. This is the pseudogap. All band calculations so far reported for various approximants are consistent with the possession of a pseudogap across E_F. The origin of the pseudogap has been ascribed to the interaction of the Fermi surface with the Brillouin zone consisting of many equivalent zone planes (see Section 9.7).

The formation of the pseudogap across E_F has also been experimentally confirmed in thermally stable quasicrystals, metastable quasicrystals and many approximants through photoemission spectroscopy [27,33], soft x-ray spectroscopy [34] and electronic specific heat measurements [35]. As a representative, we show in Fig. 15.23 combined soft x-ray emission and absorption spectra for the Al–Cu–Fe quasicrystal in comparison with those of the Hume-Rothery-type ω-phase Al_7Cu_2Fe compound and pure Al (see Section 7.8) [34]. One can clearly see that a dip or the pseudogap across E_F is evident in both

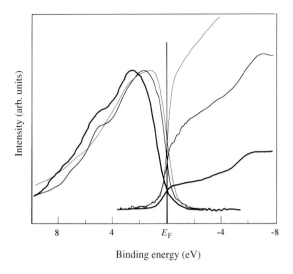

Figure 15.23. Soft x-ray Al-3p emission and Al-p absorption spectra for fcc Al (thin line), ω-phase Al$_7$Cu$_2$Fe compound (medium line) and Al$_{62}$Cu$_{25.5}$Fe$_{12.5}$ quasicrystal (thick line). The emission and absorption spectra are adjusted to the same intensity at the Fermi level E_F. The ω-phase compound has the space group P4/mnc with lattice constants $a = 6.33$ Å and $c = 14.81$ Å and contains 40 atoms in its unit cell. [Courtesy Dr E. Belin-Ferré]

quasicrystal and the ω-phase compound but not in pure Al.[25] The depth of the pseudogap is much deeper in the quasicrystal than in the ω-phase compound. Another interesting point to be noted is that the calculated density of states consist of many spiky peaks with a width of the order of 50–100 meV but they are apparently absent in the measured spectra. Indeed, the presence of spiky peaks has not been experimentally confirmed even from high-resolution photoemission spectroscopy [27, 33] and EELS measurements [36].

Electrons residing near E_F would be pushed into higher binding energies when the pseudogap is formed, thereby leading to a reduction in the electronic energy of the system. This is essentially the Hume-Rothery mechanism for the stabilization of these complex electron compounds, as discussed in Section 9.7. Hence, we believe that the formation of the pseudogap contributes to the stabilization of a quasicrystal and that its depth in the quasicrystal is deeper than

---

[25] The combined soft x-ray emission and absorption spectra are apparently split at the Fermi level even for free-electron-like pure Al. For the x-ray processes, the Fermi level of, say, pure Al is set at the inflexion point of the emission and absorption edges. These two edges result from the convolution product of the Fermi–Dirac distribution function, which at E_F separates occupied from unoccupied states, and the Lorentzian distribution of the inner level involved in the transition, and, as result, are given in the form of arctangent curves. Since this inflexion point appears at half the maximum intensity of the arctg function, an apparent dip results at E_F even in pure Al. One can conclude the presence of a pseudogap at E_F, only when the intensity at E_F of an Al-based system is lower than half the maximum intensity of pure Al.

that in the approximants or Hume-Rothery type compounds because of the possession of higher symmetries and increased multiplicities of equivalent zone planes in the former [31].

15.13 Electron transport properties in quasicrystals and approximants

As has been described above, the Fermi surface of the quasicrystal is heavily perturbed by the Brillouin zone and the pseudogap is formed in the vicinity of E_F. This certainly gives rise to a substantial effect on the electron transport properties. In a crystal, one is well aware that the resistivity decreases with increasing perfection of the crystallinity. This is a natural consequence of a decrease in structural imperfections. In contrast to a crystal, the resistivity of a quasicrystal increases as the quasicrystallinity is increased by heat treatment. This is a clear signature of the non-periodic nature of the quasilattice.

The electrical conductivity in a quasicrystal may be still described in terms of equation (10.54). Because of the non-periodicity of lattice potentials, the mean free path of the conduction electron should be short in quasicrystals. In addition, the density of states $N(E_F)$ at E_F is substantially reduced owing to the formation of the pseudogap. It is also noted that the hybridization effect between the Al-3p and the d states of the transition metal element, such as Fe, Pd and Re, is so strong that it causes bonding and antibonding states near E_F. This is coupled with the zone folding effect and yields small energy dispersions, particularly in the vicinity of E_F, which, in turn, result in an enhancement in the effective mass or a reduction in the Fermi velocity v_F. A substantial enhancement in resistivity in some MI-type quasicrystals is brought about by a simultaneous reduction in both $N(E_F)$ and v_F in equation (10.54).

Figure 15.24 shows the temperature dependence of the electrical resistivity of an amorphous Al–Mg–Pd sample obtained by melt quenching. It increases substantially upon transformation to the quasicrystalline phase at about 600 K. Indeed, the electronic specific heat coefficient ($\gamma = 0.42\,\text{mJ/mol K}^2$) in the quasicrystalline phase was found to be lower than that ($\gamma = 0.78\,\text{mJ/mol K}^2$) in the amorphous phase. This is due to the formation of the pseudogap at E_F and explains the rapid increase in resistivity upon transformation to the quasicrystalline phase (see exercise 15.2).

We emphasized in Section 15.6 that the short-range structure significantly affects the electronic states near E_F and, in turn, the electron transport properties in amorphous alloys. The short-range order develops as a result of the directional covalent bondings between Al and transition metal atoms and forms icosahedral clusters in quasicrystals and their approximants. Thus, it is of great interest to examine how the short-range structure in quasicrystals and

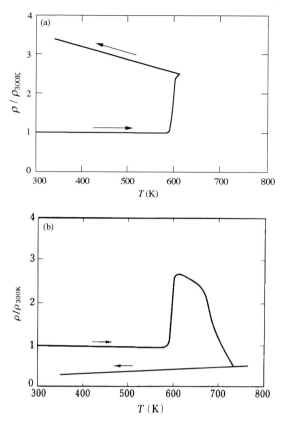

Figure 15.24. Temperature dependence of electrical resistivity of the amorphous $Al_{52}Mg_{18}Pd_{30}$ alloy heated up to (a) 608 K and (b) 790 K. The resistivity is normalized with respect to that at 300 K. The quasicrystalline phase can be achieved by lowering the temperature as shown in (a). The resistivity drops substantially upon crystallization above 680 K. The resistivities at 300 K for the amorphous and quasicrystalline phases are 220 and 780 μΩ-cm, respectively. [U. Mizutani *et al.*, *J. Phys.: Condensed Matter* **6** (1994) 7335]

approximants influences the electron transport properties. The short-range structure can be reliably determined for the approximants. For example, the edge length of the icosahedral cluster in various 1/1-1/1-1/1 approximants has been determined by the Rietveld method [30]. Figure 15.25 shows that the resistivity at 300 K sharply increases with decreasing edge length of the icosahedral cluster. A shortening of the edge length is most likely caused by an increase in the hybridization effect, which enhances the bonding and antibonding states near E_F and, in turn, contributes to an increase in resistivity.

The temperature dependence of the resistivity over the range 2–300 K for some representative non-magnetic quasicrystals is depicted in Fig. 15.26. We see that the data follow well the universal behavior shown in Fig. 15.12 for

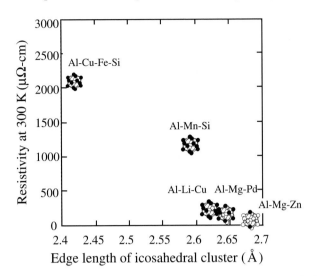

Figure 15.25. Resistivity at 300 K as a function of edge length of the icosahedral cluster in various 1/1-1/1-1/1 approximants. Solid circle indicates Al atoms and open circle other constituent atoms like Zn, Pd and Cu. [H. Yamada *et al.*, *Proc. of 6th Int. Conf. on Quasicrystals*, edited by S.Takeuchi and T.Fujiwara, (World Scientific, Singapore, 1998) pp. 664–667]

non-magnetic amorphous alloys in group (IV) and (V). First, we consider the low-resistivity regime, where the ρ–T types always change in the sequence $(a) \to (b) \to (c)$ with increasing resistivity in the same way as that in amorphous alloys. As mentioned in Section 15.8.2, the Baym–Meisel–Cote theory was able to interpret the data successfully for low-resistivity amorphous alloys in group (V). Here it is recalled that the integration over the scattering wave number K in equation (15.23) is limited to the range $0 \leq K \leq 2k_F$. This is reasonable for amorphous alloys because of the absence of long-range order. In the case of quasicrystals and their approximants, the structure factor consists of a series of sharp Bragg peaks and its information will need to be included up to high values of K far beyond $2k_F$. In other words, multiple scattering must be important. It is, therefore, surprising that the ρ–T behavior in low-resistivity quasicrystals and their approximants is essentially the same as that in low-resistivity amorphous alloys. All we can say, at the moment, is that the systematic change in the ρ–T types from (a) to (c) reflects the process of decreasing mean free path toward an average atomic distance and that all three ρ–T types must be described within the framework of the Boltzmann transport equation without invoking the weak localization effect.

It is worthy of noting that the resistivity of an Al–Mg–Zn quasicrystal of a high quality is 150 $\mu\Omega$-cm and its ρ–T type is (c) in spite of the possession of

15.13 Electron transport properties in quasicrystals and approximants

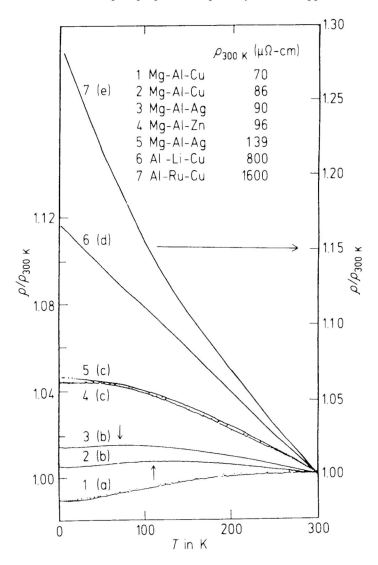

Figure 15.26. Temperature dependence of resistivity in various quasicrystals. [U. Mizutani, "Materials Science and Technology", edited by R. W. Cahn, P. Haasen and E. J. Kramer, (VCH, Germany, 1994), vol.3B *Electronic and Magnetic Properties of Metals and Ceramics, Part II*, volume editor K. H. J. Buschow, pp. 97–157]

the pseudogap at E_F [31, 35]. The absence of the weak localization effect for the pseudogap system with $g < 1$ is different from the case for amorphous alloys, where the Mott equation (15.37) holds. In other words, the mean free path effect must be still effective for low-resistivity quasicrystals and their approximants of ρ–T types (a) to (c), indicating that the electron experiences more coherent potentials than that in amorphous alloys.

The ρ–T curve of type (e) is observed in high-resistivity MI-type quasicrystals [27]. The Al–Pd–Mn and Al–Cu–Ru MI-type quasicrystals possess resistivities ranging from 0.01 to 0.1 Ω-cm. Here the weak localization effect coupled with the enhanced electron-electron interaction dominates in the same way as in high-resistivity amorphous alloys. However, the resistivity of the Al–Pd–Re quasicrystal of a high quality reaches a value as high as 1 Ω-cm at 4.2 K [27]. The temperature dependence of the electrical conductivity of a high quality Al–Pd–Re quasicrystal is particularly unique and cannot be described in terms of type (d) or (e). Instead, a power law obeying $\sigma \propto T^\alpha$ ($0.3 < \alpha < 0.7$) has been reported below 10 K [37]. The scattering mechanism for such high-resistivity quasicrystals has not yet been well clarified.

The ρ–T characteristic of many 1/1-1/1-1/1 approximants is mostly of type (a), indicating that the Boltzmann transport mechanism dominates and that the weak localization effect is absent. No matter how their crystallinity is improved, their residual resistivity apparently remains finite because of the presence of inherent chemical disordering [30]. The observation of type (a) suggests that, due to the restoration of lattice periodicity of 12–14 Å, the conduction electron propagates in these approximants in potentials which are more coherent than those in quasicrystals. It is of great interest to note that there exist some MI-type 1/1-1/1-1/1 approximants, in which the resistivity exceeds 1000 $\mu\Omega$-cm at 300 K while the ρ–T type is still (a). Since the mean free path Λ_F is longer than a, the Mott conductivity formula (15.37) should be replaced by

$$\sigma = g^2 S_F^{\text{free}} e^2 \Lambda_F / 12\pi^3 \hbar \qquad (15.39)$$

for both quasicrystals and approximants of the ρ–T types (a) to (c). An increase in resistivity above 1000 $\mu\Omega$-cm with the absence of the weak localization effect is possible, if the approximant possesses a small g-parameter around 0.3 so that the g^2-dependent electronic structure effect overwhelms the mean free path effect which guarantees type (a) [38]. In contrast, the electron transport behavior of the 2/1-2/1-2/1 approximant, which possesses a lattice constant of about 23 Å, is no longer distinguishable from that in quasicrystals.

Finally, we note that the temperature dependence of the Hall coefficient in quasicrystals and their approximants is generally stronger than that in amorphous alloys. Since both electrons and holes coexist in quasicrystals and their approximants, the discussion based on the two-band model may be appropriate (see Section 11.8). However, the interpretation of the Hall coefficient in extremely high-resistivity quasicrystals is still far from clear.

15.14 Electron conduction mechanism in the pseudogap systems

We have seen that the pseudogap is formed across E_F through different mechanisms. As mentioned in Section 15.9, an expansion of volume at high temperatures under high pressures caused the pseudogap to develop in liquid mercury. In quasicrystals and their approximants, simultaneous contacts of the Fermi surface with many equivalent Brillouin zone planes are responsible for its formation. In amorphous M_xX_{100-x} alloy systems, where M and X stand for a metal and metalloid element, respectively, the metal–insulator transition occurs when the concentration of a metallic element M is decreased below some critical value. We know that the energy gap opens at $x=0$ corresponding to the pure element X, such as semiconducting amorphous Si or Ge. Hence, the transition from a metal to an insulator with decreasing metal content can be viewed as the process of deepening the pseudogap across the Fermi level E_F. We treat both amorphous M_xX_{100-x} alloys and quasicrystals as being typical of pseudogap systems and discuss their characteristic features on a ρ–γ_{exp} diagram. At the end of this section, we briefly comment on the data in pseudogap systems in different "families" on the ρ–γ_{exp} diagram, which include heavy fermion and strongly correlated electron systems.

15.14.1 Mott conductivity formula for the pseudogap system

Mott [20] elaborated the electron conduction mechanism for a system with the pseudogap across E_F and formulated the conductivity for a Fermi gas at absolute zero from the Kubo–Greenwood formula in combination with the tight-binding approximation:

$$\sigma_0 = \rho_0^{-1} = \frac{\pi e^2 a^5 z I^2}{\hbar}[N(E_F)]^2, \tag{15.40}$$

where a is an average atomic distance, z is the coordination number of the constituent atom and I is the hopping integral defined as

$$I = \int \psi_i^* H \psi_j d\mathbf{r}, \tag{15.41}$$

where ψ_i is the wave function at the site i and H is the tight-binding Hamiltonian of the electron. The hopping integral in equation (15.41) depends on the degree of overlap of wave functions over the nearest neighbor atoms: it is small when the overlap is small. Equation (15.40) indicates that the resistivity is inversely proportional to the square of $N(E_F)$ or the measured electronic specific heat coefficient γ_{exp} in the pseudogap system. Equation (15.40) has provided a theoretical basis for the validity of equations (15.37) or (15.38) derived

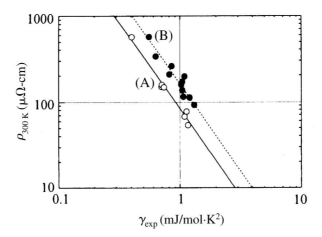

Figure 15.27. Resistivity at 300 K against the measured electronic specific heat coefficient γ_{exp} on a log–log scale for amorphous alloys in two different families: (○) Ag–Cu–Ge, Ag–Cu–Si, Mg–Zn–Sn sp electron alloys and (●) Al–Si–Ni alloys. They fall on the lines (A) and (B), respectively. [From ref. 35.]

more intuitively by Mott. Equation (15.40) may be hereafter referred to as the Mott conductivity equation.

From the experimental point of view, the validity of equation (15.40) must be tested by choosing a pseudogap system where the resistivity is high enough to assure the electron mean free path to be constrained by an average atomic distance. For this reason, non-periodic systems like amorphous alloys and quasicrystals of the ρ–T type (d) or (e) are best chosen. Further care is directed to the choice of the resistivity in equation (15.40). Ideally the residual resistivity ρ_0 is chosen. However, when the resistivity exceeds about 1000 $\mu\Omega$-cm, the quantum interference effect, which is not taken into account in equation (15.40), significantly perturbs the resistivity value at low temperatures. To avoid this difficulty, we use the resistivity value at 300 K in place of ρ_0 in the present discussion. Indeed, the choice of the resistivity at 300 K or at 4.2 K does not matter, as long as the resistivity is lower than 1000 $\mu\Omega$-cm. This is because the TCR is always less than 10 %.

Figure 15.27 shows the resistivity at 300 K against the measured electronic specific heat coefficient on a log–log scale for amorphous M_xX_{100-x} alloys in two different families [35]. The first family includes the amorphous $(Ag_{0.5}Cu_{0.5})_{100-x}Ge_x$ ($20 \leq x \leq 60$), $(Ag_{0.5}Cu_{0.5})_{77.5}Si_{22.5}$ and $Mg_{70}Zn_{30-x}Sn_x$ ($x = 0, 4, 6$) alloys in group (V). Their density of states at E_F are scarcely affected by the d states. The second family includes the Al-rich amorphous $Al_{90-x}Ni_{10}Si_x$ ($10 \leq x \leq 30$) and $Al_{85-x}Ni_{15}Si_x$ ($15 \leq x \leq 35$) alloys, in which only a small amount of the Ni-3d states constantly coexist with the sp electrons at E_F while the pseudogap is deepened with increasing Si concentration x. Thus, they are

15.14 Electron conduction mechanism in the pseudogap systems

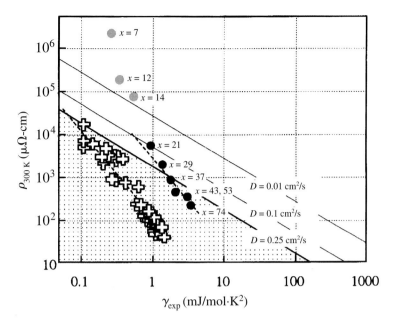

Figure 15.28. Resistivity at 300 K against γ_{exp} value on a log–log scale for quasicrystals and their approximants (⊕) and amorphous V_xSi_{100-x} alloys (●; metallic and ◐: insulator). The slope of the dashed lines is -2, whereas that of the solid lines with different diffusion coefficients is -1. The thick line refers to the line with $D = 0.25$ cm^2/s. The shaded area below it corresponds to the metallic regime. [Reproduced from refs. 35 and 39.]

selected such that the hopping integral I of equation (15.41) in the first family is larger than that in the second.

It is clear from Fig. 15.27 that the data for the sp electron amorphous alloys constitute a line (A) with a slope of -2 in excellent agreement with equation (15.40) and that the data in the second family form another parallel straight line (B), which is slightly displaced to higher resistivities relative to the first, as is expected from a reduction in I in equation (15.41). Such lines are referred to as Mott-lines.

15.14.2 Family of quasicrystals and their approximants

Sets of data for the resistivity at 300 K and the electronic specific heat coefficient in a large number of icosahedral quasicrystals and their approximants are plotted in Fig. 15.28 on a log–log diagram [35]. All data points are found to be fitted well to a Mott-line of the type mentioned above.[26] This is a bit surprising,

---

[26] Data for RT-type quasicrystals are less scattered and are fitted well to line (A) in Fig. 15.27. Data for MI-type quasicrystals are more scattered and fall on a line slightly above line (B). The larger scatter of the data points for the MI-type quasicrystals is due partly to the collection of data from different sources in the literature [35].

since the ρ–T types (a) to (c) have been observed in low-resistivity quasicrystals and approximants as discussed in Section 15.13 and, hence, the mean free path effect must be effective. But, the experimental confirmation for the presence of the Mott-line even for the low-resistivity quasicrystals and their approximants indicates that the mean free path effect is too small to be reflected on the log–log diagram. It is also inferred that the spiky peaks characteristic of the calculated DOS (see Fig. 15.22) must be rounded or smeared in real quasicrystals. The chemical disordering inherent in the atomic structure of approximants and quasicrystals may also be responsible for it [30]. It is noted that such chemical disordering effects cannot be taken into account in band calculations.

A metal–insulator transition line (MI-line) may be drawn as a rough guide on the ρ–γ_{exp} diagram. The MI-line refers to the boundary on the ρ–γ_{exp} diagram, above which no metallic data appear. As discussed in Section 15.9, we have assumed the electron diffusion coefficient D to take its possible minimum value of $0.25\,cm^2/s$. The *MI*-line in Fig. 15.28 is indeed the line with $D = 0.25\,cm^2/s$ and is the same as the high-resistivity limiting curve drawn in Fig. 15.16. The metallic regime below the MI-line is shaded in Fig. 15.28. For quasicrystals and approximants, the data point corresponding to the highest resistivity with the lowest electronic specific heat coefficient refers to that of the Al–Pd–Re quasicrystal of a high quality. This data point is found to fall very close to the MI-line. It is not possible to judge from Fig. 15.28 if an insulator in this family is achieved only in the limit of a diminishing carrier concentration or crosses the MI-line at a finite value of γ_{exp}.

15.14.3 Family of amorphous alloys in group (IV)

The change in the atomic and electronic structures and electron transport properties across the metal–insulator transition was systematically studied in the amorphous V_xSi_{100-x} ($7 \leq x \leq 74$) alloys [39]. A decrease in the V concentration lowers the 3d density of states at E_F and eventually leads to a semiconducting amorphous Si. Thus, this is typical of the pseudogap system. From Fig. 15.29, the conductivity for alloys with $x > 20$ can be interpreted in terms of the quantum interference effect characteristic of a disordered metallic system whereas those with $x < 15$ in terms of the variable-range hopping model applicable to an insulating regime (see Section 15.8.5). Therefore, the metal–insulator transition occurs in the composition range 15–20 at.%V in this system.

Figure 15.30 shows the V concentration dependence of the electronic specific heat coefficient for amorphous V_xSi_{100-x} alloys. Obviously, the density of states at E_F remains finite even in the insulating regime of $x < 15$. Indeed, the V-3d states remain finite across E_F over the whole concentration range, as

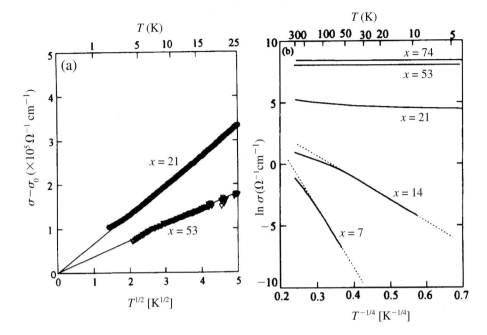

Figure 15.29. Temperature dependence of conductivity in the amorphous $V_x Si_{100-x}$ alloys. (a) The data for $x=21$ and 53 are fitted to equation (15.27). (b) The $T^{-1/4}$-dependence of conductivity holds for alloys with $x=7$ and 14. [From ref. 39.]

revealed by V-Lα soft x-ray spectroscopy.[27] The fully localized electronic states at E_F in the range $x<15$ are ascribed to the V-3d states hybridized with Si-3p states and are responsible for the observed variable-range hopping conduction.

The resistivity at 300 K and the electronic specific heat coefficient in the amorphous $V_x Si_{100-x}$ alloys are incorporated in Fig. 15.28. Here the resistivity at 300 K is again employed to circumvent the difficulty associated with the quantum interference effect in the metallic side. The data for $x>20$ fall on a straight line with a slope of -2 in excellent agreement with the Mott conductivity equation. However, we realize that the line is substantially shifted to the right relative to the Mott-line drawn through the data points for amorphous alloys in Fig. 15.27 and for quasicrystals and approximants in Fig. 15.28. It can also be noted in Fig. 15.28 that the data points with $x<15$, marked with symbol (●), deviate substantially from the extrapolated line. They should not be plotted on an equal footing, since they are insulators possessing finite localized states at E_F. Thus, we say that, in sharp contrast to the case in quasicrystals,

---

[27] The V-Lα spectrum is obtained by measuring the radiation emitted upon the transition from the V-3d band to its 2p level and provides information about the V-3d valence band structure (see Section 7.8).

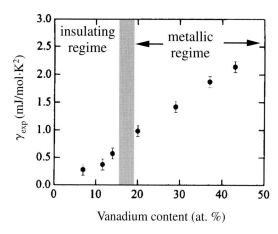

Figure 15.30. Electronic specific heat coefficient as a function of V concentration in amorphous V_xSi_{100-x} alloys. The metal–insulator transition occurs across the hatched area. [From ref. 39.]

the amorphous V–Si pseudogap system enters into an insulating regime by crossing the MI-line at a finite γ_{exp} value on the ρ–γ_{exp} diagram.

15.14.4 Family of "unusual" pseudogap systems

The metal–insulator transition, being characterized by the formation of a pseudogap at the Fermi level, is encountered in many different systems. Figure 15.31 is a summary of the ρ–γ_{exp} diagram on a log–log scale, in which the data for different pseudograp systems are plotted together with those of quasicrystals and amorphous V_xSi_{100-x} [39] and Ti_ySi_{100-y} [40] alloys. It is clear that the data for quasicrystals and these amorphous alloys in the metallic regime obey well the Mott conductivity formula (15.40), as evidenced from excellent line fitting of the data points with a slope of −2 on the log–log diagram.

There exist metallic systems, in which the datasets (ρ, γ_{exp}) definitely fail to obey the Mott conductivity formula. The data for $Sr_{1-z}La_zTiO_3$ [41] being typical of a strongly correlated electron system, are included in Fig. 15.31 for the composition range $0.5 \leq z \leq 0.95$ (see open squares). The $z = 0.95$ sample is still in the metallic regime while $LaTiO_3$ with $z = 1.0$ is an insulator. The marginally metallic $z = 0.95$ sample falls slightly above the MI-line of $D = 0.25 \, cm^2/s$ in the ρ–γ_{exp} diagram. However, the data points in the metallic regime obviously no longer follow the Mott conductivity formula (15.40).

Nishino et al. [42] found that the DO_3-type Fe_2VAl intermetallic compound exhibits semiconductor-like temperature dependence of the electrical resistivity, as shown in Fig. 15.32, and that its resistivity of $850 \, \mu\Omega$-cm at 300 K

15.14 Electron conduction mechanism in the pseudogap systems

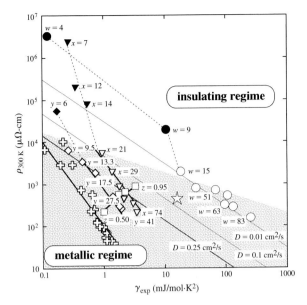

Figure 15.31. Resistivity at 300 K against γ_{exp} value on a log–log scale for ($\diamond$, $\blacklozenge$) amorphous $Ti_y Si_{100-y}$ alloys [40], ($\square$) $Sr_{1-z}La_z TiO_3$ [41], ($\star$) $Fe_2 VAl$ [42] and ($\circ$, $\bullet$) amorphous $Ce_w Si_{100-w}$ alloys [43, 44]. The data for ($\triangledown$, $\blacktriangledown$) amorphous $V_x Si_{100-x}$ alloys [39] and ($\oplus$) quasicrystals and approximants are reproduced from Fig. 15.28. The composition in the respective systems is marked near the data point. Open and filled symbols refer to samples characterized as metallic and insulating states, respectively. The data for amorphous $M_x Si_{100-x}$ (M = Ti and V) alloys in the metallic regime as well as quasicrystals are fitted to lines with the slope of -2, consistent with the Mott conductivity formula (15–40). The shaded area represents the experimentally determined metallic regime.

increases to 3000 $\mu\Omega$-cm at 4.2 K, though photoemission spectroscopy measurements revealed a sharp Fermi cutoff. This system is also classified as a pseudogap system, since E_F is located in a deep pseudogap, as shown in Fig. 15.33. Nevertheless, the electronic specific heat coefficient is deduced to be 14 mJ/mol K^2 and its large enhancement is ascribed to the spin fluctuations unique to a marginally magnetic alloy.[28] The set of (ρ_{300K}, γ_{exp}) data for this compound falls in the region close to the $D = 0.1$ cm^2/s line in the ρ–γ_{exp} diagram, well above the MI-line of $D = 0.25$ cm^2/s, even though it must be located in the metallic regime.

Finally, we discuss briefly the data for amorphous $Ce_w Si_{100-w}$ alloys [43, 44] which are classified as belonging to the heavy fermion system. The uniqueness of this system is the possession of an extremely large electronic specific heat

---

[28] The Curie temperature decreases with increasing x and disappears at $x = 0.33$ in the ternary alloys $(Fe_{1-x}V_x)_3 Al$. Indeed, the $Fe_2 VAl$ compound exhibits a strong temperature dependence of the magnetic susceptibility at least down to 4.2 K and, hence, is classified as group (III).

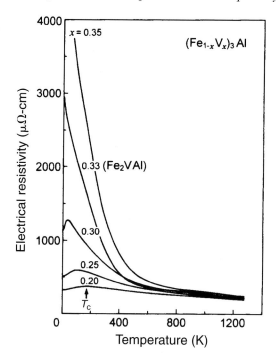

Figure 15.32. Temperature dependence of resistivity in $(Fe_{1-x}V_x)_3Al$ with $0.2 \leq x \leq 0.35$. Samples with $x \geq 0.33$ exhibit a semiconductor-like behavior. T_C indicates the Curie temperature. [From ref. 42.]

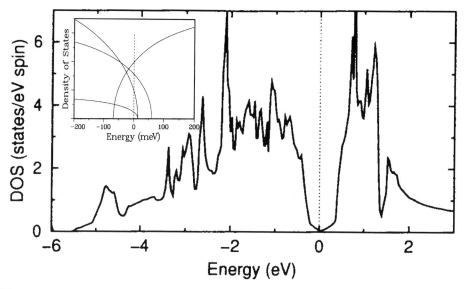

Figure 15.33. Calculated density of states of the Fe_2VAl compound. The Fermi level falls in the minimum of the pseudogap. Inset shows the density of states near the Fermi level on an expanded scale. [R. Weht and W. E. Pickett, *Phys. Rev. B* **58** (1998) 6855]

coefficient exceeding 100 mJ/mol·K² but an electrical resistivity fairly comparable to that found in amorphous V_xSi_{100-x} [39] and Ti_ySi_{100-y} [40] alloys. The Ce-4f electrons at the Fermi level are believed to play a key role both in enhancement of the electronic specific heat coefficient and in electron conduction. With decreasing Ce concentration, however, the electronic specific heat coefficient does decrease down to about 20 mJ/mol·K² before the transition to the insulating regime, which occurs when the Ce concentration is reduced to 9 at.%. The transition is clearly seen in Fig. 15.31. It is clear that the data for the amorphous Ce_wSi_{100-w} alloys neither obey the Mott conductivity formula nor fall below the MI-line of $D = 0.25\,cm^2/s$.

All data points marked by open symbols in Fig. 15.31 have to be regarded as a metal. The region encompassing open symbols is shaded. It is of great interest to note that the metallic regime apparently extends well above the MI-line of $D = 0.25\,cm^2/s$, when the value of γ_{exp} becomes large. Further studies are certainly needed to explore the electron transport mechanism in the neighborhood of the experimentally determined metal–insulator transition line in the region of γ_{exp} values exceeding 10 mJ/mol·K².

Exercises

15.1 The Debye–Waller factor can explain the negative TCR observed in the ρ–T types (b) and (c) of amorphous alloys, whose resistivities are in the range 60–150 $\mu\Omega$-cm (see Section 15.8.2). But it does not play a significant role in the case of a crystal (see Section 10.12). Discuss why this is so.

15.2 The electrical resistivity of the Al–Mg–Pd amorphous ribbon sample jumps sharply upon its transformation to the quasicrystalline phase, as shown in Fig. 15.24. The electronic specific heat coefficient γ_{exp} for the amorphous and quasicrystalline phases was experimentally determined as 0.78 and 0.42 mJ/mol K², respectively. Given the resistivity of the amorphous phase to be 220 $\mu\Omega$-cm, estimate the resistivity in the quasicrystalline phase by using the Mott conductivity equation (15.37). Assume that the mean free path of both phases is equally constrained by an average atomic distance a.

Appendix

Values of selected physical constants

quantity	symbol	value	SI	CGS
velocity of light	c	2.99792458	10^8 m·s^{-1}	10^{10} cm·s^{-1}
electron mass	m	9.1093897	10^{-31} kg	10^{-28} g
electron charge	e		$1.60217733 \times 10^{-19}$ C	4.80296×10^{-10} esu
specific charge of the electron	e/m		1.7588196×10^{11} C·kg^{-1}	$5.27645886 \times 10^{17}$ esu·g^{-1}
Planck's constant	h	6.6260755	10^{-34} joule·s	10^{-27} erg·s
	$\hbar$	1.05457266	10^{-34} joule·s	10^{-27} erg·s
quantum flux	$h/2e$	2.0678	10^{-15} tesla·m^2 = Wb	10^{-7} gauss·cm^2
Bohr magneton	$\mu_B = \mu_0 e\hbar/2m$		1.165×10^{-29} Wb·m	
			9.27314×10^{-24} joule·tesla^{-1}	9.27314×10^{-21} erg·gauss^{-1}
Avogadro's number	N_A	6.0221367×10^{23} mole^{-1}		
Boltzmann's constant	k_B	1.380658	10^{-23} joule·K^{-1}	10^{-16} erg·deg^{-1}
gas constant	R_0	8.314510	1 joule·mole^{-1}·deg^{-1}	10^7 erg·mole^{-1}·deg^{-1}
standard volume of perfect gas at 0°C, 1 atm	V_0	22.41410	10^{-3} m^3·mole^{-1}	10^3 cm^3·mole^{-1}
Bohr radius	a_0	5.291 77249	10^{-11} m	10^{-9} cm
permittivity of free space	ε_0	—	$10^7/4\pi c^2 = 8.854 \times 10^{-12}$ fared·m^{-1}	1
permeability of free space	μ_0	—	$4\pi \times 10^{-7} = 1.257 \times 10^{-6}$ henry·m^{-1}	1

Conversions: 1 eV = $1.60217733 \times 10^{-19}$ joule = $1.60217733 \times 10^{-12}$ erg = 7.35×10^{-2} Ry
1 eV = 8.0655410×10^5 m^{-1} = 8.0655410×10^3 cm^{-1}
1 Ry = 13.6058 eV
1 eV = 1.160445×10^4 K
1 cal = 4.186 joule
1 nm = 10 Å = 10^{-9} m = 10^{-7} cm

Principal symbols (by chapter)

Chapter 1

λ : wavelength
p: momentum
v: velocity of electron
m: mass of electron
$(-e)$: electronic charge
σ: electrical conductivity
n: number of electrons per volume
τ: relaxation time
k_B: Boltzmann constant
T: absolute temperature
Λ: mean free path of electron
R: gas constant
T_c: superconducting transition temperature

Chapter 2

n: principal quantum number
ℓ: azimuthal or orbital angular momentum quantum number
m: magnetic quantum number
s: spin quantum number
W_F: average kinetic energy per electron
ε_I: ionization energy
ε_c: cohesive energy
r_0: equilibrium interatomic distance
$\psi(x,y,z)$: wave function of electron
m: mass of electron
$\hbar$: Planck constant divided by 2π
E: energy eigenvalue of electron

k_x, k_y, k_z, and k: wave number of electron $k = \sqrt{k_x^2 + k_y^2 + k_z^2}$
λ: wavelength of electron
L: edge length of a metal cube
n_x, n_y, n_z: arbitrary integers
V: volume of a system
$\mathbf{k}$: wave vector of free electron
$\mathbf{p}$: electron momentum
k_F: Fermi radius
N_0: total number of electrons per mole
N_A: Avogadro number
E_F: Fermi energy
$N(E)$: electron density of states
N: total number of electrons per volume V
e/a: number of valence electrons per atom
Ω: volume per atom
T_F: Fermi temperature

Chapter 3

E_i: energy of electron in i-th sphere
Z_i: number of states available for electrons with energy E_i
N_i: number of electrons with energy E_i
ω_i: number of distinguishable ways in distributing N_i electrons over Z_i states with energy E_i
W: total number of distinguishable ways
N: total number of electrons per volume V
E: total energy of conduction electron system
k_B: Boltzmann constant
$f(E, T)$: Fermi–Dirac distribution function at temperature T
$E_F(T)$: Fermi energy at temperature T
$E_F(0)$: Fermi energy at absolute zero
$F(E)$: physically meaningful arbitrary function
ΔQ: heat input to a sample under an adiabatic condition
ΔT: temperature increment due to the heat input ΔQ
U_{el}: internal energy of the conduction electron system
$N(E_F(0))$: density of states at the Fermi level at 0 K
γ: electronic specific heat coefficient
C_{el}: electronic specific heat
T_F: Fermi temperature
N_A: Avogadro number
n_0: valency of the constituent atom; e/a is alternatively used

R: gas constant
α: lattice specific heat coefficient
Θ_D: Debye temperature
C: specific heat
γ_{exp}: experimentally derived electronic specific heat coefficient
γ_F: electronic specific heat coefficient in the free-electron model
m^*_{th}: thermal effective mass
γ_{band}: electronic specific heat coefficient derived from band calculations
$\boldsymbol{\mu}$: magnetic moment of conduction electron
H: magnetic field
θ: angle between magnetic moment $\boldsymbol{\mu}$ and applied field **H**
$p(\theta)d\theta$: probability of finding the magnetic moment at angles between θ and $\theta+d\theta$
M: component of magnetization parallel to **H**
$L(\alpha)$: Langevin function

$$\alpha = \frac{\mu H}{k_B T}$$

χ: magnetic susceptibility
m_s: spin quantum number of electron
μ_B: Bohr magneton
e/a: number of electrons per atom
A: atomic weight
d: mass density
v: velocity of conduction electron
p: electron momentum
E_0: work needed to remove to infinity an electron at the lowest energy state in the valance band
ϕ: work function
J: emission current density
n: number of electrons per unit volume
A: pre-exponential factor in Richardson–Dushman equation

Chapter 4

$f(\mathbf{r})$: arbitrary periodic function
a: lattice constant
A_n: Fourier coefficient of the periodic function $f(x)$
$\mathbf{l}_{l_x l_y l_z}$ or **l**: lattice vector
l_x, l_y, l_z: component of the lattice vector
$\mathbf{g}_{n_x n_y n_z}$ or **g**: reciprocal lattice vector
$\mathbf{a}_x, \mathbf{a}_y, \mathbf{a}_z$: basic vectors in real space

$\mathbf{b}_x, \mathbf{b}_y, \mathbf{b}_z$: basic vectors in reciprocal space
L: edge length of a cube sample
(hkl): Miller indices
d: interplanar distance
d_N: distance from an origin to the lattice plane
$\mathbf{k}$: wave vector of the incident x-ray beam
$\mathbf{k}'$: wave vector of the reflected x-ray beam
θ: glancing angle of the incident x-ray beam to the crystal plane
M or m: mass of atom
u_{la} or $u(l)$: displacement of the l-th atom
β: force constant
ξ: amplitude of lattice vibrations
q: wave number of lattice vibrations
ω: angular frequency of lattice vibrations
s: sound velocity
c: elastic stiffness constant
ρ: density
N: total number of atoms in a crystal
v: frequency of light wave
R: gas constant
$\mathbf{q}$: wave vector of lattice wave
$\omega_\mathbf{q}$: angular frequency of lattice wave
Z_i: number of states available for particles with energy E_i
N_i: number of particles with energy E_i
W: total number of distinguishable ways
N: total number of particles per volume V
E: total energy of a system
k_B: Boltzmann constant
ζ: chemical potential
S: entropy
U: internal energy
p: pressure
$n(E, T)$: Bose–Einstein distribution function at temperature T
$n_\mathbf{q}(T)$: Planck distribution function for phonons
$U_{lattice}(T)$: internal energy due to lattice vibrations at temperature T
$D(\omega)$: phonon density of states or frequency spectrum
$N(q)$: number of lattice modes enclosed by a sphere with the radius q
q_D: Debye radius
ω_D: Debye frequency
Θ_D: Debye temperature

Θ_E: Einstein temperature
ξ_0 and η_0: displacement of atoms

Chapter 5

$V(x)$: ionic potential function
$2A$: amplitude of the cosine-type ionic potential
a: lattice constant
$\psi(x)$: wave function of electron
E: energy eigenvalue
ξ: dimensionless space variable defined as $\xi = \pi x/a$
ε: dimensionless energy defined as $\varepsilon = 8mEa^2/h^2$
η: dimensionless amplitude of the ion potential defined as $\eta = 8mAa^2/h^2$
k: wave number of the Bloch wave
$\mathbf{l}$: lattice vector in a crystal defined as $\mathbf{l} = l_x\mathbf{a}_x + l_y\mathbf{a}_y + l_z\mathbf{a}_z$
$u_\mathbf{k}(\mathbf{r})$: periodic function in the Bloch wave function
$\mathbf{k}$: wave vector of the Bloch wave
$\mathbf{g}$ or $\mathbf{g}_n$: reciprocal lattice vector
$\hbar\mathbf{k}$: crystal momentum of the Bloch electron
V_0: potential height of the Kronig–Penney model
α: wave number defined as $\hbar\alpha = \sqrt{2mE}$ in the Kronig–Penney model
β: wave number defined as $\hbar\beta = \sqrt{2m(V_0 - E)}$ in the Kronig–Penney model
a and b: potential width of the Kronig–Penney model
P: parameter defined as $\lim_{\substack{b\to 0 \\ \beta\to\infty}}(\beta^2 ab/2) = P$ in the Kronig–Penney model
$V_\mathbf{n}$: Fourier coefficient of the periodic potential $V(\mathbf{r})$
$A_\mathbf{n}$: Fourier coefficient of the Bloch wave function
$E_\mathbf{n}$: unperturbed energy of the Bloch electron defined as $E_\mathbf{n} = \hbar^2(\mathbf{k} - \mathbf{g}_n)^2/2m$
ΔE_{100}: energy gap across the {100} zone planes
$E_-(k)$ and $E_+(k)$: energies of the Bloch electron defined by equation (5.40)
n: band index
λ: wavelength of the Bloch wave
d: interplanar distance
$\mathbf{i}$, $\mathbf{j}$ and $\mathbf{k}$: unit vectors in cartesian coordinates
$\mathbf{a}_x$, $\mathbf{a}_y$, and $\mathbf{a}_z$: primitive translation vectors in real space
$\mathbf{b}_x$, $\mathbf{b}_y$, and $\mathbf{b}_z$: primitive translation vectors in reciprocal space
V_B: volume of the first Brillouin zone
c and a: lattice constants of hcp lattice
$\mathbf{a}_1$, $\mathbf{a}_2$ and $\mathbf{a}_3$: primitive translation vectors of hcp lattice in real space
$\mathbf{b}_1$, $\mathbf{b}_2$ and $\mathbf{b}_3$: primitive translation (basic) vectors of hcp lattice in reciprocal space
n or e/a: number of electrons per atom

Chapter 6

Γ: origin in reciprocal space
$E_-(k_N)$: energy at the point N in the {110} zone planes in the bcc lattice
e/a: number of electrons per atom
E_c: energy at the bottom of condution band
E_v: energy at the top of valence band
E_d: donor level
E_a: acceptor level
ΔE_g: energy gap

Chapter 7

F: Lorentz force
E: electric field
B($=\mu_0$**H**): magnetic field
$(-e)$: electronic charge
c: speed of light
v: velocity of electron
r: position vector of electron
k: wave vector of electron
n: positive integer including zero
A: vector potential
S: area of the closed orbit of electron in real space
$A(\varepsilon)$: area of the closed orbit of electron in reciprocal space
ω_c: cyclotron frequency, defined as $\omega_c = eB/m$
θ: angle between the two gamma-rays in the positron annihilation experiment
$\rho(\mathbf{p})$: density of electrons with the momentum **p**
$N(p_z)$: coincidence rate in the positron annihilation experiment
$A(p_z)$: area of the cross-section of the Fermi sphere normal to the p_z-axis
p_z: Fermi momentum or the Fermi cut-off
E_1: energy of incident photon
E_2: energy of photon after scattering
E_{el}: energy of electron
β: ratio of the velocity of electron over the light speed, v/c
$J(p_z)$: probability of the scattered electron with momentum p_z
E_{kin}: kinetic energy of photoelectron
$h\nu$: energy of incident photon
ϕ: work function
E_B: binding energy of the electronic state in a solid
$N_i(E)$: initial density of states at the energy E
$N_f(E+h\nu)$: final density of states at the energy $E+h\nu$

$f(E,T)$: Fermi–Dirac distribution function at temperature T

$\sigma_{opt}(E,h\nu)$: optical transition probability or an average cross-section for all states at the energy E

$P_t(E,h\nu)$: electron transport function

$P_e(E,h\nu)$: escape function

$n(E_{kin},h\nu)$: number of photoelectrons

$S_A(E_{kin})$: analyzer sensitivity

$R_A(E_{kin},h\nu)$: total resolution function

V_{acc}: acceleration voltage

E_i: electron energy of the initial state

E_f: electron energy of the final state

ϑ: exit angle of photoelectron

φ: exit angle of photoelectron

$\mathbf{K}$: wave vector of a photoexcited electron in a crystal

$\mathbf{K}^{pe}$: wave vector of a photoelectron in vacuum

$\mathbf{g}$: reciprocal lattice vector

$\mathbf{k}$: wave vector of the Bloch electron

U_o: inner potential

$K^{pe}_\perp(i)$: wave number corresponding to the peak i in the measured photoemission spectrum in the normal emission mode

$E_{kin}(i)$: kinetic energy of the photoelectron corresponding to the peak i

$E_i(i)$: initial energy of the electron corresponding to the peak i in the normal emission mode

$k_\perp(i)$: wave vector of the Bloch electron corresponding to the peak i in the normal emission mode

λ: wavelength of soft x-ray

d: grating space of a crystal in the spectrometer

E_c: energy of the core level

$I(\omega)$: transition probability per unit time for the spontaneous emission

ψ_i: electron wave function at the initial state

ψ_c: electron wave function at the final core state

$\psi_{n,\mathbf{k}}(r)$: wave function of the valence electron specified by the wave vector $\mathbf{k}$ and band index n

N: number of unit cells in a crystal

$R_\ell(E_{n\mathbf{k}},r)$: radial wave function

$Y_{\ell m}(\theta,\varphi)$: spherical harmonic function specified by the azimuthal quantum number ℓ and magnetic quantum number m

$n_s(E)$, $n_p(E)$ and $n_d(E)$: s-, p- and d-partial density of states

$b_{n\mathbf{k},\ell m}$: expansion coefficient

$M_{pK}(E)$: transition matrix element

$\Delta\sigma(\omega)$: increase in the photoabsorption cross-section at the threshold

α: fine structure constant

$|I\rangle$: initial state in a whole system
$|F\rangle$: final state in a whole system
$V(\mathbf{r})$: interaction potential of incident electron with electrons in a solid
$d\Omega$: solid angle
Z_i: atomic number
$\mathbf{R}_i$: position vector of nucleus i
$\mathbf{r}_j$: position vector of electron j in a solid
$\mathbf{q}$: momentum transfer vector equal to $\mathbf{q} = \mathbf{k}_0 - \mathbf{k}$
$S(\mathbf{q},E)$: dynamical structure factor
$I(E)$: intensity of incident electron
$P(E)$: energy-dependent matrix element

Chapter 8

N: total number of electrons per unit volume
Ψ: total wave function in a system consisting of electrons and nuclei
V_{ee}: Coulomb potential energy due to electron–electron interaction
V_{en}: Coulomb potential energy due to electron–nucleus interaction
$(-e)$: electronic charge
$\mathbf{r}_i$: position vector pointing to i-th electron
Z_α: atomic number of a nucleus
$\mathbf{R}_\alpha$: position vector to α-th nucleus
N_n: number of nuclei per unit volume
ξ_i: spin coordinate of i-th electron
V_{nn}: nuclear potential energy
$E_{0,i}$: ground-state energy of i-th isolated atom
$\psi_i(\mathbf{r}_i)$: wave function of i-th electron
$\chi_i(\xi_i)$: spin function of i-th electron
Ψ_H: Hartree wave function
$V_{H,i}$: Hartree potential
ε_i: one-electron energy of i-th electron
Ψ_{HF}: Hartree–Fock wave function
$\delta_{\chi_i\chi_j}$: Kronecker-delta
$V_x(\mathbf{r}_i)$: exchange potential
ε_0: ground-state energy per electron in a homogeneous electron gas
ε_{kin}: kinetic energy per electron in a homogeneous electron gas
ε_X: exchange energy per electron in a homogeneous electron gas
ε_C: correlation energy per electron in a homogeneous electron gas
R_s: radius of a sphere containing one electron
ρ: electron density in a homogeneous electron gas
a_0: Bohr radius

Principal symbols

$\rho(\mathbf{r})$: electron density at the position $\mathbf{r}$
$V_{ion}(\mathbf{r})$: ionic potential at the position $\mathbf{r}$
$F[\rho]$: $T_s[\rho] + E_{XC}[\rho]$ in the LDF theory
$T_s[\rho]$: kinetic energy of electron in non-interacting system
$E_{XC}[\rho]$: exchange and correlation energy
$\varepsilon_{XC}(\rho)$: exchange and correlation energies in a homogeneous electron gas
$\mu_{XC}(\rho(\mathbf{r}))$: chemical potential defined as $d(\rho\varepsilon_{XC}(\rho))/d\rho$
$\mu_X(\rho(\mathbf{r}))$: exchange energy in a homogeneous electron system
$\mu_{X\alpha}(\rho(\mathbf{r}))$: Slater's exchange energy
$V(\mathbf{r})$: effective one-electron potential consisting of the Hartree potential and $\mu_{XC}(\rho(\mathbf{r}))$
$\mathbf{l}$: lattice vector in a crystal
$\mathbf{k}$: Bloch wave vector
$\psi_\mathbf{k}(\mathbf{r})$: Bloch wave function
$u_\mathbf{k}(\mathbf{r})$: periodic function in the Bloch wave function
$\mathbf{g_n}$: reciprocal lattice vector
$U_a(\mathbf{r})$: potential of a free atom
$\phi(\mathbf{r})$: atomic orbital wave function
$\mathbf{R}_n$: position vector pointing to the nearest neighbor lattice sites
$\mathbf{i}, \mathbf{j}$ and $\mathbf{k}$: unit vectors in cartesian coordinates in real space
α and γ: parameters in the tight-binding approximation
$X_\mathbf{k}(\mathbf{r})$: OPW function
Ω: volume per atom
$\mu_{\mathbf{k},j}$: coefficient in the OPW function
$C(\mathbf{k}+\mathbf{g_n})$: OPW expansion coefficient
U_{mn}: Fourier component of the potential in the OPW method
R_M: radius in the empty-core model
A_0: constant energy in the empty-core model
$V_\mathbf{q}$: Fourier component of bare Coulomb potential
$v(\mathbf{r}-\mathbf{l})$: muffin-tin potential specified by the lattice vector $\mathbf{l}$
ℓ: azimuthal quantum number
m: magnetic quantum number
$Y_{\ell m}(\theta,\phi)$ or $Y_{\ell m}(\hat{r})$: spherical harmonic function
$R_\ell(r)$: radial wave function in a spherically symmetric potential
$\chi_\mathbf{k}(E,\mathbf{r})$: APW function
$j_\ell(kr)$: spherical Bessel function of order ℓ
$\theta_\mathbf{k}$ and $\phi_\mathbf{k}$: polar angles of the wave vector $\mathbf{k}$
$C(\mathbf{k}+\mathbf{g_n})$: APW expansion coefficient
F_{mn}: Fourier component in the APW method
$\chi(\mathbf{r})$: unperturbed wave function in the multiple scattering theory
$\psi_n^o(\mathbf{r})$: wave function propagating outward from all the scatterers
$\psi_n^i(\mathbf{r})$: wave function incident to the n-th scatterer

$G(\kappa, \mathbf{r} - \mathbf{r}')$: Green function
κ: kinetic energy E in atomic units
$\mathscr{L}$: arbitrary linear operator
$G(\kappa, \mathbf{k}; \mathbf{r} - \mathbf{r}'')$: structure Green function
Λ: functional in the KKR method
L_ℓ: logarithmic derivative of the radial wave function $R_\ell(r)$ at the muffin-tin radius
$S_{\ell'm',\ell m}$ or $S^k_{\ell'm',\ell m}$: structure factor
$\underline{\underline{H}}$: Hamiltonian matrix
$\underline{\underline{O}}$: overlap matrix
$\mathbf{a}$: expansion coefficient vector
$\hat{r}$: polar angles (θ, φ) of the position vector $\mathbf{r}$
r_0: radius of the Wigner–Seitz sphere or atomic sphere
$\chi_{\ell m}(E, \mathbf{r})$: muffin-tin orbital
$p_\ell(E)$: function appearing in the muffin-tin orbital
$a^{jk}_{\ell m}$: expansion coefficient of the muffin-tin orbital
$V_{MT}(r)$: muffin-tin potential in the LMTO method
V_{MTZ}: muffin-tin zero
a: radius of the muffin-tin sphere
η_ℓ: phase shift of the ℓ-th partial wave
κ: wave number defined as $\sqrt{E - V_{MTZ}}$ in the LMTO method
$j_\ell(\kappa r)$: spherical Bessel function
$n_\ell(\kappa r)$: spherical Neumann function
$R_\ell(E_v, r)$: partial radial wave function at energy E_v
$\dot{R}_\ell(E_v, r)$: first derivative of the partial radial wave function at energy E_v
$\psi_{bonding}(E_B, \mathbf{r})$: bonding state with energy E_B
$\psi_{antibonding}(E_A, \mathbf{r})$: antibonding state with energy E_A
$\phi_{\ell m}(\mathbf{r})$: atomic orbital in diatomic molecule

Chapter 9

Z_1: valency of the atom
ΔZ: valence difference between host and impurity atoms
λ: Thomas–Fermi screening parameter
$U(r)$: impurity potential
$\rho(r)$: density of conduction electrons
E_F: Fermi energy
$\sigma(\theta)$: differential cross-section
$\mathbf{K}$: scattering vector
k_F: Fermi radius
Λ: mean free path of conduction electron
$\Delta\rho$: resistivity increment due to impurity scattering

v_F: Fermi velocity
n: number of electrons per unit volume
N_{imp}: number of impurity atoms per unit volume
θ: scattering angle of conduction electron
x: solute concentration in at.%
U: internal energy of the conduction electron system
$N(E)$: density of states of electrons
$v_X(\mathbf{r})$: muffin-tin potential associated with ion X at the position $\mathbf{r}$
$\mathbf{l}_n$: lattice vector for the n-th atom
$V_0(\mathbf{r})$: total potential in virtual crystal approximation
$v_{av}(\mathbf{r})$: averaged potential in an alloy
c_x: concentration of atom X
$\psi_{0,\mathbf{k}}(\mathbf{r})$: wave function in the virtual crystal approximation
$V_{CPA}(E,\mathbf{r})$: total potential in coherent potential approximation
$w(E,\mathbf{r})$: effective potential
$G(\kappa,\mathbf{r}-\mathbf{r}')$: Green function
$\psi^i(\mathbf{r})$: incident wave function
$\psi^s(\mathbf{r})$: scattered wave function
$T_n(E,\mathbf{r}',\mathbf{r}'')$: t-matrix of potential of n-th ion
$\tau_X(E,\mathbf{r},\mathbf{r}')$: t-matrix of effective potential of ion X

Chapter 10

v_F: Fermi velocity
$\mathbf{v}_i$: velocity of i-th electron
$\mathbf{v}_D$: drift velocity
n: number of conduction electrons per unit volume
$\mathbf{E}$: electric field
τ: relaxation time
$\mathbf{J}$: electrical current density
σ: electrical conductivity
ρ: electrical resistivity
μ: mobility
R_H: Hall coefficient
ε or $\varepsilon(k)$ or $\varepsilon_n(\mathbf{k})$: energy of conduction electron in n-th band
ω: angular frequency of electron
$\mathbf{v}_k$: group velocity of wave packet
$\Psi_k(x,t)$: time-dependent wave function of the conduction electron
$\Psi_k(x)$: time-independent wave function
$\Psi_k(\mathbf{r},t)$: time-dependent Bloch wave function
$\mathbf{g}$: reciprocal lattice vector

m^*_{ij}: effective mass tensor
H: magnetic field
$(+e)$: charge of hole
$f(\mathbf{r},\mathbf{k},t)$: steady-state electron distribution function at time t
$f_0(\varepsilon_\mathbf{k},T)$: Fermi–Dirac distribution function at temperature T
$\phi(\mathbf{r},\mathbf{k})$: form of deviation of distribution function from $f_0(\varepsilon_\mathbf{k},T)$
ζ: chemical potential
$Q(\mathbf{k},\mathbf{k}')$, $Q(\mathbf{k}'\to\mathbf{k})$ and $Q(\mathbf{k}\to\mathbf{k}')$: transition probability
$v_{k_\perp}$: velocity component perpendicular to the Fermi surface
S_F: area of the Fermi surface
$N(\varepsilon)d\varepsilon$: electron density of states in the energy range ε to $\varepsilon+d\varepsilon$
Λ_F: mean free path of the conduction electron at the Fermi level
ρ_lattice: resistivity due to lattice vibrations
ρ_imp: resistivity due to impurities and defects
$\rho_{300\mathrm{K}}/\rho_{4.2\mathrm{K}}$: residual resistivity ratio (RRR or 3R)
$U(\mathbf{r})$: impurity potential
$a_n(\mathbf{r}-\mathbf{l})$: Wannier function at lattice site **l**
$\psi_{\mathbf{k},n}$: Bloch wave function in n-th band
N: number of atoms or ions per volume V
$V_0(\mathbf{r})$: periodic potential
$f_n(\mathbf{l})$: envelope function
$U_\mathrm{p}(\mathbf{r})$: pseudopotential
$\mathbf{u_l}$: displacement vector of atom from its equilibrium lattice site **l**
$\mathbf{R_l}$: position vector of atom at the lattice site **l** at finite temperature
K: scattering vector
θ: scattering angle
$S(\mathbf{K})$: static structure factor
$U_\mathrm{p}(\mathbf{K})$: atomic form factor
$a(\mathbf{K})$: interference function or static structure factor
Ω_0: volume per atom
n_imp: number of impurities per volume
$\sigma(\theta)$: differential scattering cross-section
λ: wavelength of conduction electron
E_i: energy of the initial phonon state $|i\rangle$
E_f: energy of the final phonon state $|f\rangle$
ε_i: energy of the initial electron state **k**
ε_f: energy of the final electron state **k**'
ω: angular frequency of phonon
$Q(\mathbf{K},\omega)$: transition probability of the electron upon emission or absorption of phonons
$a(\mathbf{K},\omega)$: dynamical structure factor

$\langle\cdots\rangle_T$: thermal average of a system in equilibrium with heat bath at temperature T
$\beta = 1/k_B T$
q: phonon wave vector
$n(\omega)$ or n_s: Planck distribution function
$\theta_{kk'}$: angle between velocity vectors $\mathbf{v}_k$ and $\mathbf{v}_{k'}$
$2W_{ll'}(\mathbf{K})$: Debye–Waller factor
M: mass of the constituent atom
$\mathbf{e}_{\mathbf{q},j}$: j-th polarization vector of the mode **q**
$a_\mathbf{q}^+$: phonon creation operator
$a_\mathbf{q}$: phonon annihilation operator
$a^{(1)}(K,\omega)$: dynamical structure factor due to one-phonon normal process
Θ_D: Debye temperature
q_D: Debye radius

Chapter 11

κ: thermal conductivity
U: thermal current density or flow of heat
ΔT: temperature gradient
W: thermal resistivity
C: specific heat of carriers per unit volume
v: average particle velocity
Λ: mean free path of particle
α: thermal resistivity coefficient due to lattice vibrations
β: thermal resistivity coefficient due to impurities
J: electrical current density
E: electric field
L_{EE}: coefficient defined as $\mathbf{J} = L_{EE}\mathbf{E}$
L_{ET}: coefficient defined as $\mathbf{J} = L_{ET}\nabla T$
L_{TE}: coefficient defined as $\mathbf{U} = L_{TE}\mathbf{E}$
L_{TT}: coefficient defined as $\mathbf{U} = L_{TT}\nabla T$
ζ: chemical potential or Fermi level
$\varepsilon(\mathbf{k})$: energy of electron of wave vector **k**
$\mathbf{v}_\mathbf{k}$: velocity of electron of wave vector **k**
$f(\mathbf{r},\mathbf{k})$: steady-state electron distribution function
$\sigma(\varepsilon)$: electrical conductivity at energy ε
$\sigma(T)$: electrical conductivity at temperature T
Θ_D: Debye temperature
L_0: limiting Lorenz number
L: measured Lorenz number
Q: absolute thermoelectric power or the Seebeck coefficient

τ: relaxation time of conduction electron
T_F: Fermi temperature
σ_T: Thomson coefficient
s: electronic entropy density
Π: Peltier coefficient
$C_{lattice}$: lattice specific heat per unit volume
Z: figure of merit
B: magnetic flux density
ϕ: form of deviation of distrubution function from $f_0(\varepsilon_k, T)$
σ_{ij}: conductivity tensor
$\alpha = \omega_c \tau$
$\omega_c = (+e)B/m$
ρ_{ij}: resistivity tensor
R_H: Hall coefficient
$\mathbf{v}_D$: drift velocity
q: electric charge of carriers
k_F: Fermi wave number
A: atomic weight
d: mass density
e/a: number of carriers per atom
$\Delta \rho_{xx}$: magnetoresistance defined as $[\rho_{xx}(B) - \rho_{xx}(0)]$
σ_i: conductivity of the i-th carrier
$\beta_i = q_i \tau_i / m_i$
q: wave vector of phonon or electromagnetic wave
D: electric displacement
μ: permeability
$\varepsilon(\mathbf{q}, \omega)$ or $\varepsilon(\omega)$: dielectric constant
ω: angular frequency of electromagnetic wave
$\mathbf{D}^{tot}$: total electric displacement
$\hat{\varepsilon}(\omega)$: complex dielectric constant
$\sigma(\omega)$: optical conductivity
$\mathbf{J}^{tot}$: total current density
$\hat{\sigma}(\omega)$: complex optical conductivity
E_y(incident): y-component of incident electric field
E_y(trans): y-component of transmitted electric field
n: refractive index
κ: extinction coefficient
$\hat{n}$: complex refractive index
m^*_{opt}: optical effective mass
$\sigma_1(\omega)$: real part of the complex conductivity

$\sigma_2(\omega)$: imaginary part of the complex conductivity
ω_p: plasma frequency
$\varepsilon_1(\omega)$: real part of dielectric constant
$\varepsilon_2(\omega)$: imaginary part of dielectric constant
E_{in}: electric field of incident radiation
E_{ref}: electric field of reflected radiation
R: reflectance or reflectivity
n_{eff}: effective number of electrons per atom
D: diffusion coefficient
$n(x)$: number density of particles
F: external force
μ: mobility

Chapter 12

B: magnetic flux density
T_c: superconducting transition temperature
H_c: critical magnetic field
E: electric field
$\mathbf{H}_{ext}$: external magnetic field
M: magnetization
μ_0: permeability in a vacuum
χ: magnetic susceptibility
m_s: mass of superconducting electron
$\mathbf{v}_s$: drift velocity of superconducting electron
q_s: charge of superconducting electron
$\mathbf{J}_s$: superconducting current density
n_s: number of superconducting electrons per unit volume
D: displacement current
λ_L: London penetration depth
$G_s(T,\mathbf{H})$: free energy of superconducting state at temperature T and magnetic field **H**
$G_n(T,\mathbf{H})$: free energy of normal state at temperature T and magnetic field **H**
S_s: entropy of superconducting state
S_n: entropy of normal state
C_s: specific heat of superconducting state
C_n: specific heat of normal state
$\mathbf{p}_s$: average momentum of superconducting electron
A: vector potential
ϕ: scalar potential
ψ: order parameter of superconducting state

η: phase of superconducting wave function

ν_0: threshold frequency

Δ: energy gap in superconducting state

M: mass of isotope

Θ_D: Debye temperature

$M_{k_1,k_1'}$: matrix element associated with phonon mediated electron–electron interaction

$\hbar\omega_q$: energy of emitted phonon

ω_D: Debye frequency

n: number of conduction electrons per unit volume

S: quantum number of spin angular momentum

S_z: quantum number of z-component of spin angular momentum

$\Phi(\mathbf{r}_1,\mathbf{r}_2)$: wave function of Cooper pair

$\psi_0(\mathbf{r}_1,\mathbf{r}_2,\cdots,\mathbf{r}_{n_0})$: wave function in BCS ground state

$N(\varepsilon_F)$: density of states at the Fermi level

V: matrix element representing the strength of the net attractive interaction in BCS theory

$\Phi_\mathbf{P}$: Cooper pair wave function traveling with momentum $\mathbf{P}$

H_{c1}: lower critical magnetic field

H_{c2}: upper critical magnetic field

ξ: coherence length

ξ_0: coherence length in a perfectly pure metal

λ_e: electron mean free path

κ: Ginzburg–Landau constant

$\mathbf{J}_i$: transport current

$\mathbf{J}_H$: screening current

J_c: critical current density

$\mathbf{F}$: Lorentz force per unit length of magnetic flux core

$\mathbf{F}_p$: pinning force per unit length of magnetic flux core

$\Delta\eta$: difference in the phase of Cooper pair wave function

i_c: maximum superconducting tunneling current; Josephson critical current

i_s: superconducting tunneling current

V: DC voltage

R: resistance

C: capacitance

T: torque

θ: deflection angle

M: moment of inertia of pendulum

l: length of pendulum arm

κ: viscous damping

I: measuring current

i: circular current in SQUID ring

ϕ_a: applied magnetic flux passing through SQUID ring
ϕ_0: quantum magnetic flux
ΔM_H: width of magnetization hysteresis curve at magnetic field H

Chapter 13

T_C: Curie temperature
T_N: Néel temperature
T_f: spin freezing temperature
l or $\mathbf{l}_i$: orbital angular momentum of *i*-th electron
ℓ or ℓ_i: quantum number of the orbital angular momentum of *i*-th electron
m_ℓ: quantum number of the *z*-component of orbital angular momentum **l**
s or $\mathbf{s}_i$: spin angular momentum of *i*-th electron
s or s_i: quantum number of spin angular momentum of *i*-th electron
m_s: quantum number of *z*-component of spin angular momentum **s**
L: resultant orbital angular momentum of an atom or free ion
L: quantum number of resultant orbital angular momentum of an atom or free ion
M_L: quantum number of *z*-component of resultant orbital angular momentum **L**
S: resultant spin angular momentum of an atom or free ion
S: quantum number of resultant spin angular momentum of an atom or free ion
M_S: quantum number of *z*-component of resultant spin angular momentum **S**
J: total angular momentum equal to $\mathbf{L} \pm \mathbf{S}$
J: quantum number of total angular momentum
H_i: internal magnetic field
μ_B: Bohr magneton
n: magnetic moment at 0 K in units of μ_B
J: exchange integral
ε_s: energy in singlet ground state
ε_t: energy in triplet ground state
$\mathbf{r}_i$: position vector of *i*-th electron
$\mathbf{R}_i$: position vector of *i*-th ion
$\phi_i(\mathbf{r}_j)$: *i*-th atomic orbital wave function of electron at position $\mathbf{r}_j$
H^{spin}: spin Hamiltonian
$\mathbf{S}_i$: resultant spin operator at *i*-th atom
J_{ij} or *J*: exchange integral between *i*- and *j*-th atoms
U: total exchange energy
z: number of nearest neighbor atoms
δ: vector pointing to nearest neighbor atom
ω: angular frequency of spin waves
k: wave vector of spin waves
$n(\omega, T)$: Planck distribution function at temperature *T*

$M(T)$: magnetization at temperature T
$D(\omega)$: density of states for spin waves or magnons
Q: number of atoms in a unit cell
$N\uparrow$: number of spin-up electrons per unit volume
$N\downarrow$: number of spin-down electrons per unit volume
M: magnetization
N: total number of electrons per unit volume
ε_X: exchange energy
2Δ: shift of spin-up band relative to spin-down band
$\Delta\varepsilon_{kin}$: increase in kinetic energy of valence electrons due to electron transfer
$N(\varepsilon_F)$: density of states at the Fermi level
e/a: electrons per atom
ε_F: Fermi level
$\psi(r,\theta,\varphi)$: wave function of scattered electron due to a single magnetic impurity
$R(r)$: radial wave function
$Y_{\ell m}(\theta,\varphi)$: spherical harmonic function of ℓ-th partial wave
$V(r)$: impurity potential
$U(r)$: impurity potential defined as $U(r)=(2m/\hbar^2)V(r)$
r_0: distance at which the potential is truncated
$\mathfrak{R}$: radius of metal sphere
$f(\theta,\varphi)$: angular function of outgoing wave
$j_\ell(r)$: spherical Bessel function of ℓ-th partial wave
$n_\ell(r)$: spherical Neumann function of ℓ-th partial wave
$J_\ell(r)$: Bessel function of ℓ-th partial wave
η_ℓ: phase shift of ℓ-th partial wave
γ_ℓ: value of $d\ln R_\ell/dr$ in the range $r \le r_0$
r_{max}: position at first maximum of spherical Bessel function
θ: scattering angle of conduction electron
$\sigma(\theta)$: differential scattering cross-section
$P_\ell(\cos\theta)$: Legendre polynomials of ℓ-th order
σ: total scattering cross-section
σ_{tr}: transport cross-section
N_{imp}: number of impurities per unit volume
n: number of conduction electrons per unit volume
Λ: mean free path of conduction electron
v_F: Fermi velocity
m: mass of conduction electron
$\Delta n(k)\delta k$: change in density of states in range k to $k+\delta k$
$e\Delta Z$: excess nuclear charge
$\Delta n(\varepsilon)\delta\varepsilon$: change in density of states in energy range ε to $\varepsilon+d\varepsilon$
$\phi(r)$: screened impurity potential

Principal symbols

k_F: Fermi radius
ε_{3d}: energy at center of 3d band
ε_0: energy at center of virtual bound state
Δ: width of the virtual bound state
$\Delta\rho$: increment in electrical resistivity upon 1 at.% addition of 3d-transition metal element M to fcc Al
χ: magnetic susceptibility
U: Coulomb repulsive energy
τ_{sf}: relaxation time of localized spin fluctuations
τ_Δ: relaxation time of electron escaping from virtual bound state to a continuum
τ: relaxation time associated with thermal fluctuations at temperature T
τ_F: relaxation time associated with the Fermi temperature
T_{sf}: characteristic temperature associated with spin fluctuations
H_{sd}: s–d interaction
$\mathbf{R}_n$: position vector of n-th magnetic impurity atom
$\mathbf{s}$: spin angular momentum of conduction electron
$\mathbf{S}_n$: spin angular momentum of n-th magnetic impurity atom
$J(\mathbf{r} - \mathbf{R}_n)$: exchange integral between $\mathbf{s}$ and $\mathbf{S}_n$
ρ_{spin}: resistivity due to s–d interaction
c: concentration of magnetic impurity
$2D$: width of flat valence band
ρ_0: residual resistivity
T_K: Kondo temperature
$F(2k_F R_{nm})$: RKKY function
$\mathbf{I}_n$: nuclear spin angular momentum of n-th atom
A: hyperfine coupling constant between nuclear spins
R_{nm}: distance between two nuclei
$\rho_{//}$: longitudinal magnetoresistance
$\rho_\perp$: transverse magnetoresistance
$\Delta\rho/\rho_{//}$: FAR or ferromagnetic anisotropy of resistivity
c_A: concentration of the element A
$\rho_A^\uparrow$: residual resistivity caused by scattering of spin-up electron by atom A
$\rho_A^\downarrow$: residual resistivity caused by scattering of spin-down electron by atom A
α: ratio of $\rho_{//}^\downarrow$ over $\rho_{//}^\uparrow$
ρ_H: Hall resistivity
E_y: transverse electric field
R_0: normal Hall coefficient
R_s: anomalous Hall coefficient
M_z: spontaneous magnetization along z-direction
μ_0: permeability in vacuum

Chapter 14

m^*: effective mass of quasiparticle

$f(\mathbf{p},\mathbf{r})$: distribution function

$\mathbf{r}_i$: position vector of i-th electron

$\mathbf{R}_i$: position vector of i-th proton

$\psi(\mathbf{r}_1,\mathbf{r}_2)$: molecular orbital wave function of two electrons in a hydrogen molecule

H_i: one-electron Hamiltonian for i-th electron in the field of two protons

$\psi(\mathbf{r})$: one-electron wave function in the field of two protons

ε: energy of the electron in the field of two protons

$\psi_s(\mathbf{r}_1,\mathbf{r}_2)$: symmetric orbital wave function of two electrons in a hydrogen molecule

E_s: lowest energy corresponding to singlet state of a hydrogen molecule

E_t: next-lowest energy corresponding to triplet state of a hydrogen molecule

$\psi_t(\mathbf{r}_1,\mathbf{r}_2)$: antisymmetric orbital wave function of two electrons in a hydrogen molecule

$\phi_i(\mathbf{r})$: 1s atomic orbital wave function of i-th electron in a hydrogen atom

$c^+_{i\uparrow}$: operator to create a spin-up electron at site i

$c_{j\uparrow}$: operator to annihilate a spin-up electron at site j

$n_{i\uparrow}$: number of spin-up electrons at site i

U: on-site Coulomb energy

t: transfer integral or hopping matrix element

χ: electronegativity

Δ: charge transfer energy or resonance energy

W: band width

T_c: superconducting transition temperature

c_o: optimum carrier concentration

T^*: spin gap or pseudogap temperature

Chapter 15

a: diameter of atom

$g(r)$: pair distribution function

ρ_0: average number density

$\mathbf{k}_0$: wave vector of incident wave

$\mathbf{k}$: wave vector of scattered wave

θ: scattering angle

$\mathbf{K}$: scattering vector

λ: wavelength of x-ray

N: number of atoms in a solid

$I(\mathbf{K})$: intensity of scattered x-rays

f: atomic scattering factor or scattering amplitude

$S(\mathbf{K})$: structure factor

$\rho(\mathbf{r})$: radial density function
K_p: wave number corresponding to the first peak of the structure factor
r_0: average atomic distance
f_A: atomic scattering factor of atom A
$A(\mathbf{K})$: amplitude of scattered x-rays
$S_{\alpha\beta}(\mathbf{K})$: Ashcroft–Langreth partial structure factor for the atom pair α–β
c_α: concentration of α atoms
$\rho_{\alpha\beta}(r)$: number of β atoms found within a shell of thickness δr, radius r from the atom α at the origin
$g_{\alpha\beta}(r)$: partial pair distribution function for the atom pair α–β, distance r apart
$\delta_{\alpha\beta}$: Kronecker delta
$a_{\alpha\beta}$: Faber–Ziman partial structure factor for the atom pair α–β
T_x: crystallization temperature
T_g: glass transition temperature
T_m: melting temperature
TCR: temperature coefficient of electrical resistivity defined as $[=(1/\rho)(d\rho/dT)]$
$\mathbf{R}_i$: position vector of i-th ion
$U_p(\mathbf{r}-\mathbf{R}_i)$: pseudopotential of i-th ion at position $\mathbf{R}_i$
$U(\mathbf{r})$: total ionic potential
$2k_F$: Fermi diameter
$a(K)$: interference function
$U(K)$: Fourier component of the total ionic potential
Θ_D: Debye temperature
$a(K,\omega)$: dynamical structure factor at frequency ω and wave number K
Ω_0: volume per atom
ρ_0: residual resistivity
$\Delta\rho(T)$: resistivity due to the inelastic electron–phonon interaction
$e^{-2W(T)}$: Debye–Waller factor
$N(E_F)$: density of states at the Fermi level
W: band width
τ_0: elastic lifetime
τ_i: inelastic lifetime
A: scattering amplitude
θ: phase change upon scattering
σ: electrical conductivity
R_H: Hall coefficient
γ_{exp}: measured electronic specific heat coefficient
$\psi(\mathbf{r})$: wave function of the localized electron
a: characteristic radius
Δ_{ij}: energy difference between two localized states
P: probability of hopping

ν_{ph}: jumping frequency of electron due to phonons
R: distance between two localized states
D: electron diffusion coefficient
Λ_F: mean free path of conduction electron
v_F: Fermi velocity
g: ratio of density of states at the Fermi level over the corresponding free-electron value
$N(E_F)^{free}$: free-electron density of states at the Fermi level
S_F^{free}: area of the free-electron Fermi sphere
τ: golden ratio equal to $\tau = (1+\sqrt{5})/2$
a: average atomic distance
z: coordination number of the constituent atom
I: hopping integral

Hints and answers

Chapter Two

2.1 Insert $\psi(x)=\sqrt{2/L}\sin(\pi n_x x/L)$ into $p_x=\int_0^L \psi^*(x)p_x\psi(x)dx = \int_0^L \psi^*(x)\left(-i\hbar\dfrac{\partial}{\partial x}\right)\psi(x)dx$. This is easily shown to be zero.

2.2 An average volume per electron is given by $\Omega=(4\pi/3)r^3$. Insert this into equation (2.21).

2.3 Note that a quantized interval in the k_z-direction is equal to $2\pi/L_z=2\pi/10$ nm^{-1}, which is 10^6 times larger than that given by $2\pi/L_x=2\pi/L_y=2\pi/10^7$ nm^{-1} in the k_x- and k_y-directions. The cross-section of the Fermi sphere in the $k_y k_z$-plane passing through $k_x=0$ is shown in Fig.2A.1. It can be seen from the figure that the allowed states marked by dots form a series of planes parallel to the $k_x k_y$-plane with the interval $\Delta k_z=2\pi/L_z$. Let us assign a sequential number to planes in the $k_z\geq 0$ axis, starting from zero up to $(n+1)$. Note that the n-th plane is below the Fermi surface but the $(n+1)$-th plane is above it. Thus one obtains the relation

$$n\cdot\Delta k_z \leq k_F < (n+1)\cdot\Delta k_z \tag{2A.1}$$

or

$$\dfrac{\hbar^2(\Delta k_z)^2}{2m}n^2 \leq E_F < \dfrac{\hbar^2(\Delta k_z)^2}{2m}(n+1)^2. \tag{2A.2}$$

Consider the cross-section of the Fermi sphere cut through the l-th plane. A cross-section becomes a circle with radius r_l given by

$$r_l^2 = k_F^2 - (l\cdot\Delta k_z)^2. \tag{2A.3}$$

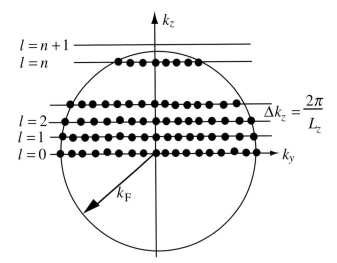

Figure 2A.1. Cross-section of the Fermi sphere in the $k_y k_z$-plane passing through $k_x = 0$. The allowed states are marked by dots. They form a series of planes parallel to the $k_x k_y$-plane and are numbered sequentially.

Suppose that N_l electrons are on the l-th plane. Then, the proportional relation

$$\pi r_l^2 : N_l = \left(\frac{2\pi}{L_x}\right) \cdot \left(\frac{2\pi}{L_y}\right) : 2 \tag{2A.4}$$

holds on the l-th plane. From equations (2A.3) and (2A.4), we obtain

$$N_l = \left(\frac{L_x L_y}{2\pi}\right)(k_F^2 - l^2 \cdot \Delta k_z^2). \tag{2A.5}$$

The total number of electrons enclosed by the Fermi sphere can be calculated as

$$N = N_0 + 2\sum_{l=1}^{n}\left(\frac{S}{2\pi}\right)\{k_F^2 - l^2 \cdot \Delta k_z^2\}, \tag{2A.6}$$

where N_0 is the number of electrons on the $k_x k_y$-plane or 0-th plane and $S = L_x L_y$. Note that the factor 2 appears as a result of the contribution from the bottom half of the Fermi sphere. Equation (2A.6) can be easily summed to

$$N = \left(\frac{S}{2\pi}\right)k_F^2(2n+1) - \left(\frac{S}{\pi}\right)(\Delta k_z)^2 \frac{1}{6}n(n+1)(2n+1). \tag{2A.7}$$

Now we remove the suffix F in k_F and assume k to be a variable. Using the relation $E = \hbar^2 k^2/2m$, we can rewrite equation (2A.7) in the form of

$$N = \left(\frac{mS}{\pi\hbar^2}\right)(2n+1)E - \left(\frac{S}{\pi}\right)(\Delta k_z)^2 \frac{1}{6}n(n+1)(2n+1). \tag{2A.8}$$

According to equation (2A.2), the lower limit of the Fermi energy is given by

$$E_F^{\min} = \left\{\frac{\hbar^2(\Delta k_z)^2}{2m}\right\}n^2. \tag{2A.9}$$

Hence, an insertion of equation (2A.9) into equation (2A.2) results in $N_{\min}$, which is expressed as

$$N_{\min} = \left(\frac{S}{6\pi}\right)(\Delta k_z)^2(2n+1)(2n-1)n. \tag{2A.10}$$

The upper limit of the Fermi energy is simply obtained by replacing n by $(n+1)$. Therefore, the following inequality is derived:

$$\left(\frac{S}{6\pi}\right)(\Delta k_z)^2(2n+1)(2n-1)n \leq N < \left(\frac{S}{6\pi}\right)(\Delta k_z)^2(2n+3)(2n+1)(n+1). \tag{2A.11}$$

Since its volume is 1×10^{-6} cm^3, the total number of electrons in the sodium thin film can be calculated as $N = 2.542 \times 10^{16}$ from its atomic weight of 22.98 g. In addition, we know that $S = 1$ cm^2 and $\Delta k_z = 2\pi/L_z = 2\pi \times 10^{-6}$ cm^{-1}. By inserting these numerical values into equation (2A.11), we obtain

$$(2n+1)(2n-1)n \leq 12\,137 < (2n+3)(2n+1)(n+1). \tag{2A.12}$$

An integer satisfying equation (2A.12) is easily found to be $n = 14$. We can calculate from equation (2A.2) the Fermi energy as $2.94 < E_F < 3.38$ eV for $n = 14$. Compare this with the value of $E_F^{\text{free}} = 3.24$ eV for bulk sodium metal (see Table 2.2).

The density of states for a two-dimensional metal can be obtained by differentiating equation (2A.8) with respect to E:

$$N_{2D}(E) = \left(\frac{S}{\pi}\right)(2n+1), \tag{2A.13}$$

where atomic units with $m = \hbar = 1$ are used. Equation (2A.2) can be rewritten as

$$\frac{2\pi^2}{L_z^2}n^2 \leq E < \frac{2\pi^2}{L_z^2}(n+1)^2. \tag{2A.14}$$

On the other hand, as shown in equation (2.22), the density of states for a three-dimensional metal can be expressed in atomic units as

$$N_{3D}(E) = \left(\frac{L_x L_y L_z}{\pi^2}\right)\sqrt{2E}. \tag{2A.15}$$

For the sake of simplicity, we take $L_x L_y = \sqrt{\pi}$ and $L_z = \sqrt{2\pi}$. Then, one

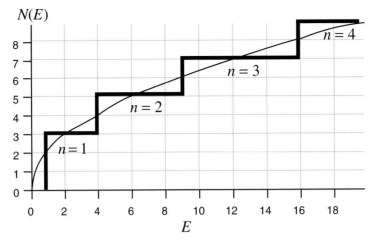

Figure 2A.2. Density of states curves for two- and three-dimensional free-electron systems.

obtains $N_{3D}(E) = 2\sqrt{E}$ and $N_{2D} = (2n+1)$ with the condition $n^2 \leq E < (n+1)^2$. Both curves are shown in Fig.2A.2.

2.4 According to Table 2.2, the volume Ω per atom in alkali metals is much larger than that in noble metals. This is caused by the difference in the crystal structure between them: noble metals crystallize into the close-packed fcc structure, alkali metals into the bcc structure with a relatively low packing density. A large value of Ω results in a higher Fermi energy through equation (2.21). In addition, the 3d-band is situated in the middle of the valence band in all three noble metals (see Section 6.4). We will learn that this contributes to raising the Fermi energy in noble metals. It is worthwhile mentioning that the difference in the electronic structure between noble metals and alkali metals is reflected in the bonding strength and, in turn, in physical properties like the melting point: 371 K for Na, 336.3 K for K, 1358 K for Cu, 1235K for Ag and 1338 K for Au.

Chapter Three

3.1 The Pauli paramagnetic susceptibility is calculated as

$$\chi = \mu_B^2 N(E_F(0)) \left\{ 1 + \frac{\pi^2}{6}(k_B T)^2 \left[\frac{1}{N(E_F)} \left(\frac{d^2 N(E_F)}{dE^2} \right) - \left(\frac{1}{N(E_F)} \frac{dN(E_F)}{dE} \right)^2 \right]_{E=E_F(0)} \right\}.$$

(3A.1)

A ratio of the second term over the first one in the curly bracket turns out to be

$$\frac{\frac{\pi^2}{6}(k_B T)^2 \left| \left[\left(\frac{d^2 N(E_F)}{dE^2} \right) - \frac{1}{N(E_F)} \left(\frac{dN(E_F)}{dE} \right)^2 \right]_{E=E_F(0)} \right|}{N(E_F(0))}. \quad (3A.2)$$

Now we use equation (2.22) for the density of states, i.e., $N(E) = C\sqrt{E}$ in the free-electron model. The relations $N'(E) = C/2\sqrt{E}$ and $N''(E) = -C/4E\sqrt{E}$ are inserted into equation (3A.2). It is reduced to $\pi^2(k_B T)^2/12[E_F(0)]^2$. An insertion of the Fermi energy of 7 eV for pure Cu and the thermal energy $k_B T = 0.025$ eV at 300 K leads to the ratio of the order of 1×10^{-5}. Hence, the contribution of the second term over the first one is merely 0.01% for Cu at room temperature.

Chapter Four

4.1 The results are shown in Fig. 4A.1. Atoms are periodically arranged with the lattice constant $a = 2$. The displacement of each atom is shown by a bold vertical bar. The solid curve represents the lattice wave with the wave vector in the first Brillouin zone and a dotted curve the lattice wave with the wave vector larger than the former by the reciprocal lattice vector. Both waves describe the same displacements of atoms.

4.2(a) The internal energy and specific heat in the Einstein model are deduced to be

$$U_{\text{lattice}} = \frac{3N\hbar\omega_0}{\exp\left(\frac{\hbar\omega_0}{k_B T}\right) - 1} \quad (4A.1)$$

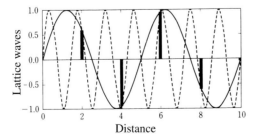

Figure 4A.1. Lattice waves with wave vectors q and $q + g$, where g is the reciprocal lattice vector. Bold bar represents the displacement of atoms.

and
$$C_{\text{lattice}} = \frac{3Rx^2}{(e^x - 1)(1 - e^{-x})}, \quad (4A.2)$$
where $x = \Theta_E/T$.

Chapter Five

5.1 The Bloch wave can be expanded as

$$\psi_{\mathbf{k}}(\mathbf{r}) = \exp(i\mathbf{k}\cdot\mathbf{r})u_{\mathbf{k}}(\mathbf{r}) = \exp(i\mathbf{k}\cdot\mathbf{r})\sum_{n=-\infty}^{\infty} A_n \exp(-i\mathbf{g}_n\cdot\mathbf{r}) \quad (5A.1)$$

since the function $u_{\mathbf{k}}(\mathbf{r})$ is periodic with the period a. If we replace the wave vector $\mathbf{k}$ by $\mathbf{k} = \mathbf{k}' + \mathbf{g}_m$, where $\mathbf{g}_m$ is the reciprocal lattice vector, then we obtain

$$\psi_{\mathbf{k}}(\mathbf{r}) = e^{i(\mathbf{k}'+\mathbf{g}_m)\cdot\mathbf{r}} u_{\mathbf{k}}(\mathbf{r}) = e^{i\mathbf{k}'\cdot\mathbf{r}} \sum_{n=-\infty}^{\infty} A_n e^{i(\mathbf{g}_m-\mathbf{g}_n)\cdot\mathbf{r}}$$

$$= e^{i\mathbf{k}'\cdot\mathbf{r}} \sum_{n'=-\infty}^{\infty} A_{n'} e^{i\mathbf{g}_{n'}\cdot\mathbf{r}} = e^{i\mathbf{k}'\cdot\mathbf{r}} u_{\mathbf{k}'}(\mathbf{r}) = \psi_{\mathbf{k}'}(\mathbf{r}). \quad (5A.2)$$

5.2 Only the diagonal elements remain in the determinant of equation (5.34), when the ionic potential is zero. Hence, it is immediately solved as

$$\cdots \left[E - \left(k - \frac{2\pi}{a}\right)^2\right] \cdot [E - k^2] \cdot \left[E - \left(k + \frac{2\pi}{a}\right)^2\right] \cdots = 0. \quad (5A.3)$$

Equation (5A.3) constitutes a series of parabolas centered at $k = (2\pi/a)n$ with $n = 0, \pm 1, \pm 2, \ldots$. This is illustrated in Fig. 5A.1. The reduced E–k relations appear successively in the first zone $-\pi/a < k \leq \pi/a$. Multi-valued E–k relations are labelled by the band indices, as shown in Fig. 5A.1. The Bragg condition is satisfied at the wave number corresponding to the intersection of the successive parabola. Note that no deviation from the free-electron model occurs at the intersection $k = (\pi/a)n$.

5.3 The relation $\mathbf{k}^2 = (\mathbf{k} - \mathbf{g}_n)^2$ is rewritten as

$$2\mathbf{k}\cdot\mathbf{g}_n = |\mathbf{g}_n|^2. \quad (5A.4)$$

As mentioned in Section 4.3, the interplanar distance d is expressed as $d = 2\pi/|\mathbf{g}_n|$. We assume that the wave vector $\mathbf{k}$ makes an angle θ with $\mathbf{g}_n$ normal to the set of the lattice planes. Now equation (5A.4) can be rewritten as $2\mathbf{k}\cdot\mathbf{g}_n = 2|\mathbf{k}|\cdot|\mathbf{g}_n|\cos[(\pi/2) - \theta] = |\mathbf{g}_n|^2$. This turns out to be $2d\sin\theta = \lambda$, if the relation $\lambda = 2\pi/k$ is inserted. If this is due to the m-th order reflections, the interplanar distance d_m must be replaced by $d_m = d/m$. Now we obtain the Bragg condition $2d\sin\theta = m\lambda$.

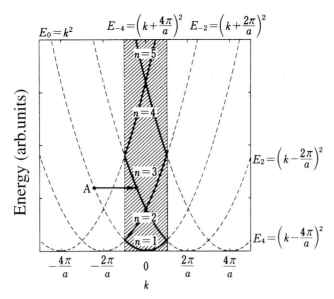

Figure 5A.1. *E–k* relation of conduction electrons in one-dimensional periodic empty-lattice with lattice constant *a*. The hatched area represents the first Brillouin zone. *n* refers to the band index.

5.4(a) The volume V_B of the first Brillouin zone for a hcp metal is $V_B = 16\pi^3/\sqrt{3}a^2 c$ and the number of electrons N accommodated in the first Brillouin zone for a hcp metal with volume V is $N = (2V/8\pi^3)(16\pi^3/\sqrt{3}a^2 c) = 4V/\sqrt{3}a^2 c$. Since the total number of atoms in a hcp metal with volume V is equal to $N = 4V/\sqrt{3}a^2 c$, we obtain the number of electrons per atom accommodated in the first Brillouin zone of a hcp metal to be equal to $(4V/\sqrt{3}a^2 c)/(4V/\sqrt{3}a^2 c) = 1$.

(b) We divide the Jones zone, Fig. 5.17(c), into two truncated hexagonal pyramids and a hexagonal prism. With reference to Fig. 5.17(a), let us denote the distance from the center Γ or A to the midpoint of each edge in the bottom- and top-plane of the truncated pyramid as L_{bottom} and L_{top}, respectively. L_{bottom} is obviously equal to $\Gamma M = 2\pi/\sqrt{3}a$, whereas L_{top} is calculated to be $[(2\pi/\sqrt{3}a) - (\sqrt{3}a\pi/2c^2)]$ from the proportional relation. Then the volume of the truncated hexagonal pyramid is given by

$$V_{\text{trunc.pyramid}} = \left(\frac{1}{3}\right)(S_{\text{bottom}} h_{\text{bottom}} - S_{\text{top}} h_{\text{top}}) = \left(\frac{4c}{3a}\right)(L_{\text{bottom}}^3 - L_{\text{top}}^3), \quad (5A.5)$$

where $S_{\text{bottom (top)}} = 6L_{\text{bottom (top)}}^2/\sqrt{3}$ is the area of the bottom- and top-hexagonal planes and $h_{\text{bottom (top)}} = (2c/\sqrt{3}a)L_{\text{bottom (top)}}$ is the height of the corresponding pyramids. Now $V_{\text{trunc.pyramid}}$ is calculated as

$$V_{\text{trunc.pyramid}} = \left(\frac{4c}{3a}\right)(L_{\text{bottom}} - L_{\text{top}})[(L_{\text{bottom}} - L_{\text{top}})^2 + 3L_{\text{bottom}}L_{\text{top}}]$$

$$= \frac{8\pi^3}{\sqrt{3}a^2c}\left\{1 - \frac{3}{4}\left(\frac{a}{c}\right)^2\left[1 - \frac{1}{4}\left(\frac{a}{c}\right)^2\right]\right\}, \quad (5A.6)$$

where $L_{\text{bottom}} - L_{\text{top}} = \sqrt{3}a\pi/2c^2$ is inserted.

The volume of the Jones zone is given by the sum of the two truncated hexagonal pyramids and the first Brillouin zone:

$$V_{\text{Jz}} = 2V_{\text{trunc.pyramid}} + V_{\text{1st zone}}$$

$$= \frac{16\pi^3}{\sqrt{3}a^2c}\left\{1 - \frac{3}{4}\left(\frac{a}{c}\right)^2\left[1 - \frac{1}{4}\left(\frac{a}{c}\right)^2\right]\right\} + \frac{16\pi^3}{\sqrt{3}a^2c} = \frac{16\pi^3}{\sqrt{3}a^2c}\left\{2 - \frac{3}{4}\left(\frac{a}{c}\right)^2\left[1 - \frac{1}{4}\left(\frac{a}{c}\right)^2\right]\right\}.$$

The number of electrons N accommodated in the Jones zone for an hcp metal with a volume V is, therefore, derived as

$$N = \frac{4V}{\sqrt{3}a^2c}\left\{2 - \frac{3}{4}\left(\frac{a}{c}\right)^2\left[1 - \frac{1}{4}\left(\frac{a}{c}\right)^2\right]\right\}. \quad (5A.7)$$

Finally, the number of electrons per atom accommodated in the Jones zone is deduced to be

$$e/a = \left\{2 - \frac{3}{4}\left(\frac{a}{c}\right)^2\left[1 - \frac{1}{4}\left(\frac{a}{c}\right)^2\right]\right\}. \quad (5A.8)$$

Chapter Seven

7.1 Equation (7Q.1) can be rewritten as

$$\frac{1}{2m}\left(\frac{\hbar}{i}\frac{\partial}{\partial x}\right)^2\psi + \frac{1}{2m}\left(\frac{\hbar}{i}\frac{\partial}{\partial y} - exB\right)^2\psi + \frac{1}{2m}\left(\frac{\hbar}{i}\frac{\partial}{\partial z}\right)^2\psi = E\psi \quad (7A.1)$$

or

$$\frac{\partial^2\psi}{\partial x^2} + \left(\frac{\partial}{\partial y} - \frac{ieB}{\hbar}x\right)^2\psi + \frac{\partial^2\psi}{\partial z^2} + \frac{2mE}{\hbar^2}\psi = 0 \quad (7A.2)$$

We seek for the solution in the form of $\psi(x, y, z) = \exp[i(\beta y + k_z z)]u(x)$. Then, (7A.2) is reduced to the form:

$$\frac{\partial^2 u}{\partial x^2} + \left[\frac{2mE'}{\hbar^2} - \left(\beta - \frac{eB}{\hbar}x\right)^2\right]u(x) = 0 \quad (7A.3)$$

with

$$E' = E - \frac{\hbar^2}{2m}k_z^2.$$

Equation (7A.3) is further rewritten as

$$-\frac{\hbar^2}{2m}\frac{\partial^2 u(x)}{\partial x^2} + \left[\frac{eB}{m}x - \frac{\hbar\beta}{m}\right]^2 u(x) = E'u(x) \tag{7A.4}$$

Equation (7A.4) represents the one-dimensional harmonic oscillator with the angular frequency $\omega_H = \frac{eB}{m}$ centred at $x_o = \frac{1}{\omega_H}\frac{\hbar\beta}{m}$. Its eigenvalue is given by

$$E' = \left(n + \frac{1}{2}\right)\hbar\omega_H$$

or

$$E = \left(n + \frac{1}{2}\right)\hbar\omega_H + \frac{\hbar^2}{2m}k_z^2. \tag{7A.5}$$

7.2 Electrons travel in an orbit on an equi-energy surface normal to the applied field. Hence, the cross-section of the Fermi surface perpendicular to the magnetic field needs to be considered. As shown in Fig. 7A.1, the area of the cross-section of the ellipsoid at $k_z = 0$ does not change when the value of k_z deviates from the origin. This is because $\partial A(k_z)/\partial k_z = 0$ at $k_z = 0$. This means that electrons within Δk_z of $k_z = 0$ can equally participate in the dHvA oscillations. In other words, the number of participating electrons, when the magnetic field is varied, is the largest on the Landau level at $k_z = 0$.

7.3(a) $\omega_c = eB/m = 8.79 \times 10^{11}$ s and

$r = v_F/\omega_c = (1.57 \times 10^{-6})/(8.79 \times 10^{11}) = 1.79 \times 10^{-6}$ m or 1.79 μm.

Use the relation $\Lambda = 5.92 \times 10^{-5} A/\rho \cdot d$, where Λ is the mean free path in Å, A is the atomic weight in g, ρ is the residual resistivity in μΩ-cm, d is the density in g/cm³ (footnote 10, p. 475, in Chapter 15). Λ turns out to be 6 μm so that $\Lambda > r$ holds.

(b) $\Delta E = \hbar\omega_c = 0.58 \times 10^{-3}$ eV
(c) $n \cong \varepsilon_F/\hbar\omega_c = 7/0.58 \times 10^{-3} = 12\,068$

7.4 The free-electron wave function $\chi(\mathbf{p}) = \exp(i\mathbf{p}\cdot\mathbf{r})$ is inserted into equation (7.23). The integrand becomes constant and, hence, the integral is reduced to

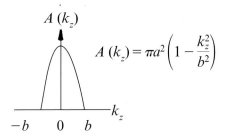

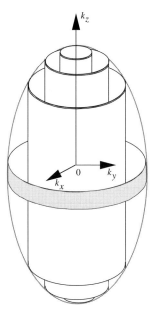

Figure 7A.1. The k_z dependence of the area of the cross-section normal to the z-axis is given by a parabola and takes its maximum at $k_z=0$. The Fermi surface is also drawn, where the electronic states contributing to the dHvA effect are marked by the shaded area. Obviously, available states are more abundant at $k_z=0$.

$$J(p_z) = \int_0^{\sqrt{p_F^2 - p_z^2}} 2\pi p\, dp = \pi(p_F^2 - p_z^2). \tag{7A.6}$$

Chapter Eight

8.1 The summation in equation (8.34) for the bcc lattice is calculated as

$$\sum_n e^{i\mathbf{k}\cdot\mathbf{R}_n} = \exp\left[i\frac{a}{2}(k_x+k_y+k_z)\right] + \exp\left[i\frac{a}{2}(-k_x+k_y+k_z)\right]$$

$$+ \exp\left[i\frac{a}{2}(k_x - k_y + k_z)\right] + \exp\left[i\frac{a}{2}(-k_x - k_y + k_z)\right]$$

$$+ \exp\left[i\frac{a}{2}(k_x + k_y - k_z)\right] + \exp\left[i\frac{a}{2}(-k_x + k_y - k_z)\right]$$

$$+ \exp\left[i\frac{a}{2}(k_x - k_y - k_z)\right] + \exp\left[i\frac{a}{2}(-k_x - k_y - k_z)\right]$$

$$= 2\cos\left(\frac{ak_z}{2}\right)\left\{\exp\left[i\frac{a}{2}(k_x + k_y)\right] + \exp\left[-i\frac{a}{2}(k_x + k_y)\right]\right.$$

$$\left. + \exp\left[i\frac{a}{2}(k_x - k_y)\right] + \exp\left[-i\frac{a}{2}(k_x - k_y)\right]\right\}$$

$$= 4\cos\left(\frac{ak_z}{2}\right)\left\{\cos\left[\frac{a}{2}(k_x + k_y)\right] + \cos\left[\frac{a}{2}(k_x - k_y)\right]\right\}$$

$$= 8\cos\left(\frac{ak_x}{2}\right)\cos\left(\frac{ak_y}{2}\right)\cos\left(\frac{ak_z}{2}\right)$$

8.2 Equation $\nabla^2\phi(r) = 4\pi\rho(\mathbf{r})$ is known as the Poisson equation in electrostatics. The electrostatic field $\phi(\mathbf{r}-\mathbf{r}')$, when a point charge is located at the position $\mathbf{r}=\mathbf{r}'$, is obviously given by $\phi(\mathbf{r}-\mathbf{r}') = (1/4\pi)(1/|\mathbf{r}-\mathbf{r}'|)$. This means that $\phi(\mathbf{r}-\mathbf{r}') = (1/4\pi)(1/|\mathbf{r}-\mathbf{r}'|)$ is the solution of equation $\nabla^2\phi(r) = \delta(\mathbf{r}-\mathbf{r}')$. By using this relation, we obtain

$$\nabla^2 G(\kappa, \mathbf{r}-\mathbf{r}') = -\frac{1}{4\pi}\nabla^2\left[\frac{\exp(i\kappa|\mathbf{r}-\mathbf{r}'|)}{|\mathbf{r}-\mathbf{r}'|}\right]$$

$$= -\exp(i\kappa|\mathbf{r}-\mathbf{r}'|)\nabla^2\left(\frac{1}{4\pi|\mathbf{r}-\mathbf{r}'|}\right) - \frac{1}{4\pi|\mathbf{r}-\mathbf{r}'|}\nabla^2[\exp(i\kappa|\mathbf{r}-\mathbf{r}'|)]$$

$$= -\exp(i\kappa|\mathbf{r}-\mathbf{r}'|)\delta(\mathbf{r}-\mathbf{r}') - \frac{\kappa^2\exp(i\kappa|\mathbf{r}-\mathbf{r}'|)}{4\pi|\mathbf{r}-\mathbf{r}'|}$$

$$= \delta(\mathbf{r}-\mathbf{r}') - \kappa^2 G(\kappa, \mathbf{r}-\mathbf{r}'),$$

where the relation $f(x)\delta(x-a) = f(a)\delta(x-a)$ is used.

Chapter Ten

10.1 According to equation (10.14b), the wave packet is explicitly written as

$$\psi(x,t) = C\int_0^\infty \exp[-a^2(k-K)^2 + i\{kx - (\hbar k^2/2m)t\}]dk. \quad (10\text{A}.1)$$

The argument of the exponential function is rewritten as

$$-a^2\xi^2 + i\{kx - (\hbar k^2/2m)t\} = -\alpha^2\xi^2 - 2\beta\xi - \gamma, \qquad (10A.2)$$

where $\xi = k - K$, $\alpha = a^2 + (i\hbar t/2m)$, $\beta = \dfrac{i}{2}[-x + (\hbar Kt/m)]$ and $\gamma = iK(-x + (\hbar Kt/2m))$. Equation (10A.1) is calculated as

$$\psi(x,t) = \dfrac{c\sqrt{\pi}}{\sqrt{a^2 + \dfrac{i\hbar t}{2m}}} \exp\left[-\dfrac{\left(x - \dfrac{\hbar Kt}{m}\right)^2}{4\left(a^2 + \dfrac{i\hbar t}{2m}\right)} + iK\left(x - \dfrac{\hbar Kt}{2m}\right)\right]. \qquad (10A.3)$$

Equation (10A.3) indicates that the probability density at $t=0$ is equal to $|\psi(x,t)|^2 = (c^2\pi/a^2)e^{-x^2/2a^2}$ and the wave packet is represented by the Gaussian function centered at $x=0$. The uncertainty in position is equal to $\Delta x = a\sqrt{2}$ from its half-width. It is also seen from equation (10A.1) that the wave packet is broadened over $\Delta k = 1/a$ about $k = K$. Hence, the uncertainty relation $\Delta x \cdot \Delta p = \sqrt{2}\hbar$ holds. The probability density of the wave packet at time t becomes

$$|\psi(x,t)|^2 = \dfrac{|c|^2\pi}{\sqrt{a^4 + \dfrac{\hbar^2 t^2}{4m^2}}} \exp\left[-\dfrac{a^2\left(x - \dfrac{\hbar Kt}{m}\right)^2}{2\left(a^4 + \dfrac{\hbar^2 t^2}{4m^2}\right)}\right], \qquad (10A.4)$$

indicating that the center of the wave packet is moved to the position $x = \hbar Kt/m$. We see, therefore, that the group velocity of the wave packet is $v = \hbar K/m$ in good agreement with the value calculated from its definition $v = \nabla_k \omega(k)|_{k=K}$. One can easily check that the wave packet continues to travel without altering the area in the Gaussian distribution.

10.2 An insertion of the Bloch wave function into the Wannier function leads to

$$a(\mathbf{r}-\mathbf{l}) = \dfrac{1}{\sqrt{N}} \sum_{\mathbf{k}} e^{-i\mathbf{k}\cdot\mathbf{l}} \left(\dfrac{1}{\sqrt{N}}\right) u(\mathbf{r}) e^{i\mathbf{k}\cdot\mathbf{r}} = \dfrac{u(\mathbf{r})}{N} \sum_{\mathbf{k}} e^{-i\mathbf{k}\cdot(\mathbf{l}-\mathbf{r})}. \qquad (10A.5)$$

Since the Brillouin zone of a simple cubic lattice is bounded by $-\pi/d < k_i \leq \pi/d$ ($i = x, y$ and z), the Wannier function at the origin $\mathbf{l}=0$ is calculated as

$$a(\mathbf{r}) = \dfrac{u(\mathbf{r})}{N} \sum_{\mathbf{k}} e^{i\mathbf{k}\cdot\mathbf{r}} = \dfrac{u(\mathbf{r})}{N} \int_{-\frac{\pi}{d}}^{\frac{\pi}{d}} e^{ik_x x} dk_x \int_{-\frac{\pi}{d}}^{\frac{\pi}{d}} e^{ik_y y} dk_y \int_{-\frac{\pi}{d}}^{\frac{\pi}{d}} e^{ik_z z} dk_z$$

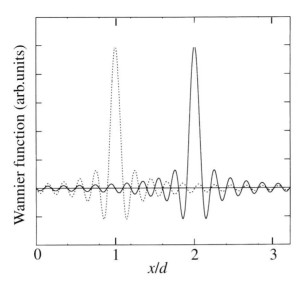

Figure 10A.1. Two Wannier functions centered at the lattice sites $x/d=1$ and 2 along x-axis in a simple cubic lattice with the lattice constant d. They are orthogonal to each other.

$$= \frac{2u(\mathbf{r})}{N} \frac{\sin\left(\frac{\pi}{d}x\right)}{\frac{\pi}{d}x} \cdot \frac{\sin\left(\frac{\pi}{d}y\right)}{\frac{\pi}{d}y} \cdot \frac{\sin\left(\frac{\pi}{d}z\right)}{\frac{\pi}{d}z}. \qquad (10\text{A}.6)$$

The integration of the product of the two Wannier functions at two different sites $l=m$ and n over the whole space is zero as shown below:

$$\int_{-\infty}^{\infty} a^*(x-md)a(x-nd)dx = \int_{-\infty}^{\infty} \left(\frac{\sin\left[\frac{\pi}{d}(x-md)\right]}{\frac{\pi}{d}(x-md)}\right) \cdot \left(\frac{\sin\left[\frac{\pi}{d}(x-nd)\right]}{\frac{\pi}{d}(x-nd)}\right) dx$$

$$= \left(\frac{d}{\pi}\right) \int_{-\infty}^{\infty} \left(\frac{\sin(x-m\pi)}{x-m\pi}\right) \cdot \left(\frac{\sin(x-n\pi)}{x-n\pi}\right) dx$$

$$= \left(\frac{d}{\pi^2}\right) \left\{ \int_{-\infty}^{\infty} \frac{\sin^2 x}{x-m\pi} dx - \int_{-\infty}^{\infty} \frac{\sin^2 x}{x-n\pi} dx \right\} = 0.$$

The two Wannier functions located at two different sites are shown in Fig. 10A.1.

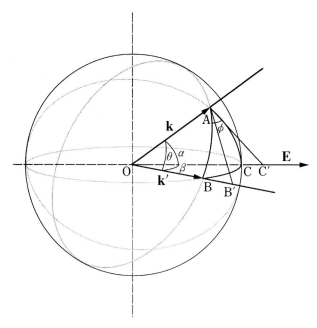

Figure 10A.2. Relation of the electric field **E** and wave vectors **k** and **k**′ in the spherical triangle ABC. [Courtesy Dr K. Ogawa]

10.3 We consider the spherical triangle shown in Fig. 10A.2. The points A, B and C are on the surface of the unit sphere and the vectors **k**, **k**′ and **E** are along OA, OB and OC directions, respectively. The angles **k**∧**k**′, **k**∧**E** and **k**′∧**E** are denoted as θ, α and β, respectively. According to the cosine rule in spherical trigonometry,

$$\cos\beta = \cos\alpha\cos\theta + \sin\alpha\sin\theta\cos\phi \qquad (10\text{A}.7)$$

Two lines AB′ and AC′ are drawn so as to be perpendicular to the vector **k** and to intersect with the lines along the vectors **k**′ and **E**, respectively. The cosine rule for the triangle $\triangle$OB′C′ and $\triangle$AB′C′ leads to

$$\text{B}'\text{C}'^2 = \text{OB}'^2 + \text{OC}'^2 - 2\text{OB}'\cdot\text{OC}'\cos\beta \qquad (10\text{A}.8)$$

and

$$\text{B}'\text{C}'^2 = \text{AB}'^2 + \text{AC}'^2 - 2\text{AB}'\cdot\text{AC}'\cos\phi. \qquad (10\text{A}.9)$$

A combination of equations (10A.8) and (10A.9) results in

$$2\text{OB}'\cdot\text{OC}'\cos\beta = (\text{OB}'^2 - \text{AB}'^2) + (\text{OC}'^2 - \text{AC}'^2) + 2\text{AB}'\cdot\text{AC}'\cos\phi$$
$$= 2\text{OA}^2 + 2\text{AB}'\cdot\text{AC}'\cos\phi. \qquad (10\text{A}.10)$$

Equation (10A.10) is rewritten as

$$\cos\beta = \frac{OA}{OB'}\cdot\frac{OA}{OC'} + \frac{AB'}{OB'}\cdot\frac{AC'}{OC'}\cos\phi = \cos\theta\cos\alpha + \sin\theta\sin\alpha\cos\phi. \quad (10A.11)$$

The relations $\mathbf{v_k}\parallel\mathbf{k}$ and $\mathbf{v_{k'}}\parallel\mathbf{k'}$ hold for the spherical Fermi surface. If we take the electric field $\mathbf{E}$ along the x-direction, we have

$$1 - \frac{\mathbf{v_{k'}}\cdot\mathbf{E}}{\mathbf{v_k}\cdot\mathbf{E}} = 1 - \frac{k'_x}{k_x} = 1 - \frac{k\cos\beta}{k\cos\alpha}, \quad (10A.12)$$

where $\mathbf{E}$ is the electric field. An insertion of equation (10A.11) into (10A.12) leads to

$$1 - \frac{\mathbf{v_{k'}}\cdot\mathbf{E}}{\mathbf{v_k}\cdot\mathbf{E}} = 1 - \frac{\cos\theta\cos\alpha + \sin\theta\sin\alpha\cos\phi}{\cos\alpha} = 1 - \cos\theta - \sin\theta\tan\alpha\cos\phi. \quad (10A.13)$$

Since the relations $1/\tau = \int[1-(\mathbf{v_{k'}}\cdot\mathbf{E}/\mathbf{v_k}\cdot\mathbf{E})]Q(\theta)d\mathbf{k'}$ and $|\mathbf{k}|=|\mathbf{k'}|=k$ hold in elastic scattering, the scattering probability is calculated to be

$$\frac{1}{\tau} = \int\left(1 - \frac{\mathbf{v_{k'}}\cdot\mathbf{E}}{\mathbf{v_k}\cdot\mathbf{E}}\right)Q(\theta)\sin\theta d\theta d\phi = \int(1-\cos\theta-\sin\theta\tan\alpha\cos\phi)Q(\theta)\sin\theta d\theta d\phi$$

$$= \int(1-\cos\theta)Q(\theta)\sin\theta d\theta d\phi.$$

10.4 By comparing the first and second lines in equation (10.88), one can explicitly write the dynamical structure factor $a(\mathbf{K},\omega)$ as

$$a(\mathbf{K},\omega) = \frac{1}{N}\left\langle\sum_f\langle i|\sum_{l'}e^{-i\mathbf{K}\cdot\mathbf{R}_{l'}}|f\rangle\cdot\langle f|\sum_l e^{i\mathbf{K}\cdot\mathbf{R}_l}|i\rangle\cdot\delta\left(\frac{\varepsilon_f-\varepsilon_i}{\hbar}-\omega\right)\right\rangle_T$$

$$= \frac{1}{N}\int\frac{dt}{2\pi}e^{i\omega t}e^{-i(\varepsilon_f-\varepsilon_i)t/\hbar}\left\langle\sum_f\langle i|\sum_{l'}e^{-i\mathbf{K}\cdot\mathbf{R}_{l'}}|f\rangle\cdot\langle f|\sum_l e^{i\mathbf{K}\cdot\mathbf{R}_l}|i\rangle\right\rangle_T, \quad (10A.14)$$

where the delta function $\delta(\omega) = \int_{-\infty}^{\infty}e^{i\omega t}(dt/2\pi)$ is replaced by its integration form. Since $\mathbf{R}_l = \mathbf{l}+\mathbf{u}_l$, we get

$$a(\mathbf{K},\omega) = \frac{1}{N}\sum_{l,l'}e^{-i\mathbf{K}\cdot(\mathbf{l}-\mathbf{l'})}\int\frac{dt}{2\pi}e^{i\omega t}e^{-i(\varepsilon_f-\varepsilon_i)t/\hbar}\left\langle\sum_f\langle i|e^{-i\mathbf{K}\cdot\mathbf{u}_{l'}}|f\rangle\cdot\langle f|e^{i\mathbf{K}\cdot\mathbf{u}_l}|i\rangle\right\rangle_T. \quad (10A.15)$$

Furthermore, the relation

$$e^{-i(\epsilon_f - \epsilon_i)t/\hbar} \langle i | e^{-i\mathbf{K}\cdot\mathbf{u}_l} | f \rangle = \langle i | e^{iH_0 t} e^{-i\mathbf{K}\cdot\mathbf{u}_l(0)} e^{iH_0 t} | f \rangle = \langle i | e^{-i\mathbf{K}\cdot\mathbf{u}_l(t)} | f \rangle \quad (10A.16)$$

holds in the Heisenberg representation. Equation (10A.15) is reduced to equation (10.89), since the relation $\Sigma_f \langle i | e^{-i\mathbf{K}\cdot\mathbf{u}_{l'}(t)} | f \rangle \cdot \langle f | e^{i\mathbf{K}\cdot\mathbf{u}_l(0)} | i \rangle = \langle i | e^{-i\mathbf{K}\cdot\mathbf{u}_{l'}(t)} e^{i\mathbf{K}\cdot\mathbf{u}_l(0)} | i \rangle$ holds [7, 8].

10.5 The dynamical structure factor is explicitly written as

$$a(\mathbf{K}, \omega) = \left(\frac{1}{N}\right) \sum_{i,f} \frac{e^{-\beta\epsilon_i}}{Z} \left| \langle f | \sum_l e^{i\mathbf{K}\cdot\mathbf{R}_l} | i \rangle \right|^2 \cdot \delta\left(\frac{\epsilon_f - \epsilon_i}{\hbar} - \omega\right), \quad (10A.17)$$

where Z is the partition function defined as $\Sigma_j e^{-\beta\epsilon_j}$ and $\left|\langle f | \Sigma_l e^{i\mathbf{K}\cdot\mathbf{R}_l} | i \rangle\right|^2$ represents the probability amplitude of the transition from the state $|\mathbf{k}, i\rangle$ to $|\mathbf{k}', f\rangle$. Now let us replace the arguments $\mathbf{K}$ and ω in $a(\mathbf{K}, \omega)$ by $-\mathbf{K}$ and $-\omega$. Then, we obtain

$$a(-\mathbf{K}, -\omega) = \left(\frac{1}{N}\right) \sum_{i,f} \frac{e^{-\beta\epsilon_i}}{Z} \langle i | \sum_{l'} e^{+i\mathbf{K}\cdot\mathbf{R}_{l'}} | f \rangle \cdot \langle f | \sum_l e^{-i\mathbf{K}\cdot\mathbf{R}_l} | i \rangle \cdot \delta\left(\frac{\epsilon_f - \epsilon_i}{\hbar} + \omega\right)$$

$$= \left(\frac{e^{-\beta\hbar\omega}}{N}\right) \sum_{i,f} \frac{e^{-\beta\epsilon_f}}{Z} \langle f | \sum_l e^{-i\mathbf{K}\cdot\mathbf{R}_l} | i \rangle \cdot \langle i | \sum_{l'} e^{+i\mathbf{K}\cdot\mathbf{R}_{l'}} | f \rangle \cdot \delta\left(\frac{\epsilon_i - \epsilon_f}{\hbar} - \omega\right)$$

$$= \left(\frac{e^{-\beta\hbar\omega}}{N}\right) \sum_{f,i} \frac{e^{-\beta\epsilon_f}}{Z} \left| \langle i | \sum_l e^{i\mathbf{K}\cdot\mathbf{R}_l} | f \rangle \right|^2 \cdot \delta\left(\frac{\epsilon_i - \epsilon_f}{\hbar} - \omega\right) = e^{-\beta\hbar\omega} a(\mathbf{K}, \omega),$$

(10A.18)

where $\left|\langle i | \Sigma_l e^{i\mathbf{K}\cdot\mathbf{R}_l} | f \rangle\right|^2$ represents the probability amplitude of the transition from the state $|\mathbf{k}', f\rangle$ to $|\mathbf{k}, i\rangle$ and the detailed balance condition assures $\left|\langle f | \Sigma_l e^{i\mathbf{K}\cdot\mathbf{R}_l} | i \rangle\right|^2 = \left|\langle i | \Sigma_l e^{i\mathbf{K}\cdot\mathbf{R}_l} | f \rangle\right|^2$.

10.6 Equation (10.93) is rewritten by substituting $f_0(\mathbf{k}) + \phi(\mathbf{k})$ for $f(\mathbf{k})$ with a subsequent use of the identity $e^{-\beta\hbar\omega} f_0(\epsilon_\mathbf{k} - \hbar\omega)[(1 - f_0(\epsilon_\mathbf{k})] = f_0(\epsilon_\mathbf{k})[1 - f_0(\epsilon_\mathbf{k} - \hbar\omega)]$:

$$\left(-\frac{\partial f_0}{\partial \varepsilon}\right)\mathbf{v_k}\cdot(-e)\mathbf{E} = -\left(\frac{2\pi}{N}\right)\int_{-\infty}^{\infty} d\omega \sum_{\mathbf{k'}} |U_p(\mathbf{K})|^2 a(\mathbf{K}, \omega)\, \delta(\varepsilon_{\mathbf{k'}} - \varepsilon_{\mathbf{k}} + \hbar\omega)$$

$$\times \{\phi(\mathbf{k})[1 + e^{-\beta(\varepsilon_\mathbf{k}-\mu)}][1 - f_0(\mathbf{k'})] - \phi(\mathbf{k'})[1 + e^{\beta(\varepsilon_{\mathbf{k'}}-\mu)}]f_0(\mathbf{k})\}, \quad (10\text{A}.19)$$

where the term involving $\phi(\mathbf{k})\phi(\mathbf{k'})$ is ignored as higher-order terms. All $\phi(\mathbf{k})$s appearing in the right-hand side are replaced by $\phi(\mathbf{k}) = \tau_\mathbf{k} \mathbf{v_k}\cdot(-e)\mathbf{E}[\partial f_0(\mathbf{k})/\partial\varepsilon]$. The quantity in the curly bracket in equation (10A.19) is then reduced to

$$\left\{\tau_\mathbf{k}[\mathbf{v_k}\cdot(-e)\mathbf{E}]\left(\frac{\partial f_0(\mathbf{k})}{\partial\varepsilon}\right)[1 + e^{-\beta(\varepsilon_\mathbf{k}-\mu)}][1 - f_0(\mathbf{k'})] - \tau_{\mathbf{k'}}[\mathbf{v_{k'}}\cdot(-e)\mathbf{E}]\left(\frac{\partial f_0(\mathbf{k'})}{\partial\varepsilon}\right)\right.$$

$$\left.\times[1 + e^{\beta(\varepsilon_{\mathbf{k'}}-\mu)}]f_0(\mathbf{k})\right\} = -\left\{f_0(\mathbf{k})[1 - f_0(\mathbf{k'})]\{\tau_\mathbf{k}[\mathbf{v_k}\cdot(-e)\mathbf{E}] - \tau_{\mathbf{k'}}[\mathbf{v_{k'}}\cdot(-e)\mathbf{E}]\}\right\}/k_\mathrm{B}T,$$
$$(10\text{A}.20)$$

where the relation $f_0(\mathbf{k})[1 - f_0(\mathbf{k})] = k_\mathrm{B}T[-\partial f_0(\mathbf{k})/\partial\varepsilon_\mathbf{k}]$ is used. The right-hand side of equation (10A.19) is reduced to

$$\frac{2\pi}{k_\mathrm{B}NT}\int_{-\infty}^{\infty} d\omega \sum_{\mathbf{k'}} |U_p(\mathbf{K})|^2 a(\mathbf{K}, \omega)\{\tau_\mathbf{k}[\mathbf{v_k}\cdot(-e)\mathbf{E}] - \tau_{\mathbf{k'}}[\mathbf{v_{k'}}\cdot(-e)\mathbf{E}]\}$$

$$\times \delta(\varepsilon_{\mathbf{k'}} - \varepsilon_\mathbf{k} + \hbar\omega)f_0(\mathbf{k})[1 - f_0(\mathbf{k'})],$$

which becomes

$$\frac{2\pi}{k_\mathrm{B}NT}\int_{-\infty}^{\infty} d\omega \sum_{\mathbf{k'}} |U_p(\mathbf{K})|^2 a(\mathbf{K}, \omega)\{\tau_\mathbf{k}[\mathbf{v_k}\cdot(-e)\mathbf{E}] - \tau_{\mathbf{k'}}[\mathbf{v_{k'}}\cdot(-e)\mathbf{E}]\}$$

$$\times \delta(\varepsilon_{\mathbf{k'}} - \varepsilon_\mathbf{k} + \hbar\omega)\left(-\frac{\partial f_0}{\partial\varepsilon_\mathbf{k}}\right)\hbar\omega n(\omega), \quad (10\text{A}.21)$$

where the relations $f_0(\varepsilon_\mathbf{k})(1 - f_0[\varepsilon_\mathbf{k} - \hbar\omega]) = -[f_0(\varepsilon_\mathbf{k}) - f_0(\varepsilon_\mathbf{k} - \hbar\omega)]n(\omega)$ and $f_0(\varepsilon_\mathbf{k} - \hbar\omega) \approx f_0(\varepsilon_\mathbf{k}) - \hbar\omega[\partial f_0(\varepsilon_\mathbf{k})/\partial\varepsilon_\mathbf{k}]$ are inserted and $n(\omega)$ is the Planck distribution function. The Boltzmann transport equation (10.93) is now reduced to equation (10.94).

10.7 In the dynamical structure factor given by equation (10.89), $e^{-i\mathbf{K}\cdot\mathbf{u}_{l'}(t)}$ and $e^{i\mathbf{K}\cdot\mathbf{u}_l(0)}$ do not commute with each other. Here we need to use the relation $e^{A+B} = e^A e^B e^{-\frac{1}{2}[A,B]}$ known as the Baker–Hansdorff theorem, where the two operators A and B are commuted with each other and satisfy the relation $[A,[A,B]] = [B,[A,B]] = 0$ [A. Messiah, *Quantum Mechanics*, (North-Holland,

Amsterdam, 1958) vol. 1, p. 442]. Then, the thermal average in equation (10.89) can be calculated as

$$\langle e^{-i\mathbf{K}\cdot\mathbf{u}_{\mathbf{l}'}(t)} e^{i\mathbf{K}\cdot\mathbf{u}_{\mathbf{l}}(0)} \rangle_T = \langle e^{i\mathbf{K}\cdot[\mathbf{u}_{\mathbf{l}}(0)-\mathbf{u}_{\mathbf{l}'}(t)]} e^{\frac{1}{2}[i\mathbf{K}\cdot\mathbf{u}_{\mathbf{l}}(0),-i\mathbf{K}\cdot\mathbf{u}_{\mathbf{l}'}(t)]} \rangle_T$$

$$= e^{(K^2/2)[\mathbf{u}_{\mathbf{l}}(0),\mathbf{u}_{\mathbf{l}'}(t)]} \langle e^{i\mathbf{K}\cdot(\mathbf{u}_{\mathbf{l}}(0)-\mathbf{u}_{\mathbf{l}'}(t))} \rangle_T$$

$$= e^{(K^2/2)[\mathbf{u}_{\mathbf{l}}(0),\mathbf{u}_{\mathbf{l}'}(t)]} e^{-\frac{1}{2}\langle[\mathbf{K}\cdot(\mathbf{u}_{\mathbf{l}}(0)-\mathbf{u}_{\mathbf{l}'}(t))]^2\rangle_T}$$

$$= e^{(K^2/2)[\mathbf{u}_{\mathbf{l}}(0),\mathbf{u}_{\mathbf{l}'}(t)]} e^{\{-\frac{1}{2}\langle(\mathbf{K}\cdot\mathbf{u}_{\mathbf{l}}(0))^2\rangle_T -\frac{1}{2}\langle(\mathbf{K}\cdot\mathbf{u}_{\mathbf{l}'}(t))^2\rangle_T + \frac{1}{2}\langle(\mathbf{K}\cdot\mathbf{u}_{\mathbf{l}}(0))(\mathbf{K}\cdot\mathbf{u}_{\mathbf{l}'}(t)) + (\mathbf{K}\cdot\mathbf{u}_{\mathbf{l}'}(t))(\mathbf{K}\cdot\mathbf{u}_{\mathbf{l}}(0))\rangle_T \}}$$

$$= e^{(K^2/2)[\mathbf{u}_{\mathbf{l}}(0),\mathbf{u}_{\mathbf{l}'}(t)]} e^{-\frac{1}{2}\{\langle(\mathbf{K}\cdot\mathbf{u}_{\mathbf{l}}(0))^2\rangle_T + \langle(\mathbf{K}\cdot\mathbf{u}_{\mathbf{l}'}(t))^2\rangle_T\}} e^{+\frac{1}{2}\langle -[\mathbf{K}\cdot\mathbf{u}_{\mathbf{l}}(0),\mathbf{K}\cdot\mathbf{u}_{\mathbf{l}'}(t)]+2(\mathbf{K}\cdot\mathbf{u}_{\mathbf{l}'}(t))(\mathbf{K}\cdot\mathbf{u}_{\mathbf{l}}(0))\rangle_T}$$

$$= e^{(K^2/2)[\mathbf{u}_{\mathbf{l}}(0),\mathbf{u}_{\mathbf{l}'}(t)]} e^{-\frac{1}{2}\{\langle(\mathbf{K}\cdot\mathbf{u}_{\mathbf{l}}(0))^2\rangle_T + \langle(\mathbf{K}\cdot\mathbf{u}_{\mathbf{l}'}(t))^2\rangle_T\}} e^{-(K^2/2)[\mathbf{u}_{\mathbf{l}}(0),\mathbf{u}_{\mathbf{l}'}(t)]} e^{\langle(\mathbf{K}\cdot\mathbf{u}_{\mathbf{l}'}(t))(\mathbf{K}\cdot\mathbf{u}_{\mathbf{l}}(0))\rangle_T},$$

(10A.22)

where the relation $\langle e^{i\mathbf{K}\cdot(\mathbf{u}_{\mathbf{l}}(0)-\mathbf{u}_{\mathbf{l}'}(t))} \rangle_T = e^{-\frac{1}{2}\langle[\mathbf{K}\cdot(\mathbf{u}_{\mathbf{l}}(0)-\mathbf{u}_{\mathbf{l}'}(t))]^2\rangle_T}$ is used. As is shown below, this relation holds, if lattice vibrations are treated in the harmonic oscillator approximation.

Since the Hamiltonian is expressed as $H=(a^+a+\frac{1}{2})\hbar\omega$ in the harmonic oscillator approximation, we can prove the relation $\langle e^{i\xi(\gamma a+\gamma^* a^+)} \rangle_T = \exp[-(\xi^2/2)|\gamma|^2 \times \coth(\hbar\omega/2k_BT)]$ [A. Messiah, *Quantum Mechanics*, (North-Holland, Amsterdam, 1958) vol. 1, p. 450–451]. In a similar way, we can prove the relation

$$\langle(\gamma a+\gamma^* a^+)^2\rangle_T \equiv \sum_{n=0}^{\infty}\left(\frac{e^{-[n+(1/2)]\hbar\omega/k_BT}}{Z}\right)\langle n|(\gamma a+\gamma^* a^+)^2|n\rangle = |\gamma|^2(1-y)\sum_{n=0}^{\infty}y^n(2n+1)$$

$$=|\gamma|^2\frac{1+y}{1-y}=|\gamma|^2\coth\left(\frac{\hbar\omega}{2k_BT}\right), \quad (10A.23)$$

where $y=e^{-\hbar\omega/k_BT}$ and $Z=e^{-\hbar\omega/2k_BT}/(1-e^{-\hbar\omega/k_BT})=y^{1/2}/(1-y)$. The second line in equation (10A.23) is obtained by using the relations $1/(1-y)=\sum_{n=0}^{\infty}y^n$ and $1/(1-y)^2=\sum_{n=0}^{\infty}(n+1)y^n$ and, hence, $\sum_{n=0}^{\infty}y^n(2n+1)=(y+1)/(1-y)^2$. A combination with the relation obtained above proves the relation $\langle e^{i\xi(\gamma a+\gamma^* a^+)} \rangle_T = e^{-(\xi^2/2)\langle(\gamma a+\gamma^* a^+)^2\rangle_T}$.

Since the first and third exponentials in equation (10A.22) are canceled to zero, we obtain the relation

$$\langle e^{-i\mathbf{K}\cdot\mathbf{u}_{\mathbf{l}'}(t)} e^{i\mathbf{K}\cdot\mathbf{u}_{\mathbf{l}}(0)} \rangle_T = e^{-\frac{1}{2}\{\langle(\mathbf{K}\cdot\mathbf{u}_{\mathbf{l}}(0))^2\rangle_T + \langle(\mathbf{K}\cdot\mathbf{u}_{\mathbf{l}'}(t))^2\rangle_T\}} e^{\langle(\mathbf{K}\cdot\mathbf{u}_{\mathbf{l}'}(t))(\mathbf{K}\cdot\mathbf{u}_{\mathbf{l}}(0))\rangle_T}. \quad (10A.24)$$

The argument $\frac{1}{2}\{\langle(\mathbf{K}\cdot\mathbf{u}_l(0))^2\rangle_T + \langle(\mathbf{K}\cdot\mathbf{u}_{l'}(t))^2\rangle_T\}$ is the Debye–Waller factor and is time-independent. Thus, the dynamical structure factor is explicitly written as

$$a(\mathbf{K},\omega) = \frac{1}{N}\sum_{l,l'} e^{-i\mathbf{K}\cdot(\mathbf{l}-\mathbf{l'})} \int_{-\infty}^{\infty} \frac{dt}{2\pi} e^{i\omega t} e^{-2W_{ll'}(\mathbf{K})} \exp\langle(\mathbf{K}\cdot\mathbf{u}_{l'}(t))(\mathbf{K}\cdot\mathbf{u}_l(0))\rangle_T.$$

10.8 An insertion of equation (10.100) into equation (10.98) for an isotropic system results in

$$W(K) = \frac{1}{2NM}\left\langle \sum_l \sum_{\alpha,\beta} K_\alpha K_\beta \sum_{\mu,\nu} \frac{\hbar}{2\sqrt{\omega_\mu \omega_\nu}} e_l^\alpha(\omega_\mu) e_l^\beta(\omega_\nu)(a_\mu^+ + a_\mu)(a_\nu^+ + a_\nu)\right\rangle_T, \quad (10A.25)$$

where $e_l^i(\omega_\mu)$ is the l-component ($l = x, y, z$) of the i-th polarization vector of the mode ω_μ. Since $\langle(a_\mu^+ + a_\mu)(a_\nu^+ + a_\nu)\rangle_T = \coth(\hbar\omega_\mu/2k_BT)$ holds, we get

$$W(K) = \frac{1}{4NM}\sum_l \sum_{\alpha,\beta} \sum_\mu K_\alpha K_\beta e_l^\alpha(\omega_\mu) e_l^\beta(\omega_\mu) \coth(\hbar\omega_\mu/2k_BT)\frac{1}{\omega_\mu}. \quad (10A.26)$$

An average over the three independent directions of the vector $\mathbf{K}$ results in

$$= \frac{1}{4NM}\frac{K^2}{3}\sum_\mu \frac{\coth(\hbar\omega_\mu/2k_BT)}{\omega_\mu}\sum_l \sum_\alpha e_l^\alpha(\omega_\mu) e_l^\alpha(\omega_\mu)$$

$$= \frac{1}{4NM}\frac{K^2}{3}\sum_\mu \frac{\coth(\hbar\omega_\mu/2k_BT)}{\omega_\mu} = \frac{1}{4NM}\frac{K^2}{3}\int_{-\infty}^{\infty} d\omega \sum_\mu \delta(\omega - \omega_\mu)\frac{\coth(\hbar\omega_\mu/2k_BT)}{\omega_\mu}$$

$$= \frac{1}{4M}\frac{K^2}{3}\int_{-\infty}^{\infty} d\omega D(\omega) \frac{\coth(\hbar\omega/2k_BT)}{\omega},$$

where $D(\omega)$ is the phonon density of states per atom. The application of the Debye model to $D(\omega)$ leads to

$$W(T) = W(0) + \left(\frac{2\pi^2}{3}\right) W(0)\left(\frac{T}{\Theta_D}\right)^2 + \ldots \quad (T \ll \Theta_D) \quad (10A.27)$$

$$W(T) = 4W(0)\left(\frac{T}{\Theta_D}\right) + \ldots \quad (T \geq \Theta_D), \quad (10A.28)$$

where $W(0) = 3(\hbar K)^2/8 Mk_B\Theta_D$.

Chapter Eleven

11.1 The ratio $(-L_{ET}/L_{EE})$ is the thermoelectric power and is reduced to equation (11.23) in the free-electron model, where $\sigma(\varepsilon) \propto \varepsilon^{3/2}$ holds. Thus, the correction term is calculated as

$$\frac{L_{TE}L_{ET}}{L_{EE}L_{TT}} = \frac{(L_{ET})^2 T}{L_{EE}L_{TT}} = \left(\frac{L_{ET}}{L_{EE}}\right)^2 \cdot \frac{\frac{\pi^2}{3}(k_B T)^2 \cdot \frac{1}{T(-e)} \left.\frac{\partial \sigma(\varepsilon)}{\partial \varepsilon}\right|_{\varepsilon=\zeta} T}{\left[\frac{\pi^2 k_B^2 T \sigma}{3(-e)^2}\right]}$$

$$= \left[\frac{T}{T_F(-e)}\right](-e)\left.\frac{\partial \ln \sigma(\varepsilon)}{\partial \varepsilon}\right|_{\varepsilon=\zeta} T \propto \left(\frac{T}{T_F}\right)^2 \quad (11A.1)$$

11.2 The x- and y-components of the current density are given by

$$J_x = \frac{(-e)}{4\pi^3}\iiint v_x \phi(\mathbf{k}) dk_x dk_y dk_z = \frac{(-e)\hbar}{4\pi^3 m}\iiint k_x (ak_x + bk_y) dk_x dk_y dk_z$$

$$= \frac{(-e)\hbar}{4\pi^3 m}\iiint (ak_x^2 + bk_x k_y) dk_x dk_y dk_z, \quad (11A.2)$$

$$J_y = \frac{(-e)}{4\pi^3}\iiint v_y \phi(\mathbf{k}) dk_x dk_y dk_z = \frac{(-e)\hbar}{4\pi^3 m}\iiint k_y (ak_x + bk_y) dk_x dk_y dk_z$$

$$= \frac{(-e)\hbar}{4\pi^3 m}\iiint (ak_y k_x + bk_y^2) dk_x dk_y dk_z, \quad (11A.3)$$

An insertion of equation (11.31) into the x-component in equation (11.30) yields

$$-(-e)E v_x \frac{\partial f_0}{\partial \varepsilon} = \frac{\phi}{\tau} - \frac{(-e)}{\hbar} B\left(\frac{\hbar}{m}\right)(bk_x - ak_y)$$

and is rewritten as

$$-(-e)\tau E v_x \frac{\partial f_0}{\partial \varepsilon} = (a - b\omega_c \tau)k_x + (b + a\omega_c \tau)k_y. \quad (11A.4)$$

Similarly, we obtain the following relation from the y- and z-components in equation (11.30)

$$0 = (a - b\omega_c \tau)k_x + (b + a\omega_c \tau)k_y. \quad (11A.5)$$

Equation (11A.5) should hold for any k_x and k_y. This is possible if $a = \omega_c \tau b$ and $b = -\omega_c \tau a$. Multiplication by k_x of both sides of equation (11A.4) yields

$$-(-e)\tau E\left(\frac{m}{\hbar}\right)v_x v_x \frac{\partial f_0}{\partial \varepsilon} = (a - b\omega_c \tau)k_x^2 + (b + a\omega_c \tau)k_x k_y. \quad (11A.6)$$

Further multiplication by $[(-e)/4\pi^3]\iiint dk_x dk_y dk_z$ of both sides yields

$$-\left(\frac{(-e)^2}{4\pi^3}\right)\left(\frac{m}{\hbar}\right)\tau E\iiint v_x v_x \frac{\partial f_0}{\partial \varepsilon} dk_x dk_y dk_z = \frac{(-e)}{4\pi^3}\iiint (ak_x^2 + bk_x k_y)\, dk_x dk_y dk_z$$

$$-\frac{(-e)}{4\pi^3}\omega_c \tau \iiint (bk_x^2 - bk_x k_y)\, dk_x dk_y dk_z.$$

The quantity in the left-hand side can be calculated in the same way as discussed in Section 10.7 and the equation above is rewritten as

$$\left[\frac{(-e)^2 \tau v_F S_F}{12\pi^3 \hbar}\right] E = \frac{(-e)}{4\pi^3}\left(\frac{\hbar}{m}\right)\iiint (ak_x^2 + bk_y k_x)\, dk_x dk_y dk_z$$

$$-\frac{(-e)}{4\pi^3}\omega_c \tau \left(\frac{\hbar}{m}\right)\iiint (bk_x^2 - bk_y k_x)\, dk_x dk_y dk_z.$$

By using equations (10.52) and (11A.2), we have

$$\left[\frac{n(-e)^2\tau}{m}\right] E = J_x - \frac{(-e)}{4\pi^3}\omega_c \tau \left(\frac{\hbar}{m}\right)\iiint (bk_x^2 - ak_y k_x)\, dk_x dk_y dk_z$$

$$= J_x - \frac{(-e)}{4\pi^3}(\omega_c \tau)^2 \left(\frac{\hbar}{m}\right)\iiint \frac{(bk_x^2 - ak_y k_x)}{\omega_c \tau}\, dk_x dk_y dk_z. \quad (11A.7)$$

An insertion of the relations $a = \omega_c \tau b$ and $b = -\omega_c \tau a$ yields

$$= J_x + \frac{(-e)}{4\pi^3}(\omega_c \tau)^2 \left(\frac{\hbar}{m}\right)\iiint (ak_x^2 + bk_y k_x)\, dk_x dk_y dk_z$$

$$= J_x + (\omega_c \tau)^2 J_x. \quad (11A.8)$$

The ratio of J_x/E_x is the xx-component of the conductivity tensor. Hence, we obtain $\sigma_{xx} = [n(-e)^2 \tau/m]\,\{1/[1 + (\omega_c \tau)^2]\}$. Similarly, the xy-component σ_{xy} of the conductivity tensor can be calculated by starting from the equation obtained by multiplying both sides of equation (11A.4) by k_y.

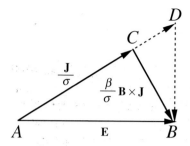

Figure 11A.1. Configurations of vectors **E** and **J**/σ in the Hall measurements. Vectors **E** and **J**/σ are both in the plane of the page and the magnetic field **B** is perpendicular to it.

11.3 A right-angle $\triangle ABC$ is constructed as shown in Fig. 11A.1 by using the three vectors **J**/σ, **E** and $\beta \mathbf{B} \times \mathbf{J}/\sigma$. $\triangle ABD$ is drawn so as to fill $\overline{DB} \perp \overline{AB}$. By using the proportional relations, we can easily find $\overline{CD} = (\beta^2 B^2)/\sigma)\mathbf{J}$ and $\overline{DB} = \beta \mathbf{B} \times \mathbf{E}$. Since the relation $\overline{AD} = \overline{AB} - \overline{DB}$ holds in $\triangle ABD$, we have $(\mathbf{J}/\sigma)(1 + \beta^2 B^2) = \mathbf{E} - \beta \mathbf{B} \times \mathbf{E}$.

11.4 In our configurations $\mathbf{J} = (J, 0, 0)$ and $\mathbf{B} = (0, 0, B)$, equation (11.42) is reduced to

$$J = (\sigma_1 + \sigma_2)E_x + (\sigma_1 \beta_1 + \sigma_2 \sigma_2)E_y B \qquad (11A.9)$$

$$0 = (\sigma_1 + \sigma_2)E_y - (\sigma_1 \beta_1 + \sigma_2 \beta_2)E_x B \qquad (11A.10)$$

$$0 = (\sigma_1 + \sigma_2)E_z, \qquad (11A.11)$$

where B^2 terms are neglected. An insertion of equation (11A.9) into equation (11A.10) yields

$$E_y = \frac{\sigma_1 \beta_1 + \sigma_2 \beta_2}{(\sigma_1 + \sigma_2)^2} JB = \frac{\sigma_1^2 R_1 + \sigma_2^2 R_2}{(\sigma_1 + \sigma_2)^2} JB, \qquad (11A.12)$$

where B^2 terms are again neglected.

11.5 The velocity correlation function is differentiated with respect to time:

$$\dot{\phi}(t) = \langle v(0)\dot{v}(t) \rangle. \qquad (11A.13)$$

The insertion of equation (11.78) into equation (11A.13) results in

$$\dot{\phi}(t) = -\frac{\langle v(0)v(t) \rangle}{\tau} + \frac{\langle v(0)F(t) \rangle}{m}.$$

Since the correlation between $F(t)$ and velocity $v(0)$ must be zero, we have

$$\dot{\phi}(t) = -\frac{\varphi(t)}{\tau}. \qquad (11A.14)$$

The solution of equation (11A.14) is obviously given by

$$\phi(t) = \phi(0)\exp\left(-\frac{t}{\tau}\right) = \langle v(0)^2\rangle \exp\left(-\frac{t}{\tau}\right).$$

The diffusion constant D in equation (11.83) turns out to be

$$D = \int_0^\infty \langle v(0)v(t)\rangle dt = \int_0^\infty \phi(t)dt = \langle v(0)^2\rangle \int_0^\infty \exp\left(-\frac{t}{\tau}\right)dt = \langle v(0)^2\rangle \tau. \qquad (11\text{A}.15)$$

According to the equipartition law of energy, we have $m\langle v(0)^2\rangle/2 = k_B T/2$ and, hence, obtain the Einstein relation $D = \langle v(0)^2\rangle \tau = (\tau/m)k_B T$.

Chapter Twelve

12.1 By taking the rotation of $\mathbf{B} = (-m_s/n_s q_s^2)\text{rot}\mathbf{J}_s$ from equation (12.13) with subsequent use of equation (12.8) and $\text{div}\mathbf{J}_s = 0$ from the equation of continuity, we immediately obtain $\mu_0 \mathbf{J}_s = (-m_s/n_s q_s^2)\text{rot rot}\mathbf{J}_s = (-m_s/n_s q_s^2)\nabla^2 \mathbf{J}_s$.

12.2 Equation (7.4) is applied to the present case:

$$\oint \left(\frac{2m\mathbf{J}_s}{n_s q_s} + q_s \mathbf{A}\right) \cdot d\mathbf{s} = nh, \qquad (12.\text{A}.1)$$

where the mass of the superconducting paired electrons is $2m$, its current density is $\mathbf{J}_s$ and n is a positive integer. Equation (7.5) is now written as

$$\oint q_s \mathbf{A} \cdot d\mathbf{s} = q_s \iint \text{rot}\mathbf{A}dS = q_s \iint \mathbf{B}dS = q_s \phi_a, \qquad (12\text{A}.2)$$

where ϕ_a is the applied magnetic flux passing through the ring and S represents the area in real space of the closed orbit of the paired electrons. Hence, equation (12A.1) becomes

$$\frac{2m}{n_s q_s}\oint \mathbf{J}_s \cdot d\mathbf{s} + q_s \phi_a = nh. \qquad (12\text{A}.3)$$

As is clear from Exercise 12.1, the superconducting current $\mathbf{J}_s$ flows only in the surface layer characterized by the penetration depth. Thus, the line integral in equation (12A.3) inside a superconductor is zero. Thus, we obtain

$$\phi_a = \frac{nh}{q_s} = n\phi_0, \qquad (12\text{A}.4)$$

where $\phi_0 = h/q_s$.

Chapter Thirteen

13.1 (a) For example, we have

$$\int \psi_1^* V(x,y,z) \psi_2 d\tau = \int |f(r)|^2 xy(Ax^2 + By^2 + Cz^2) d\tau, \quad (13A.1)$$

which is an odd power of x and, hence, must be zero. Instead, the diagonal term, such as $\int \psi_1^* V \psi_1 d\tau = \int |f(r)|^2 (Ax^4 + By^2 + Cz^2) d\tau$, is finite. The three energy levels are reduced to $E_1 = A\alpha + (B+C)\beta$, $E_2 = B\alpha + (C+A)\beta$ and $E_3 = C\alpha + (A+B)\beta$, where

$$\alpha = \int x^4 |f(r)|^2 d\tau = \int y^4 |f(r)|^2 d\tau = \int z^4 |f(r)|^2 d\tau$$

and

$$\beta = \int x^2 y^2 |f(r)|^2 d\tau = \int y^2 z^2 |f(r)|^2 d\tau = \int z^2 x^2 |f(r)|^2 d\tau$$

(b) Since $L_z = (\hbar/i)[x(\partial/\partial y) - y(\partial/\partial x)]$, we have

$$\int \psi_1^* L_z \psi_1 d\tau = \frac{\hbar}{i} \int |f(r)|^2 \left(x \frac{\partial}{\partial y} - y \frac{\partial}{\partial x} \right)(Ax^4 + By^2 + Cz^2) d\tau, \quad (13A.2)$$

which is again an odd function of the variables x and y and thus the integral is reduced to zero.

More generally we say that a non-degenerate wave function is obtained, if the admixture of the wave functions with the two z-components of an opposite sign makes the resulting wave function real. The discussion above is typical of this example. As another example, the admixture of two running waves $(e^{im_\ell \phi} \pm e^{-im_\ell \phi})/\sqrt{2}$ results in either $\sin m_\ell \phi$ or $\cos m_\ell \phi$, both of which become real again. The orbital angular momentum operator is expressed as $\mathbf{L} = \mathbf{r} \times \mathbf{p} = (\hbar/i)[\mathbf{r} \times \mathrm{grad}]$ and always involves the imaginary part. The operator for the z-component of the angular momentum also involves the imaginary part. The expectation value of the orbital angular momentum for the non-degenerate wave function is given by

$$\langle \mathbf{L} \rangle = \int \psi^*(\mathbf{r}) \mathbf{L} \psi(\mathbf{r}) d\tau = \int \psi^*(\mathbf{r}) \mathbf{L} \psi^*(\mathbf{r}) d\tau = \left[\int \psi^*(\mathbf{r}) \mathbf{L} \psi(\mathbf{r}) d\mathbf{r} \right]^* = -\langle \mathbf{L} \rangle, \quad (13A.3)$$

where $\psi^*(\mathbf{r}) = \psi(\mathbf{r})$ is used. Therefore, we always have $\langle \mathbf{L} \rangle = 0$, as long as the wave function is real.

13.2 The density of states of magnons $D(\omega)$ is given by $(1/2\pi)^3 4\pi k^2 (dk/d\omega) d\omega$. An insertion of the dispersion relation $\hbar\omega = (2JSa^2)k^2$ yields

$$D(\omega) = \frac{1}{4\pi^2} \left(\frac{\hbar}{2JSa^2} \right)^{3/2} \omega^{1/2}. \quad (13A.4)$$

The calculation of $\sum_k n_k$ is straightforward:

$$\sum_k n_k = \int_0^\infty n(\omega, T)D(\omega)d\omega = \frac{1}{4\pi^2}\left(\frac{\hbar}{2JSa^2}\right)^{3/2} \cdot \int_0^\infty \frac{\omega^{1/2}}{e^{\beta\hbar\omega} - 1}d\omega$$

$$= \frac{1}{4\pi^2}\left(\frac{k_BT}{2JSa^2}\right)^{3/2} \cdot \int_0^\infty \frac{x^{1/2}}{e^x - 1}dx.$$

The definite integral has the value of $(0.0587)(4\pi)^2$. The number of atoms per unit volume is given by Q/a^3, where Q is 1, 2 and 4 for simple cubic, bcc and fcc lattices, respectively. The magnetization $M(0)$ at absolute zero is equal to NS. Hence, magnetization at finite temperatures $M(T)$ is calculated as

$$M(T) = NS - \sum_k n_k = M(0)\left(1 - \sum_k n_k/NS\right) = M(0)\left[1 - \frac{0.0587}{SQ}\left(\frac{k_BT}{2JS}\right)^{3/2}\right]. \quad (13A.5)$$

13.3 The numbers of spin-up and spin-down electrons must be equal to $N/2$ before introduction of the exchange energy Δ. Hence, we have the relation $N/2 = \frac{1}{2}\int_0^{\varepsilon_F^o} N(\varepsilon)d\varepsilon$, where $N(\varepsilon)$ is the total density of states. The spin-down band is shifted to higher energies by Δ while the spin-up band to lower energies by Δ. If we ignore charge transfer, the Fermi level would be shifted to $\varepsilon_F^o + \Delta$ and $\varepsilon_F^o - \Delta$ for spin-down and spin-up bands, respectively. Indeed, the solution of equations $N/2 = \frac{1}{2}\int_\Delta^x N(\varepsilon - \Delta)d\varepsilon$ and $N/2 = \frac{1}{2}\int_{-\Delta}^x N(\varepsilon + \Delta)d\varepsilon$ is easily found to be $x = \varepsilon_F^o + \Delta$ and $x = \varepsilon_F^o - \Delta$, respectively. These are shown by the dashed lines in Fig. 13.5. There is no net change in the total energy, since we have

$$U = \frac{1}{2}\left(\int_\Delta^{\varepsilon_F^o + \Delta} \varepsilon N(\varepsilon - \Delta)d\varepsilon + \int_{-\Delta}^{\varepsilon_F^o - \Delta} \varepsilon N(\varepsilon + \Delta)d\varepsilon\right) - \int_0^{\varepsilon_F^o} \varepsilon N(\varepsilon)d\varepsilon = 0. \quad (13A.6)$$

So far we have no approximation. However, charge transfer should take place to bring the two Fermi levels to coincide, as illustrated in Fig. 13A.1(a). The kinetic energy increases upon the transfer of electrons. To discuss the kinetic energy, however, we should use Fig. 13A.1(b) in place of Fig. 13A.1(a). This is because the kinetic energy of both spin-up and spin-down electrons must be zero at the bottom of each band. Note in Fig. 13A.1(b) that the Fermi level after the charge transfer is shown by a thick line and is located in a different position for the spin-up and spin-down electrons. In other words, the

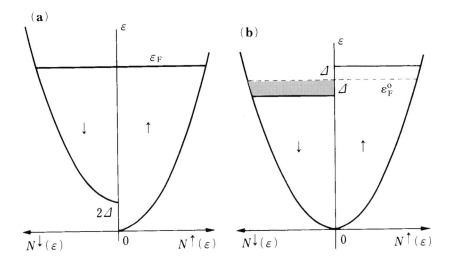

Figure 13A.1. (a) Shift of the spin-up and spin-down sub-bands due to the exchange energy. To evaluate the kinetic energy, (a) must be redrawn as (b), since the kinetic energy for spin-up and spin-down electrons is zero at the bottom of the band. The spin-down electrons in the shaded area are transferred to the spin-up band to establish a new Fermi level.

new Fermi level ε_F is formed after the transfer of spin-down electrons in the shaded area in (b) to the unoccupied state above ε_F^o in the spin-up band. The number of electrons in the shaded area is equal to $N[\varepsilon_F^o + (\Delta/2)]\Delta$ and its average energy is $[\varepsilon_F^o - (\Delta/2)]$. The average energy will increase to $[\varepsilon_F^o + (\Delta/2)]$ after the transfer to the spin-up band. The increase in the kinetic energy is, therefore, given by

$$\Delta \varepsilon_{kin} = \Delta \left\{ N\left(\varepsilon_F^o + \frac{\Delta}{2}\right) \cdot \left(\varepsilon_F^o + \frac{\Delta}{2}\right) - N\left(\varepsilon_F^o - \frac{\Delta}{2}\right) \cdot \left(\varepsilon_F^o - \frac{\Delta}{2}\right) \right\}$$

Its Taylor expansion leads to

$$\Delta \varepsilon_{kin} \approx \Delta^2 \left\{ N(\varepsilon_F^o) + \varepsilon_F^o \left(\frac{\partial N(\varepsilon)}{\partial \varepsilon}\right)_{\varepsilon = \varepsilon_F^o} \right\} \approx \Delta^2 N(\varepsilon_F^o), \quad (13A.7)$$

where the second term in the curly bracket is neglected.

The magnetization at absolute zero can be calculated from Fig. 13A.1(a) as

$$M(0) = \int_0^{\varepsilon_F} N(\varepsilon) d\varepsilon - \int_{2\Delta}^{\varepsilon_F} N(\varepsilon - 2\Delta) d\varepsilon = \int_0^{\varepsilon_F} N(\varepsilon) d\varepsilon - \int_0^{\varepsilon_F - 2\Delta} N(\varepsilon) d\varepsilon$$

$$= \left(\int_0^{\varepsilon_F} N(\varepsilon)d\varepsilon\right) - \left(\int_0^{\varepsilon_F} N(\varepsilon)d\varepsilon + \int_{\varepsilon_F}^{\varepsilon_F - 2\Delta} N(\varepsilon)d\varepsilon\right) \approx 2\Delta \cdot N(\varepsilon_F). \quad (13A.8)$$

Combining equations (13A.7) and (13A.8) yields equation (13.11) in the text. Note that the difference between $N(\varepsilon_F)$ and $N(\varepsilon_F^\circ)$ is only of the order of $\Delta/2\varepsilon_F$.

13.4 The differential cross-section can be expressed as

$$\sigma(\theta) = |f(\theta)|^2 = \frac{1}{k^2}\left|\sum_{\ell=0}^{\infty}(2\ell+1)e^{i\eta_\ell}\sin\eta_\ell P_\ell(\cos\theta)\right|^2. \quad (13A.9)$$

The Legendre functions are orthogonal with each other and satisfy the relations:

$$\int_0^\pi P_\ell(\cos\theta)P_{\ell'}(\cos\theta)\sin\theta d\theta = \int_{-1}^1 P_\ell(w)P_{\ell'}(w)dw = 0 \quad (\ell \neq \ell')$$

$$= \frac{2}{2\ell+1} \quad (\ell = \ell') \quad (13A.10)$$

The total cross-section σ given by $\sigma = 2\pi\int_0^\pi \sigma(\theta)\sin\theta d\theta$ is calculated by inserting equation (13A.9):

$$\sigma = 2\pi\int_0^\pi \sigma(\theta)\sin\theta d\theta = \frac{2\pi}{k^2}\sum_{\ell=0}^{\infty}\int_{-1}^1 (2\ell+1)^2\sin^2\eta_\ell P_\ell^2(w)dw$$

$$= \frac{2\pi}{k^2}\sum_{\ell=0}^{\infty}(2\ell+1)^2\sin^2\eta_\ell \cdot \frac{2}{2\ell+1} = \frac{4\pi}{k^2}\sum_{\ell=0}^{\infty}(2\ell+1)\sin^2\eta_\ell. \quad (13.22)$$

An insertion of equation (13A.7) into the transport cross-section given by $\sigma_{tr} = 2\pi\int_0^\pi \sigma(\theta)(1-\cos\theta)\sin\theta d\theta$ yields

$$\sigma_{tr} = \frac{2\pi}{k^2}\int_{-1}^1 \left[\sum_{\ell=0}^{\infty}(2\ell+1)e^{i\eta_\ell}\sin\eta_\ell P_\ell(w)\right]^*(1-w)\left[\sum_{\ell=0}^{\infty}(2\ell+1)e^{i\eta_\ell}\sin\eta_\ell P_\ell(w)\right]dw,$$

$$(13A.11)$$

where $w = \cos\theta$. The Legendre functions satisfy the following relations:

$$(\ell+1)P_{\ell+1} = (2\ell+1)wP_\ell - \ell P_{\ell-1},$$

$$\int_{-1}^1 wP_\ell^2 dw = \frac{1}{2\ell+1}\int_{-1}^1 [(\ell+1)P_{\ell+1}P_\ell + \ell P_{\ell-1}P_\ell]dw = 0$$

$$\int_{-1}^{1} wP_\ell P_{\ell+1} dw = \frac{1}{2\ell+1} \int_{-1}^{1} [(\ell+1)P_{\ell+1}P_{\ell+1} + \ell P_{\ell-1}P_{\ell+1}] dw$$

$$= \frac{1}{2\ell+1} \cdot (\ell+1) \cdot \frac{2}{2\ell+3} = \frac{2(\ell+1)}{(2\ell+1)(2\ell+3)}$$

Hence, equation (13A.11) is calculated as

$$\sigma_{tr} = \frac{2\pi}{k^2} \int_{-1}^{1} \left| \sum_{\ell=0}^{\infty} (2\ell+1) e^{i\eta_\ell} \sin\eta_\ell P_\ell(w) \right|^2 dw$$

$$- \frac{2\pi}{k^2} \int_{-1}^{1} \left[\sum_{\ell=0}^{\infty} (2\ell+1) e^{i\eta_\ell} \sin\eta_\ell P_\ell(w) \right]^* w \left[\sum_{\ell=0}^{\infty} (2\ell+1) e^{i\eta_\ell} \sin\eta_\ell P_\ell(w) \right] dw. \quad (13A.12)$$

The first term in equation (13A.12) denoted as I_1 is reduced to

$$I_1 = \frac{2\pi}{k^2} \int_{-1}^{1} \left| \sum_{\ell=0}^{\infty} (2\ell+1) e^{i\eta_\ell} \sin\eta_\ell P_\ell(w) \right|^2 dw = \frac{4\pi}{k^2} \sum_{\ell=0}^{\infty} (2\ell+1) \sin^2\eta_\ell$$

$$= \frac{2\pi}{k^2} \left[\sum_{\ell=0}^{\infty} 2(\ell+1)\sin^2\eta_\ell + \sum_{\ell=0}^{\infty} 2\ell \sin^2\eta_\ell \right]$$

$$= \frac{2\pi}{k^2} \left[\sum_{\ell=0}^{\infty} 2(\ell+1)\sin^2\eta_\ell + \sum_{\ell=0}^{\infty} 2(\ell+1)\sin^2\eta_{\ell+1} \right]$$

$$= \frac{4\pi}{k^2} \sum_{\ell=0}^{\infty} (\ell+1)(\sin^2\eta_\ell + \sin^2\eta_{\ell+1}) \quad (13A.13)$$

The second term in equation (13A.12) denoted as I_2 is reduced to

$$I_2 = -\frac{2\pi}{k^2} \int_{-1}^{1} \left[\sum_{\ell=0}^{\infty} \{(2\ell+1) e^{-i\eta_\ell} \sin\eta_\ell P_\ell(w)(2\ell+3) e^{i\eta_{\ell+1}} \sin\eta_{\ell+1} P_{\ell+1}(w) \right.$$

$$\left. + (2\ell+3) e^{-i\eta_{\ell+1}} \sin\eta_{\ell+1} P_{\ell+1}(w)(2\ell+1) e^{i\eta_\ell} \sin\eta_\ell P_\ell(w) \} \right] dw$$

$$= -\frac{2\pi}{k^2} \int_{-1}^{1} \sum_{\ell=0}^{\infty} (2\ell+1)(2\ell+3) \{ e^{i(\eta_\ell - \eta_{\ell+1})} + e^{-i(\eta_\ell - \eta_{\ell+1})} \}$$

$$\times \sin\eta_\ell \sin\eta_{\ell+1} P_{\ell+1}(w) P_\ell(w) dw$$

$$= -\frac{2\pi}{k^2}\sum_{\ell=0}^{\infty}(2\ell+1)(2\ell+3)2\cos(\eta_\ell - \eta_{\ell+1})\sin\eta_\ell \sin\eta_{\ell+1} \cdot \frac{2(\ell+1)}{(2\ell+1)(2\ell+3)}$$

$$= -\frac{2\pi}{k^2}\sum_{\ell=0}^{\infty}(\ell+1)\cos(\eta_\ell - \eta_{\ell+1})\sin\eta_\ell \sin\eta_{\ell+1}$$

$$= -\frac{2\pi}{k^2}\sum_{\ell=0}^{\infty}(\ell+1)(\cos\eta_\ell \cos\eta_{\ell+1} \sin\eta_\ell \sin\eta_{\ell+1} + \sin^2\eta_\ell \sin^2\eta_{\ell+1}). \quad (13A.14)$$

The transport cross-section σ_{tr} equal to $I_1 + I_2$ is now given by

$$\sigma_{tr} = \frac{4\pi}{k^2}\sum_{\ell=0}^{\infty}(\ell+1)(\sin^2\eta_\ell + \sin^2\eta_{\ell+1} - 2\cos\eta_\ell \cos\eta_{\ell+1}\sin\eta_\ell \sin\eta_{\ell+1}$$

$$- 2\sin^2\eta_\ell \sin^2\eta_{\ell+1})$$

$$= \frac{4\pi}{k^2}\sum_{\ell=0}^{\infty}(\ell+1)(\sin\eta_\ell \cos\eta_{\ell+1} - \cos\eta_\ell \sin\eta_{\ell+1})^2 = \frac{4\pi}{k^2}\sum_{\ell=0}^{\infty}(\ell+1)\sin^2(\eta_\ell - \eta_{\ell+1}).$$

$$(13.23)$$

Chapter Fourteen

14.1 Note that the spin function is expressed as $|\uparrow\rangle = \begin{pmatrix} 1 \\ 0 \end{pmatrix}$ and $|\downarrow\rangle = \begin{pmatrix} 0 \\ 1 \end{pmatrix}$ in matrix form. The spin function of interest to us is given by

$$\chi_s = \frac{1}{\sqrt{2}}(|\uparrow\downarrow\rangle + |\downarrow\uparrow\rangle) = \frac{1}{\sqrt{2}}\left\{\begin{pmatrix} 1 \\ 0 \end{pmatrix}_1 \begin{pmatrix} 0 \\ 1 \end{pmatrix}_2 + \begin{pmatrix} 0 \\ 1 \end{pmatrix}_1 \begin{pmatrix} 1 \\ 0 \end{pmatrix}_2 \right\}. \quad (14A.1)$$

A straightforward calculation easily results in $S^2\chi_s = 2\hbar^2\chi_s$ and $S_z\chi_s = 0$.

Chapter Fifteen

15.1 The observed negative TCR in amorphous alloys is of the order of 10^{-4}–10^{-5}/K and is one or two orders of magnitude smaller than the positive TCR in a metal crystal. Hence, we are discussing a very small effect in amorphous alloys. The Debye–Waller factor arises from the dynamical structure factor $a(K,\omega)$ in the Baym resistivity formula (see equation (10.96) and Section 10.12) and appears both in the Bloch–Grüneisen formula for a crystal (follow the derivation from equations (10.104) to (10.106)) and in the

Baym–Meisel–Cote formula for amorphous alloys (see equations (15.23) to (15.25)).

The structure factor $a(K,\omega)$ at 0 K consists of a series of delta functions at reciprocal lattice vectors in a crystal whereas it is a continuous function of the wave vector K in an amorphous solid. At finite temperatures, the Debye–Waller factor in $a(K,\omega)$ reduces the intensity of the diffraction peak at the reciprocal lattice vector in crystals. But we know that this does not affect the resistivity in a crystal. However, the Debye–Waller factor reduces the whole spectrum of $a(K,\omega)$ in amorphous alloys, contributing to a decrease in resistivity with increasing temperature.

15.2 The g-parameter of the amorphous phase may be well approximated as unity because of the destruction of the Brillouin zone. Then the g-parameter of the quasicrystalline phase is given by the ratio of the measured electronic specific heat coefficients of the two phases, i.e., $g = \gamma_{QC}/\gamma_{Amo}$. The Mott conductivity formula leads to the relation $\rho_{QC}/\rho_{Amo} = 1/g^2$. We obtain $\rho_{QC} = \rho_{Amo}/g^2 = 220/(0.42/0.78)^2 = 758\,\mu\Omega$-cm. This is in good agreement with the measured resistivity of $780\,\mu\Omega$-cm (see Fig. 15.24).

References

Chapter One

1. M. Planck, *Ann. Physik*, **4** (1901) 553
2. A. Einstein, *Ann. Physik*, **17** (1905) 132
3. N. Bohr, *Phil. Mag.* **26** (1913) 1
4. J. Franck and G. Hertz, *Verhandl. Deut. Physik. Ges.* **16** (1914) 512
5. A. H. Compton, *Phys. Rev.* **21** (1923) 715; **22** (1923) 409
6. W. Pauli, *Z. Physik*, **31** (1925) 765
7. L. de Broglie, *Phil. Mag.* **47** (1924) 446
8. C. Davisson and L. H. Germer, *Phys. Rev.* **30** (1927) 705
9. W. Heisenberg, *Z. Physik*, **33** (1925) 879
10. E. Schrödinger, *Ann. Physik*, **79** (1926) 361, 489, 734
11. P. Zeeman, *Phil. Mag.*, *Ser.* 5, **43** (1897) 226
12. J. J. Thomson, *Phil. Mag.*, *Ser.* 5, **44** (1897) 293
13. P. Drude, *Ann. Physik.* **1** (1900) 566
14. H. A. Lorentz, *The Theory of Electrons and its Applications to the Phenomena of Light and Radiant Heat* (Second Edition, Dover Publications, New York, 1952)
15. W. Heisenberg, *Z. Physik* **43** (1927) 172
16. E. Fermi, *Z. Physik*, **36** (1926) 902
17. P. A. M. Dirac, *Proc. Roy. Soc.* (London) **112** (1926) 661
18. W. Pauli, *Z. Physik.* **41** (1927) 81
19. A. Sommerfeld, *Z. Physik*, **47** (1928) 1; A. Sommerfeld and H. Bethe, *Elektronentheorie der Metalle* (Springer-Verlag, 1967)
20. F. Bloch, *Z. Physik*, **52** (1928) 555
21. A. H. Wilson, *Proc. Roy. Soc.* (London) **A133** (1931) 458
22. N. F. Mott and H. Jones, *The Theory of the Properties of Metals and Alloys* (Clarendon Press, Oxford, 1936; Dover Publications, 1958)
23. A. H. Wilson, *The Theory of Metals* (Cambridge University Press, First edition 1936, Second edition 1953)
24. J. Bardeen and W. H. Brattain, *Phys. Rev.* **74** (1948) 230; *Phys. Rev.* **75** (1949) 1208
25. J. Bardeen, L. N. Cooper and J. R. Schrieffer, *Phys. Rev.* **106** (1957) 162; *Phys. Rev.* **108** (1957) 1175
26. J. G. Bednorz and K. A. Müller, *Z. Physik. B–Condensed Matter* **64** (1986) 189
27. D. Shechtman, I. Blech, D. Gratias and J. W. Cahn, *Phys. Rev. Lett.* **53** (1984) 1951

28. C. Kittel, *Introduction to Solid State Physics* (John Wiley & Sons, Inc. Seventh Edition, New York, 1996)
29. N. W. Ashcroft and N. D. Mermin, *Solid State Physics* (Saunders College, West Washington Square, Philadelphia, PA 19105 1976)
30. C. S. Barrett and T. B. Massalski, *Structure of Metals* (McGraw-Hill, New York, First edition 1943, Third edition 1966)
31. H. Smith and H. H. Jensen, *Transport Phenomena* (Clarendon Press, Oxford, 1989)
32. J. M. Ziman, *Principles of the Theory of Solids* (Cambridge University Press, First edition 1964, Second edition 1972)

Chapter Two

1. L. Pauling, *The Nature of the Chemical Bonding* (Cornell Univ. Press, Third Edition, 1960)
2. A. J. Dekker, *Solid State Physics* (Prentice Hall, Inc., 1957)
3. C. Kittel, *Introduction to Solid State Physics* (John Wiley & Sons, Inc., Sixth Edition, 1986)

Chapter Three

1. For example, A. H. Wilson, *Thermodynamics and Statistical Mechanics* (Cambridge University Press, 1966)
2. T. B. Massalski and U. Mizutani, *Prog. Mat. Sci.* **22** (1978) 151–262
3. L. Landau, *Z. Physic.* **64** (1930) 629
4. C. Kittel, *Introduction to Solid State Physics* (John Wiley & Sons, Inc., New York, Seventh Edition, 1996) pp. 417–419
5. S. Dushman, *Rev. Mod. Phys.* **2** (1930) 381

Chapter Four

1. C. Kittel, *Introduction to Solid State Physics* (John Wiley & Sons, Inc., New York, Seventh Edition, 1996)
2. M. Suganuma, T. Kondow and U. Mizutani, *Phys. Rev.* **B23** (1981) 706
3. H. B. Huntington, *Solid State Physics* vol. 7, edited by F. Seitz and D. Turnbull, (Academic Press, 1958) pp. 214–349

Chapter Five

1. C. Kittel, *Introduction to Solid State Physics* (John Wiley & Sons Inc., New York, Seventh Edition, 1996) pp. 177–179

Chapter Six

1. A. P. Cracknell, *The Fermi Surfaces of Metals* (Taylor & Francis Ltd, London, 1971)

2. M. J. G. Lee, *Proc. Roy. Soc. (London)* **A295** (1966) 440
3. A. B. Pippard, *Phil. Trans. Roy. Soc.* **A250** (1957) 325
4. D. Shoenberg, *Phil. Mag.* **5** (1960) 105
5. J. C. Slonczewski and P. R. Weiss, *Phys. Rev.* **109** (1958) 272
6. M. S. Dresselhaus, G. Dresselhaus and J. E. Fischer, *Phys. Rev.* **B15** (1977) 3180
7. N. F. Mott and H. Jones, *The Theory of the Properties of Metals and Alloys* (Clarendon Press, Oxford, 1936)
8. H. Jones, *The Theory of Brillouin Zones and Electronic States in Crystals* (North-Holland, Amsterdam, 1975)
9. C. Kittel, *Introduction to Solid State Physics* (John Wiley & Sons, Inc., Seventh Edition, New York, 1996)

Chapter Seven

1. D. Shoenberg, *Magnetic Oscillations in Metals* (Cambridge University Press, 1984)
2. I. M. Lifshitz and A. M. Kosevich, *Soviet Physics (JETP)* **2** (1956) 636
3. S. Berko, *Compton Scattering*, edited by B. Williams, (McGraw-Hill, 1977), Chapter 9, pp. 273–322
4. R. H. Stuewer and M. J. Cooper, *Compton Scattering*, edited by B. Williams, (McGraw-Hill, 1977), pp. 1–27
5. B. Feuerbacher, B. Fitton and R. F. Willis, *Photoemission and the Electronic Properties of Surface* (John Wiley & Sons, Chichester, 1978)
6. *Photoemission in Solids I*, Topics in Applied Physics, vol. 26, edited by M. Cardona and L. Ley, (Springer-Verlag, Berlin, 1978)
7. S. Hüfner, *Photoelectron Spectroscopy*, Springer Series in Solid State Physics 82, edited by M. Cardona, (Springer-Verlag, Berlin, 1995)
8. J. J. Yeh and I. Lindau, At. *Data and Nucl. Data Tables* **32** (1985) 1
9. C. N. Berglund and W. E. Spicer, *Phys. Rev.* **136** (1964) A1030
10. J. K. Lang, Y. Baer and P. A. Cox, *J. Phys. F: Metal Phys.* **11** (1981) 121
11. N. W. Ashcroft and N. D. Mermin, *Solid State Physics*, Saunders College, West Washington Square, Philadelphia, PA 19105 (1976), pp. 308–309
12. *X-ray Spectroscopy*, edited by L. V. Azaroff, (McGraw Hill, 1974)
13. C. Bonnelle, *Annual Report C, The Royal Society of Chemistry*, (1987) pp. 201–272
14. T. Fukunaga, H. Sugiura, N. Takeichi and U. Mizutani, *Phys. Rev.* **B54** (1996) 3200
15. R. D. Leapman, L. A. Grunes and P. L. Fejes, *Phys. Rev.* **B26** (1982) 614

Chapter Eight

1. J. S. Faulkner, *Prog. Mat. Sci.* **27** (1982) pp. 1–187
2. S. Raimes, *The Wave Mechanics of Electrons in Metals* (North-Holland, 1970) pp. 102–105
3. D. Bohm and D. Pines, *Phys. Rev.* **82** (1951) 625; ibid **85** (1952) 338; ibid **92** (1953) 609
4. P. Nozieres and D. Pines, *Phys. Rev.* **111** (1958) 442.
5. W. Kohn and L. J. Sham, *Phys. Rev.* **140** (1965) A1133
6. M. Schlüter and L. J. Sham, *Physics Today,* **35** (February, 1982) 36–43

7. P. Hohenberg and W. Kohn, *Phys. Rev.* **136** (1964) B864
8. V. L. Moruzzi, A. R. Williams and J. F. Janak, *Phys. Rev.* **B15** (1977) 2854
9. N. D. Lang, *Solid State Physics*, **28** (1973), edited by H. Ehrenreich, F. Seitz and D. Turnbull, pp. 225–300
10. U. von Barth and L. Hedin, *J. Phys.:Solid State Phys.* **5** (1972) 1629
11. V. Heine, *Solid State Physics* **24** (1970), edited by H. Ehrenreich, F. Seitz and D. Turnbull, pp. 1–36; M. L. Cohen and V. Heine, ibid pp. 37–248; V. Heine and D. Weaire, ibid pp. 249–463
12. W. Kohn and N. Rostoker, *Phys. Rev.* **94** (1954) 1111
13. J. M. Ziman, *Principles of the Theory of Solids* (Cambridge University Press, Second Edition, 1972)
14. O. K. Andersen, *Phys. Rev.* **B12** (1975) 3060
15. O. K. Andersen, O. Jepsen and D. Glötzel, in *Highlights of Condensed-Matter Theory*, edited by F. Bassani, F. Fumi and M. P. Tosi (North Holland, New York 1985)
16. H. L. Skriver, *The LMTO Method*, Springer Series in Solid State Physics, vol. 41 (1984), edited by M. Cardona, P. Fulde, K. von Klitzing and H. -J. Queisser

Chapter Nine

1. T. B. Massalski and U. Mizutani, *Prog. Mat. Sci.*, **22** (1978) pp. 151–262
2. H. Jones, *The Theory of Brillouin Zones and Electronic States in Crystals* (North-Holland, 1975), pp. 210–219
3. V. Heine and D. Weaire, *Pseudopotential Theory of Cohesion and Structure* Solid State Physics, vol. 24, (Academic Press, 1970) pp. 250–463
4. A. Messiah, *Quantum Mechanics* (John Wiley & Sons Inc., New York, 1986), see Chapter XIX.
5. J. S. Faulkner, *Prog. Mat. Sci.*, **27** (1982) pp. 1–187
6. N. F. Mott and H. Jones, *The Theory of the Properties of Metals and Alloys* (Clarendon Press, Oxford, 1936; Dover Publications, 1958)

Chapter Ten

1. S. Raimes, *The Wave Mechanics of Electrons in Metals* (North-Holland, 1970)
2. J. M. Ziman, *Principles of the Theory of Solids* (Cambridge University Press, 1964)
3. J. M. Ziman, *Phil. Mag.* **6** (1961) 1013
4. L. I. Schiff, *Quantum Mechanics* (McGraw-Hill, New York, 1955)
5. N. F. Mott, *Metal–Insulator Transitions* (Taylor & Francis Ltd, London, 1990)
6. M. Itoh, private communications
7. N. W. Ashcroft and N. D. Mermin, *Solid State Physics* Saunders College, West Washington Square, Philadelphia, PA 19105 (1976)
8. C. Kittel, *Quantum Theory of Solids* (John Wiley & Sons, Inc., New York, 1963) pp. 368–370
9. G. Baym, *Phys. Rev.* **135** (1964) A1691

Chapter Eleven

1. J. M. Ziman, *Principles of the Theory of Solids* (Cambridge University Press, 1964)
2. K. Mendelssohn and H. M. Rosenberg, *Solid State Physics* **12**, edited by F. Seitz and D. Turnbull, (Academic Press, New York, 1961) pp. 223–274
3. H. M. Rosenberg, *Low Temperature Solid State Physics* (Clarendon Press, Oxford, 1963)
4. C. Kittel, *Introduction to Solid State Physics* (John Wiley & Sons, Inc., New York, Seven Edition, 1996)
5. H. Sato, T. Matsuda and U. Mizutani, *Physica B* **144** (1987) 173
6. N. W. Ashcroft and N. D. Mermin, *Solid State Physics* Saunders College, West Washington Square, Philadelphia, PA 19105 (1976)
7. H. Smith and H. Hojgaard Jensen, *Transport Phenomena* (Clarendon Press, Oxford 1989)
8. G. Mahan, B. Sales and J. Sharp, *Physics Today* (March 1997) 42–47
9. F. Wooten, *Optical Properties of Solids* (Academic Press, 1972)
10. P. O. Nilsson, *Optical Properties of Metals and Alloys*, Solid State Physics, (Academic Press, 1974) vol. 29, pp. 139–234
11. P. Nozieres and D. Pines, *Phys. Rev.* **113** (1959) 1254
12. H. Ehrenreich, H. R. Philipp and B. Segall, *Phys. Rev.* **132** (1963) 1918
13. E. A. Taft and H. R. Philipp, *Phys. Rev.* **138** (1965) A197
14. R. Kubo, *J. Phys. Soc. Jpn* **12** (1957) 570
15. A. Einstein, *Ann. Physik* **17** (1905) 549
16. J. M. Ziman, *Elements of Advanced Quantum Theory* (Cambridge University Press, 1969) pp. 94–104

Chapter Twelve

1. H. Kamerlingh-Onnes, *Leiden Comm.* **122b** (1911) 124c
2. J. G. Bednorz and K. A. Müller, *Z. Physik. B – Condensed Matter* **64** (1986) 189
3. J. Bardeen, L. N. Cooper and J. R. Schrieffer, Phys. Rev. **106** (1957) 162; ibid **108** (1957) 1175
4. W. Meissner and R. Ochsenfeld, *Naturwissen*, **21** (1933) 787
5. F. London and H. London, *Proc. Roy. Soc. (London)* **A149** (1935) 71
6. F. London, *Superfluids*, vol. 1, (Dover Publications, 1960)
7. V. L. Ginzburg and L. D. Landau, *Soviet Physics, JETP*, **20** (1950) 1064
8. H. Fröhlich, *Phys. Rev.* **79** (1950) 845
9. L. Hoddeson, *MRS Bulletin*/January (1999) pp. 50–55
10. B. S. Deaver and W. M. Fairbank, *Phys. Rev. Lett.* **7** (1961) 43; R. Doll and M. Näbauer, ibid **7** (1961) 51
11. A. A. Abrikosov, *Fundamentals of the Theory of Metals* (North-Holland, 1988)
12. A. C. Rose-Innes and E. H. Rhoderick, *Introduction to Superconductivity* (Pergamon Press, Second edition, 1978)
13. B. D. Josephson, *Phys. Letters* **1** (1962) 1; ibid **1** (1962) 251
14. M. Murakami, *Prog. Mat. Sci.* **38** (1994) 311
15. C. P. Bean, *Phys. Rev. Lett.* **8** (1962) 250
16. U. Mizutani, A. Mase, H. Ikuta, Y. Yanagi, M. Yoshikawa, Y. Itoh and T. Oka, *Mat. Sci. Eng.* **B56** (1999) 400

Chapter Thirteen

1. A. H. Morrish, *The Physical Principles of Magnetism* (John Wiley & Sons, Inc., New York, 1965)
2. S. Chikazumi, *Physics of Magnetism* (John Wiley & Sons, Inc., New York, 1964)
3. R. M. White, *Quantum Theory of Magnetism* Springer Series in Solid State Sciences vol. 32, (Springer, Berlin, Heidelberg, 1983)
4. G. T. Rado and H. Suhl eds, *Magnetism* vols. I–V, (Academic Press, New York, 1963–1973)
5. E. P. Wohlfarth and K. H. J. Buschow eds., *Ferromagnetic Materials*, vols. 1–7 (North Holland, Amsterdam, 1980–1993); S. Legvold, Rare earth metals and alloys in vol. 1 (1980); I. A. Campbell and A. Fert, Transport properties of ferromagnets in vol. 3, Chapter 9, (1982)
6. T. Moriya, *Spin Fluctuations in Itinerant Electron Magnetism*, Springer Series in Solid-State Sciences vol. 56 (Springer, Berlin, Heidelberg, 1985)
7. E. C. Stoner, *Proc. Roy. Soc.* (London) **A165** (1938)372; D. H. Martin, *Magnetism in Solids* (Iliffe Books, London, 1967) pp. 227–238
8. N. F. Mott, *Adv. Phys.* **13** (1964) 325
9. L. I. Schiff, *Quantum Mechanics* (McGraw-Hill, Second edition, 1955)
10. J. Friedel, *Can. J. Phys.* **34** (1956) 1190; *Nuovo Cimento*, Suppl. **7** (1958) 287
11. J. Friedel, *Phil. Mag.* **43** (1952) 153
12. A. C. Hewson, *The Kondo Problem to Heavy Fermions* (Cambridge University Press, 1993)
13. P. W. Anderson, *Phys. Rev.* **124** (1961) 41
14. J. Kondo, *Prog. Theor. Phys.* **32** (1964) 37; *Solid State Physics*, vol. 23, edited by F. Seitz, D. Turnbull and H. Ehrenreich (Academic Press, London, 1969); Kinzoku Densi Ron (Shokabo, Tokyo 1983) (in Japanese)
15. A. J. Heeger, *Solid State Physics*, vol. 23, edited by F. Seitz, D. Turnbull and H. Ehrenreich, (Academic Press, London, 1969)
16. K. Sato, T. Fujita, Y. Maeno, Y. Onuki and T. Komatsubara, *J. Phys. Soc. Jpn* **58** (1989) 1012; A. Sumiyama, Y. Oda, H. Nagano, Y. Onuki, K. Shibutani and T. Komatsubara, ibid **55** (1986) 1294
17. M. A. Ruderman and C. Kittel, *Phys. Rev.* **96** (1954) 99
18. T. Kasuya, *Prog. Theor. Phys.* **16** (1956) 45; ibid **16** (1956) 58
19. K. Yosida, *Phys. Rev.* **106** (1957) 893
20. J. W. F. Dorleijn and A. R. Miedema, *J. Phys. F: Metal Phys.* **5** (1975) 487; ibid **5** (1975) 1543; *Phys. Letters* **55A** (1975) 118
21. M. N. Baibich, J. M. Broto, A. Fert, F. Nguyen Van Dau, F. Petroff, P. Etienne, G. Creuzet, A. Friederich and J. Chazelas, *Phys. Rev. Lett.* **61** (1988) 2472
22. M. A. M. Gijs and G. E. W. Bauer, *Advances in Physics* **46** (1997) 285
23. C. M. Hurd, *The Hall Effect in Metals and Alloys* (Plenum Press, New York, 1972)

Chapter Fourteen

1. N. F. Mott and E. A. Davis, *Electronic Processes in Non-Crystalline Materials* (Clarendon Press, Oxford, 1971) p. 121
2. A. A. Abrikosov, *Fundamentals of the Theory of Metals* (North-Holland, Amsterdam, 1988) p. 19; N. W. Ashcroft and N. D. Mermin, *Solid State Physics*,

Saunders College, West Washington Square, Philadelphia, PA 19105 (1976) p. 346
3. N. W. Ashcroft and N. D. Mermin, *Solid State Physics*, Saunders College, West Washington Square, Philadelphia, PA 19105 (1976) p. 676
4. Y. Tokura and T. Arima, *Japan. J. Appl. Phys.*, **29** (1990) 2388
5. L. F. Mattheiss, *Phys. Rev. Letters* **58** (1987) 1028
6. J. Hubbard, *Proc. Roy. Soc. (London)* **A276** (1963) 238; ibid **A277** (1964) 237; ibid **A281** (1964) 401
7. L. Pauling, *The Nature of the Chemical Bond* (Cornell University Press, 1967), p. 63
8. W. A. Harrison, *Solid State Theory* (MacGraw-Hill, 1970) pp. 76–78
9. D. B. McWhan, A. Menth, J. P. Remeika, W. F. Brinkman and T. M. Rice, *Phys. Rev.* **B7** (1973) 1920
10. J. Zannen, G. A. Sawatzky and J. W. Allen, *Phys. Rev. Letters* **55** (1985) 418
11. A. Fujimori, *Strong Correlation and Superconductivity*, Springer Series in Solid-State Sciences 89, edited by H. Fukuyama, S. Maekawa and A. P. Malozemoff, (Spring-Verlag, Berlin, 1989)
12. *Spectroscopy of Mott Insulator and Correlated Metals*, Springer Series in Solid-State Sciences 119, edited by A. Fujimori and Y. Tokura, (Spring-Verlag, Berlin, 1995)

Chapter Fifteen

1. N. F. Mott and H. Jones, *The Theory of the Properties of Metals and Alloys*. Reprinted Edition (Dover Publications, 1958)
2. D. Shechtman, I. Blech, D. Gratias and J. W. Cahn, *Phys. Rev. Lett.* **53** (1984) 1951
3. D. Levine and P. J. Steinhardt, *Phys. Rev. Lett.* **53** (1984) 2477
4. Y. Waseda, *The Structure of Non-Crystalline Materials, Liquids and Amorphous Solids* (MacGraw-Hill, New York, 1980); *Prog. Mat. Sci.* **26** (1981) 1
5. N. W. Ashcroft and D. C. Langreth, *Phys. Rev.* **156** (1967) 685; ibid **159** (1967) 500
6. T. E. Faber and J. M. Ziman, *Phil. Mag.* **11** (1965) 153
7. U. Mizutani and C. H. Lee, *Mater. Trans*, JIM **36** (1995) 210
8. W. Buckel and R. Hilsch, *Z. Physik*, **138** (1954) 109
9. P. Duwez, R. H. Willens and W. Klement Jr, *J. Appl. Phys.* **31** (1960) 1136; W. Klement Jr, R. H. Willens and P. Duwez, *Nature* **187** (1960) 869
10. R. Pond Jr and R. Maddin, *Trans. Met. Soc. AIME* **245** (1969) 2475
11. H. S. Chen and C. E. Miller, *Rev. Sci. Instrum.* **41** (1970) 1237
12. U. Mizutani, *Prog. Mat. Sci.* **28** (1983) 97; *Phys. Stat. Sol.* (b) **176** (1993) 9
13. J. Hafner, S. S. Jaswal, M. Tegze, A. Pflugi, J. Krieg, P. Oelhafen and H. -J. Güntherodt, *J. Phys. F: Met. Phys.* **18** (1988) 2583
14. N. Takeichi, H. Sato and U. Mizutani, *J. Phys.: Condensed Matter* **9** (1997) 10145
15. T. Fukunaga, H. Sugiura, N. Takeichi and U. Mizutani, *Phys. Rev.* **B54** (1996) 3200
16. H. Tanaka and M. Itoh, *Phys. Rev. Lett.* **81** (1998) 3727; H. Tanaka, *Phys. Rev.* **B57** (1998) 2168; M. Itoh, ibid **B45** (1992) 4241
17. J. M. Ziman, *Phil. Mag.* **6** (1961) 1013
18. J. M. Ziman, *Proc. of the 2nd Int. Conf. on The Properties of Liquid Metals*, edited by S. Takeuchi, (Tokyo, 1972, Taylor & Francis Ltd, London)

19. P. W. Anderson, *Phys. Rev.* **109** (1958) 1492
20. N. F. Mott, *Phil. Mag.* **13** (1966) 989; ibid **19** (1969) 835; ibid **26** (1972) 1015; ibid **B44** (1981) 265; *Metal–Insulator Transitions* (Taylor & Francis Ltd, 1990)
21. L. V. Meisel and P. J. Cote, *Phys. Rev.* **B16** (1977) 2978; ibid **B17** (1978) 4652
22. N. F. Mott, *Phil. Mag.* **26** (1972) 1249
23. G. Bergman, *Phys. Rev.* **B28** (1983) 2914
24. B. L. Altshuler and A. G. Aronov, *Electron–Electron Interactions in Disordered Systems*, edited by A. J. Efros and M. Pollak, (Elsevier Science Pub., 1985) pp. 1–153
25. N. F. Mott, *J. Non-Cryst. Solids* **1** (1968) 1; *Phil. Mag.* **19** (1969) 835; V. Ambegaokar, B. I. Halperin and J. S. Langer, *Phys. Rev.* **B4** (1971) 2612; M. Pollack, *J. Non-Cryst. Solids* **11** (1972) 1
26. U. Even and J. Jortner, *Phys. Rev. Lett.* **28** (1972) 31
27. *Physical Properties of Quasicrystals*, edited by Z. M. Stadnik, Solid State Sciences, vol. 126, (Springer-Verlag, Berlin, 1999)
28. C. Janot, *Quasicrystals, A Primer*, (Clarendon Press, Oxford 1992), p. 34
29. V. Elser and C. L. Henley, *Phys. Rev. Lett.* **55** (1985) 2883
30. U. Mizutani, W. Iwakami, T. Takeuchi, M. Sakata and M. Takata, *Phil. Mag. Lett.* **76** (1997) 349
31. T. Takeuchi and U. Mizutani, *Phys. Rev.* **B52** (1995) 9300
32. T. Fujiwara and T. Yokokawa, *Phys. Rev. Lett.* **66** (1991) 333
33. Z. M. Stadnik, D. Purdie, M. Garnier, Y. Baer, A. -P. Tsai, A. Inoue, K. Edagawa, S. Takeuchi and K. H. J. Buschow, *Phys. Rev.* **B55** (1997) 10938
34. E. Belin-Ferré and A. Traverse, *J. Phys.:Condensed Matter* **3** (1991) 2157; E. Belin-Ferré, *Mat. Res. Soc. Symp. Proc.* **553** (Materials Research Society, Boston, 1999), pp. 347–358
35. U. Mizutani, *J. Phys.:Condensed Matter* **10** (1998) 4609
36. M. Terauchi, M. Tanaka, A. P. Tsai, A. Inoue and T. Masumoto, *Phil. Mag. Lett.* **74** (1996) 107
37. M. Ahlgren, C. Gignoux, M. Rodmar, C. Berger and Ö. Rapp, *Phys. Rev.* B55 (1997) R11915; J. C. Lajaunias, *Proc. of 9th Int. Conf. on Rapidly Quenched Materials*, (Bratislava, 1996)
38. U. Mizutani, *Mat. Sci. Eng.*, **294–296** (2000) 464 presented at 7th Int. Conf. on Quasicrystals, (Stuttgart, 1999)
39. U. Mizutani, T. Ishizuka and T. Fukunaga, *J. Phys.:Condensed Matter* **9** (1997) 5333
40. A. Y. Rogatchev, T. Takeuchi and U. Mizutani, *Phys. Rev.* **B61** (2000) 10010
41. Y. Tokura, Y. Taguchi, Y. Okada, Y. Fujimori, T. Arima, K. Kumagai and Y. Iye, *Phys. Rev. Lett.* **70** (1993) 2126
42. Y. Nishino, M. Kato, S. Asano, K. Soda, M. Hayasaki and U. Mizutani, *Phys. Rev. Lett.* **79** (1997) 1909
43. T. Hihara, K. Sumiyama, H. Yamauchi, Y. Homma, T. Suzuki, K. Suzuki, *J. Phys.:Condensed Matter* **5** (1993) 8425
44. T. Biwa, M. Yui, T. Takenchi and U. Mizutani, to be published in *Mater. Trans. JIM* (2001)

Materials index

α-phase Al–Mn–Si 498
α-phase Cu–Zn 236
Ag 130, 477
Ag–Pd 246
Al 135, 146, 206, 305, 325, 327, 342, 477
Al–Cu–Fe 494, 500
Al–Cu–Ru 506
Al–Li–Cu 494
Al–Li–Cu 500
Al–Mg–Ag 494, 498
Al–Mg–Cu 498
Al–Mg–Pd 498
Al–Mg–Zn 494, 498, 504
Al–Mn, Al_{86}–Mn_{14} 451, 498
Al–Pd–Mn 494
Al–Pd–Mn 506
Al–Pd–Re 494
Al–Pd–Re 506, 510
$Al_x Mg_{39.5} Zn_{60.5-x}$ 498
amorphous systems
  (Ag–Cu)–Ge ; $(Ag_{0.5}Cu_{0.5})_{100-x}Ge_x$ 491–3, 508
  $(Ag_{0.5}Cu_{0.5})_{77.5}Si_{22.5}$ 508
  (Ca–Mg)–Al 491–2
  Al–(Cu–Y) 470
  Al–Mg–Pd 502
  $Al_{85-x}Ni_{15}Si_x$ 508
  $Al_{90-x}Ni_{10}Si_x$ 508
  $Au_{70}Si_{30}$ 463
  $Ce_w Si_{100-w}$ 513
  $Cu_{40}Y_{60}$ 470
  Mg–(Cu–Y) 470
  Mg–Zn 467
  $Mg_{70}Zn_{30}$ 469, 473
  $Mg_{70}Zn_{30-x}Sn_x$ 508
  Ni 456
  $Ni_{81}B_{19}$ 460
  $Ti_x Si_{100-x}$ 512, 515
  $V_x Si_{100-x}$ 510, 512, 515
As 141, 145
Au 130, 229, 290
Au–(V) 417
Au–Cu 228, 245
Au_3Cu 231
$AuCu_3$ 231

AuCuI 231
AuCuII 231

B 146
Be 133, 239
β′-phase Cu–Zn alloy 237
Bi 141, 148
Bi–Sr–Ca–Cu–O 376
Bi_2Te_3 312

Ca 132
Cd 133, 316, 477
Ce 171
$Ce_x La_{1-x} Cu_6$ 418
Co 386, 395, 430
Cr_2O_3 447
Cs 126
Cu 80, 130, 159, 168, 186, 229, 253, 290, 413, 477
Cu–Ga 232
Cu–Ge 232
Cu–(Mn) 417
Cu–Ni 246, 248
Cu–Zn 187, 232, 241

demagnetized Ni 420
diamond 81, 143
Dy 419

Er 419

Fe 386, 395, 399, 413
Fe–Co 398
Fe–Ni 398, 420
$FeSi_2$ 312

Ga 135, 146
Gd 171, 401, 418, 419
Ge 143
graphite 137, 325, 327

helium 15
Hg 360
  liquid 491

$HgBa_2Ca_2Cu_3O_8$ 376
Ho 419

In 135, 146, 477

K 126
KCL 14

La 171
La–Ba–Cu–O 334, 376
La_2CuO_4 440
$La_{2-x}Sr_xCuO_4$ 439, 447, 448
$(La_{1-x}Sr_x)_2CuO_4$ 376
lead 81
Li 126
liquid mercury (Hg) 491
liquid Ni 456
liquid Sn 166

Mg 133
Mn 413
Mo 386

Na 126, 291, 477 see also sodium
NaCl 14
Nb 346, 386
Nb–Ti 366, 368, 377
Nb_3Ge 376
Nb_3Sn 366, 368, 377
$Nd_{2-y}Ce_yCuO_{4-\delta}$ 448
Ni 186, 386, 395, 399, 400
  liquid 456
Ni–Co 398, 420, 425
Ni–Cu 398
Ni–Fe 425
$Ni_{1-x}Cu_x$ 397
$Ni_{1-x}Zn_x$ 397
$Ni_{97}Al_3$ 428
$Ni_{70}Co_{30}$ 422
$Ni_{97}Co_{3-x}Rh_x$ 423
$Ni_{97}Cr_{3-x}M_x$ (M = Al, Fe, Mn, Ti, etc.) 423
NiO 432, 447

ω-phase Al_7Cu_2Fe 500

P 145
Pb 137, 360, 362
Pb–PbO–Pb 368
Pd 386, 400
Pt 290, 386

Rb 126

Sb 141, 145
Sc_3In 385
Si 143, 168
Sn 360, 477
Sn(α) 143
sodium 5, 22, 26 see also Na
Sr 132
$Sr_{1-z}La_zTiO_3$ 512

Tb 419
Ti 386
Ti_2O_3 447
Tl 135
Tl–Ba–Ca–Cu–O 376
Tm 419

V 168, 386
V_2O_3 444, 447

$YBa_2Cu_3O_7$; $YBa_2Cu_3O_{7-\delta}$ 376, 447

ζ-phase Ag–Al 239, 241
ζ-phase Cu–Ge 239, 241
Zn 41, 133, 239, 316, 477
Zr 386
$ZrZn_2$ 385

Subject index

1/1-1/1-1/1 approximant 498–9, 503
2/1-2/1-2/1 approximant 498–9
$2k_F$ scattering 485
$2p_\sigma$ orbital 440–3
3/2-2/1-2/1 approximant 499
3d-transition metal oxides 444
$3d_{x^2-y^2}$ orbital of Cu site 440–3

α-phase 232
Abrikosov theory 362
absolute thermoelectric power 303
absorption edge 318
absorption of infrared electromagnetic wave 347
absorption spectra of Cu–Zn alloys 318
absorption spectrum 184, 187
absorptivity 187
AC conductivity 324
AC Josephson effect 370, 372–3
acceptor level 147
acoustic branch 78
acoustic phonons 77, 79
actinide metals 386
Al-3p partial density of states 179
Al-Kα line 163, 177, 179
Al-Kβ line spectrum 177, 179
alkali metals 126
amorphous alloy 291, 309, 386, 451, 462 ff.
amorphous alloys in group (V) 479–80, 491
amorphous film 462
amorphous metals 8
amplitude of reflectance 324
analyzer sensitivity 165
Anderson localization 485
  theory 474, 483–4
Anderson model 413
angular correlation curve 158, 242
angular-integrated photoemission spectroscopy 172
angular-resolved photoemission spectroscopy (ARPES) 172
anomalous Hall coefficient 429–30
anomalous skin effect 130

antibonding band 139
antibonding molecular orbital 221–2, 436
  wave function 439
antibonding state 222, 443, 502
antiferromagnetic insulating phase 448–9
antiferromagnetic insulator 440
antiferromagnetic metals and alloys 385
antisymmetric orbital wave function 435
antisymmetric wave function 191, 193
approximant 245, 452, 497
APW function 210
APW method 214
Ar gas sputtering 169
area of the Fermi surface 268
Ashcroft–Langreth structure factor 459
atomic size-factor effect 234
atomic % 229
atomic form factor 277
atomic orbital 200, 222
  wave functions 272
atomic scattering factor 455, 458
atomic sphere 216, 218
atomic sphere approximation (ASA) 216
atomic structure
  of amorphous metals 452
  of liquid metals 452
atomic units 212
attractive electron–electron interaction 349–50
Auger electron 180
augmented muffin-tin orbital 222
augmented plane wave (APW) function 209
  method 207
average atomic distance 279, 456, 475, 481–2, 490, 492–3, 504, 507–8
average electron concentration 397
average field approximation 399, 411, 413
average kinetic energy per electron 26
average number density 453, 455
azimuthal quantum number 10, 21, 176, 178, 209

β′-phase 232, 235
β-phase 232, 235
back scattering 278

band calculations 26 *see also* CPA band
   calculations
band index 102, 121, 178, 218
band structure 1
   effect 272, 481
band width 445
basic vector 56–7, 105, 494, 496
basis functions 200, 203, 210, 215
Baym resistivity formula 284, 287, 291, 479
Baym–Meisel–Cote theory 479, 481, 482, 486, 491,
   493, 504
bcc lattice 63
BCS ground state 355
BCS theory 7, 9, 334–5, 342, 346–7, 354, 356–7,
   360, 362
Bean critical-state model 380
belly 131
Bessel function 406
Bi2212-type superconductor 379
Bi2223-type superconductor 377, 379
black-body 69
Bloch condition 201, 210, 213
Bloch electron 92, 178, 229, 257–8
Bloch state 91–2, 102, 254, 353, 401
Bloch sum 200, 203, 210, 223
Bloch $T^{3/2}$-law 394
Bloch theorem 7, 86, 88, 91, 93, 94, 102, 199,
   245–6, 270, 279, 291, 386, 466, 482
Bloch wave 7, 91, 93, 99, 223, 256, 271, 395, 438,
   483
   function 97, 200
   vector 201, 210, 213, 217
Bloch waves 273
Bloch–Grüneisen law, 284, 288–9, 291, 349, 402,
   416, 450, 480
block layer 439, 447
Bohr magneton 391, 395, 399
Bohr radius 195
Bohr–Sommerfeld quantization rule 149
Boltzmann distribution function 32, 34
Boltzmann equipartition law 6, 40
Boltzmann factor 44
Boltzmann relation 71
Boltzmann transport equation 9, 264–5, 279, 481,
   491, 493, 504, 506
   for isotropic metal 280
bonding band 139
bonding molecular orbital 221–2, 436, 438
bonding state 222, 443, 470, 502
Born approximation 276, 475, 479
Bose–Einstein distribution function 69, 70–1
Bose–Einstein statistics 354
bound state 351, 353
boundary energy 361–2
Bragg condition 62, 103–4, 177
Bragg law 61
Bragg reflection 259–60
Bragg scattering 103, 237
Bravais lattice 57
Bremsstrahlung 170
Bremsstrahlung Isochromat Spectroscopy (BIS)
   170

Brillouin zone *see also* first Brillouin zone *and*
   second Brillouin zone
   of bcc lattice 106
   of fcc lattice 110
   of hcp lattice 113
   for two-dimensional square lattice 106
Brownian motion 329

c-axis oriented $SmBa_2Cu_3O_{7-\delta}$ superconductors
   380
carrier concentration dependence of thermoelectric
   power 309
carrier entropy 305, 309
central-field approximation 10
centrifugal potential 410
charge transfer energy 444
charge transfer-type insulator 447–8
chemical disordering 506, 510
chemical potential 33, 71, 197, 266
   gradient 298
chemical short-range order 458
cleaving 169
closed shell 386
coherence length 362, 368
coherent potential approximation (CPA) method
   246
coherent scattering intensity 455–6, 458–9
cohesive energy 14, 192, 198
collective motion
   of atoms 65
   of electrons 194
combined XPS and IPES spectra 171
compensated (metal) 133
complete orthogonal set of wave functions
   273
complete solid solution 231, 248
complex cubic γ-phase 237
complex dielectric constant 320, 323
complex optical conductivity 321–3
complex refractive index 319
Compton effect 160
conduction band 120, 144, 261
conduction electrons 14, 81
conductivity formula 488–9
   for isotropic system 269
conductivity tensor 313, 332
Cooper pair 351–60, 368
coordination number 507
core electrons 14
correlation energy 194, 197–9
correlation function 332
cosine-type periodic potential 86
cosine-type stationary wave 104
Coulomb correlation 194
Coulomb energy 412, 414, 427
Coulomb field 485
Coulomb potential energy 190, 433
Coulomb repulsive energy 441
covalent bonding 10, 15, 145, 461, 467, 470
CPA band calculations 247–8
critical current density 364–5, 379–80
critical field 335, 341, 346, 355

cross-section for inelastic scattering of incident electron 182
crystal momentum 260–1
  of the Bloch wave 93
crystal monochrometer 180
crystal spectrometer 177
crystalline electric field 389
crystallization temperature 464
CsCl-type ordered structure 232
CuO_2 plane 439, 443, 447–8
cuprate compound 432, 439, 450
cuprate oxides 9
Curie law 46, 385, 390, 416
Curie temperature 248, 385, 391, 394, 401–2, 422, 466
Curie–Weiss law 385, 391, 399, 411, 414, 430
current–voltage characteristics 347
curved quartz crystal monochrometer 163
cut-off frequency 74

DC Josephson effect 369
DC sputtering technique 462
DC voltage across juction 373
de Haas–Shubnikov effect 153
de Haas–van Alphen effect 127, 130, 140, 142, 148, 399
de Haas–van Alphen measurement 237
Debye formula for lattice specific heat 75
Debye frequency 74
Debye model 40–1, 73, 79–80, 288, 290, 295
Debye radius 74, 290
Debye sphere 74
Debye temperature 41, 74, 81, 288, 349
Debye–Waller factor 284–5, 480–1
decaganol quasicrystal 494
deflection angle 372
degenerate Fermi gas 297
degree of freedom 67, 72
delocalization of electrons 485
delta function 212, 268, 279, 285–6, 455
dense Kondo systems 418
density 66
density of states 36, 199, 409, 500 see also final density of states and initial density of states
  at the Fermi level 43, 354, 413, 416
  in transition metals 400
  curve 119, 168, 238–9, 241
  of Cu metal 131
  of graphite 140
  of Li metal 129
  of Si 144
detailed balance condition 282
deviation from Ohm's law 266
diadic 268
diamagnetism of ions 49
diamond structure 143
diatomic linear chain model 77
dielectric constant 320, 324, 327
difference in the valency between the impurity and matrix 409
differential cross-section 227, 403

differential scanning calorimetry (DSC) 464
differential scattering cross-section 279, 408
differential thermal analysis (DTA) 464
diffraction phenomena 103
diffusion 264
diffusion coefficient 330–1, 488
dilute magnetic alloy 416
dimensionless figure of merit 309
dipole approximation 180, 183
dipole selection rule 183
direct beam 456
direct transition 317
directional bonding 15, 502
disordered alloy 245
disordered fcc phase 231
dispersion 254
dispersion relation 65, 74, 78, 80, 95, 255, 320
  of conduction electrons 102
  of spin waves 394
dissipation 330
divalent liquid metals 477
divalent metals 132
donor level 146
dopant 145
Doppler effect 161
drift velocity 250, 232, 269
Drude conductivity formula 252
Drude edge 326
Drude expression
  for AC conductivity 322
  for optical conductivity 322
  for conductivity 331
Drude model 270
Drude theory 249
Drude-type spectrum 326
Dulong–Petit law 6, 40, 73, 76, 308
$d_{x^2-y^2}$ symmetry 448
dynamical effect 413
dynamical structure factor 183, 281, 284, 286, 479–80

E–$\mathbf{k}$ relation 100, 102, 242
  of Al metal 136
  of Cu metal 131, 174
  of Li metal 127
  of Na metal 127
  of Si 144
ϵ-phase 232, 235, 238
edge length of icosahedral cluster 503
EELS measurements see electron-energy-loss spectroscopy
effective electric field 298
effective mass 261, 275
  of electron 260
  of quasiparticle 433
  tensor 260
effective potential 246
Einstein model 79
Einstein relation 329–32
elastic continuum 66
elastic life time 484
elastic medium 66

elastic scattering 227, 276–8, 302, 474, 476, 484–5
  of conduction electrons 483
elastic stiffness constant 66, 80
electrical conduction process 300
electrical conductivity 5, 7, 249, 251, 254, 297, 305, 485, 493
  formula 267, 305, 490
  due to inelastic electron–phonon interaction 280
  for isotropic metals 268
  tensor 268
electrical current density 251, 296
electrical resistivity 7, 252, 309, 417
  due to electron–phonon interaction 280
  formula at finite temperatures 280
electrochemical factor 234
electromagnetic wave 186
electromotive force 303
electron compounds 235
electron concentration 234, 241–2
electron conduction in the normal state 358
electron conduction mechanism 507
  in amorphous alloys 488
electron configuration 11, 386
electron correlation effects 399
electron density 195
electron density of states 25
  at the Fermi level 299
  per unit volume 268
electron diffusion coefficient 510
electron doping 448
electron-energy-loss spectroscopy (EELS) 181–4
  measurements 501
electron Fermi surface 316
electron liquid 433
Electron Spectroscopy for Chemical Analysis (ESCA) 167
electron transport function 165
electron transport properties 466
electron–electron interaction 43, 190, 432, 434, 437, 440, 485
  enhanced 485, 506
electron–magnon interaction 401–2
electron–nucleus interaction 190
electron–phonon interaction 43, 271, 402, 416 *see also* inelastic electron–phonon interaction
electron-yield detection technique 180
electronegativity 234, 443
  difference 444
electronic entropy 312
  density 305
electronic specific heat 7, 37–8, 40–1, 238, 294
  coefficient 39, 41, 239, 299, 346, 411, 470, 486, 491, 502, 507, 509–10, 513
  measurements 500
  in the superconducting state 346
  per unit volume 305
electronic structure effect 490, 506
electronic thermal conductivity 293–4, 296, 298
electrons per atom 234, 398
emission current density 51
empty-core model 205
endothermic reaction 465

energy band 10, 11, 14
energy conservation 182
  law 160, 286, 349
  of electron 276
energy current 297
energy dependence
  of photoionization cross-section 166
  of relation time 309
energy dispersion relation 433
energy gain per electron 350–1
energy gap 1, 102–3, 116, 120–1, 144, 238, 261, 354, 357
  in superconductor state 347
  in superconductor 345
energy loss near-edge structure (ELNES) 183
energy of electromagnetic wave in a vacuum 321
energy spectrum 356
  of quasiparticles 356
energy-independent muffin-tin orbital 221, 223
entropy of superconducting state 342
envelope function 273, 275
equation of continuity 343
equi-energy surfaces 202
equipartition law 6, 72, 76, 329
equivalent circuit 370
escape function 165
Ewald sphere 61
exchange constant 416
exchange energy 194–9, 395–7
exchange integral 392–3, 395–6, 437
exchange interaction 391, 395, 400
excited state of superconductor 355–6
exothermic reaction 465
experimentally derived electronic specific heat coefficient 42
experimentally determined metal–insulator transition line 515
extended x-ray absorption fine structure (EXAFS) 179
extended zone scheme 121, 133
external torque 372
extinction coefficient 319–20
extinction rule 63, 115
extrinsic semiconductor 145

Faber–Ziman structure factor 459
failure
  of the Bloch theorem 467
  of one-electron approximation 438
  of ordinary band calculations 440
  of the Ziman theory 479
FAR effect *see* ferromagnetic anisotropy of resistivity
Fe/Cr multilayered films 425
Fe_2VAl intermetallic compound 512
Fermi cut-off 158, 513
Fermi diameter 315
Fermi edge 410, 474
Fermi energy 23, 26–7, 30, 32, 37, 225
  at a finite temperature 33
Fermi gas 433
Fermi level 33, 268, 397

Fermi liquid 432–4
Fermi momentum 158, 162
Fermi radius 23, 27, 118, 154, 227, 314, 418
Fermi sphere 23, 30, 33, 82, 116, 355
Fermi statistics 356
Fermi surface 9, 23, 116, 119, 199, 236, 249, 358, 484
  of Cu metal 130
  of electrons 124, 136–7
  of holes 124, 136
  of Na metal 127
  of Zn metal 133
Fermi surface–Brillouin zone interaction 9, 116, 242–3, 462
Fermi temperature 26, 39, 48, 298, 306, 309
Fermi velocity 227, 249, 253, 294–5, 491
Fermi wavelength 26
Fermi–Dirac distribution function 29, 33–4, 39, 51, 165, 265–6, 270, 282, 298
Fermi–Dirac statistics 6, 33, 40
Fermiology 142, 155
fermions 33
ferromagnetic anisotropy of resistivity (FAR) 421
  effect 422, 425
ferromagnetic coupling 393
ferromagnetic metals and alloys 385
ferromagnetism 390, 391
Fibonacci chain 496, 497
Fick equation 330
field cooling (FC) 336
final density of state 165–7, 400, 482
final energy state 174
fine structure constant 180
first Brillouin zone 66, 68 82, 93, 102, 105, 109, 133, 139, 143, 202, 259
  of bcc lattice 110, 126
  of fcc lattice 112–13, 130
  of hcp primitive cell 115
  of simple cubic lattice 116
first-order reflection 61
five-fold symmetry 244, 451, 494
flow of heat 292, 301
flux pinning effect 365
flux quantum 374
fluxoid 374
fluxon 374
forbidden band 96, 102
forbidden gap 144
force constant 64, 77, 80
forward scattering 278
four-dimensional space 495
Fourier transform of pseudopotential 206
Frank–Kasper compound 452
Frank–Kasper phase 498
free electron, 14, 18, 21
free energy 355
  for the normal state 341
  for the superconducting state 341–2
free volume 477
free–electron model 19, 23, 36, 43, 48, 91, 118, 133–5, 137, 158, 162, 174, 179, 193, 207, 236, 250, 254, 298, 305–6, 309, 314–15, 350

free-electron wave function 93
frequency
  of spin fluctuations 414
  of superconducting tunneling curent 373
frequency spectrum 73
frictional force 250
Friedel oscillations 410, 469
Friedel sum rule 408–9
Fröhlich theory 349, 351, 354
functional of density 196
fundamental reflections 231

g-parameter 490–3
γ-phase 232, 235, 244
gamma-rays 156–7
gas constant 39
gas evaporation method 462
Gauss theorem 149
Gaussian function 255
giant magnetoresistence (GMR) effect 425, 427–8
Ginzburg–Landau (GL) equations 345
Ginzburg–Landau constant 362
Ginzburg–Landau theory 344, 358, 362
glass transition temperature 464
golden ratio 495–7
Goldschmidt radius 458
good quantum number 21, 235, 355, 387
grating space of a crystal 177
Green function 212, 214, 247
Green theorem 214
ground state 11, 388, 437
ground-state energy 192, 198, 392
  per electron 194, 354
ground-state wave function 354
group velocity 254, 360
  of wave packet 255, 257
gun technique 463

η-phase 232, 235
half-filled band 444
Hall coefficient 9, 153, 254, 312–13, 315–16, 470, 485–6, 491–3, 506
Hall coefficient in amorphous alloys 316
Hall effect 7
  in magnetic metals 428
Hall resistivity 428
harmonic oscillator 69, 88
  approximation 284
Hartree approximation 192, 193, 195
Hartree field 193
Hartree potential 193, 197, 199
Hartree–Fock approximation 195, 432
Hartree–Fock equation 194
hcp structure 133
heat–current conversion efficiency 309
heavy fermion 418, 507, 513
Heisenberg model 393
Heisenberg uncertainty principle 6
Heitler–London approximation 434, 439
Heitler–London model 437
helium gas discharge lamp 162
Helmholtz wave equation 220

hexagonal close–packed structure 113
high-resistivity amorphous alloys 482
high-resistivity limiting curve 488–9, 510
high-T_c cuprate superconductor 439, 447
higher-order reflection 61, 103
hole doping 439
hole Fermi surface 316
hole with $d_{x^2-y^2}$ symmetry 448
holes 118, 134, 136, 140, 142, 261, 263, 314
homogeneous electron gas 194, 197–9
homonuclear diatomic molecule 221
hopping conduction 483
hopping integral 507
hopping matrix element 442
horizontal scattering process 301
Hubbard Hamiltonian 441
Hubbard model 432, 441–3
Hume-Rothery alloys 240
Hume-Rothery electron compounds 240
Hume-Rothery electron phases 235–6
Hume-Rothery mechanism 501
Hume-Rothery rule 232, 234, 241, 244
Hund rule 172, 388–9, 438
hybridization effect 132, 502–3
hydrogen molecule 392, 434–8
hyperfine coupling constant 418–19
hysteresis in irreversible magnetization curve 365

I–V characteristics 368, 372
icosahedral cluster 502
icosahedral quasicrystal 494
icosahedral symmetry 495
ideal electron gas 433
imperfections 364
impurity potential 271, 273, 403–4, 406
impurity scattering 270, 302
incomplete shell 386
indirect transition 318
individual Cooper pair wave function 354
inelastic electron–phonon interaction 280, 283, 285, 299, 309, 479–80, 483
inelastic life time 484
inelastic one-phonon normal process 288
inelastic scattering 299, 302, 474
infrared 186
initial density of states 165
inner potential 174
insulator 1
interacting electron system 433
interaction potential 182
interatomic distance 11
interatomic pair potential 467, 469–70
interband transition 186–7, 318, 326–7
interference function 277
intermediate state 361
intermetallic compound 231, 245
internal energy of electron system 38, 242
internal field 390–1
interplanar distance 59, 62
interstitial alloys 245
intraband transition 318, 327
intrinsic semiconductor 120, 144

Inverse Photoemission Spectroscopy (IPS) 169–170
Ioffe–Regel criterion 279, 490
ionic bonding 10, 14
ionization energy 14
irreducible wedge 199
irreversible magnetization curve 365
isotope effect 345, 347, 349, 351, 354–5
isotropic metal 251, 268
isotropic system 275
itinerant electron model 395–6, 398, 399–400
Itoh formula of the Hall conductivity 472

jellium model 193
Jones zone 116, 142, 238
Josephson critical current 368
Josephson current 370
Josephson device 368
Josephson effect 368, 373
Josephson junction 368, 370, 372–4
jungle gym 137

k-space 19
K-transition 179
Kβ-emission and K-absorption spectra 181
Kelvin relation 297
kinetic theory of gases 294
KKR–ASA method 216, 219 see also Korringa–Kohn–Rostoker (KKR) method
Kondo effect 9, 385, 414, 416, 418–19, 466
Kondo lattice 418
Kondo problem 416
Kondo temperature 416–17
Koopmans theorem 444
Korringa–Kohn–Rostoker (KKR) method 211, 214, 217–18
Kramers–Kronig relation 325, 327
Kronecker-delta 194, 214
Kronig–Penney model 93
Kubo formula 324, 328–9, 332
Kubo–Greenwood formula 507

L-transition 179
Landau diamagnetism 49
Landau level 151–2
Landau theory 344
Landé g-factor 387
Langevin function 45
Langevin theory 390
Laplace equation 216
large-angle scattering 301
lattice constant 11, 54, 59, 64, 77, 86, 100, 238, 394
  of Cu metal 270
  of Li metal 130
lattice specific heat 40, 41, 69, 72, 309
  coefficient 41
  per unit volume 308
lattice sum 214
lattice thermal conductivity 293
lattice vector 55, 65, 91, 199
lattice vibrations 64, 66, 271
lattice wave 65, 69, 82, 433

Laue condition 103–4
layered perovskite cuprates 432, 443
layered perovskite structure 439
LCAO model 203 *see also* linear combination of atomic orbitals
Legendre polynomials 408
lifetime of localized spin fluctuations 413
limiting Lorenz number 299
Linde law 226, 228
Linde rule 403
linear augmented plane wave (LAPW) method 440
linear chain model 64
linear combination of atomic orbitals (LCAO) 201, 221
linear-muffin-tin orbital (LMTO) method 215–16, 223 *see also* LMTO-recursion method
linear-response theory 324, 332
linearized Boltzmann transport equation 266–7, 276, 296–7, 299, 302, 312, 322, 475, 479
liquidus curve 229
LMTO-recursion method 470
local atomic structure 460
local density functional (LDF) method 195–8, 440
local electron density 198
local magnetic moment 412, 415
local spin-density functional method 198
localization effect 491
localized electron model 390, 399
localized magnetic spin 419
localized moment 413, 414, 416, 418, 428
logarithmic divergence 416
logarithmic temperature dependence of resistivity 417
London equation 340, 343–5
London penetration depth 340
London theory 9, 335, 338
long-range order 344
  of momentum 359
longitudinal magnetoresistance 420
longitudinal wave 68, 80, 287
Lorentz force 49, 148, 343, 365, 428–9
low-resistivity amorphous alloys 486
lower critical field 360
lower Hubbard band 444–5, 448

Mackay icosahedron (MI)-type quasicrystals 494
magnetic domain 393, 420
magnetic energy 362
magnetic field dependence
  of critical current density 366, 377
  of magnetization 360
magnetic field sensor 425
magnetic flux lines 363
magnetic impurities 403
magnetic impurity atom 405
magnetic moment 44, 387, 389, 390–3, 417, 448
magnetic prism-type spectrometer 182
magnetic quantum number 10, 21, 178, 209
magnetic susceptibility 46, 148
magnetization 45, 319
  curve 362, 379–80
  of Ni 397

magneto-thermal oscillation 153
magnetoresistance 7, 316–17, 420
magnon density of states 394
magnons 394, 401
main peak in the structure factor 467
many-body effect 169, 416
marginally insulating regime 474
marginally metallic regime 474
Matthiessen rule 271, 293
maximum phonon wave number 288
maximum solubility limit 232
maximum superconducting tunneling current 368
Maxwell equation 318, 336, 338, 365
Maxwell–Boltzmann distribution function 50
Maxwell–Boltzmann distribution law 29
Maxwell–Boltzmann statistics 6, 53
mean escape depth of the photoexcited electron 169
mean free path 6, 7, 227, 253, 270, 294, 332, 401, 481–2, 490–3, 504, 506, 508
  effect 481, 490, 505–6, 510
  of conduction electrons at the Fermi level 269, 279, 291, 295, 475, 502
  of electrons 362
  of phonons 308
  of sp-electrons 386, 466
mean phonon energy 354
mechanical alloying technique 462
mechanism of superconductivity 450
medium-range structure 457
Meissner effect 335, 337–8, 340, 345, 355, 358, 360
melt-processing technique 379–80
melt-quenching 462
metal 1
metal–insulator transition 312, 444, 483, 486, 488, 507, 512
metal–insulator transition line (MI-line) 510, 512–13, 515
metallic bonding 10, 14
metastable quasicrystals 494
method of Lagrangian multipliers 31, 71
Mg-Kα line 162
MI-type quasicrystals 506
Miller indices 57, 82, 97, 99, 100
minimum diffusion coefficient 488
minimum metallic conductivity 483, 491
mixed state 361–3
mobility 253
mobility edge 483
mode of lattice waves 67
molecular dynamics simulations 469–70
molecular field 395
molecular orbital 222
  method 436
moment of inertia 372
momentum conservation law 156, 160, 286, 318, 349
  of the Bloch electron 92
moment operator 91
monovalent liquid metals 477
monovalent metals 126, 130
monster 134, 136

Mott conductivity formula, Mott relation 490, 505–8, 511–12, 515
Mott insulator 447
Mott s–d scattering model 400–1, 482
Mott two-current model 422
Mott–Hubbard insulator 444, 447
Mott-line 509
muffin-tin orbital 216, 220
muffin-tin potential 207–8, 210, 213–14, 218, 245
muffin-tin radius 214
muffin-tin zero 208, 219
multiple scattering 504,
  theory 211, 218, 246
multiple-phonon process 286
multiplet structure 172
multiplicity of equivalent zone planes 241
μ-phase 235

n-dimensional hyperspace 495
n-type semiconductor 146
NaI (Tl) scintillators 156
Nd123 superconductor 379
$Nd_2Fe_{14}B$ magnet 381
near-edge x-ray absorption fine structure (NEXAFS) 179
nearest neighbour atoms 201
nearly-free-electron model 97, 99, 200, 241, 466, 473, 493
neck 131, 159, 237
neck diameter 131
negative scattering amplitude 460
negative TCR 474, 477, 481–2, 485
neutron diffraction
  measurements 399, 470
  technique 460
neutron inelastic scattering 80, 450
neutron scattering amplitude 460
neutron zero-alloy 460
NFE model 203–4 see also nearly-free-electron model
noble metals 130
non-interacting electrons 433
non-magnetic metals 386
non-periodic system 451, 466
Nordheim law 228, 231
normal Hall coefficient 429–30
normal Hall effect 429
normal modes of lattice vibrations 65
normal process 286
nuclear spin 418
number of electrons
  per atom 42, 116, 133
  per unit volume 5, 269
number of impurities per unit volume 227
number of phonons 295
number of superconducting electrons per unit volume 338
number-density function 455, 459

oblique transition 317
Ohm's law 251, 358
on-site Coulomb energy 441–2, 444–5

on-site Coulomb interaction 432
one-dimensional monatomic lattice 64, 93, 103, 105
one-dimensional periodic lattice 88, 121
one-electron approximation 9, 190–2, 198–9, 432–5, 439, 444, 485
one-electron band calculations 440
one-electron Schrödinger equation 193
one-phonon normal process 290
one-phonon process 285–6
Onsager relation 297
optical absorption spectrum 317
optical branch 79
optical conductivity 317, 319–20, 324, 327
  spectrum 325, 327
optical constant 324–5
optical effective mass of conduction electron 323
optical excitations 317
optical mode 79
optical phonons 77, 79
optical properties 9
optical reflectance 317
optical reflection 184
optical transition 167
  probability 165
optimum doped region 449–50
OPW wave function see orthogonalized plane wave (OPW) method
orbital angular momentum quantum number 10
order parameter 344–5, 362–3
ordered alloy 231
ordering energy 362
orthogonalized plane wave (OPW) method 203
overdoped region 449–50

p-band 139, 327
p–n junction 147
p-type semiconductor 147
pair distribution function 452–6
parallel space 495, 497
paramagnetic metals and alloys 385
paramagnetic susceptibility 48
paramagnetism of free-electron gas 6
partial density
  of final states 180
  of states 178, 183
partial pair distribution function 459–60
partial structure factor 458–61
partial wave 220
  method 215, 405, 408
Pauli exclusion principle 4, 11, 13, 21–2, 26, 29–30, 46, 69, 81, 161, 191, 193–5, 351, 387–8, 435, 444
Pauli paramagnetic susceptibility 434
Pauli paramagnetism 44, 48, 385, 466
Peltier coefficient 305–6
Peltier effect 297, 306–7
pendulum 370
penetration depth 338
Penrose quasilattice 498
pentavalent semimetals 141
perfect crystal at absolute zero 279

perfect diamagnetism 341, 360
periodic boundary condition 18, 23, 27, 66, 89
periodic empty-lattice 93
   model 99, 123
periodic Penrose lattice 451
periodic square-well potential 93
periodic zone scheme 121
permeability 339
perpendicular space 496–7
phase coherence 485
   of traveling wave 359
phase diagram 228–9
phase difference 368–72
   of tunneling current 374
phase of reflectance 324
phase shift 220, 368, 406, 409–10, 454
phonon annihilation operator 285
phonon creation operator 285
phonon density of states 73
phonon drag effect 306–7, 309
phonon mean free path 306
phonon scattering 270
phonon thermal conductivity 296
phonon thermal resistivity 294
phonon-assisted hopping of electrons 486
phonon–electron interaction 294, 307
phonon-mediated electron–electron interaction 351
phonon–phonon interaction 307
phonons 69–70, 81, 433
photoabsorption cross-section 180
photoelectric effect 69
photoelectrons 162, 164
photoemission
   experiment 248
   process 164, 173
   spectroscopy 447, 470, 500
   spectrum 163, 174
photoexcited electrons 164
photoinization cross-section 165, 169
photons 4, 69
physical space 495
pinning centers 365, 379
pinning force per unit length of core 365
Planck constant 3, 69
   divided by 2π 16
Planck distribution function 72–3, 283, 286, 394
plane wave 17, 87, 200, 271, 275
   approximation 272, 275
plasma frequency 323
plasma oscillation 186, 327
plasmon 194, 433
point-contact transistor 7
Poisson equation 225
polarization 68, 319
   vector 285, 287
polyvalent liquid metals 477
positive Hall coefficient 472
positive hole 263
positive ions 14
positron annihilation 155, 236
primary solid solution 232

primitive cell 114
primitive translation vector 56, 105–6, 114, 138
   in reciprocal space 108, 114
principal quantum number 10–11, 21, 150
pseudo-binary dilute alloys 423
pseudogap 245, 312, 450, 490–2, 500, 502
   system 507–8, 510
pseudopotential 99, 205, 276, 475
   method 144, 204, 243

quantized flux 364
quantized magnetic flux 364
   distribution 363
quantum interference effect 474, 485, 508, 510–11
quantum magnetic flux 374
quantum number of orbital angular momentum 387
quantum oscillations 154–5
quasi-elastic scattering 301
quasicrystal 244, 291, 451, 470, 494
quasilattice 502
quasiparticle 356, 394, 433
quasiperiodic Penrose lattice 451
quasiperiodicity 498
quenching of 3d orbital angular momentum 389

ρ–γ_{exp} diagram 507, 510, 512
radial distribution function 470
radial wave function 178, 403, 406
radius of a sphere containing one electron 195
random force 329
rare earth metals 386
rare earth-123 type superconductors 379
reciprocal lattice vector 54, 56, 63, 65, 82, 92, 99, 100, 102, 104, 175, 200, 210, 260
reciprocal space 19, 22, 56
rectification 147
reduced zone scheme 121, 134, 237
reflectance
   measurement 324
   spectrum 325
reflection spectrum 186
refractive index 319–20
regular hexagonal network 137
relaxation time 5, 227, 250, 267, 270, 278, 283, 299, 302, 307, 316, 326, 358, 400, 422
   approximation 266, 283, 296–7, 299, 313, 322, 324
repeated elastic scattering events 484
repeated zone scheme 121
repulsive centrifugal potential 403
residual resistivity 271, 302, 405, 416, 480, 483, 485
residual resistivity ratio (RRR or 3R) 271
resistivity at 300 K 509
   formula in the partial wave method 408
   due to impurities and defects 271
   due to lattice vibrations 271
   maximum 419, 481
   minimum 385, 414, 419
   of transition metals 400
   tensor 313

resonance energy 444
resonant scattering 410
resultant orbital angular momentum 387–8
resultant spin angular momentum 387–9
rhombic triacontahedron 497
rhombic triacontahedron (RT)-type quasicrystals 494
rhombohedral unit cell 141
Richardson–Dushman equation 53
Riemann zeta function 76, 291
Rietveld method 503
rigid pendulum 371
rigid-band model 245, 248, 411
RKKY interaction 418–19
Russel–Saunders nomenclature 388

s-band 139
s–d exchange constant 419
s–d exchange Hamiltonian 415
s–d interaction 401, 414–15
s–d scattering model 482
saturation magnetic moment 397
saturation magnetization 393, 398
scale invariance 497
scaling law 483
scattering amplitude 455
scattering angle 227, 277, 476
scattering cross-section 6
scattering potential 276
scattering probability 400, 415, 482
scattering vector 277, 280, 476
screened Coulomb interaction 194
screened potential 226, 403, 433
screening current 364
screening radius 225–6
second Brillouin zone 115
second-order perturbation theory 279, 350, 399
second-order phase transition 344
Seebeck coefficient 303–5
selection rule 167, 176–8
self-similarity 497
semimetal 137, 140
set of lattice planes 57, 59, 81–2
shielding surface current 341
short-range order 245, 458, 461, 467, 470, 502
short-range structure 457, 502
Si-$K\beta$ SXES spectrum 179
sign of TCR 477, 482
simple cubic lattice 68, 202
simple cubic metal 116
simple hypercubic lattice in six-dimensional space 497
simple liquid metals 279, 466, 472, 474–5
simple-harmonic oscillator 65
sine-type stationary wave 104
single-roll melt-spinning apparatus 494
single-roll quenching method 463
single-roll spinning wheel apparatus 462, 472
singlet state 353, 392–3, 416–17, 435, 437
six–dimensional lattice 451, 495
Slater determinant 194
Slater–Pauling curve 398–9

small-angle scattering 291
smeared Fermi surface–Brillouin zone effect 467
soft x-ray absorption spectroscopy (SXAS) 176, 179, 181
soft x-ray emission and absorption spectra 500
soft x-ray emission spectroscopy (SXES) 176
    spectra of Al and Al_2O_3 179
soft x-ray spectroscopy 176, 500
solidus curve 230
sound velocity 66, 73, 80
sp^2 orbital 139
space lattice 57
space-symmetries 178
specific heat 7, 355, 417 see also electronic specific heat and lattice specific heat
    in superconducting state 346
    of carriers per unit volume 294
    of superconducting state 342
spectrometer resolution 165
spherical Bessel function 209, 214, 220, 406–7
spherical harmonic function 178, 209, 216, 403
spherical Neumann function 214, 220, 406
spherical wave 220
spiky peaks 501, 510
spin configuration 388, 392
spin fluctuation 399–400, 413, 417, 513
spin freezing temperature 385, 419
spin operator 393
spin polarization 416
    of conduction electron 419
spin quantum number 10, 21
spin wave 385, 394, 401
    theory 390, 394
spin-down band 397, 400
spin-down electron 46, 198, 422, 424, 427–8, 442, 444
spin-glass 385, 418–19
spin-orbit interaction 183
spin-up band 397, 400
spin-up electron 46, 198, 422, 424, 427–8, 442, 444
splat quenching 463
spontaneous emission 177
spontaneous magnetization 390
sputtering method 462
SQUID magnetometer 376 see also superconducting quantum interference device
stability of phases 244
standard voltage 373
standing wave 260–1
static distribution of ions 475
static source of scattering 271
static structure factor 277, 285–6, 480
stationary wave 21
statistical fluctuations 330
steady state 251, 259, 264–5, 314, 328, 358
steady-state electron distribution 270
Stirling formula 31
Stoner condition 396
strength of net attractive interaction 354
strongly correlated electron system 9, 432–9, 441, 444–5, 507, 512

structural disorder 477
structure factor 214, 218, 279, 455–6
structure Green function 213
subshell photoionization cross-section 165
substitution rule 497
substitutional disordered alloy 245
sum rule 324
superconducting bulk magnet (SBM) 382
superconducting ceramic oxides 376
superconducting current density 365
superconducting electron 338
superconducting magnet 365–6, 377
superconducting motor 382
superconducting permanent magnet 380
superconducting phase 448
superconducting phenomenon 334
superconducting quantum interference device (SQUID) 373
superconducting ring with two junctions 373
superconducting transition 41, 346
    temperature 334, 354, 449
superconducting tunneling current 368, 370–2
superconducting wire 366, 377
superconductivity 7, 334–5
supercooled liquid 464
supercooling 464
superfluidity 351
superlattice 231
    reflections 231
    structure 245
superstructure 231
surface energy 198
symmetric orbital wave function 435
synchrotron radiation 162–3, 180

$T^{-1/4}$-law 488
t-matrix 247
$T^{3/2}$-law
T^5-law 291, 300
tail cancellation 218
Taylor theorem 34
temperature coefficient of resistivity (TCR) 252, 473
temperature dependence
  of conductivity 486
  of electrical impurity resistivity 419
  of electronic thermal resistivity 295–6
  of free volume 464
  of Hall coefficient 316
  of magnetic susceptibility 391
  of resistivity in amorphous alloys 483
  of resistivity of simple liquid metals 472
  of the Fermi energy 36
  of the Pauli paramagnetism 385
  of thermoelectric power 305, 309
temperature gradient 293, 297, 300, 303
temperature-dependent magnetic susceptibility 466
tensor 268
tetravalent metals 137
thermal average 281
thermal conduction process 300

thermal conductivity 5, 7, 9, 293–4, 297, 309
thermal current 293, 307
thermal current density 293, 296–7
  of phonon 308
thermal effective mass 43
thermal energy at room temperature 33
thermal positron 156
thermal resistivity 293
thermally stable quasicrystals 494
thermionic emission 7, 50–1
thermocouple 305
thermodynamics of superconductor 341
thermoelectric device materials 309
thermoelectric power 9, 153, 302–3, 304–6, 309, 417
  due to phonon drag 308
Thomas–Fermi approximation 225, 403, 407
Thomas–Fermi screening parameter 226
Thomson coefficient 305
three-dimensional physical space 451, 495, 497
three-dimensional quasiperiodicity 494
tight-binding approximation 437, 483, 507
tight-binding Hamiltonian 507
tight-binding linear muffin-tin orbital particle source method 472
tight-binding method 200–1, 271, 436, 438
time-averaged electrostatic field 193
total electronic thermal resistivity 293
total reflection 327
total resolution function 165
total scattering cross-section 408
total thermal conductivity 295
transfer integral 442
transistor action 147
transition probability 266, 275, 278, 280–1, 475
transmission electron microscope 182
transport cross-section 408
transport current 364
transverse magnetoresistance 317, 420
transverse wave 68, 80, 287
traveling wave 19
triplet state 392–3, 435, 437
trivalent metals 135
tunneling effect 368
tunneling experiment 347
tunneling for quasiparticles 369
two-band model 316–17, 506
two-current model 422–3, 424
two-dimensional lattice model 138
two-dimensional quasiperiodicity 494
two-wave approximation 99, 104
type-I superconductors 360
type-II superconductors 345, 360, 364–5

ultraviolet photoemission spectroscopy (UPS) 162
ultraviolet rays 162
Umklapp phonon–phonon interaction 294
Umklapp process 286
underdoped region 449–50
upper critical field 360, 364
upper Hubbard band 444–5, 448

valence band 13, 120, 144, 261
valence electrons 13–14
  per atom 26
van der Waals bonding 10
van der Waals force 15
van der Waals interaction 137
van Hove singularity 129, 241, 467
variable-range hopping conduction 511
variable-range hopping model 486, 488, 510
variational principle 192, 194, 196–7, 211, 213, 215, 223
VCA potential *see* virtual crystal approximation (VCA) model
vector potential 149, 343, 345
vertical scattering process 301
vertical transition 317
virtual bound state 405, 410–11, 413
virtual crystal approximation (VCA) model 246
virtual phonon 350
viscous damping 372
volume per atom 26, 106, 110, 116, 203, 238, 278

Wannier function 272–3, 275
wave number dependence of the pseudopotential 476
wave packet 254–5
wave vector 17, 21, 91

wavelength 18
  of incident x-ray 454
weak ferromagnetism 385
weak localization 474, 483, 485, 492–3, 504, 506
weak periodicity 467
weight % 229
Wiedemann–Franz law 5, 299, 301–2
Wigner–Seitz cell 108, 208, 213
  of bcc lattice 112
  of fcc lattice 109
Wigner–Seitz sphere 216
window 495
work function 50, 163, 198

x-ray absorption near-edge structure (XANES) 179
x-ray diffraction 59, 454, 470
x-ray photoemission spectroscopy (XPS) 163
Xα method 199
XPS valence band 168, 248

Zeeman effect 4
zero field cooling (ZFC) 337
zero-phonon process 285
zero-point motion 15
ζ-phase 232, 235, 238–9
Ziman resistivity formula, Ziman theory 275, 278, 466, 472, 474–7, 491